# Achieving sustainable cultivation of rice

## Volume 2: Cultivation, pest and disease management

It is widely recognised that agriculture is a significant contributor to global warming and climate change. Agriculture needs to reduce its environmental impact and adapt to current climate change whilst still feeding a growing population, i.e. become more 'climate-smart'. Burleigh Dodds Science Publishing is playing its part in achieving this by bringing together key research on making the production of the world's most important crops and livestock products more sustainable.

Based on extensive research, our publications specifically target the challenge of climate-smart agriculture. In this way we are using 'smart publishing' to help achieve climate-smart agriculture.

Burleigh Dodds Science Publishing is an independent and innovative publisher delivering high quality customer-focused agricultural science content in both print and online formats for the academic and research communities. Our aim is to build a foundation of knowledge on which researchers can build to meet the challenge of climate-smart agriculture.

For more information about Burleigh Dodds Science Publishing simply call us on +44 (0) 1223 839365, email info@bdspublishing.com or alternatively please visit our website at www.bdspublishing.com.

**Related titles:**
Achieving sustainable cultivation of rice Volume 1: Breeding for higher yield and quality
Print (ISBN 978-1-78676-024-1); Online (ISBN 978-1-78676-026-5, 978-1-78676-027-2)

Rice insect pests and their management
Print (ISBN 978-1-78676-196-5); Online (ISBN 978-1-78676-198-9, 978-1-78676-199-6)

Achieving sustainable cultivation of wheat Volume 1: Breeding, quality traits, pests and diseases
Print (ISBN 978-1-78676-016-6); Online (ISBN 978-1-78676-018-0, 978-1-78676-019-7)

Achieving sustainable cultivation of wheat Volume 2: Cultivation techniques
Print (ISBN 978-1-78676-020-3); Online (ISBN 978-1-78676-022-7, 978-1-78676-023-4)

Achieving sustainable cultivation of maize Volume 1: From improved varieties to local applications
Print (ISBN 978-1-78676-008-1); Online (ISBN 978-1-78676-010-4, 978-1-78676-011-1)

Achieving sustainable cultivation of maize Volume 2: Cultivation techniques, pest and disease control
Print (ISBN 978-1-78676-012-8); Online (ISBN 978-1-78676-014-2, 978-1-78676-015-9)

Chapters are available individually from our online bookshop: https://shop.bdspublishing.com

climate•SMART•publishing
IN AGRICULTURAL SCIENCE

BURLEIGH DODDS SERIES IN AGRICULTURAL SCIENCE

NUMBER 04

# Achieving sustainable cultivation of rice

## Volume 2: Cultivation, pest and disease management

Edited by Professor Takuji Sasaki,
Tokyo University of Agriculture, Japan

burleigh dodds
SCIENCE PUBLISHING

Published by Burleigh Dodds Science Publishing Limited
82 High Street, Sawston, Cambridge CB22 3HJ, UK
www.bdspublishing.com

Burleigh Dodds Science Publishing, 1518 Walnut Street, Suite 900, Philadelphia, PA 19102-3406, USA

First published 2017 by Burleigh Dodds Science Publishing Limited


Library of Congress Control Number: 2016962110

British Library Cataloguing in Publication Data
A catalogue record for this book is available from the British Library

ISBN 978-1-78676-028-9 (print)
ISBN 978-1-78676-030-2 (online)
ISBN 978-1-78676-031-9 (online)
ISSN 2059-6936 (print)
ISSN 2059-6944 (online)

Typeset by Deanta Global Publishing Services, Chennai, India
Printed by Lightning Source

# Contents

Series list xi

Acknowledgements xv

Introduction xvi

**Part 1 Rice cultivation techniques**

1 Advances in irrigation techniques for rice cultivation 3
*D. S. Gaydon, CSIRO Agriculture, Australia*
- 1 Introduction 3
- 2 Water-saving measures 4
- 3 Scale-dependency of water productivity and water savings 5
- 4 Aerobic rice as a water-saving measure 7
- 5 Alternate wetting and drying (AWD) as a water-saving measure 10
- 6 Saturated soil culture (SSC) as a water-saving measure 13
- 7 Case study: water-saving irrigation in southeast Australia 15
- 8 Future trends and conclusion 17
- 9 Where to look for further information 19
- 10 References 19

2 Advances in nutrient management in rice cultivation 25
*Bijay-Singh, Punjab Agricultural University, India and V. K. Singh, Indian Agricultural Research Institute, India*
- 1 Introduction 25
- 2 Real-time site-specific nitrogen management in rice using non-invasive optical methods 27
- 3 Site-specific nutrient management for intensive rice-cropping systems 33
- 4 Controlled-release and slow-release nitrogen fertilizers 36
- 5 Urease and nitrification inhibitors 39
- 6 Deep placement of nitrogen fertilizers 42
- 7 Phosphorus and potassium 44
- 8 Micronutrients 46
- 9 Integrated plant nutrient management based on organic resources and mineral fertilizers 47
- 10 Summary and future trends 50
- 11 Where to look for further information 51
- 12 References 52

3 Sustainable rice cultivation in coastal saline soils: a case study 69
*Sukanta K. Sarangi and Buddheswar Maji, ICAR-Central Soil Salinity Research Institute, India*
- 1 Introduction 69
- 2 The Sundarbans region of India 72

3 Management of coastal acid saline soils 77
4 Salt- and flood-tolerant rice varieties for coastal saline soils 78
5 Nursery management for wet season rice 82
6 Rice field management after transplanting 85
7 Management practices for dry season rice 86
8 Managing toxic gases from coastal rice fields 90
9 Removing excess soluble salts from the root zone 91
10 Integrated nutrient management (INM) for coastal saline soils 91
11 Reducing salinity build-up through mulching/cover crops 95
12 Management practices for increasing income from rainfed lowlands 96
13 Contingency plan for cyclone affected area 96
14 Recommended practices for cultivation of *kharif* and *rabi* rice 98
15 Alternative livelihood options 99
16 Future trends and conclusion 99
17 Where to look for further information 100
18 References 100

4 Dry-seeded and aerobic rice cultivation 105
*T. Parthasarathi, Ben-Gurion University of the Negev, Israel; M. Kokila, D. Selvakumar, V. Meenakshi and A. Kowsalya, Tamil Nadu Agricultural University, India; K. Vanitha, Tamil Nadu Rice Research Institute, Tamil Nadu Agricultural University, India; A. Tariq, University of Copenhagen, Denmark; A. Surendran, Rutgers University, USA; and Eli Vered, Netafim Irrigation Ltd, Israel*
1 Introduction 105
2 Dry direct seeded rice cultivation 105
3 Good agricultural practices (GAP) in dry direct seeding 107
4 Aerobic rice cultivation 113
5 Good agricultural practices (GAP) in aerobic rice cultivation 113
6 Conclusion and future trends 118
7 Where to look for further information 119
8 References 119

5 Processing rice straw and husks as co-products 121
*Nguyen Van Hung, Carlito Balingbing, James Quilty, Bjoern Ole Sander, Matty Demont and Martin Gummert, International Rice Research Institute (IRRI), The Philippines*
1 Introduction: rice by-products 121
2 Availability of rice straw and husks 123
3 In-field rice residue processing and management 126
4 Off-field rice residue processing and management 129
5 Using rice residues for energy production 133
6 Non-energy uses of rice residues 140
7 Summary 144
8 Future trends 146
9 Where to look for further information 148
10 References 148

**Part 2 Overall management of rice cultivation**


6 Yield gap analysis towards meeting future rice demand 157
*Kazuki Saito, Pepijn van Oort, Ibnou Dieng, Jean-Martial Johnson, Abibou Niang, Kokou Ahouanton and Amakoe Delali Alognon, Africa Rice Center, Benin; Atsuko Tanaka, JICA and Africa Rice Center, Benin; Kalimuthu Senthilkumar and Elke Vandamme, Africa Rice Center, Tanzania; Cyriaque Akakpo, INRAB, Benin; Zacharie Segda, CNRST/INERA, Burkina Faso; Ibrahim Bassoro and Delphine Mapiemfu Lamare, IRAD, Cameroon; Moundibaye D. Allarangaye, ITRAD, Chad; Henri Gbakatchetche, CNRA, Côte d'Ivoire; Belay A. Bayuh, EIAR, Ethiopia; Famara Jaiteh, NARI, The Gambia; Ralph K. Bam, CSIR-CRI; Wilson Dogbe, CSIR-SARI, Ghana; Keita Sékou, IRAG, Guinea; Raymond Rabeson, FOFIFA, Madagascar; Niananköro Kamissoko, IER, Mali; Illiassou Maïga Mossi, INRAN, Niger; Oladele S. Bakare, NCRI, Nigeria; Fanny L. Mabone, INERA, DR Congo; Elie R. Gasore, RAB, Rwanda; Idriss Baggie, SLARI, Sierra Leone; Geophrey J. Kajiru, Ministry of Agriculture, Tanzania; Komlan A. Ablede, ITRA, Togo; and David Nanfumba, NARO, Uganda*
1 Introduction 157
2 Studies on quantification of yield gaps 160
3 Studies to identify causes of yield gaps 169
4 Future trends and conclusion 174
5 Where to look for further information 178
6 References 178

7 Developments in the system of rice intensification (SRI) 183
*Norman Uphoff, Cornell University, USA*
1 Introduction: understanding the protean nature of SRI 183
2 The evolution of SRI: from practices to principles to fundamentals 184
3 The spread of SRI to other countries and crops and overall effects 187
4 Contributions of SRI to achieving climate-smart agriculture 194
5 Developments in SRI methodology 198
6 Assessing SRI yields 200
7 SRI in relation to other production systems 202
8 Summary and future trends 205
9 Where to look for further information 206
10 References 206

8 Assessing the sustainability impacts of rice cultivation 213
*Wyn Ellis, Sustainable Rice Platform, Thailand*
1 Introduction 213
2 Trends in the production and consumption of rice 214
3 Sustainability concerns: resource use and climate change impacts 217
4 Assessing sustainability in rice systems: introduction to the Sustainable Rice Platform (SRP) 220
5 Application of the SRP standard for sustainable rice cultivation 225
6 Conclusions 229
7 Where to look for further information 229

8 Acknowledgements 229
9 References 230

**Part 3 Rice pests**

9 Rice insect pests: biology and ecology 235
*E. A. Heinrichs, University of Nebraska-Lincoln, USA; F. Nwilene, The Africa Rice Center, Nigeria; M. Stout, Louisiana State University, USA; B. Hadi, International Rice Research Institute (IRRI), The Philippines; and T. Freitas, Universidade Federal Rio Grande do Sul, Brazil*
1 Introduction 235
2 Biology and ecology of rice-feeding insects: root and stem feeders 237
3 Biology and ecology of rice-feeding insects: stem borers 240
4 Biology and ecology of rice-feeding insects: rice gall midges 245
5 Biology and ecology of rice-feeding insects: leafhoppers and planthoppers 249
6 Biology and ecology of rice-feeding insects: foliage feeders 255
7 Biology and ecology of rice-feeding insects: panicle feeders 260
8 References 264

10 Management of rice insect pests 271
*E. A. Heinrichs, University of Nebraska-Lincoln, USA; F. Nwilene, The Africa Rice Center, Nigeria; M. Stout, Louisiana State University, USA; B. Hadi, International Rice Research Institute (IRRI), The Philippines; and T. Freitas, Universidade Federal Rio Grande do Sul, Brazil*
1 Introduction 271
2 Overview of concepts and options for integrated pest management (IPM) of rice 272
3 Cultural practices in rice IPM 273
4 Promoting natural enemies of rice pests: conservation biological control 276
5 Augmentative biological control 279
6 Selective use of insecticides 280
7 Dissemination mechanisms for rice IPM 283
8 References 285

11 Plant protection products in rice cultivation: critical issues in risk assessment and management to promote sustainable use 293
*Maura Calliera and Ettore Capri, Università Cattolica Sacro Cuore di Piacenza, Italy*
1 Introduction 293
2 Scientific modelling for the risk assessment of PPPs 294
3 The role of ethical, social and behavioural issues in PPP risk assessment 298
4 The European SUD and IPM 300
5 Farmer involvement in IPM 302
6 Conclusion 303
7 Where to look for further information 304
8 References 305

12 Integrated pest management for sustainable rice cultivation: a holistic approach 309
*F. G. Horgan, University of Technology Sydney, Australia*
1 Introduction 309
2 Emerging rice production landscapes and 'crop health syndromes' 310
3 Increasing rice production 312
4 Increasing water productivity 315
5 Reducing labour costs 316
6 Increased availability and marketing of agrochemicals 317
7 Pest management for rice: scientific approaches 320
8 Knowledge of the rice production system 322
9 Biocides, traps and barriers 322
10 Transgenic and herbicide-resistant rice varieties 324
11 Genetic approaches to resistance and tolerance 324
12 Enhancing natural regulation 326
13 Integrating management practices 329
14 Education and extension 329
15 Future trends and conclusion 331
16 Where to look for further information 332
17 References 333

13 Control of rodent pests in rice cultivation 343
*P. R. Brown, CSIRO Agriculture and Food, Australia; B. Douangboupha, National Agricultural and Forestry Research Institute, Lao PDR; N. M. Htwe, International Rice Research Institute, Myanmar; J. Jacob, Julius Kühn Research Institute, Germany; L. Mulungu, Sokoine University of Agriculture, Tanzania; N. T. My Phung, Provincial Department of Agriculture and Rural Development, Vietnam; G. R. Singleton, International Rice Research Institute, The Philippines and University of Greenwich, UK; A. M. Stuart, International Rice Research Institute, The Philippines; and Sudarmaji, Assessment Institute for Agricultural Technology of Yogyakarta, Indonesia*
1 Introduction 343
2 Rodent impacts on rice 345
3 Rodent pests of rice and their biology 348
4 Rodent management methods 351
5 Case studies of management of rats in rice 360
6 Other vertebrate pests in rice 366
7 Future trends and conclusion 366
8 Where to look for further information 368
9 References 369

14 Integrated weed management techniques in rice 377
*Simerjeet Kaur and Gulshan Mahajan, Punjab Agricultural University, India; and Bhagirath S. Chauhan, The University of Queensland, Australia*
1 Introduction 377
2 Weeds in rice crops: the problem 378
3 The need for IWM in rice production 380
4 IWM: an overview 380

5 IWM techniques: preparation and control of the growing environment 382
6 IWM techniques: establishing the rice crop 384
7 IWM techniques: managing the rice crop 388
8 The integration of weed management techniques 392
9 Summary 394
10 Where to look for further information 395
11 References 395

Index 399

# Series list

| Title | Series number |
|---|---|
| **Achieving sustainable cultivation of maize - Vol 1**<br>From improved varieties to local applications<br>Edited by: Dr Dave Watson, CGIAR Maize Research Program Manager, CIMMYT, Mexico | 001 |
| **Achieving sustainable cultivation of maize - Vol 2**<br>Cultivation techniques, pest and disease control<br>Edited by: Dr Dave Watson, CGIAR Maize Research Program Manager, CIMMYT, Mexico | 002 |
| **Achieving sustainable cultivation of rice - Vol 1**<br>Breeding for higher yield and quality<br>Edited by: Prof. Takuji Sasaki, Tokyo University of Agriculture, Japan | 003 |
| **Achieving sustainable cultivation of rice - Vol 2**<br>Cultivation, pest and disease management<br>Edited by: Prof. Takuji Sasaki, Tokyo University of Agriculture, Japan | 004 |
| **Achieving sustainable cultivation of wheat - Vol 1**<br>Breeding, quality traits, pests and diseases<br>Edited by: Prof. Peter Langridge, The University of Adelaide, Australia | 005 |
| **Achieving sustainable cultivation of wheat - Vol 2**<br>Cultivation techniques<br>Edited by: Prof. Peter Langridge, The University of Adelaide, Australia | 006 |
| **Achieving sustainable cultivation of tomatoes**<br>Edited by: Dr Autar Mattoo, USDA-ARS, USA & Prof. Avtar Handa, Purdue University, USA | 007 |
| **Achieving sustainable production of milk - Vol 1**<br>Milk composition, genetics and breeding<br>Edited by: Dr Nico van Belzen, International Dairy Federation (IDF), Belgium | 008 |
| **Achieving sustainable production of milk - Vol 2**<br>Safety, quality and sustainability<br>Edited by: Dr Nico van Belzen, International Dairy Federation (IDF), Belgium | 009 |
| **Achieving sustainable production of milk - Vol 3**<br>Dairy herd management and welfare<br>Edited by: Prof. John Webster, University of Bristol, UK | 010 |
| **Ensuring safety and quality in the production of beef - Vol 1**<br>Safety<br>Edited by: Prof. Gary Acuff, Texas A&M University, USA & Prof.James Dickson, Iowa State University, USA | 011 |
| **Ensuring safety and quality in the production of beef - Vol 2**<br>Quality<br>Edited by: Prof. Michael Dikeman, Kansas State University, USA | 012 |
| **Achieving sustainable production of poultry meat - Vol 1**<br>Safety, quality and sustainability<br>Edited by: Prof. Steven C. Ricke, University of Arkansas, USA | 013 |
| **Achieving sustainable production of poultry meat - Vol 2**<br>Breeding and nutrition<br>Edited by: Prof. Todd Applegate, University of Georgia, USA | 014 |
| **Achieving sustainable production of poultry meat - Vol 3**<br>Health and welfare<br>Edited by: Prof. Todd Applegate, University of Georgia, USA | 015 |

**Achieving sustainable production of eggs - Vol 1** 016
Safety and quality
Edited by: Prof. Julie Roberts, University of New England, Australia

**Achieving sustainable production of eggs - Vol 2** 017
Animal welfare and sustainability
Edited by: Prof. Julie Roberts, University of New England, Australia

**Achieving sustainable cultivation of apples** 018
Edited by: Dr Kate Evans, Washington State University, USA

**Integrated disease management of wheat and barley** 019
Edited by: Prof. Richard Oliver, Curtin University, Australia

**Achieving sustainable cultivation of cassava - Vol 1** 020
Cultivation techniques
Edited by: Dr Clair Hershey, formerly International Center for Tropical Agriculture (CIAT), Colombia

**Achieving sustainable cultivation of cassava - Vol 2** 021
Genetics, breeding, pests and diseases
Edited by: Dr Clair Hershey, formerly International Center for Tropical Agriculture (CIAT), Colombia

**Achieving sustainable production of sheep** 022
Edited by: Prof. Johan Greyling, University of the Free State, South Africa

**Achieving sustainable production of pig meat - Vol 1** 023
Safety, quality and sustainability
Edited by: Prof. Alan Mathew, Purdue University, USA

**Achieving sustainable production of pig meat - Vol 2** 024
Animal breeding and nutrition
Edited by: Prof. Julian Wiseman, University of Nottingham, UK

**Achieving sustainable production of pig meat - Vol 3** 025
Animal health and welfare
Edited by: Prof. Julian Wiseman, University of Nottingham, UK

**Achieving sustainable cultivation of potatoes - Vol 1** 026
Breeding, nutritional and sensory quality
Edited by: Prof. Gefu Wang-Pruski, Dalhousie University, Canada

**Achieving sustainable cultivation of oil palm - Vol 1** 027
Introduction, breeding and cultivation techniques
Edited by: Prof. Alain Rival, Center for International Cooperation in Agricultural Research for Development (CIRAD), France

**Achieving sustainable cultivation of oil palm - Vol 2** 028
Diseases, pests, quality and sustainability
Edited by: Prof. Alain Rival, Center for International Cooperation in Agricultural Research for Development (CIRAD), France

**Achieving sustainable cultivation of soybeans - Vol 1** 029
Breeding and cultivation techniques
Edited by: Prof. Henry Nguyen, University of Missouri, USA

**Achieving sustainable cultivation of soybeans - Vol 2** 030
Diseases, pests, food and non-food uses
Edited by: Prof. Henry Nguyen, University of Missouri, USA

**Achieving sustainable cultivation of sorghum - Vol 1** 031
Genetics, breeding and production techniques
Edited by: Prof. Bill Rooney, Texas A&M University, USA

**Achieving sustainable cultivation of sorghum - Vol 2** 032

Sorghum utilisation around the world
Edited by: Prof. Bill Rooney, Texas A&M University, USA

**Achieving sustainable cultivation of potatoes - Vol 2** 033
Production and storage, crop protection and sustainability
Edited by: Dr Stuart Wale, Potato Dynamics Ltd, UK

**Achieving sustainable cultivation of mangoes** 034
Edited by: Professor Víctor Galán Saúco, Instituto Canario de Investigaciones Agrarias (ICIA), Spain & Dr Ping Lu, Charles Darwin University, Australia

**Achieving sustainable cultivation of grain legumes - Vol 1** 035
Advances in breeding and cultivation techniques
Edited by: Dr Shoba Sivasankar et al., CGIAR Research Program on Grain Legumes, ICRISAT, India

**Achieving sustainable cultivation of grain legumes - Vol 2** 036
Improving cultivation of particular grain legumes
Edited by: Dr Shoba Sivasankar et al., CGIAR Research Program on Grain Legumes, ICRISAT, India

**Achieving sustainable cultivation of sugarcane - Vol 1** 037
Cultivation techniques, quality and sustainability
Edited by: Prof. Philippe Rott, University of Florida, USA

**Achieving sustainable cultivation of sugarcane - Vol 2** 038
Breeding, pests and diseases
Edited by: Prof. Philippe Rott, University of Florida, USA

**Achieving sustainable cultivation of coffee** 039
Breeding and quality traits
Edited by: Dr Philippe Lashermes, Institut de Recherche pour le Développement (IRD), France

**Achieving sustainable cultivation of bananas - Vol 1** 040
Cultivation techniques
Edited by: Prof. Gert Kema, Wageningen University, The Netherlands & Prof. André Drenth, University of Queensland, Australia

**Global Tea Science** 041
Current status and future needs
Edited by: Dr V. S. Sharma, Formerly UPASI Tea Research Institute, India & Dr M. T. Kumudini Gunasekare, Coordinating Secretariat for Science Technology and Innovation (COSTI), Sri Lanka

**Integrated weed management** 042
Edited by: Emeritus Prof. Rob Zimdahl, Colorado State University, USA

**Achieving sustainable cultivation of cocoa - Vol 1** 043
Genetics, breeding, cultivation and quality
Edited by: Prof. Pathmanathan Umaharan, Cocoa Research Centre – The University of the West Indies, Trinidad and Tobago

**Achieving sustainable cultivation of cocoa - Vol 2** 044
Diseases, pests and sustainability
Edited by: Prof. Pathmanathan Umaharan, Cocoa Research Centre – The University of the West Indies, Trinidad and Tobago

**Water management for sustainable agriculture** 045
Edited by: Prof. Theib Oweis, Formerly ICARDA, Lebanon

**Improving organic animal farming** 046
Edited by: Dr Mette Vaarst, Aarhus University, Denmark & Dr Stephen Roderick, Duchy College, Cornwall, UK

**Improving organic crop cultivation** 047
Edited by: Prof. Ulrich Köpke, University of Bonn, Germany

**Managing soil health for sustainable agriculture - Vol 1** 048
Fundamentals
Edited by: Dr Don Reicosky, USDA-ARS, USA

**Managing soil health for sustainable agriculture - Vol 2** 049
Monitoring and management
Edited by: Dr Don Reicosky, USDA-ARS, USA

**Rice insect pests and their management** 050
E. A. Heinrichs, Francis E. Nwilene, Michael J. Stout, Buyung A. R. Hadi & Thais Freitas

**Improving grassland and pasture management in temperate agriculture** 051
Edited by: Prof. Athole Marshall & Dr Rosemary Collins, University of Aberystwyth, UK

**Precision agriculture for sustainability** 052
Edited by: Dr John Stafford, Silsoe Solutions, UK

# Acknowledgements

We wish to acknowledge the following for their help in reviewing particular chapters:

- Chapter 2: *Dr G. A. K. Kumar, Central Rice Research Institute (CRRI), India*
- Chapter 13: *Dr Gary Witmer, USDA-APHIS, USA*

# Introduction

Rice (*Oryza sativa* L.) is a principal food for over 3 billion people (around half the global population), providing up to 23% of calorie requirements worldwide, and is the staple food for the 70% of the world's poor living in Asia. Rice is cultivated in more than 100 countries across the world on around 154 million hectares annually, equal to 11% of the world's cultivable land. It has been predicted that global demand for rice will increase from 479 million tons in 2014/15 to as much as 551 million tons by 2029/30. At present, the increase in rice production rate (at under 1% per year) is lower than the increase in population of 1.5% per year. The rice community therefore faces the challenge of increasing rice production by 25-50% over the next thirty years, in the face of a rising population, plateauing rice yields, pressures on land and other resources available for rice cultivation and an unstable, changing climate. The two volumes of *Achieving sustainable cultivation of rice* summarize some of the key research addressing this challenge. This volume (Volume 2) focuses on developments in cultivation techniques and pest management in making production more efficient and sustainable. Volume 1 reviews developments in breeding to improve yields as well as research on improving nutritional and other aspects of quality.

## Part 1 Rice cultivation techniques

The majority of the world's rice is grown under irrigated conditions in which the fields are flooded from planting to harvest. Rice land receives 35–45% of all the world's irrigation water (which itself uses around 70% of all the world's developed water resources). In view of the importance of rice and growing concerns about future water scarcity, achieving water savings in irrigated rice production has become one of the key research challenges for the rice-growing world.

As Chapter 1 points out, there are a number of potential research avenues being pursued to achieve water savings without sacrificing rice yields. These avenues include:

- breeding short-duration, high-yielding modern rice varieties which require less water due to less time growing in the field (discussed in Chapters 3–5 in Volume 1 of *Achieving sustainable cultivation of rice*)
- breeding and cultivation of varieties which yield better in harsh environments (such as higher salinity or drought conditions – discussed, e.g., in Chapter 3 in this volume and Chapter 6 in Volume 1)
- precision agriculture, site-specific nutrient management (SSNM) and improved crop establishment methods to increase yields and minimize wasted resources (reviewed, e.g., in Chapters 2, 6 and 7 in this volume)
- improved irrigation techniques which minimize unproductive water losses in the cropping system (reviewed in Chapter 1)

Essentially the aim of any water-saving technique is to increase irrigated water productivity (WP), the amount of rice produced per unit of irrigation water applied. At the field level, water is either stored in the soil matrix, transpired to the atmosphere via the crop, lost by evaporation directly from the soil or water surface, or lost from the system via a number of other possible mechanisms (such as deep drainage below the reach of crop roots, lateral flows out of the field which include surface runoff or seepage under the bunds, etc.). Any attempts to save water must take account of factors such as evapotranspiration (ET) from soil and plants which takes water out of the system.

In the Indian state of Punjab, for example, declining water table depth has been a major concern for decades due to over-exploitation of water resources (particularly pumping of groundwater). Numerous water-saving practices have been suggested to save irrigation water and contribute to arresting the decline (such as laser land levelling, alternate wetting and drying (AWD) techniques, conservation agriculture practices, raised beds, delayed rice transplanting or shorter-duration rice varieties). These irrigation water-saving techniques can achieve water savings at the field scale. However, it has been suggested that irrigation water-saving options which primarily reduce ET (such as residue retention to reduce soil evaporation) are most likely to succeed in addressing the problem in Punjab, where evaporative losses are the true losses in this system at the regional scale which are driving the water table decline.

Chapter 1 discusses three water-saving techniques: aerobic rice systems, AWD, and saturated soil culture (SSC). These can be ranked according to levels of water availability, starting with aerobic rice systems. Further up the spectrum towards fully ponded lowland rice systems (which may be rainfed or irrigated) are interim irrigated practices such as AWD and SSC.

Aerobic rice is a term introduced by the International Rice Research Institute (IRRI) for high-yielding rice grown under non-flooded conditions in non-puddled and unsaturated (aerobic) soil. In an aerobic rice system, the crop can be dry direct seeded or transplanted and soils are kept aerobic throughout the growing season. Supplemental irrigation is applied as necessary, for example using sprinkler irrigation, furrow irrigation on beds, and drip and subsurface irrigation. The use of surface mulches in aerobic rice systems has been noted to increase irrigation water productivity. Studies have suggested that, by reducing water use during land preparation and limiting seepage, percolation and evaporation, aerobic rice had 51% lower total water use and 32–88% higher water productivity, expressed as gram of grain per kilogram of water, than flooded rice. When specific aerobic cultivars are grown under aerobic conditions and compared with specific lowland cultivars grown under lowland conditions, the reported WP gains are even higher. The WUE of the aerobic varieties under aerobic conditions can be 164–188% higher than that of lowland cultivated rice varieties. A key problem in aerobic rice cultivation is yield decline resulting from continuous cropping of aerobic rice, due to factors including build-up of pests such as nematodes and the loss of soil organic carbon reserves. These problems can be offset by techniques such as crop rotation, better nitrogen (N) management and cultivar improvement. Dry-seeded and aerobic cultivation of rice are both discussed in Chapter 4 in this volume.

An established irrigation water-saving technology for rice is AWD, based on the fact that high yields in rice can be achieved without continuous flooding. It has been demonstrated over many years in numerous environments that rice can extract adequate water from the soil around roots even in the absence of standing water in the field. In AWD, once the crop is established, the pond depth is allowed to fall to a threshold depth below the surface of the soil for a certain period before the next irrigation is applied. As chapter 1 shows, AWD improves water productivity (WP) universally when applied within recommended guidelines. However, research varies in terms of AWD's impact on yields. AWD is unlikely to be a key solution to the crisis of falling water tables in regions such as the Indian

Northwest, where technologies to decrease ET are the main objective. However, in areas where deep percolation losses are effectively lost from the farmer's irrigation system, AWD water savings may be a real solution to water-shortage problems.

SSC is an irrigation practice which involves keeping the rice-growing soil saturated but not ponded throughout the growth period, thereby reducing the hydraulic head of the water, and in turn reducing the percolation rate and water lost to deep drainage. SSC is the typically nominated water management method in the system of rice intensification (SRI), which is reviewed in detail in Chapter 7 in this volume. Studies show a wide range of water savings of between 9 and 40%, depending on factors such as soil permeability. The literature also indicates that, in combination with AWD, SSC has the potential to increase water savings to over 50%.

Chapter 1 concludes with a case study of the use of AWD in growing japonica rice in temperate southeast Australia. It shows how research has been able to establish the optimal level of moisture stress between irrigation periods to save water without compromising yields. Locally targeted research to fully understand the interactions between crop, management and environmental constraints, and then to subsequently use this understanding to optimize practical advice to farmers, represents the greatest challenge to researchers in the field of irrigation water savings.

It has been estimated that rice accounts for up to 15% of total fertilizer use, a figure that rises to 20% in the case of the use of nitrogen fertilizer. Fertilizers account for 20 to 25% of total production costs in some rice systems. Due to large field-to-field variability in soil nutrition, the efficient use of fertilizers is not possible with the traditional reliance on broad-based blanket recommendations for their use. A key challenge for more efficient and sustainable rice cultivation is therefore a more targeted use of fertilizers. Chapter 2 provides a comprehensive review of the ways this is being achieved for nitrogen (N), phosphorus (P) and potassium (K).

SSNM for rice was developed in Asia by the IRRI as a plant-based approach for applying fertilizer to optimally match the needs of the rice crop in a specific field and season. The major focus of SSNM is on managing both temporal variability in plant nutrient status during the growing season and managing field-specific spatial variation in nutrient requirements. Given its importance, a particular focus is on N requirements, with requirements for P and K also taken into consideration. The chapter also discusses the targeted application of micronutrients such as zinc, iron, manganese and boron. Modelling has been used to predict soil nutrient supply and plant uptake at differing stages of crop growth. To ensure that supply of N matches the crop need at critical growth stages, the estimated total fertilizer N requirement by rice crop is then apportioned among multiple application points during the growing season ('fixed-time/adjustable dose').

A key element in achieving SSNM is the development of diagnostic tools that can assess 'real-time' requirements of rice for N and other nutrients. The chapter reviews the development and use of non-invasive, optical methods based on leaf greenness, absorbance and/or reflectance of light by leaves. These include leaf colour charts (LCC); chlorophyll meters; ground-based sensors; and digital, aerial and satellite imaging techniques. The chapter discusses the principles and application of chlorophyll meters, LCC and optical sensors in detail. Both 'real-time' and 'fixed-time/adjustable dose' methods can be used to target N, P and K application more precisely.

In addition to investigating better control of the timing and amount of fertilizer, research has also looked at how it can be better delivered. One of the reasons for the low fertilizer use efficiency of water-soluble N fertilizers such as urea, ammonium sulphate and ammonium carbonate in rice is the imbalance between the timing and degree that fertilizer is delivered to plants and the actual demands of the crop. Slow- and controlled-release N fertilizers increase N use efficiency and rice yields whilst reducing fertilizer losses because they synchronize N availability with plant demand. Slow-release fertilizers consist of compounds of generally low water solubility, which become available on enzymatic hydrolysis by urease or other biological catalysts. The N release patterns, rates and duration of slow-release fertilizers are thus strongly dependent on soil properties and environmental conditions such as moisture and temperature. Controlled-release fertilizers consist of highly soluble urea pills or granules coated with a water-insoluble material such as sulphur or polyolefin that control the rate, pattern and duration of N release. Besides the advantages of controlled-released fertilizers in reducing N losses to the environment and increasing fertilizer N use efficiency in rice, the rate of N application or the number of applications during the growing season can also be reduced, which has the added advantage of saving labour costs.

There is considerable interest in the use of urease and nitrification inhibitors in rice systems because urea is the most important conventional N fertilizer for rice and N loses via NH3 volatilization and nitrification–denitrification significantly contribute to low fertilizer use efficiency. The short-lived effect of the inhibitor is a limitation when urea is applied to flooded rice and further research is needed to improve our understanding of when urease and nitrification inhibitors can be used most cost-effectively. Placement of fertilizer nutrients, particularly deep placement in the soil, is also crucial to increase rice yield because it minimizes N losses. The chapter reviews research on the effective use of urea supergranules (USG) and band placement of liquid urea using mechanical injection techniques.

Finally, as the chapter shows, organic manures have regained importance as components of integrated plant nutrient management (IPNM) strategies. A return to managing soil fertility through the combination of organic resources and mineral nutrient inputs has become one of the key features of sustainable soil management. The basic concept underlying IPNM is the maintenance and improvement of soil health for more sustainable, long-term crop productivity with fertilizers used as a supplement to nutrients supplied by different organic sources available at the farm to meet the nutrient requirement of the crops. The maintenance and/or improvement of organic matter in soil is central to the philosophy of IPNM, using inputs such as farmyard manure, crop residues, sewage sludge or food industry wastes. During the last two decades, many studies of rice cultivation in South Asia have shown the positive impact on rice yields from integrated management of different organic and mineral fertilizer inputs. Future challenges include reducing the costs and simplifying SSNM procedures for rice farmers without losing their ability to target fertilizer use more effectively as well as refining IPNM to integrate different sources of nutrition.

As mentioned in Chapter 1, a major challenge in rice cultivation is cultivation in less favourable environments, given both the limits to productivity gains in established irrigated rice environments and the likely impact of climate change. Chapter 3 reviews how to optimize rice cultivation in saline, flood-prone coastal areas, using West Bengal in India as an example. It builds on topics in both Chapters 1 and 2, for example irrigation and nutrient management as well as improving soil health through conservation agriculture techniques. It also demonstrates the importance of an integrated and holistic approach from the use of appropriate varieties to better cultivation, nutrient and irrigation management practices.

The chapter first explores the development of improved, salt-tolerant varieties using mutagenic techniques to cross local landraces as well as the shift to growing salt- and flood-tolerant varieties during the wet season whilst using salt-tolerant short-duration varieties during the dry season. It also explores the role of improved agronomic practices such as raising healthy seedlings through proper nursery management techniques such as optimal seeding density and nutrition. It also reviews better methods for planting, whether transplanting of aged seedlings during the wet season or the use of drum seeders for early, more efficient crop establishment during the dry season. Weed management is also covered, a subject discussed in detail in Chapter 14.

Building on Chapter 2, the chapter also reviews the use of IPNM using combined organic and inorganic sources of nutrition, including farmyard manure and biofertilizers such as *Azotobacter*, *Azospirillum* and blue green algae. It also discusses the role of cover crops such as the legume *Sesbania aculeata* and *Azolla*, a free-floating freshwater fern which fixes nitrogen in the soil and inhibits salt movement into rice plants, as well as inorganic nutrition such as deep placement of USG. The chapter discusses other aspects of conservation agriculture such as mulching during dry season using rice husk and straw or organic farm waste, for example, as well as techniques for leaching of excess soluble salts from the root zone of crops and ways of adapting SRI techniques (a topic discussed in Chapter 7).

Direct seeding (also known by the acronym DSR (direct-seeded rice)) is an alternative to more traditional transplanting techniques particularly suited to rainfed regions. Seeds are sown on a dry or moist non-puddled soil surface, and then incorporated by ploughing or harrowing. Key issues in successful dry direct seeding include effective land levelling (e.g. to ensure effective irrigation) and good crop establishment (to ensure optimum plant density), for example, by optimizing planting date, seed treatment, seed rate and planting depth. Other issues include precise water management (especially in the early stages of growth to ensure good rooting and seedling growth) and effective weed management (to avoid competition with young rice plants). The chapter reviews good agricultural practices (GAP) in DSR, from cultivar selection and soil preparation to seed treatment and sowing, irrigation and fertilization, thinning, weed management and harvesting.

As highlighted in Chapter 1, aerobic cultivation is a water-saving production system in which rice is grown in well-drained, non-puddled and non-saturated soil conditions. The main advantages of aerobic cultivation include reduced water use, lower labour requirements and costs, the need for fewer seeds, improved efficiency in fertilizer use and preservation of soil texture and quality. Challenges facing aerobic rice cultivation include lower yields compared to traditional transplanting and competition from weeds. The chapter summarizes GAP techniques to address these issues.

The final chapter in Part 1 looks at the handling of by-products from rice production. As Chapter 5 points out, each kg of milled rice results in roughly 1.4 kg of rice straw and 0.28 kg of husk. Rice straw is separated from the grains after the plant is threshed whilst rice husk is a by-product in the milling process. The intensification of rice cropping systems with the increased use of combine harvesters and shorter breaks between crops has both significantly increased the volume of these by-products, made collection more difficult and complicated the traditional use of straw as a soil amendment. This has resulted in an increase in straw burning with damaging consequences for soil health and the environment. The chapter looks at ways of overcoming these problems such as speeding up straw decomposition during breaks between crops and the development of rice straw baling systems to collect straw more easily. The chapter also reviews ways of using the by-products, for example to generate fuel, heat or electricity through thermal, chemical or bio-based processes. It includes illustrative case studies on rice husk combustion for paddy drying, power generation from rice straw combustion, conversion of rice straw to biochar for use as a soil amendment, cultivation of mushroom on rice straw and the use of rice straw for livestock feed.

# Part 2 Overall management of rice cultivation

The range of techniques discussed in Part 1 can only be effective if there are methods for accurately assessing yield gaps and their causes, setting targets for improvement and measuring improvements in productivity. Chapter 6 starts by reviewing differing definitions of yield, starting with 'potential yield', defined as the maximum yield that can be obtained from a crop in a given environment. As the chapter points out, there are other yield benchmarks, namely 'economic yield', 'experimental yield' and 'best farmers' yield' which lie between potential yield (or water-limited potential yield) and average on-farm yield.

The chapter provides a valuable overview and comparison of existing yield gap studies such as the results of the Global Yield Gap Atlas (GYGP) programme for rice, what they show and their degree of reliability. It looks first at studies focusing on the quantification of yield gaps, then describes studies concentrating on identifying the causes of yield gaps. Examples are mainly from sub-Saharan Africa, where yields have been historically low, though research here mirrors similar results from South Asia. The chapter discusses the various methods for investigating causes, including crop simulation models, use of survey data and farm experiments. Research has shown a broad range of factors contributing to yield gaps, including:

- Biophysical (weather, soil, water, pest pressure, weeds);
- Technical/management (labour availability, timing of operations, tillage, variety or seed selection, management of water, nutrients, weeds and pests, and harvesting);
- Socio-economic (social and economic status, farmers' cultural traditions, commitments and obligations, knowledge, family size and farm profitability);
- Institutional and policy (government policy, rice price, credit, input supply, land tenure, markets, research, development and extension);
- Technology learning and linkages (competence and resources of extension staff; knowledge and skills; linkages among public, private and NGO extension staff).

One illustration comes from the Senegal River Valley which identified how delayed sowing resulted in a yield reduction of around 1 t/ha and the reasons behind this delay such as availability of credit, machinery and irrigation water. The chapter concludes by reflecting on the ways in which different approaches can be used and how they can complement each other. It concludes with a discussion of the challenges to achieving better quantification of yield gaps and their causes, and the implications of yield gap studies for sustainable agricultural development to meet future rice demand. As it points out, rapid advances are being made in remote sensing technology for estimating various parameters such as soil quality, cropping season and water availability. Combined with the expected increase in sources of reliable weather data and improved bias correction for satellite-based weather data, advances in technology will enable the development of crop models of potential and water-limited yield that are applicable to a wide range of spatial scales.

Chapter 7 reviews one of the best-known and controversial systems of rice production: SRI which, from its origins in Madagascar some 30 years ago, is now used in over 50 countries. SRI involves practices such as:

- transplanting single, young, widely spaced seedlings, typically in square patterns
- keeping soil mostly wet but not flooded, either with small daily applications of water or with AWD techniques (as discussed in Chapter 1)
- controlling weeds mechanically, both to protect the crop and aerate the soil
- enriching the soil with organic matter

The principles underlying SRI can be summarized as:

- early crop establishment with healthy plants and minimal competition between plants to promote growth of root systems and canopies
- promotion of soil health/fertility
- management of water to avoid anaerobic soil conditions

SRI has allowed a significant reduction in use of inputs such as water and fertilizer. It can be seen as sharing a common foundation with conservation agriculture. Written by one of the world's foremost authorities on SRI, the chapter reviews the origins and spread of SRI as well as more recent developments such as the use of direct seeding for crop establishment and mechanization of SRI operations. The chapter also provides a detailed assessment of the types of yield achieved by SRI in different conditions and provides a careful review of the sometimes conflicting results on how it compares with other types of rice cultivation practice.

The final chapter in Part 2 picks up on resource efficiency and sustainability issues explored in the chapters in Part 1 as well as other chapters such as Chapter 7 on SRI. As noted earlier, rice cultivation consumes significant amounts of irrigation water, fertilizers and other inputs such as pesticides. Sustainability issues in rice cultivation include:

- resource use efficiency (land, water, agrochemicals, labour)
- climate change impacts
- impacts on ecosystem services
- soil impacts (e.g. salinization)

Flooded rice cultivation (irrigated, rainfed and deep-water cultivation) also represents an important source of atmospheric methane and, to a lesser extent, nitrous oxide. Rice cultivation not only has a significant environmental impact but is, itself, vulnerable to the effects of climate change. This makes it critical to make rice cultivation more sustainable as it seeks to improve yields.

As Chapter 8 indicates, voluntary sustainability standards (VSS) have emerged as an important tool to promote more sustainable practices in agri-food systems. However, progress in rice has been slow, in part because, with a low-value bulk commodity such as rice, compliance costs cannot easily be borne by small, resource-poor farmers. In addition, with less than 2% of global rice production destined for high-value markets, there has been less consumer pressure for sustainability initiatives.

This has led organizations such as the United Nations Environment Programme and the IRRI to launch the sustainable rice platform (SRP) in 2011 to promote adoption of sustainable climate-smart best practices in rice cultivation. Such practices include:

- AWD and other ways of improving irrigation (discussed in Chapter 1)
- Multi-stress-tolerant varieties (reviewed, e.g., in Chapter 6 in Volume 1)
- SSNM (discussed in Chapter 2)
- Methods for sustainable intensification of rice cropping systems (such as SRI) (reviewed in Chapter 7)
- Biofortification initiatives (surveyed in Part 2 of Volume 1)
- Low-cost post-harvest technologies to manage residues and reduce food losses (discussed in Chapter 5)

A set of quantitative indicators (SRP performance indicators for sustainable rice cultivation) has been developed by SRP stakeholders to provide a system for objective measurement of the benefits and impacts of adopting good environmental practices in such areas as greenhouse gas emissions, water, nutrient and pesticide use efficiency. This allows farmers to work towards the SRP standard for sustainable rice cultivation. This was launched in 2015 as the world's first VSS for rice, serving to both define what constitutes sustainability and promote improvement. By incorporating a scoring system, the standard allows for stepwise improvement in order to encourage and reward progress towards full compliance. The current standard is currently undergoing an extensive multi-country field validation programme in Brazil, Cambodia, China, India, Indonesia, Myanmar, Pakistan, the Philippines, Thailand, Uganda and Vietnam. The findings of the validation programme will be used to revise the standards and indicators, and then define targets and voluntary/mandatory levels of compliance for each requirement, with the aim of providing a viable path to improving sustainability.

## Part 3 Rice pests

The final part of the book reviews key pests which limit rice yields and measures for their management. As Chapter 9 points out, the rice plant is an ideal host for many insect species. There are over 800 insect species which damage rice in one way or another, although the majority of these species do not cause a significant problem. In tropical Asia, about 20 species are of major importance. In Africa, 15 species of insects are considered serious pests of rice whilst about 20 species cause significant damage in the Americas. To develop effective pest management strategies, it is essential to properly identify and to understand the biology and ecology of major insect pests. This chapter utilizes the unique knowledge and expertise of leading rice entomologists from Africa, Asia and the Americas to provide the first global overview of rice insect pests. The chapter reviews insects based on the following feeding types: root and stem feeders, stem borers, rice gall midges, leafhoppers and planthoppers, foliage feeders and panicle feeders. In each case, the chapter provides a summary of each type together with a case study of a particular insect. The case studies cover the rice water weevil, rice striped borer, Asian rice gall midge, brown planthopper, African rice hispa and rice bug. The discussion of each insect includes geographical distribution, plant hosts other than rice, description and biology, plant damage and ecology.

With Chapter 9 as a context, Chapter 10 reviews techniques for managing insect pests. The chapter suggests a four-phase framework for the integrated management of insect pests, starting with measures to prevent damaging levels of pests (phases 1 and 2) and thus minimize the need for curative actions (phases 3 and 4). There are two main types of preventative strategy: 1) cultural practices that directly target insect pest populations (primary practices) and 2) crop husbandry practices with secondary effects on pest populations (secondary practices). Practices that directly target insect pest populations include mechanical and physical removal of insect pest populations.

Among cultural practices for pest management, deployment of host-resistant plants (a primary cultural practice) and fertilizer management (a secondary cultural practice) form basic building blocks in developing a rice ecosystem with reduced vulnerability against insect pests. As Chapter 10 points out, much work has been done to identify resistance genes, especially against hemipteran pests, and breed them into elite varieties. The chapter refers to several general deployment strategies to improve the durability of resistance genes against pests and diseases: gene rotation, field mixture of resistance genes and gene pyramiding. In general, fertilizer management affects rice insect pest populations by modifying rice plants' suitability for and attractiveness towards herbivorous insects as well as their enemies.

Phases 3 to 4 (which seek to remove rather than prevent pests) cover chemical and biological control of pests. High usage of pesticides potentially damages populations of predators which prey on insect pests and can result in resistance among pests. Conservation biological control aims to remedy this situation by limiting insecticide use, promoting selective insecticides and altering crop habitats to allow better support for natural enemy populations. Various habitat management schemes, for example, intercropping and introduction of wildflower strips or plants repellent to herbivores, have been shown to disrupt pests, enhance natural enemies and reduce crop damage. In contrast to conservation biological control, the aim of which is to improve the impact of indigenous natural enemy community, augmentative biological control typically aims for a limited-time increase in specific natural enemy species to temporally suppress pest populations and avoid economic damage in yield. Augmentative biological control is usually achieved by field releases of predators, parasitoids or pathogens with expected effects to last either season-long (inoculative biological control) or within a short period of time during the season (inundative biological control).

Finally, Chapter 10 reviews ways of disseminating these management practices through farmer field schools (FFS) and mass media campaigns (MMC). FFS and MMC complement each other by creating a nucleus of farmers with in-depth understanding of agro-ecosystem management and the ability to conduct their own field experiments, while mass communicating simpler pest management messages towards a larger farming audience.

As Chapter 10 shows, even when seeking to keep their use to a minimum, insecticides remain a component of pest management alongside other methods of control. This theme is reiterated in Chapter 11 which argues that the use of plant protection products (PPPs) such as insecticides, herbicides and fungicides is still a key component in ensuring the sustainability of rice cultivation. It is essential to evaluate the risks these products pose to human health and the environment, and to determine what measures are required to keep the risks within acceptable limits. Chapter 11 reviews how the risks posed by PPPs are currently evaluated, focusing on the importance of multi-tier assessments and the range of available simulation models, and considers where current risk assessment practices require improvement. Methods of managing the risks associated with PPPs and encouraging more sustainable rice cultivation are then discussed, using the example of the European Sustainable Use Directive which emphasizes the importance of integrated pest management (IPM) and how IPM strategies can be implanted in practice by farmers.

Chapter 12 builds on Chapters 9 through 11 by looking in more detail at key issues and challenges in implementing IPM strategies in practice in rice cultivation. As the chapter points out, rice cultivation in many parts of the world has continued to depend significantly on pesticide use. However, the trend towards an increasing availability of pesticides has not resulted in correspondingly lower pest and weed incidences or sustained yield improvements. Instead, it has led to area-wide problems of resistant pests and weeds, sustained outbreaks, for example, of planthoppers and leaffolders, problems in managing other pests such as snails, and concerns about the wider impact of pesticides on environmental and human health.

A key reason for the continued threat from pests is that the efficiency of natural enemies such as spiders and predatory wasps is reduced by insecticides, either because they are killed by the chemicals or because their behaviours are altered. Increased pest reproduction and reduced mortality leads to exponential population growth rates in high-fecundity species such as planthoppers and leaffolders, a phenomenon known as 'resurgence'. Many scientific reports have described pesticide-induced resurgence in a range of rice pest insects including planthoppers, stemborers and leaffolders, as well as paddy-dwelling species of health concern such as mosquitoes.

The chapter identifies three main ingredients for successful IPM application. The first of these is knowledge of the rice production system, its component species (including the rice plant) and the nature of interactions between these species. It is also essential to understand how rice ecosystems, particularly their pest and weed populations, are regulated through negative feedback loops which can lead, for example, to problems of resurgence. The chapter introduces the concept of 'crop health syndromes' which emphasizes the need both to understand the complexity of host–plant interactions and, in consequence, to develop a holistic approach in pest management. As an example, several herbivores including insects, snails and rodents are associated with paddy field weeds, sometimes facilitating the weeds by reducing shade from the rice canopy. Weeds may in turn lead to increased damage to rice from birds, as has been noted from recent studies in Africa. In different conditions, certain 'pests' may become beneficial 'allies' because of their predatory or herbivore roles, for example, apple snails may increase rice production costs by damaging young seedlings in flooded rice systems, but can play a major role in reducing weed biomass during later crop stages. IPM can only be effective if these interactions are properly understood.

An example of the importance of understanding of these interactions can be seen in research on the way cultural practices such as use of fertilizers affects pests and their management. As discussed in Chapter 10, field and laboratory studies have indicated higher levels of damage from insects, snails and weeds to rice produced under high nitrogen conditions. The structure of weed communities (relative abundance and species richness) may also significantly change under high nitrogen conditions. Farmers who adopt high-yield varieties, particularly hybrid rice varieties, will often use higher amounts of fertilizer to meet the yield potential of the varieties. Hybrid rice has therefore been associated with higher incidences of several pests and diseases, including stemborers and planthoppers.

Building on a full understanding of pest ecology, the second key component of a successful IPM strategy is a set of tools to boost natural regulation of pests and avoid triggering damaging crop health syndromes. A precondition for using IPM techniques successfully is effective detection and monitoring of pests. As an example, the chapter highlights the use of light traps to determine the extent, direction and climatic conditions that generate planthopper migrations and to act as an early warning system for rice farmers before planthoppers alight in their fields. Recent research has attempted to develop technologies to better detect pests including using remote sensing applications. It will be necessary to continually update damage thresholds and estimated yield losses as climates and crop management practices change and because different rice varieties often have specific responses to pest attack.

The chapter provides a detailed review of research into different IPM tools. These range from developing and evaluating botanical pesticides such as those derived from the neem plant, to the development of synthetic pheromones to disrupt the life cycles of pests such as striped stemborers and ricebugs, or repellents such as methyl anthranilate, pulegone and caffeine. Biocontrol of insects and snails using ducks and fish has also gained recent attention as part of integrated farming systems. Building on Chapter 10, the chapter reviews recent research on the use of augmentative biological control for the control of rice field insects, nematodes and storage pests. These studies have focused on the development of rearing and application methods for *Trichogramma* spp. to control stemborers for example, and the use of bacteria such as *Pseudomonas fluorescens* to control insects, nematodes and rice diseases. Among the advantages of using fungal and bacterial biocontrol agents are the ease with which they can be reared and the facility with which they can be applied in the field. However, they do exert selection pressure for pest adaptation if used in excess.

An important element of IPM is to encourage predators of insect pests. Ecological engineering research in Asia has mainly focused on creating flower strips and vegetable patches as a source of alternative foods and refuge for the natural enemies of planthoppers and stemborers. Recent reports have indicated that such systems can successfully increase rice yields, reduce pest damage, reduce or eliminate pesticide inputs, increase the biodiversity and function of predators including predatory birds, and increase farm profits and food production by using the space available on rice bunds (levees) to grow vegetables.

Another element in an IPM arsenal is the development of pest-resistant varieties. In a recent study, for example, researchers from South and Southeast Asia evaluated the effectiveness of several donor varieties with known resistance genes against the brown planthopper. Their results indicated that most genes were now ineffective, but that traditional varieties and landraces containing two or more resistance genes remained largely effective. This supports the idea that pyramiding resistance genes can improve the durability of anti-herbivore resistance. Following the development of Bt-transgenic crops with resistance against chewing insects, transgenic resistance has become a major focus of research into reducing stemborer and leaffolder damage in rice. Transgenic rice expressing lectins, protease inhibitors and venoms have also been considered to combat rice planthoppers and other rice pests. More recently, research into RNAi technology has gained some attention for controlling planthoppers and stemborers.

However, it is clear that more research is required to bring the different 'technologies' together into functioning and sustainable rice production systems. Research into ecological engineering and biocontrol has rarely incorporated innovations in pest-resistant varieties whilst breeding initiatives often produce a resistant variety without considering how it can best be deployed within an overall IPM programme. Future research might benefit from more successfully integrating such diverse components of pest management.

The third and final key component of successful IPM is effective communication of knowledge and guidance of farmers towards best management practices. IPM is therefore as much about communication networks, effective teaching and extension activities, and agricultural policy as it is about ecology or technology. Research suggests that the main reasons for non-adoption of IPM include knowledge-related issues (lack of training, little technical support or limited availability of information), structural issues (lack of government policy and support, or limited funding) and implementation issues (prohibitive costs, limited access to inputs or over-complex systems). As an example, a lack of adequate extension support and of basic ecological knowledge means that farmers often initiate pesticide applications when they observe 'insects' or damage. However, they cannot distinguish pests from beneficial organisms, and cannot project from their field observations to realistic damage potentials. The result is inappropriate and ineffective pesticide application that may even do more harm than good.

Fortunately, there are initiatives to educate and support farmers in IPM implementation, such as the 'three reductions, three gains' (3R3G) campaign promoted through the Ministry of Agriculture and Rural Development in Vietnam which was successful in reducing pesticide inputs. The campaign promoted a reduction in seeding rates, fertilizers and pesticides to improve rice profitability. The chapter highlights other initiatives such as knowledge banks, digital information repositories with simplified retrieval interfaces to help users understand rice crop management and pest problems. In some cases specialized knowledge 'apps' have been made available to farmers through smart phones to help in pest identification and management, for example the Rice Knowledge Bank set up by the Bangladesh Rice Research Institute.

As Chapter 13 points out, rodents damage rice at all stages of crop development and are significant pests in all rice cropping systems: irrigated lowland, rainfed lowland, upland rainfed and deep-water/tidal wetlands. In some countries such as Indonesia rodents are considered the most important pre-harvest pest of rice, causing around 15% losses annually. In intensive lowland irrigated rice cropping systems rodents typically cause relatively low but regular losses each year. Probably of greater concern for smallholder farmers, however, are the episodic outbreaks that can lead to losses of 50-100% of the rice crop. In addition, post-harvest losses from rodents can be as much as 10-14%. Surveys of rice farmers highlight their concerns about rodents as a pest they find particularly hard to control.

There are a wide range of management options used for rodent management in rice systems. These can be categorized as physical, chemical and biocontrol techniques. The ultimate aim of a rodent management programme is to reduce the impact of damage rather than necessarily to eliminate pests completely. Chemical methods include acute and anticoagulant rodenticides. It is important to note that, for several years, no new rodenticidal compounds have been registered for use in crop protection. This has put greater emphasis on combining existing compounds as well as improving delivery mechanisms for rodenticides such as pulse baiting and pre-baiting to allow rats and mice to get used to feeding at a known site. As in the case of insects, knowledge of life cycles, foraging behaviour and food preferences is vital to the successful use of rodenticides. As an example, baits placed within rice fields are likely to be more attractive and more readily accepted during the tillering stage of the rice crop because food availability is low and pest rodents move into the rice field from the surrounding habitats during this time. However, the use of rodenticides can be limited by rodent behaviours such as neophobia (fear of new objects), conditioned aversion to the bait base or rodenticide, and the development of resistance. Other chemical methods include fumigation and the use of repellent compounds.

Limits in the use of chemical techniques have highlighted the importance of other approaches. Physical methods include trapping and barriers, including trap barrier systems or community trap barrier systems. Methods for the biocontrol or rodents include sterility control, reproductive inhibitors, parasites and diseases, though, in many cases, there has been limited research and commercial development. An example is the trialling of a rodent-specific parasite *Sarcocystis singaporensis* in rice fields in Thailand which resulted in reduced yield loss and demonstrated positive benefit-cost ratios similar to those obtained with conventional control techniques.

An important aspect of control is habitat management, including rice bunds (the small earthen banks surrounding rice field plots) which provide an important potential habitat for rodents. Effective management can be achieved through treatment of non-crop areas which are sources of habitat for rodents, including bunds or undisturbed weedy areas which provide burrows and nesting sites. This can be managed by slashing the weeds to reduce nesting habitat and increase predation risk. At a broad scale, crop synchrony is a highly effective form of management. For rodent species that exhibit a peak in breeding activity that coincides with the ripening stage of rice crops, crop synchrony limits the duration of this breeding period by reducing food availability. Habitat management can also be used to encourage predators such as owls.

Based on IPM principles, ecologically based rodent management (EBRM) is essentially a combination of the control methods mentioned above, but conducted in a way that takes proper account of the ecology and biology of the particular rodent pest species as well as the particular rice agro-ecosystem. To develop a successful rodent management strategy, it is important to identify the rodent species of concern and understand its ecology in the specific ecosystem. Key issues include the timing of control (to take advantage of stage of breeding and population abundance) and undertaking management over sufficiently large areas to reduce the chance of reinvasion after treatment.

Chapter 13 concludes with case studies of rodent control programmes in Vietnam, Laos and Tanzania. In the case of Vietnam, for example, farmers have traditionally relied heavily on the use of rodenticides and more informal techniques such as electrocution and spreading sump oil mixed with insecticides onto flooded rice fields to manage the rodent problem. These methods both have had limited success and have raised concerns about safety and environmental impact. To develop better programmes for rodent control, farming communities were trained and supported in implementing EBRM through "training-of-trainers" of local extension staff. A range of community-based rodent control options were trialled which were found to be relatively inexpensive to implement and resulted in reductions in yield losses. These include community actions such as synchronized cropping, rat control campaigns at key times, field hygiene and the use of community trap barrier systems when damage is expected to be high. After implementing EBRM, rodent damage was reduced by 33-50%, rice yields were increased by 2-5%, rodenticide use was reduced by 62-90% and the use of electrocution was reduced by 95%. Key lessons from implementing such programmes include the need to have good coordination between civic and government agencies to enable farmer participation, to have effective leadership of farmer groups and for management to be conducted early in the growth of the rice crop before rodent populations commence breeding.

Looking ahead, the chapter points out that improved monitoring systems and forecasting systems are urgently needed for EBRM. Most monitoring is conducted too late after damage has already occurred and there are very few predictive models available. There is strong interest in developing automated remote sensing equipment for rodents (in all crops, not just in rice). The challenge is making such equipment reliable and cheap, and not too data intensive. There is also interest in developing mobile phone applications for monitoring of rodents and provision of advice on better rodent control.

The chapter also reviews other vertebrate pests of rice such as birds. In some countries in Africa, birds are considered to be the second most important pest of rice after weeds. Preventative measures used to manage bird pests in rice include synchronous cropping, weed control (as weeds attract birds during the early grain filling stage) and nest destruction. Protective measures include the use of repellent substances, protecting fields or nurseries with nets or wires, and bird scaring techniques. However, aside from weed management, which has been shown to reduce bird visitation rates to rice fields, evidence that clearly demonstrate the effectiveness of these methods is limited.

As Chapter 13 points out, there are opportunities to explore the synergistic effects of management strategies with multiple benefits for rice production, for example concurrent weed and rodent management. It has also been shown that effective management of weeds also helps in managing insect pests and diseases. As Chapter 14 shows, rice grain yield losses caused by grasses can reach as high as 40%, with DSR systems particularly vulnerable. Continuous use of a particular cropping practice, such as repeated use of a particular herbicide, can inadvertently contribute to a shift in the dominance and distribution of weeds. Controlling weeds with only one or two techniques gives some weeds a chance to adapt to those practices. The use of herbicides with the same mechanism of action, year after year, has thus resulted in herbicide-resistant weed chemotypes. Repeated herbicide use may also result in weed flora shifting towards more persistent perennials, as well as a build-up of herbicide residues in the soil and in the harvested rice. Hundreds of weed species have now become resistant to major classes of herbicide, and many species are resistant to multiple herbicide classes. Regions that have practiced direct seeding over several decades have been particularly affected. As an example, weeds have become a major cause of yield decline for rice farmers in Sri Lanka who struggle to control grasses and sages using available products.

The chapter discusses the key components of integrated weed management (IWM), from preparation and control of the growing environment to establishing and managing the rice crop. Techniques which can be incorporated into an IWM strategy include choice of cultivar, tillage, crop rotation, water management, use of fertilizers and herbicides, as well as mechanical and biological control of weeds. Besides controlling existing weed problems, IWM puts greater emphasis on preventing weed reproduction and establishment in the first place, and on minimizing weed competition with the crop. Preventative techniques start with cultivar selection, measures to prevent seed contamination and appropriate land preparation techniques. As an example, some allelopathic rice cultivars can inhibit weed growth. Precision land levelling or using a laser levellers to create regular, sloping fields, for example, improves water management which can have a significant effect on both weed growth and bio-efficacy of herbicides. Scouting fields routinely and keeping non-cropped areas free of weeds can limit the persistence of weeds from one crop to the next.

Chapter 14 then covers establishment methods and planting patterns such as the use of stale seedbeds. Maintaining narrow row spacing and high seeding rates, for example, allows less space for weeds to become established. Some cultivation techniques such as conservation agriculture may promote the germination and emergence of weed seeds that stay on or near the soil surface. Herbicide application or mechanical weeding may be needed to eradicate these before sowing or planting. Cover crops like *Azolla* and mulches can also reduce weed problems in conservation agriculture by preventing weed seed germination or by suppressing the growth of emerging seedlings. Including crop rotation in conservation agriculture can also be a successful approach to reducing weed pressure. Well-targeted crop fertilization, especially nitrogen, is a promising agronomic practice in reducing weed interference in crops by helping crops to outperform weeds. However, under high weed pressure, crop fertilization can be counterproductive since nutrient absorption can be higher in weeds than in crop plants.

The chapter concludes by looking at mechanical and other methods for weed control. It covers developments in biological control such as *Hirschmanniella spinicaudata*, a rice root nematode which controls most upland rice weeds; *Bactra verutana,* a moth which destroys *Cyperus rotundu; Altica cyanea* (the steel blue beetle), which completely destroys *Ludwigia parviflora*; and the rust fungus *Puccinia canaliculata*, which controls *Cyperus esculentus* L. It then reviews the range of non-selective or selective, and pre-, early post- or post-emergence herbicides and ways they can best be deployed as part of an IWM programme.

## Summary

Each chapter identifies potential avenues for further research. One common thread is the development of better monitoring techniques, whether to measure plant nutritional status, water losses from crops, pest or weed populations. The chapters in Volume 2 highlight a number of other common themes. They demonstrate the importance of the research community in understanding and explaining fundamental processes such as plant and pest ecosystems, and then using this understanding to design technologies and systems to ensure more resource-efficient, pest-free and sustainable rice cultivation.

However, as many chapters show, it is equally important then to translate these technologies and systems into practical programmes which farmers can use on the ground. This involves a range of processes such as localizing programmes so that they are tailored to local conditions, reducing costs so that technologies are affordable and translating programmes into a series of practical steps that farmers can implement easily. As Chapters 1, 10, 12 and 13 show, this also requires effective training and dissemination activities together with access to resources. As Chapter 8 demonstrates, farmers also need appropriate standards to work and the incentives to make improvements which may only benefit them in the long term.

A second major theme is the interconnectedness of many initiatives to improve rice cultivation, and potential synergies from integrating different approaches. One example is improving the efficiency of fertilizer management discussed in Chapter 2 which also has an impact on pest and weed management (as shown in Chapters 10, 12, 13 and 14). Chapter 3 provides an excellent example of the way different management practices can be combined to allow rice to be cultivated successfully and sustainably in a more challenging environment. Another aspect of such interconnectedness is how to navigate potential trade-offs between different programmes, for example reconciling aspects of conservation agriculture with effective weed management or rodent control. This links to a third common theme: the focus on conservation agricultural techniques. Anticipated to some degree by SRI, conservation agriculture, with its emphasis on understanding and optimizing soil health, has implications for improving water and nutrient management as well as control of insect pests.

Part 1

# Rice cultivation techniques

Chapter 1

# Advances in irrigation techniques for rice cultivation

*D. S. Gaydon, CSIRO Agriculture, Australia*

1 Introduction
2 Water-saving measures
3 Scale-dependency of water productivity and water savings
4 Aerobic rice as a water-saving measure
5 Alternate wetting and drying (AWD) as a water-saving measure
6 Saturated soil culture (SSC) as a water-saving measure
7 Case study: water-saving irrigation in southeast Australia
8 Future trends and conclusion
9 Where to look for further information
10 References

## 1 Introduction

Increasing scarcity of fresh water resources is a growing concern in many major food-producing areas of the world, including some of the major rice-growing areas in Asia and Africa (Godfray et al., 2010; Rijsberman, 2006). Projected future increases in temperatures are also likely to exacerbate the issue of crop water stress (IPCC, 2014). This, combined with the imperative to produce more food for a growing global population, is driving a focus globally on increasing water productivity in irrigated agriculture, or, put more simply, growing more food with less water (Keating et al., 2010).

Rice is one of the key elements of global food security, being the staple food for most of the world's population, particularly people in low and lower-middle-income countries. Nearly 90% of the global rice crop is produced by Asian farmers (Devendra and Thomas, 2002), and some of the facts associated with production of this food source are truly staggering. For example, rice farming is the single largest use of land for producing food on Earth (with some 696 million tons produced in 2010 from around 158 million hectares), yet only 7% of all rice is exported from its country of origin. Rice is produced along Myanmar's Arakan Coast, where the growing season average rainfall is greater than 5100 mm; in places like Australia's Riverina and parts of Africa where the growing season rainfall may be less than 100 mm; at the Upper Sindh in Pakistan with average season

http://dx.doi.org/10.19103/AS.2016.0003.14

temperatures of 33°C; and in parts of Japan where the mean growing season temperature falls as low as 17°C (GRISP, 2013).

With such huge variation in rice-growing environments and landscapes, not surprisingly there are a wide range of cultivation and irrigation practices. The majority of the world's rice is grown under irrigated conditions in which the fields are flooded from planting to harvest. Rice land receives 35–45% of all the world's irrigation water (which itself uses some 70% of all the world's developed water resources) (Bouman, 2013; Barker et al., 1998). In view of the importance of rice to humanity and the current and growing concerns about future water scarcity, achieving water savings in irrigated rice production has become one of the key research challenges for the rice-growing world.

There are a number of potential research avenues being pursued to achieve such water savings without sacrificing rice yields. Essentially the aim is to increase irrigated water productivity – the amount of rice produced per unit of irrigation water applied. These avenues include (i) breeding of short-duration, high-yielding modern rice varieties which require less water due to less time in the field (Khush, 1995); (ii) breeding of varieties which yield better in harsh environments (salinity, submergence, heat, cold, drought) (Zhang, 2007); (iii) breeding of varieties better suited to new water-saving irrigation practices (for example deep-rooted genes for more water-stressed conditions) (iv) precision agriculture, site-specific N management and improved crop establishment methods to increase yields and minimize wasted resources (Y. Yao et al., 2012; Peng et al., 2010; Kumar and Ladha, 2011); (v) land-forming and Conservation Agriculture practices which seek to improve soil health and system sustainability (Hobbs et al., 2008); and (vi) improved irrigation techniques which minimize unproductive water losses in the cropping system. In the future, successful water-saving practices will require continual adaptation to climatic changes and other emerging challenges, such as growing pressure to utilise marginal lands (Borlaug, 2007).

Other chapters in this publication are addressing the first four of the above avenues, and many of these focus on increasing water productivity by increasing yields as the primary focus. Although in reality many of these solutions work together in a system context (Tuong et al., 2005), this chapter will restrict itself to consideration of the key techniques for improving rice water productivity through enhanced irrigation practices and aiming to reduce irrigation water use in rice cultivation.

# 2 Water-saving measures

Before moving into a literature review on irrigation methods for water savings in rice production, a consideration of the definition of 'water productivity' (WP) is warranted to avoid confusion. A general definition of WP is:

**WP** = **rice produced** (amount or value)/**water used** (mm)

From the perspective of a crop physiologist or a breeder, the 'water used' component would refer to the amount of water transpired by the crop, leading to the 'transpired water productivity' (WPT – a purely physiological term). However from the perspective of the farmer or a field-scale irrigation planner, the 'water used' is a broader term, referring to the total irrigation water applied to the crop. They are interested in extracting as much value as possible out of every drop of available irrigation water, and maximising the 'irrigated water productivity' (WPI – which encompasses water used in crop transpiration, evaporation from the soil or pond surface, and also any losses of water from the field in the form of seepage under/through bunds, deep drainage and runoff). Further, at a regional scale the water resources planner will be more interested in the productivity of all regional water resources (WPTWI) including diverted or pumped irrigation water, rainfall, plus any deep drainage or runoff waters which are re-couped somewhere else in the system and re-used (Humphreys et al., 2010; Tuong et al., 2005).

As this review is on advances in irrigation techniques for rice cultivation, the focus will primarily be on methods and practices for saving irrigation water at the field scale (WPI); however considerations at the larger regional scale (WPTWI) will be reflected upon when relevant.

# 3 Scale-dependency of water productivity and water savings

The scale-dependency of water productivity and water savings has long been recognised (Molden et al., 2003). Water applied at a *field scale* is stored in the soil matrix, transpired to the atmosphere via the crop, lost by evaporation directly from the soil or water surface, or lost from the system via a number of other possible mechanisms (deep drainage below the reach of crop roots, lateral flows out of the field which include surface runoff or seepage under the bunds, etc.). Water applied in excess of crop needs results in an increasing proportion of these losses (Fig. 1; Fereres and Soriano, 2007). Technologies such as Conservation Agriculture (Hobbs, 2007) may lead to decreased water losses due to an increased capacity of the soil to store water (improved structure), and reductions in evaporation via use of surface mulches.

There is a general positive relationship between crop yield and water transpired by the crop (Perry et al., 2009); hence farmers may seek to minimise such losses from their fields in order to maximise their WPI. However, at a *larger scale*, the 'losses' of an individual farmer may be recouped by other users in the system (or even the farmer himself) and productively used, therefore not really constituting true losses. An example of this phenomenon is applied irrigation water lost as deep drainage from a field, entering the regional water table, to be once again pumped to the surface and used for irrigation and productive crop transpiration either at the same location or further afield (Humphreys et al., 2010; Paydar et al., 2009).

In Punjab, India, for example, declining water table depth has been a major concern for decades, with over-exploitation of water resources (particularly groundwater pumping) often being blamed (Jalota et al., 2007). Numerous water-saving practices have been mooted to save irrigation water and contribute to arresting the decline (such as laser land levelling, replacing rice with other crops, alternate wetting and drying (AWD) water management in rice, zero-till wheat, raised beds, delayed rice transplanting, shorter duration rice varieties, etc.). These irrigation water-saving practices may achieve water savings at the *field scale,* however those savings a likely due to reductions in deep drainage in many instances, with little effect on evapotranspiration (ET). Because deep drainage is returned to the water-table and not lost from the system in this region, reducing deep drainage alone is hence unlikely to have the required impact on water-table decline. It is important to note that reducing deep drainage may have major additional benefits in Punjab and is not irrelevant – for example reduced energy consumption for groundwater pumping and reduced pollution of the groundwater via nutrient leaching – however, this alone will not address the key environmental crisis of water-table decline. The irrigation water-saving options which primarily reduce ET are the ones that are most likely to succeed in addressing the problem in Punjab (Humphreys et al., 2010) as evaporative losses are the *true* losses in this system at the regional scale, and are driving the water-table decline. In other hydrogeological systems, the picture may be quite different – deep drainage losses may be genuine losses, un-recoupable by irrigators, and hence any savings in both ET and deep drainage may constitute true irrigation water savings in those systems.

Another aspect of scale which needs addressing in a review like this is the concept of water-limited versus land-limited production environments (Gaydon et al., 2012a). Water-savings are not a universally desirable aim in production systems – water savings do not in themselves allow the growth of more rice on *a given field* (Bouman et al., 2001). Rice is very sensitive to water stress, and attempts to reduce water inputs may result in yield reduction (Tuong et al., 2005). If the water savings from one field cannot be diverted to produce additional rice at another field (in other words, expanding the area of production), then an overall reduction in rice production may result. Field-level WP and yield can only be increased concurrently by water-saving irrigation plus raising crop yields through varietal improvement or better agronomy (fertiliser management, weed and pest control, etc.). Total rice production increases from water savings can occur by using the water saved

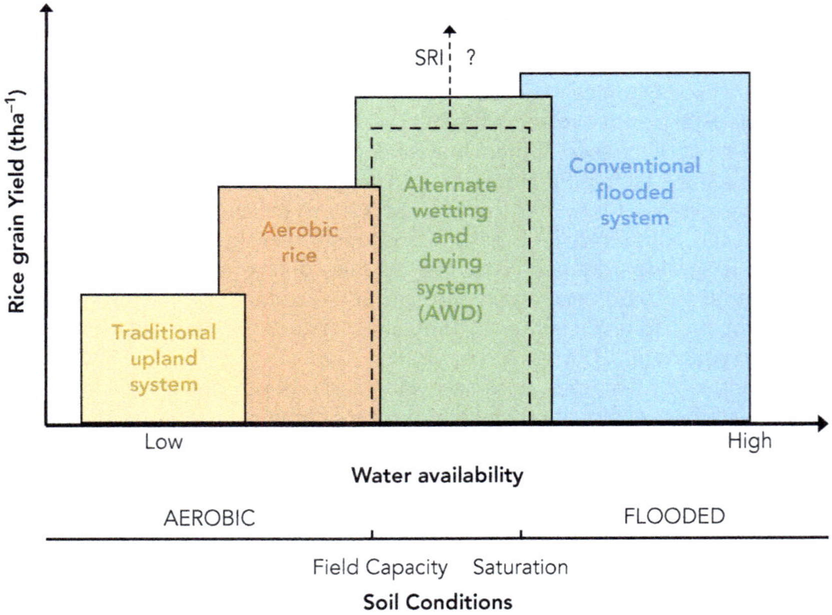


**Figure 1** Schematic presentation of irrigation system and rice yields as a function of water availability (Adapted from Farooq et al., 2009 and Tuong et al., 2005, with modification).

in one location to irrigate new land in another, and this can only occur in water-limited environments when land area is not the limiting factor (Gaydon et al., 2012a). The other possibility is that irrigation water saved during one crop could be used to better irrigate a second (or third) crop as part of a rotation on the same field, providing any of those subsequent crops were originally water-limited. These crops could be additional rice crops, or else non-flooded crops like wheat, maize or legumes.

So in conclusion, the Punjab example illustrates that water-savings at a field scale may not necessarily flow on to water savings at higher scales. The most scale-independent improvements in irrigation efficiency are those which reduce ET (Humphreys et al., 2010), although in certain hydrogeological circumstances reduction in deep drainage and lateral water losses may also be true savings. It is therefore important to understand the hydrogeology of the region of interest, as this will define the criteria on which the effectiveness of irrigation water savings can be judged. Also, water savings and increased water productivity at the field scale are not a means to produce more rice, unless the water saved can be diverted to additional irrigated land or crops. This implies that water savings are not a desirable aim in land-limited production environments (Gaydon et al., 2012a), and that additional cropping land (or other water-limited crops in the same rotation) are pre-requisites to realise the benefits of any water-savings achieved (Bouman, 2007a).

In the following sections we will review the current status and knowledge for a number of water-saving irrigation techniques (Fig. 1), and assess them in view of these perspectives. Figure 1 illustrates these practices on a scale of increasing water availability for the growing rice crop. At the low end of the spectrum, traditional upland systems are rainfed, often on poor soils, rarely ponded and rice yields are generally low. Increased water productivity in these systems may be achieved via agronomic practices such as residue retention to reduce soil evaporation, but since they are not irrigated, they will not be considered further in this chapter. Aerobic rice systems are irrigated but not ponded, with a focus on water economy. Moving further up the spectrum towards fully ponded lowland rice systems (which may be rainfed or irrigated), we encounter interim irrigated practices such as alternate wetting and drying (AWD) and the system of rice intensification (SRI). These also aim for water savings over fully ponded systems, but seek higher yields than aerobic systems, with less water savings. More complete descriptions of these systems, together with assessment of benefits and risks, are discussed further in the following sections.

# 4 Aerobic rice as a water-saving measure

## 4.1 Description

Aerobic rice is a term introduced by the International Rice Research Institute (IRRI) for high-yielding rice grown under non-flooded conditions in non-puddled and unsaturated (aerobic) soil. The crop is responsive to nutrient supply, can be rainfed or irrigated, and tolerates (occasional) flooding (Bouman and Tuong, 2001). On the sliding water-availability scale figure (Fig 1.) it can be seen that aerobic rice fits in between AWD and pure rainfed upland rice. In an aerobic rice system, the crop can be dry direct-seeded or transplanted and soils are kept aerobic throughout the growing season. Supplemental irrigation is applied as necessary (Nie et al., 2012). A key component of success in aerobic rice system is selection of appropriate cultivars (Wang et al., 2002), as root exploration and drought tolerance are key selection criteria. For example under declining soil moisture (measured as a % of the available soil water) most crops maintain transpiration up to around 30% of available soil water after which the rate declines (Loomis and Connor, 1992); in rice, there appears to be a linear decline from around 70% of available soil water (Lilley and Fukai, 1994). Recent varieties bred specifically for aerobic conditions are targeted at approaching the behaviour of other non-rice crops (wheat, maize, etc.).

There are a range of ways in which irrigation water may be applied in aerobic rice systems. These include sprinkler irrigation (Kahlown et al., 2007), furrow irrigation on beds (Choudhury et al., 2009) and drip and subsurface irrigation (Parthasarathi et al., 2014).

It has been stated that weeds are the most severe constraints to widespread adoption of aerobic rice (Rao et al., 2007; Jabran and Chauhan, 2014); however, rice yield decline reported in a number of studies after years of continuous aerobic rice cultivation clearly represents another severe constraint which needs to be addressed (Kreye et al., 2009).

## 4.2 Benefits

Growing rice under continuously unsaturated soil conditions can maximise water-use efficiency and minimise both labour requirements and greenhouse-gas emissions (Kato et al., 2014). In temperate environments, aerobic rice culture regularly produces grain yields greater than 9 t $ha^{-1}$ (e.g., in yields in the United States under centre-pivot sprinklers reach 10 t $ha^{-1}$, with up to 11.4 t $ha^{-1}$ achieved in central Japan). However, yields remain at less than 8 t $ha^{-1}$ in the tropics (Kato et al., 2014).

The harshness (heat, evaporative demand) of the environment is pertinent to how well rice responds to aerobic irrigation, with large yield reductions in sprinkler irrigated rice of 35–70% reported in temperate Australia (Muirhead et al., 1989) compared with conventional flooded yields of $10^{+}$ t $ha^{-1}$. Associated water savings were similar percentages resulting in negligible changes to WP; however, it must be noted that the varieties used in the experiment were not adapted to aerobic conditions and were typical local lowland rice varieties (Humphreys et al., 1989). In Brazil, aerobic rice cultivars with high grain yields of 5–7 t $ha^{-1}$ have been developed (Castañeda et al., 2002). While in northern China, the grain yields of 8 t ha-1 and even higher have been achieved using high-yielding aerobic rice cultivars under appropriate management practices (Wang et al., 2002).

By reducing water use during land preparation and limiting seepage, percolation, and evaporation, aerobic rice had 51% lower total water use and 32–88% higher water productivity, expressed as gram of grain per kilogram of water, than flooded rice (Bouman et al., 2005). Castañeda et al. (2003) reported that rice yields under aerobic conditions were 2.4–4.4 t $ha^{-1}$ (14–40% lower than under flooded conditions), but that water use decreased relatively more than yield, resulting in irrigated water productivity increases by 20–40% (in one case even 80%) over that under flooded conditions. Water savings averaged 73% of irrigation water for land preparation and 56% during the crop growth period, compared with conventional flooded production (Castañeda et al., 2003).

The above studies were all conducted with the same cultivars used for flooded and aerobic cultivation. When specific aerobic cultivars are employed under aerobic conditions and compared with specific lowland cultivars under lowland conditions, the reported WP gains are even higher. The WUE of the aerobic varieties under aerobic conditions can be 164–188% higher than that of a lowland cultivated rice varieties (Farooq et al., 2009).

The use of surface mulches in aerobic rice systems has been noted to increase irrigation water productivity (Qin et al., 2006), significantly increasing the leaf area per plant, main root length, tap root length and root dry weight per plant of crop, in addition to reducing soil evaporation losses. In an experiment in China comparing flooded rice irrigation with aerobic rice under different surface mulching scenarios, it was found that WP was increased in the aerobic rice system by 307–321% under surface mulching, but only by 98–138% under the no-mulch treatments (Zhang et al., 2008). Obviously there are socio-economic aspects to the use of mulch in small-holder farming systems which must be taken into consideration (requirements for livestock feed, etc.); however the biophysical benefits of mulching in aerobic rice systems, from a grain production and irrigation water perspective, appear clear.

There is also evidence that the intermittent irrigation characteristic of aerobic (or AWD) rice systems may be a useful tool to fight the build-up of arsenic in flooded rice cultivation of boro rice production in Bangladesh (Roberts et al., 2010). This appears to be largely driven by the disruption of saturated soil redox conditions under aerobic or AWD management.

Labour use is also saved in aerobic rice because more labour is required for land preparation such as puddling, transplanting and irrigation activities in flooded rice (Wang et al., 2002; Farooq et al., 2009).

Less methane emissions are expected under aerobic than under flooded conditions, but with higher nitrous oxide emissions (Wassmann et al., 2000). Weller et al. (2016) considered the combined global warming potential (GWP) due to both $CH_4$ and $N_2O$ and reported the GWP of conventional flooded paddy rice as 3868 ± 2137 kg $CO_2$-eq $ha^{-1}$ $season^{-1}$, compared with 2029 ± 1050 kg $CO_2$-eq $ha^{-1}$ $season^{-1}$ for an aerobic rice system, over a 4-year trial. This represents a reduction in global warming potential of 47% for aerobic rice irrigation. However, there are two confounding elements in this evaluation: (i) when this is scaled according to grain yield the reduction was less, but (ii) the aerobic rice system also comprised one fully-flooded (wet season) crop per year and one aerobic crop. The results are also significantly affected by soil type and climate. At this stage it is fair to say there is little published evidence that aerobic rice systems present better GWP outcomes than AWD systems; however, enhanced performance compared with conventionally flooded systems has been clearly demonstrated.

## 4.3 Risks

With continuous aerobic rice, there have been concerns reported regarding the so-called 'soil sickness'. Yield decline resulting from continuous cropping of aerobic rice has been identified as a major constraint to the widespread adoption of aerobic rice technology (Nie et al., 2012). The decline also appears to be greater in the dry season than in the wet (Peng et al., 2006); however the magnitude of yield decline after continuous cropping of aerobic rice depends strongly on the number of seasons that aerobic rice was continuously cropped, soil properties, climates, rice cultivars, and management practices (Kreye et al., 2009).

It has been suggested that growth-limiting factors other than water or N affected the crop in these situations, from the stage of panicle initiation onwards. Root knot nematodes and micronutrients have been nominated as possible major constraints to crop development in these continuous aerobic rice situations (Kreye et al., 2009). Strategies to overcome continuous cropping obstacle of aerobic rice have been developed. These strategies include crop rotation, N management practice, soil acidification, and cultivar improvement (Nie et al., 2012). The problem is not unknown in other cropping systems without appropriate measures including rotations and crop-breaks (Barker and Koenning, 1998; Thompson et al., 2012; Grabau and Chen, 2016), and breaking up the lifecycle of the pest (for example nematodes) is the aim. A detailed review of the literature on options for managing yield decline in aerobic rice systems is found in Nie et al. (2012).

Loss of soil organic carbon reserves has also been identified as a risk in aerobically irrigated rice systems. The introduction of aerobic phases in lowland rice systems has been found to result in decreased organic C and total N due to low C sequestration in soils (Witt et al., 2000). A recent study by Kraus et al. (2016) indicated that under land management change which sees flooded paddy rice replaced by an aerobic or upland crop, approx. 2.8–3.4 t C $ha^{-1}$ $yr^{-1}$ residue incorporation after harvest is required to achieve stable soil organic carbon (SOC) stocks in the new system. This suggests that aerobic rice irrigation systems may need to include some principles of conservation agriculture (e.g. residue retention) to maintain long-term productivity.

Aerobic rice may also be a risky practice in saline-prone environments, as the reduced water application will likely result in reduced flushing of salts below the root zone, increasing the potential for greater build-up of salts and consequent crop salt stress.

# 5 Alternate wetting and drying (AWD) as a water-saving measure

## 5.1 Description

An established irrigation water-saving technology for rice is *alternate wetting and drying* (AWD) developed earlier but document by IRRI in the 1970s (Bouman and Tuong, 2001; Belder et al., 2004; Sandhu et al., 1980). This technology is based on the fact that high yields in rice can be achieved without continuous flooding. Once the crop is established, the pond depth is allowed to fall to a threshold depth below the surface of the soil surface for a certain period before the next irrigation is applied. In 2001, the IRRI developed a set of simplified guidelines for AWD technology using a field water tube as the tool to monitor the water level below the soil surface. Irrigation is applied (to a depth of around 5 cm) when the perched water table falls to 15 cm below the soil surface. The field water tube (Fig. 2a and b) is a perforated tube that can be fabricated with materials such as polyvinyl chloride (PVC) pipe, bamboo, plastic bottles, or even tin cans, with diameter ranging from 10 to 20 cm (Bouman et al., 2007b).The threshold of 15 cm is called 'safe AWD', as this will not cause any significant yield decline. In more technical terms, irrigation is applied when soil water tension increases to 10 kPa at a depth of 15 cm (Sudhir-Yadav et al., 2011a). Safe AWD is now recommended to farmers in several Asian countries, including Vietnam, India, Bangladesh and the Philippines (Lampayan et al., 2015; Palis et al., 2014).

It has been demonstrated over many years now in numerous environments that rice can extract adequate water from the soil around roots, even in the absence of standing water in the field (Lampayan et al., 2015). In some circumstances there are constraints on which phases of the crop growth can be irrigated in this way (see following case study from Australia's Riverina (Dunn and Gaydon, 2011)); however, in many environments, the majority of crop growth period can be irrigated using AWD, with the sensitive period around flowering exempted (flooding of 5cm applied for 1–2 weeks).

## 5.2 Benefits

Whereas practically all studies report significant decreases in water use (mm/ha) with AWD in comparison with continuous flooding, some studies have actually reported an increase in rice grain yield (Zhang et al., 2008; Won et al., 2005; Yang et al.,2007), whereas others report a decrease in rice grain yield (Tabbal et al., 2002; Belder et al., 2004). In 2001 Bouman and Tuong asserted that AWD would always result in some degree of yield decline due to the inevitability of incurring some level of rice crop water stress; however, subsequent studies have indicated this is not always the case.

In a 4-year Chinese experiment, Feng et al. (2007) demonstrated that rice yields under flooded conditions were around 8000 kg $ha^{-1}$ with 900 mm total (rain, irrigation) water input. Irrigation water savings of 40–70% were achieved without any yield loss by applying AWD. In a different region of China, Belder et al. (2004) reported that the AWD technology can reduce field water use by 15–20% without significantly affecting yield. In Hubei province, another experiment showed that AWD reduced irrigation water without significantly impacting grain yields and increased the mean water productivity by 16.9% compared with continuously flooded irrigation (Tan et al., 2013). Sudhir-Yadav et al. (2011a,b) demonstrated the feasibility of reducing irrigation amount (30–50%) while maintaining yield by replacing puddled transplanted rice (PTR) with direct-seeded rice (DSR) using AWD in the Indian Punjab, provided that soil tension is kept lower than 20 kPa at 20 cm depth. However, in the Philippines, in an area with a relatively deep groundwater table, intermittent irrigation every 8 or 10 days reduced the rice yield on average by 25% over that of the control and the average water input by 60% that of the control. Water productivity, however, still nearly doubled the continuously flooded value (Tabbal et al., 2002). But with 20 days of continuous drought, yields declined so much that even water productivities declined to 82% of the value under continuous flooding. In experiments in Yangzhou, Eastern China, the grain yield was increased by 9.3–9.5% under AWD with moderate soil drying between anthesis and maturity (limit of −25 kPa at 15–20cm depth), while it was reduced by 7.5–7.8% under a more severe form of AWD over the same period (limit of −50kpa), compared with that under conventional flooding. Water applied to the moderate AWD treatment was 44% and to the severe AWD treatment was 25% of the amount applied to the conventionally flooded treatment. The moderate AWD also significantly improved milling, appearance and cooking qualities, while the severe AWD decreased these qualities, once again in comparison with continuously flooded production. The authors concluded that a moderate wetting drying regime during the grain-filling phase holds great promise to increase both yield quantity and quality and also to save water (Zhang et al., 2008). The authors also suggested that soil drying to −25 kPa is notably beneficial during the grain-filling period, but remarked that during their experiments relatively cool conditions prevailed during this crop growth phase and questioned whether the same results would be achieved under warmer tropical conditions.

So this raises the hypothesis that climate may explain discrepancies, although it is likely more complex than that. In summary, the discrepancies in the water savings/yield trade-offs with AWD are probably due to variations in any number of things: soil characteristics, depth to water-table and other local hydrological aspects, timing of the applied water stress, and also varietal differences. It is also possible that different climates during key crop stages are influential. Although some varieties are likely to perform better under AWD than others, it appears that benefits are available to all. In Hubei, China 2009–10, Super Hybrid rice developed for high-yielding lowland production was compared under continuous flooding (CF) and AWD. AWD saved 24% and 38% irrigation water compared with CF in 2009 and 2010, respectively. There was insignificant difference in grain yield between AWD and CF. These results suggest that high-yielding varieties developed for the continuously flood-irrigated rice system could still produce high yield under safe AWD experienced in this study. 'Super' hybrid rice varieties do not necessarily require more water input to produce high grain yield. (Yao et al., 2012). It is also likely that varietal traits including drought tolerance and deep rooting will be advantageous in AWD systems and can act to reduce any observed yield losses resulting from water stress.

It seems clear that AWD improves WP universally when applied within recommended guidelines; however as illustrated above, the specific impacts on and trade-offs between irrigation water use and rice grain yields will vary. Whether AWD will solve water shortage problems is also not clear-cut. Bouman et al. (2007) showed that most of the irrigation water savings from AWD are caused by reduced deep drainage (or percolation) rates. This will reduce groundwater recharge and may lead to decreased opportunities for groundwater irrigation in some regions (e.g. NW India, Humphreys et al., 2010). In Punjab, Aurora et al. (2006) showed 350 mm of water savings from AWD in PTR (25–30% of continuously flooded water use), but ET was only reduced by about 30 mm. From this perspective of water savings, AWD is unlikely to be a key solution to the crisis of falling water tables in the North-West India, where technologies to decrease in ET are the main objective (Humphreys et al., 2010). However, in areas where deep percolation losses are effectively lost from the farmer's irrigation system, AWD water savings may be a real solution to any water-shortage problems.

In addition to water saving advantages, the AWD practice was reported to reduce methane ($CH_4$) emissions from rice fields (Hosen, 2007) due primarily to reduced time in a flooded state under reduced soil conditions. Among many agricultural practices, water management has been recognised as one of the most promising approaches to reduce $CH_4$ emission (Wassmann et al., 2000; Li et al., 2002). Midseason drainage is a common practice used by farmers to control the unproductive tillers. Field studies have shown that midseason drainage reduces $CH_4$ emissions by above 10% (Wassmann et al., 2000). In Chinese experiments, the $CH_4$ emissions under $AWD_{15}$ (15cm threshold for irrigation) and $AWD_{30}$ were more than 30% and 60% lower than under CF, respectively (Liang et al., 2016). However reductions in $CH_4$ are only part of the emissions picture when changing from CF to AWD – the increased drying and re-wetting of soil associated with AWD irrigation also increases nitrification (on soil drying) and denitrification (on soil wetting), resulting in the release of $N_2O$ and other gaseous N products, which have higher global warming potential per mass than $CH_4$.

*Carbon Dioxide equivalents* are a universally accepted method of integrating the net effect of gaseous C and N products on global warming potential (Robertson et al., 2000). Several studies are now available which explore the global warming potential of different rice water management systems, and the literature reveals a common conclusion (Linquist et al., 2015). For example, carbon dioxide equivalents of $CH_4$ and $N_2O$ emissions from paddy fields under AWD irrigation were found to be 788 kg $CO_2$ $ha^{-1}$, a reduction of 61.4% compared with those from conventional flooded irrigation (Yang et al., 2012). In summary, N emission are increased, but C emissions are decreased under AWD water management, with the net impact being reduced carbon dioxide equivalents. It can therefore be concluded that AWD is an effective technique for mitigating the global warming potential of irrigated rice production.

There is evidence available that the investment to develop and disseminate the AWD technology has had a high rate of return, with benefit-cost ratio of 7:1. The evidence of economic benefits at the farm level when aggregated will more than compensate for the total research investments made to develop and disseminate the AWD technology (Lampayan et al., 2015).

## 5.3 Risks

Although AWD can substantially reduce irrigation water use at the field scale, there is the risk that excessive AWD drying in puddled clay-based soils can inadvertently create the opposite outcome due to development of deep cracks extending below the plow-pan, which actually increase total crop water use from rapid bypass flow upon re-wetting, and before the cracks self-seal (Bouman and Tuong, 2001). In Hubei province, China, the deep drainage under AWD was reduced by 15.3% in 2007 and 8.3% in 2008 compared to CF (Tan et al., 2103). However, the cumulative percolation of the first 5 days after irrigation in AWD plots was significantly larger than that in CFI plots. The authors concluded from this that bypass flow across the plow-pan, facilitated by the opening of cracks in the unstructured (puddled) soil during the drying phase, allows significantly increased water losses in the period before the cracks 'self-close'. This potentially decreases the water-saving effectiveness of AWD and increases the potential for $NO_3$-N loading to the groundwater.

Similar to aerobic rice irrigation, AWD may also be a risky practice in saline-prone environments, with increased potential for greater build-up of salts and consequent crop salt stress due to reduced leaching of salts (compared with continuously flooded systems).

# 6 Saturated soil culture (SSC) as a water-saving measure

## 6.1 Description

Saturated Soil Culture (SSC) is an irrigation practice which involves keeping the rice-growing soil saturated but not ponded throughout the growth period, thereby reducing the hydraulic head of the water, and in turn reducing the percolation rate and water lost to deep drainage. Evaporative losses are likely to be similar to conventional flooded practice (Borrell et al., 1997).

SSC is the typically nominated water management method in the System of Rice Intensification (SRI) (initially described in Stoop et al., 2002; Uphoff et al., 2002), at least up until panicle initiation, after which a very thin pond of 1–2 cms is recommended. Numerous authors have either debunked (Sheehy et al., 2004; McDonald et al., 2006) or supported (Hussain et al., 2004; Sinha et al., 2007) the remarkable claims of yield and productivity increases available through SRI; however for the purposes of this chapter, water savings available through employing SSC in comparison with conventional flooded practice are reported as positive. This section will not attempt to delve into issues relating to SRI grain production potential – only to examine potential water savings through implementation of SSC irrigation practices.

## 6.2 Benefits

From a large Indian, Philippine and Japanese dataset, Bouman and Tuong concluded that under saturated soil conditions water savings were on average 23% (±14%) with yield reductions of only 6% (±6%), compared with production through maintaining a 5–10 cm pond throughout crop growth. This was noted as the most promising option to save water and increase WP without decreasing land productivity too much. According to Bouman and Tuong (2001), although substantial increases in WP were possible when applied following 'safe AWD' protocols, AWD always results in reduced land productivity. Overall production increases were then only possible by following a philosophy of 'spreading the water' over wider cropping areas (Gaydon et al., 2012a), or else diverting saved water to another crop in the rotation on the same field.

In a two-season experiment in Northern Australia on a permeable soil, Borrell et al. (1997) found that there was no significant difference in rice yield or rice quality between SSC and conventional flooded practice; however, there was approximately 32% saving in irrigation water use from SSC in both seasons. Water productivities (defined as weight of grain produced per mm applied water) were increased by up to 26%.

In a Chinese investigation comparing SRI to conventional flooding, Zhao et al. (2010) reported that irrigation use was reduced by 57.2% through SRI methods. Since they also recorded significant gains in rice yield with SRI, an impressive increase in irrigated WP of 194.9% was achieved; however the latter stages of their SRI crops were irrigated in an AWD pattern every 3–7 days. The authors also noted that SRI may be worth further study, solely on the basis of possible water savings – irrespective of the sometimes fierce debate over the impact of SRI on grain yield potential. Thakur et al. (2016), in summarising the current state of information on SRI, reported that water savings are possible up to 40% using pure SSC irrigation methods (i.e. without AWD).

Although claims on the increasing water productivity of SRI-practices, including reduced water use, are found in many publications, there is a surprising lack of studies quantifying the magnitude of these changes under farmer conditions (Berkhout et al., 2015). Adusumilli et al. (2011) reported irrigation water savings of 52% for SRI over conventional practice, and Gathorne-Hardy et al. (2013) indicated a 55% water saving, for an increase in WP of 139%; however, all of these figure represent a mix of SSC (prior to panicle initiation) and AWD (till maturity).

In summary, the pure SSC culture studies report irrigation water-savings ranging between 9 and 40%. This range is most likely due to the range of soil permeability used in experiments. The literature also indicates that in combination with AWD, SSC has the potential to increase water-savings higher, up to 52–57.2%.

Reduced total global warming potential associated with SRI technology has been reported (1.1 kg $CO_2$ equivalents per kg paddy rice, compared with 2.8 for conventional flooded practice; Gathorne-Hardy et al., 2013). However, once again, the water management employed in this system is a mix of SSC and AWD.

## 6.3 Risks

There is a risk of increased rice floret sterility in cold weather, particularly relevant in temperate climates. SSC provides no thermal buffering protection to the rice crop from the water mass, such as can be achieved in fully ponded production (Williams and Angus, 1994). In these ponded systems, the developing rice floret may be submerged into warmer water to avoid low air temperatures. Authors also note increased weed control problems over conventional flooded practice which require additional inputs in terms of herbicide and/or labour (Borrell et al., 1997; Ly et al., 2012; Gathorn-Hardy et al., 2016)

# 7 Case study: water-saving irrigation in southeast Australia

In temperate southeast Australia, japonica rice is traditionally grown under fully irrigated conditions over a long growing period (close to six months, starting October and going through till April). Consequently grain-filling occurs over an extended time, and grain yields are among the highest in the world (12+ tonnes/ha). Since the mid-1980s the average water productivity (WP) of the total New South Wales (NSW) rice crop has effectively doubled (Humphreys et al., 2006). The breeding and introduction of high-yielding semi-dwarf cultivars, together with reduced field water use, have largely driven this improvement; however, several climate-specific adaptations in management practice have also been influential. The climate constraints in this region are significantly different from much of the tropical rice-growing world – the average rice crop in southeast Australia requires around 1200 mm of water throughout its growing cycle, yet on average only about 350 mm of that comes from rainfall (Humphreys et al., 2006). The rest is supplied via irrigation from either diverted surface water or groundwater pumping. During the growing period the crop experiences hot, dry weather with daytime temperatures often rising above 40°C and night-time temperatures falling below 15°C on a regular basis. Historically, the risk of the crop being exposed to night-time temperatures less than the damaging 17°C threshold during the sensitive microspore period is around 40% – a very high production risk due to low-temperature floret sterility (Williams and Angus, 1994) with the potential for dramatic yield reductions. The application of deep water (20–25 cm) during this period is a recommended method adopted by farmers to protect the pollen against low temperature (Humphreys et al., 2006). Australian rice growers are regarded as rapid adopters of improved technology (McDonald, 1994; Clampett et al., 2000), and the deep-water flooding technology at microspore has had an almost universal uptake by industry.

Historically, water shortages were never a problem for Australian rice producers, receiving at least 100% of their requirements every season from as far back as 1912 (at the commencement of the irrigation schemes) right up until the mid-1990s (McIntyre et al., 2011; Gaydon et al., 2012a). At this point, a mix of political and climatic factors led to unprecedented restrictions in production due to low water availability and extreme variability of supply in the intervening period till the present (Gaydon et al., 2012a). This has driven a growing interest in water-saving technology.

There was thought to be little opportunity for increasing water productivity from the current flooded rice-growing practice (Humphreys et al., 2006), so farmers and researchers looked to AWD and realised that one opportunity to save water may be through reducing the length of time the crop is flooded. The period when rice must be flooded in deep water for cold-temperature protection during the microspore period presented a clear constraint (Williams and Angus, 1994). Also, flooding is required during the subsequent reproductive and grain-filling period to meet the crop water requirements during the very high crop growth rates (250–300 kg/ha/d) that occur between panicle initiation (PI) and flowering. The high potential evapotranspiration demand in this very dry environment over this period could result in crop water deficit stress if a non-ponded culture was used (Humphreys et al., 2005) and result in significant yield losses. The early phase of crop growth from establishment to the time when the crop reaches full canopy cover offered particular potential for reducing water use using AWD. During this period the plant canopy is small and evaporation from the free water surface constitutes up 40% of total evaporation loss in continuously flooded rice (Simpson et al., 1992).

In the early 1980s AWD was first trialled in Australia. Heenan and Thompson (1984) recorded water savings of 23% through AWD every seven days prior to the establishment of continuous flooding at PI, achieving equivalent grain yields to conventionally flooded practice. Subsequent studies on different sites also demonstrated an increase in water productivity, with AWD ranging from 0.06 to 0.23 kg/m$^3$ compared to the normal flooding regime (Thompson and Griffin, 2006). However, all these early studies with AWD in Australia involved regular irrigation intervals with little moisture stress on the crop during the unflooded period. It was thought that greater water savings may be achieved with less frequent irrigations, but the optimal level of moisture stress between irrigation was unknown.

Dunn and Gaydon (2011) therefore conducted experiments investigating water savings from a range of treatments, flush irrigated at different cumulative evapotranspiration frequencies prior to the start of continuous flooding, which was delayed until just prior to panicle initiation. They demonstrated increased input water productivity from higher levels of imposed crop water stress during the initial non-flooded period. Irrigation intervals at 160 mm cumulative ET (equivalent to around 10 day AWD intervals) significantly improved input water productivity above that of the conventional drill sown treatment (by 17%). Irrigating at 80 mm intervals resulted in a significant but lesser improvement (9%). Notable from these experiments were the vigour with which the crops in the highly stressed AWD treatments bounced back once they received full flooding at PI – it was difficult to visually distinguish them from the conventional treatments during grain filling, and yield effects were minimal. However delaying the application of continuous flooding extended the period of crop growth and the authors concluded that it may be desirable to move sowing forward 7–10 days so that pollen microspore could still occur at the safest time in regard to cold temperature damage. A subsequent modelling study extending this experimental work and focusing on long-term risk found that establishing the crop up to a month earlier with the harsher AWD imposition could minimise the low-temperature risk and maximise system WP, but indicated that projected future climate changes by 2030 could see the optimum crop establishment time reverting to the original optimum establishment time (Gaydon et al., 2012b).

This case study is an example of how environmental constraints can limit the options for applying water-savings irrigation technology, but how targeted research combining field trials and modelling can allow the development of smart technology which takes such environmental constraints into account. Locally targeted research to fully understand the interactions between crop, management and environmental constraints and to subsequently use this understanding to optimise practical advice to farmers represents the greatest challenge to researchers in the field of irrigation water-savings. Participatory engagement with farmers allows researchers to gain insights into system constraints which may limit farmer options. Once these types of understandings are obtained, field experimentation in combination with cropping systems modelling offers researchers insights into systems performance and risks over larger timescales, and indeed in potential future climates. This is likely to be of growing importance to future researchers in the field of irrigation and crop production, as we move into climate conditions for which past experiences can yield us limited insights.

# 8 Future trends and conclusion

The call to 'grow more rice with less water' (Guerra et al., 1998) clearly requires more efficient use of water in irrigation and increased rice WP through advances in irrigation techniques; however this review has revealed that such a goal is more complex than it first seems. Water-saving irrigation increases water productivity, but in rice generally results in decreased yield (Tuong et al., 2005). There are some examples of concurrent water savings and yield increases (Liang et al., 2016), but in general water-savings do not produce more rice with less water in the same field. Field-level water productivity and yield can best be increased concurrently by water-saving irrigation plus raising crop yields through varietal improvement or better agronomy (fertiliser management, weed and pest control, etc.). Total rice production increases from water savings alone can occur by using the water saved in one locality to irrigate new land in another, or by diverting saved water onto another sequential water-limited crop (either another rice crop or non-flooded crop) in rotation on the same field. If this is not done or is not possible, a strategy of saving water at the field level potentially threatens total rice production at large (Bouman and Tuong, 2001). This implies that water savings alone are not a desirable aim for land-limited production environments, but require additional available land and crops in which to expand production to realise benefits (Gaydon et al., 2012a; Bouman, 2007a).

Water savings at a field scale may not necessarily flow on to water savings at higher scales. The most scale-independent improvements in irrigation efficiency are those which reduce ET (Humphreys et al., 2010), although in certain hydrogeological circumstances reduction in deep drainage and lateral water losses may also be true savings. It is therefore important to understand the hydrogeology of the region of interest, as this will define the criteria on which the effectiveness of irrigation water savings can be judged.

A review of current literature has revealed that the AWD rice irrigation technology is capable of achieving 15–70% water savings over conventional flooded irrigation practice, without any significant yield reductions. The savings strongly correlate with soil percolation rate rather than evaporative demand, hence varying greatly geographically as indicated by this large per cent savings range. This 'Safe AWD' has been widely extended within the rice-growing world with a demonstrably high return on research investment. For some environment-variety combinations, significant increases in crop yields have even been demonstrated, resulting in even greater WP gains. Saturated soil culture (SSC) irrigation such as those demonstrated by SRI spans the gap between AWD and conventional practice, but seems more limited in its application due to more intense management requirements and reduced water savings.

In regions where water availability is lower, aerobic rice systems may be the preferred irrigation practice with greater yield reductions than AWD, but with potential for even larger irrigation water use savings. Whether this necessarily leads to greater WP than AWD systems may depend on variety, soil, and environmental conditions. But to realise the advantages of these aerobic (or AWD) WP gains, there must be extensive cropping lands available in which to 'spread the saved water' and increase cropping area, or else to more intensively irrigate another previously water-stressed crop in the rotation. Aerobic rice is not the system of choice for land-limited cropping systems without significant water shortages. Also, it is impossible to escape from the fact that rice is still a drought-sensitive crop, which prefers high-moisture content soils. There are limits for stressing rice crops in aerobic irrigation beyond which WP plummets due to great reductions in grain yields. In many cases where irrigation water is more limiting, it may be more productive to focus on other dry-season crops like wheat, maize, legumes.

The basic biophysical advantages of rice irrigation water-savings technologies detailed in this chapter are now quite well understood from experimental studies under a range of conditions and environments. However, less is known about potential savings extrapolated to wider geographical areas incorporating a range of soil-types, water availabilities and climates, or the impact of regional socio-economic constraints when these techniques are employed within cropping systems management practices like Conservation Agriculture (Hobbs, 2007). Hence a future trend in irrigation research for rice-based systems will surely be integrating existing knowledge using tools like crop models, water-resources and climate models, together with global information systems (GIS) to draw useful conclusions at a regional or policy scale regarding initiatives to increase water productivity in irrigation. A hydrological characterisation and mapping of Asia's rice area is needed to assess the extent and magnitude of potential water savings, because the gains vary with soil type and climate (Belder et al., 2004).

Cropping systems philosophies like Conservation Agriculture also raise new questions about proven rice irrigation practices. For example, what is the potential to use AWD with unpuddled soils (less cracking during drying phase than puddled soils, due to maintenance of soil structure) to reduce the risk of bypass flow transporting $NO_3$-N to the ground water (Tan et al., 2013)? What are greenhouse gas emission and soil organic carbon implications of land management change options? Some studies have started to investigate these broader cropping system issues (e.g. Kraus et al., 2016; Kadiyala et al., 2014) and these are regarded a pertinent future trends in these irrigated systems.

The *crop-scale* questions explored by experiments of the past will now morph into *systems-scale questions*. For example, how best to irrigate rice in rotation with other non-flooded crops and pastures? To which on-farm enterprises are the water resources best allocated? Is it more profitable for the farmer to optimise his rotational system, rather than optimise the performance of any one crop? Should irrigation water savings from one crop be employed on another subsequent crop in the same field, or should the area of cultivation be expanded? Is limited irrigation water best used on a rice crop, or on another type of crop (wheat, maize etc.)? Which option gives the best outcomes for farmer profitability, or the best outcomes for regional food security?

As food production pressure increases in coming years, there will be an increasing pressure to produce food from marginal lands and harsh environments. Questions for irrigation scientists in rice-based systems may include, 'How are we best to employ brackish irrigation water to expand available supply and hence crop-able area?' 'What is the best strategy to mix available fresh water supplies with other water supplies of lesser quality?' and 'When should they be used on the crop?'

Climate change adaptation will obviously figure strongly in future irrigation science research trends. What will be the future climate impacts on irrigation water requirements? On irrigation water availability? On irrigation water quality? How will these affect the crop yields, water savings and water productivities possible via AWD and aerobic technologies, and how will they increase the risk or variability of returns?

Cropping systems and water-resources models are useful tools to explore future scenarios, provided they are well parameterised, calibrated and validated locally for the region of interest. They can be used to extend what we learn from a limited number of years in field experiments and to give insights into system performance and risk over much longer timescales and in different climatic conditions. For many of these questions raised for the future of irrigation in rice system, well-validated cropping systems models, when used judiciously together with field experimentation, will allow irrigation scientists of the future to explore the effects of future conditions over broad expanses of the rice-growing world and evaluate performance of different strategies and technologies. It is difficult to see how traditional field-based agronomy and irrigation studies can yield the required information on how best to irrigate rice-based systems into the future without the concurrent use of cropping systems and water-resources models. The field-experimentation/modelling nexus is likely to be a growing trend in rice irrigation research as we prepare to feed humanity in the coming centuries.

# 9 Where to look for further information

The following references provide enhanced detail on many of the concepts raised in this chapter. Additionally, the reader may gain valuable information from the IRRI website (http://irri.org/resources ) and a range of freely available resource publications therein. The International Water Management Institute (IWMI) also has publicly accessible resources related to water management and water savings options (http://www.iwmi.cgiar.org/publications/latest/).

# 10 References

Adusumilli, R. and Bhagya Laxmi, S., 2011. Potential of the system of rice intensification for systemic improvement in rice production and water use: the case of Andhra Pradesh, India. *Paddy and Water Environment* 9, 89–97.

Alberto, M. C. R., Wassmann, R., Hirano, T., Miyata, A., Kumar, A., Padre, A. and Amante, M., 2009. CO 2/heat fluxes in rice fields: comparative assessment of flooded and non-flooded fields in the Philippines. *Agricultural and Forest Meteorology* 149(10), 1737–50.

Anthofer, J., 2004. *The Potential of the System of Rice Intensification (Sri) For Poverty Reduction in Cambodia.* Conference on International Agricultural Research for Development, Berlin, 5–7 October.

Arora, V. K., 2006. Application of a rice growth and water balance model in an irrigated semi-arid subtropical environment. *Agricultural Water Management* 83, 51–7.

Barker, K. R. and Koenning, S. R., 1998. Developing sustainable systems for nematode management. *Annual Review of Phytopathology* 36(1), 165–205.

Barker, R., Dawe, D., Tuong, T. P., Bhuiyan, S. I. and Guerra, L. C. (1998). The outlook for water resources in the year 2020: Challenges for research on water management in rice production. In 'Assessment and Orientation Towards the 21st Century', *Proceedings of 19th Session of the International Rice Commission*, 7–9 September 1998, FAO, pp. 96–109.

Belder, P., Bouman, B. A. M., Cabangon, R., Guoan, L., Quilang, E. J. P., Li, Y., Spiertz, J. H. J. and Tuong, T. P. 2004. Effect of water-saving irrigation on rice yield and water use in typical lowland conditions in Asia. *Agricultural Water Management* 65,193–210.

Berkhout, E., Glover, D. and Kuyvenhoven, A., 2015. On-farm impact of the System of Rice Intensification (SRI): Evidence and knowledge gaps. *Agricultural Systems* 132, 157–66.

Borlaug, N., 2007. Feeding a hungry world. *Science* 318(5849), 359.

Borrell, A., Garside, A. and Fukai, S., 1997. Improving efficiency of water use for irrigated rice in a semi-arid tropical environment. *Field Crops Research* 52(3), 231–48.

Bouman, B. A. M. and Tuong, T. P., 2001. Field water management to save water and increase its productivity in irrigated lowland rice. *Agricultural Water Management* 49(1), 11–30.

Bouman, B. A. M., 2007. A conceptual framework for the improvement of crop water productivity at different spatial scales. *Agricultural Systems* 93(1), 43–60.

Bouman, B. A. M., Lampayan, R. M. and Tuong, T. P., 2007b. *Water Management in Irrigated Rice: Coping with Water Scarcity. International Rice Research Institute*, Los Baños, Philippines, p. 54.

Bouman, B. A. M., 2013. Bas Bouman's blog – Global Rice Science Partnership, IRRI http://irri.org/blogs/bas-bouman-s-blog-global-rice-science-partnership/does-rice-really-use-too-much-water (accessed 29 April 2016).

Castañeda, A. R., Bouman, B. A. M., Peng, S. and Visperas, R. M., 2002. The potential of aerobic rice to reduce water use in water-scarce irrigated lowlands in the tropics: opportunities and challenges. In Bouman, B. A. M., Hengsdijk, H., Hardy, B., Bindraban, P. S., Tuong, T. P. and Ladha, J. K. (Eds), *Water-Wise Rice Production*, 8–11 April 2002, International Rice Research Institute, Los Baños, Philippines, pp. 165–76.

Castañeda, A. R., Bouman, B. A. M., Peng, S. and Visperas, R. M., 2003. The potential of aerobic rice to reduce water use in water-scarce irrigated lowlands in the tropics. In Bouman, B. A. M., Hengsdijk, H., Hardy, B., Bindraban, P. S., Tuong, T. P. and Ladha, J. K. (Eds), *Water-Wise Rice Production*, 8–11 April 2002 at IRRI Headquarters in Los Baños, Philippines, International Rice Research Institute, Los Baños, Philippines.

Choudhury, B. U., Bouman, B. A. M. and Singh, A. K., 2007. Yield and water productivity of rice–wheat on raised beds at New Delhi, India. *Field Crops Research* 100(2), 229–39.

Clampett, W. S., Williams, R. L. and Lacy, J. M., 2000. Major achievements in closing yield gaps of rice between research and farmers in Australia. In *Proceedings of the Expert Consultation on Yield Gap and Productivity Decline in Rice Production*, 5–7 September 2000. FAO, Rome, Italy, pp. 411–28.

Devendra, C. and Thomas, D., 2002. Smallholder farming systems in Asia. *Agricultural Systems* 71(1–2), 17–25.

Dunn, B. W. and Gaydon, D. S., 2011. Rice growth, yield and water productivity responses to irrigation scheduling prior to the delayed application of continuous flooding in south-east Australia. *Agricultural Water Management* 98(12), 1799–807.

Farooq, M., Kobayashi, N., Wahid, A., Ito, O. and Basra, S. M., 2009. Strategies for producing more rice with less water. *Advances in Agronomy* 101, 351–88.

Gathorne-Hardy, A., Reddy, D. N. and Harriss-White, B., 2013. A Life Cycle Assessment (LCA) of greenhouse gas emissions from SRI and flooded rice production in SE India. *Taiwan Journal of Water Conservancy* 61, 120–5.

Gathorne-Hardy, A., Reddy, D. N., Venkatanarayana, M. and Harriss-White, B., 2016. System of Rice Intensification provides environmental and economic gains but at the expense of social sustainability – A multidisciplinary analysis in India. *Agricultural Systems* 143, 159–68.

Gaydon, D. S., Meinke, H. and Rodriguez, D., 2012a. The best farm-level irrigation strategy changes seasonally with fluctuating water availability. *Agricultural Water Management* 103, 33–42.

Gaydon, D. S., Meinke, H. and Dunn, B. W., 2012b. Maximizing water productivity through model-aided design of delayed continuous flood irrigation in Australian rice production, Chapter 7 in Gaydon, D. S., 2012. Living with Less Water: Development of viable adaption options for Riverina irrigators, Ph.D thesis, Wageningen University, Wageningen, the Netherlands, p. 225. ISBN: 978-94-6173-232-3. http://library.wur.nl/WebQuery/wda/1987914.

Godfray, H. C. J., Beddington, J. R., Crute, I. R., Haddad, L., Lawrence,D., Muir, J. F., Pretty, J., Robinson, S., Thomas, S. M. and Toulmin, C., 2010. Food security: The challenge of feeding 9 billion people. *Science* 327, 812–18.

Grabau, Z. J. and Chen, S., 2016. Influence of long-term corn–soybean crop sequences on soil ecology as indicated by the nematode community. *Applied Soil Ecology* 100, 172–85.
GRiSP (Global Rice Science Partnership). 2013. *Rice Almanac*, 4th edition, Los Baños (Philippines), International Rice Research Institute, p. 283.
Guerra, L. C., Bhuiyan, S. I., Tuong, T. P. and Barker, R., 1998. *Producing More Rice With Less Water From Irrigated Systems*. SWIM Paper 5. IWMI/IRRI, Colombo, Sri Lanka, p. 24.
Hobbs, P. R., Sayre, K. and Gupta, R., 2008. The role of conservation agriculture in sustainable agriculture. *Philosophical Transactions of the Royal Society of London B: Biological Sciences*, 363(1491), 543–55.
Heenan, D. P. and Thompson, J. A., 1984. Growth, grain yield and water use of rice grown under restricted water supply in New South Wales. *Australian Journal of Experimental Animal Husbandry* 24, 104–9.
Hobbs, P. R., 2007. Conservation agriculture: what is it and why is it important for future sustainable food production? *Journal of Agricultural Science* 145(2), 127–37.
Hosen, Y., 2007. Whether water saving reduces the global warming potential of irrigated paddies. *Proceedings of 8th Conference of the East and Southeast Asian Federation of Soil Science*, Tsukuba, Japan, pp. 8–13.
Humphreys, E., Lewin, L. G., Khan, S., Beecher, H. G., Lacy, J. M., Thompson, J. A., Batten, G. D., Brown, A., Russell, C. A., Christen, E. W. and Dunn, B. W., 2006. Integration of approaches to increasing water use efficiency in rice-based systems in southeast Australia. *Field Crops Research* 97(1), 19–33.
Humphreys, E., Kukal, S. S., Christen, E. W., Hira, G. S. and Sharma, R. K., 2010. 5 Halting the Groundwater Decline in North-West India – Which Crop Technologies will be Winners? *Advances in Agronomy* 109(5), 155–217.
Humphreys, E., Muirhead, W. A., Melhuish, F. M., White, R. J. G. and Blackwell, J. B., 1989. The growth and nitrogen economy of rice under sprinkler and flood irrigation in southeast Australia. II. Soil moisture and mineral N transformations. *Irrigation Science* 10, 201–13.
Husain, A. M., Chowhan, G., Barua, P., Uddin, A. F. M. and Rahman, A. B. M, 2004. Final evaluation report on verification and refinement of the system of rice intensification (SRI) project in selected areas of Bangladesh. PETRRA-Project, IRRI, Dhaka, Bangladesh.
IPCC, 2014. Climate Change 2014: Synthesis Report. Contribution of Working Groups I, II and III to the Fifth Assessment Report of the Intergovernmental Panel on Climate Change [Core Writing Team, R. K. Pachauri and L. A. Meyer (eds.)]. IPCC, Geneva, Switzerland, p. 151.
Jabran, K. and Chauhan, B. S., 2015. Weed management in aerobic rice systems. *Crop Protection* 78, 151–63.
Kadiyala, M. D. M., Mylavarapu, R. S., Li, Y. C., Reddy, G. B. and Reddy, M. D., 2012. Impact of aerobic rice cultivation on growth, yield, and water productivity of rice–maize rotation in semiarid tropics. *Agronomy Journal* 104(6), 1757–65.
Kahlown, M. A., Raoof, A., Zubair, M. and Kemper, W. D., 2007. Water use efficiency and economic feasibility of growing rice and wheat with sprinkler irrigation in the Indus Basin of Pakistan. *Agricultural Water Management* 87(3), 292–8.
Kato, Y. and Katsura, K., 2014. Rice adaptation to aerobic soils: physiological considerations and implications for agronomy. *Plant Production Science* 17(1), 1–12.
Keating, B. A., Carberry, P. S., Bindraban, P. S., Asseng, S., Meinke, H. and Dixon, J., 2010. Eco-efficient Agriculture: Concepts, Challenges, and Opportunities. *Crop Science* 50 (Suppl. 1), S109–S119.
Khush, G. S., 1995. Modern varieties – their real contribution to food supply and equity. *Geojournal* 35(3), 275–84.
Kraus, D., Weller, S., Klatt, S., Santa Bárbara, I., Haas, E., Wassmann, R., Werner, C., Kiese, R. and Butterbach-Bahl, K., 2016. How well can we assess impacts of agricultural land management changes on the total greenhouse gas balance ($CO_2$, $CH_4$ and $N_2O$) of tropical rice-cropping systems with a biogeochemical model? *Agriculture, Ecosystems & Environment*, 224, 104–15.

Kreye, C., Bouman, B. A. M., Castañeda, A. R., Lampayan, R. M., Faronilo, J. E., Lactaoen, A. T. and Fernandez, L., 2009. Possible causes of yield failure in tropical aerobic rice. *Field Crops Research* 111(3), 197–206.

Kumar, V. and Ladha, J. K., 2011. Direct Seeding of Rice: Recent Developments and Future Research Needs. *Advances in Agronomy* 111, 297–413.

Lampayan, R. M., Rejesus, R. M., Singleton, G. R. and Bouman, B. A., 2015. Adoption and economics of alternate wetting and drying water management for irrigated lowland rice. *Field Crops Research* 170, 95–108.

Li, C. S., Qiu, J. J., Frolking, S., Xiao, X. M., Salas, W., Moore, B., Boles, S., Huang, Y. and Sass, R., 2002. Reduced methane emissions from large-scale changes in water management of China's rice paddies during 1980-2000. *Geophysical Research Letters* 29(20), 33–4.

Liang, K., Zhong, X., Huang, N., Lampayan, R. M., Pan, J., Tian, K. and Liu, Y., 2016. Grain yield, water productivity and CH 4 emission of irrigated rice in response to water management in south China. *Agricultural Water Management* 163, 319–31.

Lilley, J. M. and Fukai, S., 1994. Effect of timing and severity of water deficit on four diverse rice cultivars I. Rooting pattern and soil water extraction. *Field Crops Research* 37(3), 205–13.

Linquist, B. A., Anders, M. M., Adviento-Borbe, M. A. A., Chaney, R. L., Nalley, L. L., Da Rosa, E. F. and Kessel, C., 2015. Reducing greenhouse gas emissions, water use, and grain arsenic levels in rice systems. *Global Change Biology*, 21(1), 407–17.

Loomis, R. S. and Connor, D. J., 1992. *Crop Ecology: Productivity And Management In Agricultural Systems*. Cambridge, Cambridge University Press.

Ly, P., Jensen, L. S., Bruun, T. B., Rutz, D. and de Neergaard, A., 2012. The System of Rice Intensification: Adapted practices, reported outcomes and their relevance in Cambodia. *Agricultural Systems* 113, 16–27.

Macadam, R., Drinan, J. and Inall, N., 2002. Building Capacity for Change in the Rice Industry. Rural Industries Research and Development Corporation (RIRDC) Publication No. 02/009. Rural Industries Research and Development Corporation, Canberra, ACT, Australia.

McDonald, D. J., 1994. Temperate rice technology for the 21st century: an Australian example. *Australian Journal of Experimental Agricullture* 34, 877–88.

McDonald, A. J., Hobbs, P. R. and Riha, S. J., 2006. Does the system of rice intensification outperform conventional best management?: A synopsis of the empirical record. *Field Crops Research* 96(1), 31–6.

McIntyre, S., McGinness, H. M., Gaydon, D. and Arthur, A. D., 2011. Introducing irrigation efficiencies: prospects for flood-dependent biodiversity in a rice agro-ecosystem. *Environmental Conservation* 38(03), 353–65.

Molden, D., Murray-Rust, H., Sakthivadivel, R. and Makin, I., 2003. A water productivity framework for understanding and action. In Kijne, J. W., Baker, R. and Molden, D. (Eds), *Water Productivity In Agriculture: Limits And Opportunities For Improvement*. Wallingford UK (CABI), pp. 1–18.

Muirhead, W. A., Blackwell, J. B., Humphreys, E. and White, R. J. G., 1989. The growth and nitrogen economy of rice under sprinkler and flood irrigation in southeast Australia. I. Crop response and N uptake. *Irrigation Science* 10, 183–99.

Mushtaq, S., 2016. Economic and policy implications of relocation of agricultural production systems under changing climate: Example of Australian rice industry. *Land Use Policy* 52, 277–86.

Nie, L., Peng, S., Chen, M., Shah, F., Huang, J., Cui, K. and Xiang, J., 2012. Aerobic rice for water-saving agriculture. A review. *Agronomy for Sustainable Development* 32(2), 411–18.

Palis, F. G., Lampayan, R. M., Bouman, B. A. M., et al., 2014. Adoption and dissemination of alternate wetting and drying technology for boro rice cultivation in Bangladesh. In Kumar, A. (Ed.), *Mitigating Water-ShortageChallenges in Rice Cultiva-tion: Aerobic and Alternate Wetting and Drying RiceWater- Saving Technologies*. IRRI and Asian Development Bank, Manila.

Parthasarathi, T., Mohandass, S., Senthilvel, S. and Vered, E., 2013. Effect of various micro irrigation treatments on growth and yield response of aerobic rice. *International Agricultural Engineering Journal*, 22(4), 49.

Paydar, Z., Gaydon, D. S. and Chen, Y., 2009. A methodology for up-scaling irrigation losses. *Irrigation Science* 27 (5), 347–56.

Peng, S., Bouman, B. A. M., Visperas, R. M., Castan˜eda, A., Nie, L. and Park, H. K., 2006. Comparison between aerobic and flooded rice in the tropics: Agronomic performance in an eight-season experiment. *Field Crops Research* 96, 252–9.

Peng, S., Buresh, R. J., Huang, J., Zhong, X., Zou, Y., Yang, J., Wang, G., Liu, Y., Hu, R., Tang, Q. and Cui, K., 2010. Improving nitrogen fertilization in rice by site-specific N management. A review. *Agronomy for Sustainable Development* 30(3), 649–56.

Perry, C., Steduto, P., Allen, R. G. and Burt, C. M., 2009. Increasing productivity in irrigated agriculture: agronomic constraints and hydrological realities. *Agricultural Water Management* 96(11), 1517–24.

Qin, J., Hu, F., Zhang, B., Wei, Z. and Li, H., 2006. Role of straw mulching in non-continuously flooded rice cultivation. *Agricultural Water Management* 83(3), 252–60.

Rao, A. N., Johnson, D. E., Sivaprasad, B., Ladha, J. K. and Mortimer, A. M., 2007. Weed management in direct-seeded rice. *Advances in Agronomy* 93, 153–255.

Rijsberman, F. R., 2006. Water scarcity: fact or fiction? *Agricultural Water Management* 80(1), 5–22.

Roberts, L. C., Hug, S. J., Voegelin, A., Dittmar, J., Kretzschmar, R., Wehrli, B., Saha, G. C., Badruzzaman, A. B. M. and Ali, M. A., 2010. Arsenic dynamics in porewater of an intermittently irrigated paddy field in Bangladesh. *Environmental Science & Technology* 45(3), 971–6.

Robertson, G. P., Paul, E. A. and Harwood, R. R., 2000. Greenhouse gases in intensive agriculture: contributions of individual gases to the radiative forcing of the atmosphere. *Science* 289(5486), 1922–5.

Sandhu, B. S., Khera, K. L., Prihar, S. S. and Singh, B., 1980. Irrigation needs and yield of rice on a sandy-loam soil as affected by continuous and intermittent submergence. *The Indian Journal of Agricultural Science* 50 (6), 492–6.

Sheehy, J. E., Peng, S., Dobermann, A., Mitchell, P. L., Ferrer, A., Yang, J., Zou, Y., Zhong, X. and Huang, J., 2004. Fantastic yields in the system of rice intensification: fact or fallacy?. *Field Crops Research* 88(1), 1–8.

Sinha, S. K. and Talati, J., 2007. Productivity impacts of the system of rice intensification (SRI): A case study in West Bengal, India. *Agricultural Water Management* 87(1), 55–60.

Stoop, W. A., Uphoff, N. and Kassam, A., 2002. A review of agricultural research issues raised by the system of rice intensification (SRI) from Madagascar: opportunities for improving farming systems for resource-poor farmers. *Agricultural Systems* 71, 249–74.

Sudhir-Yadav, Gill, G., Humphreys, E., Kukal, S. and Walia, U., 2011a. Effect of water management on dry seeded and puddled transplanted rice. Part 1: Crop performance. *Field Crops Research* 120(1), 112–22.

Sudhir-Yadav, Humphreys, E., Kukal, S., Gill, G. and Rangarajan, R., 2011b. Effect of water management on dry seeded and puddled transplanted rice Part 2: Water balance and water productivity. *Field Crops Research* 120(1), 123–32.

Tabbal, D. F., Bouman, B. A. M., Bhuiyan, S. I., Sibayan, E. B. and Sattar, M. A., 2002. On-farm strategies for reducing water input in irrigated rice; case studies in the Philippines. *Agricultural Water Management* 56(2), 93–112.

Tan, X., Shao, D., Liu, H., Yang, F., Xiao, C. and Yang, H., 2013. Effects of alternate wetting and drying irrigation on percolation and nitrogen leaching in paddy fields. *Paddy and Water Environment* 11(1–4), pp. 381–95.

Thakur, A. K., Uphoff, N. T. and Stoop, W. A., 2015. Scientific Underpinnings of the System of Rice Intensification (SRI): What Is Known So Far? *Advances in Agronomy* 135, 147–79.

Thompson, J. A. and Griffin, D., 2006. Delayed flooding of rice – effect on yields and water. *IREC Farmers Newsletter* 173 (Spring), 52–3.

Thompson, J. P., Mackenzie, J. and Sheedy, G. H., 2012. Root-lesion nematode (Pratylenchus thornei) reduces nutrient response, biomass and yield of wheat in sorghum–fallow–wheat cropping systems in a subtropical environment. *Field Crops Research* 137, 126–40.

Tuong, P., Bouman, B. A. M. and Mortimer, M., 2005. More rice, less water – integrated approaches for increasing water productivity in irrigated rice-based systems in Asia. *Plant Production Science* 8(3), 231–41.

Uphoff, N., Fernandes, E. C. M., Yuan, L. P., Peng, J. M., Rafaralahy, S. and Rabenandrasana, J., 2002. Assessment of the system for rice intensification (SRI). In *Proceedings of the International Conference*, Sanya, China, 1–4 April 2002. Cornell International Institute for Food, Agriculture and Development (CIIFAD), Ithaca, NY.

Wang, H., Bouman, B. A. M., Zhao, D., Wang, C. and Moya, P. F., 2002. Aerobic rice in northern China: opportunities and challenges. In Bouman, B. A. M., Hengsdijk, H., Hardy, B., Bindraban, P. S., Tuong, T. P. and Ladha, J. K. (Eds), *Water-wise Rice Production. Proc International Workshop on Water-wise Rice Production*, 8–11 April 2002, Los Baños, Philippines. International Rice Research Institute, Los Baños, Philippines, pp. 143–54.

Wassmann, R., Neue, H. U., Lantin, R. S., Makarim, K., Chareonsilp, N., Buendia, L. V. and Rennenberg, H., 2000. Characterization of methane emissions from rice fieldsin Asia: II. Differences among irrigation, rainfed, and deepwater rice. *Nutrient Cycling in Agroecosystems* 58, 13–22.

Williams, R. L. and Angus, J. F., 1994. Deep floodwater protects high nitrogen rice crops from low temperature damage. *Australian Journal of Experimental Agriculture* 34, 927–32.

Witt, C., Cassman, K. G., Olk, D. C., Biker, U., Liboon, S. P., Samson, M. I. and Ottow, J. C. G., 2000. Crop rotation and residue management effects on carbon sequestration, nitrogen cycling and productivity of irrigated rice systems. *Plant and Soil* 225, 263–78.

Won, J. G., Choi, J. S., Lee, S. P., Son, S. H. and Chung, S. O., 2005. Water saving by shallow intermittent irrigation and growth of rice. *Plant Production Science* 8, 487–92.

Yang, J., Liu, K., Wang, Z. and Zhu, Q., 2007. Water-saving and high-yielding irrigation for lowland rice by controlling limiting values of soil water potential. *Journal of Integrative Plant Biology* 49, 1445–54.

Yang, S., Peng, S., Xu, J., Luo, Y. and Li, D., 2012. Methane and nitrous oxide emissions from paddy field as affected by water-saving irrigation. *Physics and Chemistry of the Earth, Parts A/B/C* 53, 30–7.

Yao, F., Huang, J., Cui, K., Nie, L., Xiang, J., Liu, X., Wu, W., Chen, M. and Peng, S., 2012. Agronomic performance of high-yielding rice variety grown under alternate wetting and drying irrigation. *Field Crops Research* 126, 16–22.

Yao, Y., Miao, Y., Huang, S., Gao, L., Ma, X., Zhao, G., Jiang, R., Chen, X., Zhang, F., Yu, K. and Gnyp, M. L., 2012. Active canopy sensor-based precision N management strategy for rice. *Agronomy For Sustainable Development* 32(4), 925–33.

Zhang, Q., 2007. Strategies for developing green super rice. *Proceedings of the National Academy of Sciences* 104(42), 16402–9.

Zhang, Z., Zhang, S., Yang, J. and Zhang, J., 2008. Yield, grain quality and water use efficiency of rice under non-flooded mulching cultivation. *Field Crops Research* 108(1), 71–81.

Chapter 2

# Advances in nutrient management in rice cultivation

*Bijay-Singh, Punjab Agricultural University, India and V. K. Singh, Indian Agricultural Research Institute, India*

1 Introduction
2 Real-time site-specific nitrogen management in rice using non-invasive optical methods
3 Site-specific nutrient management for intensive rice-cropping systems
4 Controlled-release and slow-release nitrogen fertilizers
5 Urease and nitrification inhibitors
6 Deep placement of nitrogen fertilizers
7 Phosphorus and potassium
8 Micronutrients
9 Integrated plant nutrient management based on organic resources and mineral fertilizers
10 Summary and future trends
11 Where to look for further information
12 References

## 1 Introduction

Rice (*Oryza sativa* L.) is the staple food for nearly half of the world's population. In 2014, global production of rice was more than 740 Mt, of which 90% was recorded in Asia (FAOSTAT, 2016). As global grain demand is projected to double by 2050, the challenge to achieve even higher rice production levels still remains. Fertilizer use is one of the major factors for the continuous increase in rice production; more than 20% of fertilizer nitrogen (N) produced worldwide is used in the rice fields of Asia. Irrigated and rain-fed lowland rice systems account for 92% of total rice production and nutrients applied as fertilizers account for 20–25% of total production costs in these rice systems. Of the total 172.2 Mt fertilizer (N + $P_2O_5$ + $K_2O$) consumed globally during 2010–11, 14.3% (24.7 Mt) was used in rice fields. Percentages for N, phosphorus (P) and potassium (K) were 15.4, 12.8 and 12.6, respectively (Heffer, 2013).

http://dx.doi.org/10.19103/AS.2016.0003.16

After the introduction of mineral fertilizers, a large body of literature has become available on nutrient management in rice and different rice-based cropping systems. The knowledge generated from these studies generally resulted in the evolution of nutrient management for rice in the form of blanket recommendations for small regions with similar climate and landform. By adopting these recommendations, although to variable extents, in different rice-growing regions of the world, farmers have achieved varying levels of N, P and K use efficiencies, which still need to be augmented further to meet the challenges of increasing rice production with reduced inputs of energy and minimal damage to the environment. Because of large field-to-field variability of soil nutrient supply, efficient use of nutrients applied as fertilizers is not possible when broad-based blanket recommendations for fertilizers are used (Adhikari et al., 1999). Among many factors that influence fertilizer use efficiency, one potentially important factor is the uncertainty in deciding the amount of fertilizer nutrient to be applied in a given field (Lobell, 2007). When blanket recommendations are followed, besides use of nutrients in excess of the requirement of the crop in many fields, another major reason of low fertilizer use efficiency is the inefficient splitting of fertilizer applications. For example, Peng and Cassman (1998) demonstrated that recovery efficiency of top-dressed urea during panicle initiation stage could be as high as 78%. The strategies for fertilizer management must also be responsive to temporal variations in crop nutrient demand to achieve supply–demand synchrony. When fertilizer applications are not synchronized with crop demand, losses of nutrients from the soil–plant system are large, leading to low fertilizer use efficiency.

Unlike P and K, management of N in rice has received more attention of researchers because (i) deficiency of N is probably the most common problem in rice and tends to be of large economic significance, (ii) proper application of N fertilizers is vital to improve crop growth and grain yields, especially in intensive agricultural systems, (iii) insufficient and/or inappropriate fertilizer N management can be detrimental to crops and the environment, (iv) supply of N to rice and losses of N to the environment are greatly influenced by management of water in rice culture and (v) no suitable soil-test method has been established and implemented for determining the N-supplying capacity for soils used to produce rice. Optimal N-management strategies aim at matching fertilizer N supply with actual crop demand, thus maximizing crop N uptake and reducing N losses to the environment. Since late 1990s, some advances in technologies and strategies have been made to further enhance fertilizer N use efficiency in rice. These include site-specific and real-time N management, non-destructive quick test of the N status of plants, new types of slow-release and controlled-release fertilizers, deep placement of fertilizer application and use of urease inhibitor and nitrification inhibitor to decrease N losses.

As for N, little improvement in fertilizer use efficiency can be expected from current blanket recommendations for fertilizer P and K in rice, and site-specific approaches are the answer (Dobermann et al., 1998). Some useful advancements in this direction are already becoming available. With widespread use of mineral fertilizers in rice, organic manures were thought of as a secondary source of nutrients. However, with increasing awareness about soil health and sustainability in agriculture, organic manures and many diverse organic materials have gained importance as components of integrated plant nutrient management (IPNM) strategies. The IPNM is a holistic approach and seeks to optimize plant nutrient supply with an overall objective of adequately nourishing rice crop as efficiently as possible, and improve and maintain the health of the soil base while minimizing potentially adverse impacts to the environment.

The recent advances in nutrient management in rice have been primarily driven by the continuing need to increase rice production. In addition, the fact that it will not be possible to continue the way the plant nutrients have been managed so far because agriculture adds globally significant and environmentally detrimental amounts of N and P to terrestrial ecosystems (Vitousek et al., 1997), at rates that may triple if past practices are used to achieve another doubling in food production (Tilman et al., 2001). The environmental impacts of agricultural practices are the costs that are typically unmeasured and often do not influence the farmer or societal choices about production methods (Tilman et al., 2002).

# 2 Real-time site-specific nitrogen management in rice using non-invasive optical methods

Substantial portions of applied N are lost due to the lack of synchrony of plant–N demand with N supply. The timing of fertilizer N application is used to best match the demand of N by crop plants with supply. In mid-1980s and 1990s, the emphasis was shifted from reducing N losses to feeding crop needs for increasing fertilizer N use efficiency (Buresh, 2007). The research was oriented towards finding means and ways to apply fertilizer N in real time using crop- and field-specific needs. Several methods based on soil tests and analyses of tissue samples were tried to predict cereal N needs during vegetative growth stages (Fox et al., 1989; Hong et al., 1990; Magdoff et al., 1990; Binford et al., 1992; Sims et al., 1995; Justes et al., 1997). These studies showed good correlations with grain yield and acceptable levels of accuracy; however, soil and tissue tests were time consuming, cumbersome and expensive. Moreover, the prospects remained bleak for accurate N prescriptions developed using soil tests before the season (Han et al., 2002). Tissue tests were also of limited value for predicting fertilizer N needs because a period of 10–14 days from sampling to receiving a fertilizer recommendation dose does not seem a practical proposition. Thus, most farmers use leaf colour as a visual and subjective indicator of the need for N fertilizer, although visual estimate of leaf colour is influenced by sunlight variability and is a non-quantitative method for determining the N needs of rice.

An important element of site-specific N management is the development and use of diagnostic tools that can assess 'real-time' N need of crop plants (Fageria and Baligar, 2005). The concept of using spectral ratio reflectance to rapidly quantify colour of intact plant leaves appears to have originated with Inada (1963). This concept is based on the assumption that spectral characteristics of radiation reflected, transmitted or absorbed by leaves can provide a better indication of plant chlorophyll content (Richardson et al., 2002). Further intensification of the efforts on investigation of leaf spectral characteristics occurred in the 1970s, along with the development of instrumentation and interest in evaluating potential uses of remote sensing (Jackson, 1986). More recently, some non-invasive optical methods based on leaf greenness, absorbance and/or reflectance of light by the intact leaf have been developed. These include chlorophyll meters, leaf colour charts (LCC), ground-based remote sensors and digital, aerial and satellite imageries. Since late 1990s, chlorophyll meter, LCC and a handheld GreenSeeker® optical sensor unit (NTech Industries Incorporation, Ukiah, CA) have been extensively tried to improve N use efficiency in cereals grown in different agro-ecosystems and regions (Varinderpal-Singh et al., 2010).

## 2.1 Chlorophyll meters

Handheld chlorophyll meters provide a fast, easy, on-site and precise way to measure the relative quantity of chlorophyll in rice leaves. For N management, the handheld Minolta SPAD-502 (also known as 'SPAD meter') is the most used chlorophyll meter. It measures relative difference in crop N status and can detect the onset of an N stress before it is visible to human eyes (Francis and Piekielek, 1999).

Research focused on improving N use efficiency using SPAD meter can be divided into two broad groups. In the first group, relationships between SPAD readings and N content of leaves have been studied. In rice, the relationship between SPAD meter reading and N content in leaves has been found to be non-linear (Chubachi et al., 1986; Peng et al., 1993; Shukla et al., 2004; Esfahani et al., 2008). Peng et al. (1993) suggested adjustment of SPAD readings for specific leaf weight to improve upon the prediction of leaf N concentration in rice. Thus, Peng et al. (1996) and Shukla et al. (2004) observed a linear correlation between SPAD values and rice leaf N concentration measured on leaf area basis for all the growth stages and lines tested. The second group of researchers has focused on evaluating the relationship between SPAD readings and the need for top dress N (Turner and Jund, 1991; Wells et al., 1993; Tuner and Jund, 1994; Maiti et al., 2004; Khurana, 2005). Two approaches have been used to guide fertilizer N applications to rice: (a) when SPAD value is less than a set critical reading (Balasubramanian et al., 1999; Bijay-Singh et al., 2002; Maiti et al., 2004; Satawathananont et al., 2004) and (b) when a sufficiency index (defined as SPAD value of the plot in question divided by that of a well-fertilized reference plot or strip) falls below 0.90 in rice (Hussain et al., 2000). Despite greater reliability of the sufficiency index or dynamic threshold value approach, the fixed threshold value approach is more practical as it does not require a well-fertilized or N-rich plot.

### 2.1.1 Fixed SPAD meter threshold value approach

Peng et al. (1996) were among the first who focused on determining a fixed critical SPAD value that rice farmers could refer to in the field. For rice cultivar IR72 grown in dry season in the Philippines, top-dressing of 30 kg N ha$^{-1}$ was recommended when SPAD value fell below the critical number of 35. In field trials in the Philippines, using 35 as the critical SPAD reading was found to result in similar yields with less N fertilizer applied (higher agronomic efficiency) compared to fixed split-timing schemes. The critical SPAD value had to be reduced to 32 during the wet season due to continuous cloud cover for most of the growing season (Balasubramanian et al., 1999). The SPAD value of 37.5 was found to be critical for rice in northwestern India (Bijay-Singh et al., 2002). It has also been suggested that different threshold SPAD values may have to be used for different varietal groups (Balasubramanian et al., 2000). For rice cultivars grown in the Indo-Gangetic plain in India, the threshold SPAD value of 37 or 37.5 has been found to be appropriate for optimum rice yields (Bijay-Singh et al., 2002; Maiti et al., 2004), whereas for rice cultivars grown in South India, the threshold SPAD value was found to be 35 (Nagarajan et al., 2004). While critical SPAD values for rice in Asia ranged between 32 and 37.5, Stevens and Hefner (1999) determined critical SPAD values of 40 and 41 for two different rice cultivars in Missouri (United States). In Texas, Turner and Jund (1994) reported the need for N in rice when SPAD values of the most recently matured leaf were less than the critical value of 40.

In majority of the irrigated, transplanted or direct-seeded rice farms across Vietnam (Son et al., 2004; Tan et al., 2004), China (Wang et al., 2001), Indonesia (Abdulrachman et al.,

2004), the Philippines (Gines et al., 2004), Thailand (Satawathananont et al., 2004) and India (Bijay-Singh et al., 2002; Maiti et al., 2004; Nagarajan et al., 2004; Khurana, 2005), chlorophyll meter-based N management led to significant increases in N use efficiency (NUE) compared to the farmers' fertilizer practices (Table 1). Despite these increases, NUE at some sites (like in Jinhua, China) remained moderate in absolute terms, and it was hypothesized that it could be further improved by synchronizing N with water management (Wang et al., 2001).

### 2.1.2 Dynamic SPAD meter threshold (sufficiency index) approach

Hussain et al. (2000) evaluated fertilizer N management in rice following the sufficiency index approach. Sufficiency index was monitored at 7–10 days interval, and whenever it was less than the critical value of 90%, 30 kg N $ha^{-1}$ was broadcasted up to the time of 50% flowering. Rice yields obtained for different cultivars were similar to those obtained in the fixed-time N application treatment but with 30 kg less N $ha^{-1}$. Similar results have been reported by Bijay-Singh (2008). A small over-fertilized plot need to be established to follow the sufficiency index approach, but it has the advantage of being self-calibrating for different soils, seasons and cultivars.

## 2.2 Leaf colour charts

LCC is a high-quality plastic strip on which a series of panels are embedded with colours based on the wavelength characteristics of leaves; the colours range from yellowish green to dark green and cover a continuum from leaf N deficiency to excessive leaf N content (Pasuquin et al., 2004). The LCCs measure leaf greenness and the associated leaf N by visually comparing light reflection from the surface of leaves and the LCC (Yang et al., 2003). These are simple, easy-to-use and inexpensive alternatives to chlorophyll meters (IRRI, 1999) and are visual and subjective indicators of plant N deficiency. Developed from a Japanese prototype (Furuya, 1987), several types of LCCs are available now. The most common ones are those developed by the International Rice Research Institute (IRRI), Zhejiang Agricultural University, China and the University of California, Davis, California.

There are two major approaches in the use of the LCC (Witt et al., 2007). The fixed splitting pattern approach provides a recommendation for the total N fertilizer requirement and a plan for splitting and timing of applications in accordance with crop growth stage, cropping season, variety used and crop establishment method. The LCC is used at critical growth stages to decide whether the recommended standard N rate would need to be adjusted up or down based on leaf colour (Witt et al., 2007; Bijay-Singh et al., 2012). In the real-time approach, a prescribed amount of fertilizer N is applied whenever the colour of rice leaves falls below a critical LCC value. Local guidelines on the LCC use have now been developed for the major irrigated rice domains.

### 2.2.1 Real-time N management in rice using fixed LCC threshold values

As shade 4 on the LCC represents greenness equivalent to SPAD value somewhere between 35 and 37, it was found to be threshold value for inbred rice varieties prevalent in the Indo-Gangetic plains in India (Bijay-Singh et al., 2002; Varinderpal-Singh et al., 2007; Yadvinder-Singh et al., 2007; Thind et al., 2010). For direct wet-seeded rice grown under

**Table 1** Effect of chlorophyll meter (CM) on fertilizer nitrogen use efficiency (NUE) in rice across different regions in Asia

| Country, region, experimental year(s), critical SPAD value | $AE_N$ † kg grain kg $N^{-1}$ FFP‡ | CM | $RE_N$ † kg N kg $N^{-1}$ FFP | CM | $PFP_N$ † kg grain kg $N^{-1}$ FFP | CM | Reference |
|---|---|---|---|---|---|---|---|
| India, Nadia, West Bengal, 2001–3, 37 | 24.3a§ | 42.4b | 0.43a | 0.55b | 56.6a | 77.3b | Maiti et al. (2004)‡ |
| Thailand, Suphan Buri, Central Plain, 1997–9, 35 | 7.4a | 9.0a | 0.22a | 0.29b | 30.1a | 33.0a | Satawathananont et al. (2004) |
| Indonesia, Sukamandi, West Java, 1997–9, 32–35 | 10.2a | 13.0b | 0.31a | 0.46b | 28.9a | 29.0a | Abdulrachman et al. (2004) |
| China, Jinhua, Zhejiang, 1997–9, 36 | 6.0a | 11.0b | 0.18a | 0.29b | 36.9a | 40.0a | Wang et al. (2001) |
| India, Aduthurai, Old Cauvery Delta, 1997–9, 35 | 13.9a | 16.0b | 0.39a | 0.46b | 32.8a | 38.0b | Nagarajan et al. (2004) |
| Vietnam, Hanoi, Red River Delta, 1997–9, 33–37 | 14.0a | 18.0b | 0.33a | 0.39b | 43.1a | 46.0b | Son et al. (2004) |
| India, Punjab, 2003 and 2004, 36–37.5 | 8.8a | 16.1b | 0.20a | 0.30b | 34.7a | 44.2b | Khurana (2005) |
| Vietnam, Omon, Mekong Delta, 1997–9, 33–37 | 15.0a | 20.0b | 0.34a | 0.44b | 45.2a | 46.0a | Tan et al. (2004) |
| The Philippines, Nueva Ecija, 1996, 35 | 18.2a | 19.7a | – | – | 48.0a | 44.7a | Balasubramanian et al. (1999) |
| India, Ludhiana, Punjab, 1999, 37.5 | 20.0a | 23.7b | 0.44a | 0.51b | – | – | Bijay-Singh et al. (2002)‡ |
| Philippines, Maligaya, Central Luzon, 1997–9, 35 | 12.0a | 15.0b | 0.32a | 0.46b | 36.9a | 34.0a | Gines et al. (2004) |
| India, Thanjavur, New Cauvery Delta, 1997–9, 35 | 13.6a | 15.0b | 0.45a | 0.46a | 27.9a | 31.0b | Nagarajan et al. (2004) |

† $AE_N$, agronomic efficiency of applied N; $RE_N$, apparent recovery efficiency of applied N; $PFP_N$, partial factor productivity of applied N.

‡ FFP, farmers' fertilizer practice in which all nutrient management was done by the farmer without any interference by the researcher. However, in some studies conducted only on research farms and not in actual farmers' fields, FFP denotes fixed-schedule N application, for example, in Maiti et al. (2004) at 100 kg N $ha^{-1}$ and in Bijay-Singh et al. (2002) at 120 kg N $ha^{-1}$.

§ For each NUE index ($AE_N$, $RE_N$ or $PFP_N$) and site, values with different letters are significantly different at the 0.05 probability level.

northwest Indian conditions, LCC shade 3 (or simply LCC 3) proved a better threshold value (Bijay-Singh et al., 2006). In northeastern India, Maiti et al. (2004) established LCC 4 as the critical value for transplanted rice. In the Upper Gangetic Plains of India, Shukla et al. (2004) established LCC 3, 4 and 5 as the critical values for *basmati*, inbred and hybrid rice cultivars.

Table 2 lists two categories of comparisons between LCC-based site-specific N management and farmers' fertilizer practice (FFP) for managing N in rice. In the first category, the most commonly observed effect of following LCC-based N management is the production of rice yield similar to that with FFP but with less fertilizer N application. In the second category, an increase in grain yield with a reduction in N fertilizer use was observed by following the LCC method (Table 2). Increase in partial factor productivity in all the comparisons listed in Table 2 may also occur due to retention of increasing proportion of N inputs in soil organic and inorganic N pools. By using LCC, Thind et al. (2010) observed saving in total fertilizer N application along with significantly higher grain yields than with blanket fertilizer recommendation. In Bangladesh, Alam et al. (2005, 2006a,b) observed that the use of LCC for N management in transplanted rice significantly increased NUE (>35%), average grain yield (5 to >12%) and profits (>19%) across villages and seasons over the farmers' practices. Haque et al. (2003) observed significantly large increases in PFP with a saving of 19–37 kg N $ha^{-1}$ under LCC N management treatment than the farmers' fertilizer practice for nine different rice genotypes. The grain yield increases with LCC were, however, non-significant. It is important to note that in the LCC method, neither the total amount of N to be applied nor the numbers of splits for N applications to be made are fixed. Balasubramanian (2002) observed that both parameters vary depending on indigenous N supply and/or crop requirement.

Adoption of LCC for managing N by farmers is not likely without the promise of adequate economic returns (Dobermann and Cassman, 2004) because in the end, the economic benefit holds the key to the success or failure of a technology. Ladha et al. (2005) placed the use of LCC in the very high benefit:cost ratio category. The LCC is an ideal tool to optimize N use, irrespective of the source of N applied – organic fertilizers, bio fertilizers or chemical fertilizers. It is very effective in avoiding over-application of N fertilizers (Bijay-Singh et al., 2002) that ensures minimal environmental degradation.

### 2.2.2 Fixed-time variable rate dose approach for LCC-based N management in rice

Many a times, farmers prefer less frequent monitoring of leaf colour, as they are strongly accustomed to applying fertilizer N at growth stages as per blanket recommendation. The fixed-time option involves monitoring of leaf colour using LCC only at the growth stages critical for adequate supply of N, such as active tillering, panicle initiation and a week before initiation of flowering. Applications of fertilizer N upwards or downwards can then be adjusted based on the leaf colour, which reflects the relative need of the crop for N at these stages. Witt et al. (2007) have described the fixed-time adjustable dose strategy in which split N application doses at active tillering and panicle initiation of transplanted rice are given based on expected yield response and leaf colour defined by IRRI–LCC as yellowish green (LCC value 3), intermediate (LCC value 3.5) and green (LCC value 4). In India, Bijay-Singh et al. (2012) worked out appropriate combination of fixed and adjustable rates of fertilizer N at critical stages of transplanted rice. A dose of 30 kg N $ha^{-1}$ at transplanting as prescriptive N management proved to be adequate for achieving high yields of rice. Corrective N management consisting of adjustable N doses

**Table 2** Comparison of leaf colour chart (LCC) method with farmers' fertilizer practice (FFP) for N management in rice in Asia

| Country, region, year, critical LCC value, type of rice, number of farms | N used, kg N $ha^{-1}$ | | Grain yield[§], Mg $ha^{-1}$ | | $AE_N$ [†, §] | | $PFP_N$ [†] | | Reference |
|---|---|---|---|---|---|---|---|---|---|
| | FFP[‡] | LCC | FFP | LCC | FFP | LCC | FFP | LCC | |
| *Same grain yield with reduced N fertilizer application following LCC* | | | | | | | | | |
| The Philippines, Maligaya, 1998, LCC-4, TPR, 11 | 78 | 33 | 3.97a | 3.87a | 9b | 20a | 51 | 117 | Balasubram-anian et al. (2003) |
| The Philippines, Maligaya, 1999, LCC-4, TPR, 11 | 74 | 46 | 4.49a | 4.68a | 12b | 19a | 91 | 102 | |
| Vietnam, Cai Lay District, 1998, LCC-3, B-WSR, 28 | 120 | 82 | 5.24a | 5.26a | – | – | 44 | 64 | |
| Vietnam, Cai Lay District, 1999, LCC-3, B-WSR, 7 | 99 | 70 | 6.34a | 6.31a | – | – | 64 | 90 | |
| India, Haryana, 2001, LCC-4, TPR, 165 | 149 | 124 | 6.36a | 6.37a | – | – | 43 | 51 | |
| Bangladesh, Gazipur, 2002, LCC-4, TPR, 9 | 72 | 46 | 4.46a | 4.56a | – | – | 62 | 102 | Haque et al. (2003) |
| India, Punjab, 2002, LCC-4, TPR, 107 | 153 | 113 | 6.0a | 6.0a | – | – | 39 | 53 | Varinderpal-Singh et al. (2007) |
| India, Punjab, 2003, LCC-4, TPR, 48 | 115 | 91 | 6.5a | 6.5a | – | – | 57 | 71 | |
| India, Punjab, 2004, LCC-4, TPR, 53 | 134 | 100 | 8.1a | 8.2a | – | – | 61 | 82 | |
| India, Punjab, 2005, LCC-4, TPR, 142 | 145 | 107 | 7.0a | 7.1a | – | – | 48 | 66 | |
| India, Punjab, 2000, LCC-4, TPR, 8 | 120 | 91 | 6.53a | 6.61a | 20.8 | 27.8 | 57 | 85 | Yadvinder-Singh et al. (2007) |
| India, Punjab, 2001, LCC-4, TPR, 8 | 120 | 85 | 7.10a | 7.04a | 15.4 | 20.7 | 60 | 94 | |
| India, Punjab, 2002, LCC-4, TPR, 11 | 126 | 78 | 6.93a | 7.12a | 11.3 | 17.8 | 52 | 83 | |
| *Increase in grain yield with reduced N fertilizer application following LCC* | | | | | | | | | |
| The Philippines, Maligaya, 1998, LCC-3, B-WSR, 6 | 151 | 125 | 4.53b | 5.15a | 6b | 14a | 30 | 41 | Balasubram-anian et al. (2003) |
| Vietnam, Huyen District, 1999, LCC-3, B-WSR, 18 | 98 | 80 | 4.63b | 4.92a | – | – | 47 | 62 | |
| India, Uttar Pradesh, 2002, LCC-4, TPR, 1 | 150 | 135 | 6.9b | 7.6a | 20.7b | 28.1a | 46 | 56 | Shukla et al. (2004) |
| Bangladesh, southwestern region, LCC-4, TPR, 33 | 149 | 100 | 3.8b | 4.1a | 10b | 16a | 25 | 41 | Alam et al. (2006b) |

[†] $AE_N$, agronomic efficiency of applied N; $PFP_N$, partial factor productivity of applied N.
[‡] FFP, farmers' fertilizer practice in which all nutrient management was done by the farmer without any interference by the researcher. However, in some studies conducted only on research farms and not in actual farmers' fields, FFP denotes fixed-schedule N application.
[§] For grain yield and NUE index of $AE_N$, at different sites, values with different letters are significantly different at the 0.05 probability level.

was worked out as application of 45, 30 or 0 kg N $ha^{-1}$ depending upon leaf colour to be <LCC shade 4, between LCC shade 4 and 5 or ≥LCC shade 5 both at maximum tillering and panicle initiation stages, and 30 kg N $ha^{-1}$ only if leaf colour is less green than LCC shade 4 at initiation of flowering.

## 2.3 Optical sensors

Chlorophyll meter and LCC do not take into account the photosynthetic rates or biomass production and the expected yields for working out fertilizer N requirements. Optical sensors measure visible and near-infrared (NIR) spectral response from plant canopies to detect the N stress (Peñuelas et al., 1994; Ma et al., 1996). Chlorophyll contained in the palisade layer of the leaf controls much of the visible light (400–720 nm) reflectance as it absorbs 60% of all the incident light in the red wavelength bands (Campbell, 2002). Reflectance of the NIR electromagnetic spectrum (720–1300 nm) depends upon structure of the mesophyll tissues which reflect as much as 60% of all incident NIR radiation (Campbell, 2002). Spectral vegetation indices such as the normalized-difference vegetation index [NDVI defined as: $(F_{NIR} - F_{Red})/(F_{NIR} + F_{Red})$, where $F_{NIR}$ and $F_{Red}$ are, respectively, the fractions of emitted NIR and red radiation reflected back from the sensed area] have proved useful for indirectly obtaining information such as photosynthetic efficiency, productivity potential and potential yield (Peñuelas et al., 1994; Thenkabail et al., 2000; Raun et al., 2001; Báez-González et al., 2002) and have been found sensitive to leaf area index, green biomass (Peñuelas et al., 1994) and photosynthetic efficiency (Aparicio et al., 2002). A handheld GreenSeeker optical sensor unit has been used for site-specific N management in rice.

Several researchers have used mid-season spectral reflectance measurements with optical/crop canopy sensors to estimate rice growth and N status of rice (Xue et al., 2004; Nguyen et al., 2006; Bajwa et al., 2010; Ali et al., 2014). Rather than using a critical NDVI value for recommending fertilizer N, the optical sensor works out the fertilizer N requirement of the crop on the basis of the difference in N uptake between estimates of yield potential with no added fertilizer N and with fertilizer N application, and an efficiency factor. Based on target yield approach and split fertilization approach, Xue et al. (2014) used GreenSeeker optical sensor for top-dressing N at panicle initiation stage of rice. Tubaña et al. (2011) also used canopy reflectance to top-dress fertilizer N at panicle initiation stage of rice. Recently, Bijay-Singh et al. (2015) found that high yields along with high N use efficiency in transplanted rice can be achieved by applying a moderate amount of fertilizer N at transplanting and enough fertilizer N to meet the high N demand during the period between active tillering and panicle initiation before applying an optical sensor-guided fertilizer N dose at panicle initiation stage of rice. Optical sensor-assisted N management resulted in similar rice grain yields as the blanket recommendation for the region, but with reduced N rates leading to greater recovery efficiency (by 5.5–21.7%) and agronomic efficiency [by 4.7–11.7 kg grain (kg N applied)$^{-1}$].

# 3 Site-specific nutrient management for intensive rice-cropping systems

Soil nutrient-supplying capacity as determined through soil-test analyses may be used to tailor fertilizer recommendations, but soil-test analyses often do not effectively account for effects of soil submergence on soil nutrient supply. Soil submergence drastically alters biological and chemical processes that influence the release of plant-available nutrients. In addition, soil tests cannot help rice farmers adjust fertilizer application to rice based on crop performance and climatic conditions during a given season. Site-specific nutrient management (SSNM) for rice as developed in Asia is a plant-based approach for applying fertilizer to optimally match the needs of the rice crop in a specific field and season (IRRI, 2007). It enables farmers to dynamically adjust fertilizer use to fill the deficit between the nutrient needs of a high-yielding crop and the nutrient supply from naturally occurring indigenous sources such as soil, crop residues, manures and irrigation water, and can achieve high nutrient use efficiencies as well as high yields. The IRRI played the lead role in developing the SSNM approach across diverse irrigated rice-growing environments in Asia. During 1997–2000, it was evaluated and refined on irrigated rice farms in eight major rice-growing areas across six countries in Asia (Dobermann et al., 2004). The SSNM concept was systematically simplified to an extent that farmers and extension workers can use it to practise plant need-based management of N, P and K in rice. For site-specific N management, the use of LCC (see Section 2.2) was included. In 2003–4, the simplified SSNM was evaluated in farmers' fields at about 20 locations in tropical and subtropical Asia (Bangladesh, China, India, Indonesia, Myanmar, Thailand, the Philippines and Vietnam), each representing an area of intensive rice farming on more than 100 000 ha with similar soils and cropping systems (Buresh, 2004).

The major focus of SSNM is on managing field-specific spatial variation in indigenous supply of N, P and K, temporal variability in plant N status during the growing season and medium-term changes in soil P and K supply resulting from actual nutrient balance. A modified Quantitative Evaluation of the Fertility of Tropical Soils (QUEFTS) model (Janssen et al., 1990; Witt et al., 1999) was used to predict soil nutrient supply and plant uptake in absolute terms in the high-yielding irrigated rice systems in Asia. It provided the relationship between grain yield and nutrient accumulation as a function of climatic yield potential and the supply of N, P and K. In the QUEFTS model, a linear relationship between grain yield and plant nutrient uptake implies that internal efficiencies are constant until yield targets reach about 70–80% of yield potential. The relationship between grain yield and nutrient uptake enters a non-linear phase as yields approach the potential yield and internal nutrient efficiencies decline. To model this in a generic sense, two boundary lines describing the minimum and maximum internal efficiencies of N, P and K in the plant across a wide range of yields and nutrient status are determined empirically. Dobermann and Witt (2004) used more than 2000 entries on the relationship between grain yield and nutrient uptake to derive these generic boundary lines of internal efficiencies. For the linear phase of the relationship between yield and nutrient uptake, the balanced N, P and K uptake requirements for 1000 kg of rice grain yield were estimated from the respective envelope functions as 14.7 kg N, 2.6 kg P and 14.5 kg K. The corresponding borderlines for describing the minimum and maximum internal efficiencies were estimated at 42 and 96 kg grain $kg^{-1}$ N, 206 and 622 kg grain $kg^{-1}$ P and 36 and 115 kg grain $kg^{-1}$ K, respectively (Witt et al., 1999). The parameters were found to be valid for any site in Asia for rice varieties with a harvest index of about 0.45–0.55. Witt et al. (2007) provided guidelines on optimal rates of N, P and K adjusted to field-specific yield levels and indigenous supply of nutrients.

To follow plant-based SSNM approach, the fertilizer N required by a crop can be estimated from the anticipated crop response to fertilizer N, which is the difference between a yield target and the yield without fertilizer N (N-limited yield). The yield target is the rice grain yield attainable by farmers with good crop and nutrient management and average climatic conditions for a given location. The N-limited yield can be determined with the nutrient omission plot technique (IRRI, 2007) as the grain yield of a crop not fertilized with N but supplied with enough quantity of other nutrients to ensure that these do not limit yield. As only a fraction of the fertilizer N applied to rice is taken up by the crop, total amount of fertilizer N required for each tonne of increase in grain yield depends on the agronomic efficiency of fertilizer N use by rice. An efficiency of fertilizer N use of 18–20 should be achievable with SSNM and good crop management in tropical Asia. In high-yielding seasons with very favourable climatic conditions, an efficiency of fertilizer N use of 25 is often achievable with good crop management.

To ensure that supply of N matches the crop need at critical growth stages, the estimated total fertilizer N requirement by rice crop is then apportioned among multiple times of application during the growing season. Since rice plants do not need much N before the tillering stage, SSNM advocates application of small to moderate amount of fertilizer N to young rice within 14 days after transplanting or 21 days after direct sowing (IRRI, 2007). In order to achieve high yield, rice plants require sufficient N at early and mid-tillering stages to achieve an adequate number of panicles (grain bunches), at panicle initiation stage to increase grain number per panicle and during the ripening phase to enhance grain filling. Although the SSNM approach as developed by IRRI advocates use of LCC for monitoring the relative greenness of a rice leaf as an indicator of the leaf N status (Witt et al., 2005b) and guides application of fertilizer N doses to rice at appropriate stages, gadgets like chlorophyll meter (see Section 2.1) can also be used for this purpose. Both the 'real-time' and the 'fixed-time/adjustable dose' N-management options (see Sections 2.2.1 and 2.2.2) can be used to apportion the fertilizer N during the growing season of the crop.

In recent years, some modifications in the amount and times of planned fertilizer N applications during the rice-growing season have been introduced in the SSNM procedure to suit farmers in different regions. Peng et al. (2006) distributed the total N at one day before transplanting, mid-tillering, panicle initiation and heading with approximate proportions of 35, 20, 30 and 15%, respectively. The actual rates of N top-dressing at mid-tillering and panicle initiation were adjusted by $\pm 10$ kg N $ha^{-1}$ according to leaf N status measured with a SPAD or LCC. Wang et al. (2007) applied 25% of the total N requirement as basal and remaining N was applied in two to three splits at critical growth stages. Generally, 30 and 35% of the total N was applied at early tillering and at panicle initiation stages, respectively, but the rates were adjusted as per chlorophyll meter or LCC readings. Hu et al. (2007) tested some modified SSNM procedures. In one of the modified SSNM strategy, the number of N applications was reduced from four to three by combining the N applications at mid-tillering and panicle initiation or eliminating N application at heading. In the other modification, the N rate of basal application was increased because some farmers feared that a decrease in basal N application rate would reduce grain yield. In yet another modification, there was N top-dressing during the first 2 weeks after transplanting of rice, which is contrary to the standard SSNM procedure. Moreover, the LCC was not used in modified SSNM approaches to adjust the rate of N top-dressing. When tested against the standard SSNM, the modified SSNM procedures performed very competitively. Das et al. (2009) tested a SSNM in rice in Eastern India based on QUEFTS model and found that it resulted in improved site-specific and balanced fertilizer management in rice.

As the amounts of P taken up by a rice crop are directly related to crop yield, the requirement of rice for fertilizer P in a given field is obtained from an estimate of an attainable yield target and a P-limited yield (IRRI, 2007). Total K needed by rice is also worked on similar lines. As for N, the yield target can be estimated from the grain yield in a fully fertilized plot with no nutrient limitations and good management. Because rice grain yield is directly related to the total amount of P taken up by rice, the P-limited yield approximates the P supplied to rice from indigenous sources such as soil, crop residue, irrigation water, organic amendments and manures. The K-limited yield similarly approximates the K supplied to rice from indigenous sources. Irrigation water can be an important indigenous source of K that is taken accounted for with K-limited yield in the SSNM approach (Buresh, 2004). The attainable yield target and P-limited yield are used with a nutrient decision support system (Witt et al., 2005a) to determine the amount of fertilizer P required to both overcome P deficiency and maintain soil P fertility. Where the soil P supply is small, 20 kg $P_2O_5$ $ha^{-1}$ is applied for each ton of target grain yield increase (difference between yield target and yield in P-limited plot). The maintenance fertilizer P rates are designed to replenish the P removed with grain and straw, assuming a low to moderate return of crop residues (Witt et al., 2007). Similarly, the attainable yield target and K-limited yield, together with an estimate of the amount of retained crop residue, are used to determine the amount of fertilizer $K_2O$ required to both overcome K deficiency and maintain soil K fertility. Where the soil K supply is small, 30 kg $K_2O$ $ha^{-1}$ is applied for each ton of target grain yield increase (difference between yield target and yield in K-limited plot). The maintenance fertilizer K rates are designed to replenish the K removed with grain and straw by considering the amount of straw returned to the field from the previous crop.

SSNM is compatible with integrated management of organic and inorganic nutrient sources, and it protects the environment by preventing excessive application of N fertilizer, which could leak from rice fields to contaminate water bodies and increase greenhouse gas emissions to the atmosphere. Pampolino et al. (2007) estimated environmental impacts and evaluated the economic benefits of refined SSNM practices at on-farm research sites in major rice-growing areas in southern India, the Philippines and southern Vietnam. The management of N through SSNM increased the fertilizer N use efficiency leading to more grain yield per unit of fertilizer N as compared to existing farmers' fertilizer practices.

# 4 Controlled-release and slow-release nitrogen fertilizers

One of the reasons for the low fertilizer use efficiency of the water-soluble N fertilizers such as urea, ammonium sulphate and ammonium carbonate in rice is the imbalance between the time and intensity that fertilizer gives off its nutrient and the demands of crop. Controlled- and slow-release N fertilizers increase N use efficiency and yield of rice crop and reduce N losses by better synchronizing N availability with plant demand. Slow-release fertilizers consist of compounds of generally low water solubility, which become available on enzymatic hydrolysis by urease or other biological catalysts. Examples are isobutylidene diurea (IBDU), oxamide and urea formaldehyde. The N-release patterns, rates and duration of slow-release fertilizers are strongly dependent on variable soil properties, in particular biological activity and external conditions such as moisture content, wetting and drying and temperature (Aarnio and Martikainen, 1995; Trenkel, 2010). Thus, release of N may not match the crop growth demand due to varying weather conditions, so that their effects on agriculture cannot be consistent and predictive (Shaviv, 2001; Trenkel, 2010). On the other hand, the controlled-release fertilizers consist of highly soluble urea prills or granules coated with a water-insoluble material such as sulphur or polyolefin that control the rate, pattern and duration of N release. Polymer-coated urea is one of the most widely used controlled-release products in arable crop production systems (Golden et al., 2009). Unlike slow-release fertilizers, performance of controlled-release fertilizers is relatively less affected by soil properties, and the N-release longevity can be controlled by using organic/inorganic polymers that are thermoplastics or resins or sulphur as a physical diffusion barrier, and altering the ratio of components in the coatings (Shaviv, 2001).

Figure 1 illustrates the mode of action of controlled-release N fertilizers vis-à-vis conventional water-soluble fertilizers, which release all N in a short period after being applied to soil with appropriate soil moisture. If applied at planting, conventional N fertilizers readily release N when the crop sink strength is low, so that N becomes vulnerable to losses through $NH_3$ volatilization, denitrification, runoff or leaching. Thus, N-release curve for conventional water-soluble fertilizers does not match the dynamic needs of crop growth. Applying conventional water-soluble N fertilizers in split doses to rice is only an attempt to supply N in synchrony with N uptake pattern of the crop. As shown in Fig. 1, the sigmoidal rate of supply of plant-available mineral N from controlled-release fertilizers is synchronized with crop demand, considerably reducing the initial exposure of the fertilizer to losses from the soil–plant system.

Several recent reviews (Chen et al., 2008; Guertal, 2009; Trenkel, 2010; Ramesh and Reddy, 2011; Gonzalez et al., 2012; Azeem et al., 2014; Timilsena et al., 2015) indicate increased interest in slow- and controlled-release materials as N fertilizers with enhanced efficiency. Besides the advantages of controlled-released fertilizers in reducing N losses to the environment and increasing fertilizer N use efficiency in rice (Pandey and Singh, 1987; Agarwal et al., 1990; Hassan et al., 1992; Kitamura and Imai, 1995; Chang and Youngdahl, 1997; Wakimoto, 2004; Yan et al., 2008; Kiran et al., 2010; Patil et al., 2010; Zhang et al., 2012; Jat et al., 2012; Ye et al., 2013; Chalk et al., 2015; Wang et al., 2015), the rate of N application or the number of applications during the growing season can be reduced,

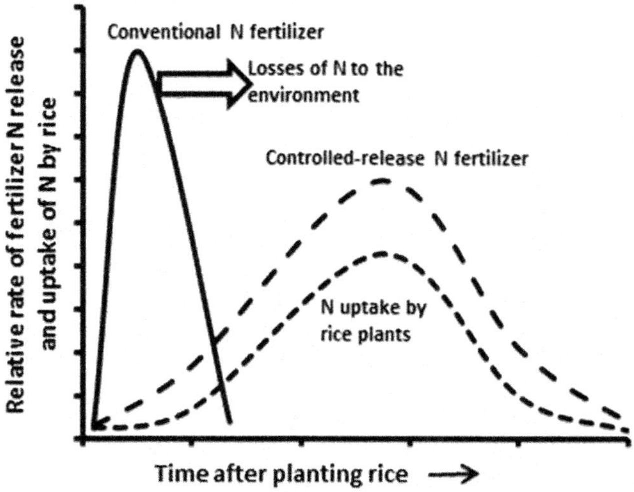


**Figure 1** Relative rate of fertilizer N release from conventional and controlled-release fertilizers and N uptake by rice plants.

which has the added advantage of savings in labour costs. Although slow- and controlled-release fertilizers such as sulphur-coated urea, IBDU and urea formaldehyde are available for more than three decades (Trenkel, 2010; Chalk et al., 2015), these could not be used on a large scale for crops like rice primarily because of prohibitive cost. In recent decades, many new controlled-release fertilizers with different kinds of polymer coatings have been developed. Unlike sulphur-coated urea that releases urea through small pinholes, resulting in a more difficult controlled N-release pattern, polymer-coated urea releases N by diffusion of urea through the swelling polymer membrane. Polymer-coated fertilizer technologies vary greatly between producers depending on the choice of the coating material and the coating process. In general, the polymer coating material represents 3–15% of the total weight of the finished product. For example, the capsule or coating film of Meister® (encapsulated urea) is 50–60 μm in thickness and approximately 10% in weight (Fujita and Shoji, 1999).

Different variants of polymer-coated urea have been designed to synchronize N release and crop N uptake with minimum side effects. These have already been tested for achieving high fertilizer N use efficiency along with high grain yield of rice in a large number of studies from all over the world (Sato et al., 1993; Shoji and Kanno, 1994; Singh et al., 1995, 2007; Blaise and Prasad, 1996; Fashola et al., 2002; Carreres et al., 2003; Wakimoto, 2004; Acquaye and Inubushi, 2004; Kondo et al., 2005; Tang et al., 2007; Yan et al., 2008; Kiran et al., 2010; Patil et al., 2010; Zhang et al., 2012; Ye et al., 2013; Wang et al., 2015). Three derivatives of polymer-coated urea with 6, 8 and 12% coating (wt/wt) at an identical N application rate were evaluated by Wang et al. (2015) during two rice-growing seasons. While the 6% polymer-coated urea could improve $^{15}N$ recovery and reduce $^{15}N$ loss, and increase grain yield slightly due to an initial $^{15}N$ burst occurring at high-field temperatures after basal fertilization, the 8 or 12% coated urea better met plant N demand from transplanting to heading, greatly enhanced $^{15}N$ recovery and decreased $^{15}N$ loss and $NH_3$ volatilization. But, unlike a significant increase of yield for 12% coated urea, 8% coated urea did not increase yield due to $^{15}N$ release and excessive $^{15}N$ uptake by plants at ripening. In India, Patil et al. (2010) found that even by applying a 50:50 mixture of polymer-coated urea and ordinary urea, total N dose can be reduced by 50% without any reduction in grain yield of rice. In China, a new variant of polymer-coated urea has been developed (Zhang et al., 2006) that uses coating of thermoplastic resin, which is made from recycled plastic films initially used for greenhouses on local vegetable farms. This material is about 70% cheaper than the new polymer-coated material (Yang et al., 2012a). Derivatives of urea coated with the cheap polymer are being produced on a large scale in China using a fluidized bed boating process. Yang et al. (2012b, 2013) could record 23–104% increase in recovery efficiency by applying a single dose of new polymer-coated urea fertilizer with varying N-release longevities compared to applying same amount of N as conventional urea.

There is significant penetration of polymer-coated urea use in rice fields in Japan. Methods for single basal applications of polymer-coated urea to no-till direct-seeded rice and to rice seedlings have already been standardized (Shoji and Kanno, 1994, 1995; Kaneta, 1995; Wakimoto, 2004). The development of sigmoidal controlled-release N fertilizers, such as Meister – a polyolefin-coated urea – has enabled farmers to use a single basal nursery application and a single basal field application. Both applications can meet the whole plant N demand throughout the growing season without any top-dressing. Table 3 shows a comparison of N use efficiency in rice as measured by $^{15}N$ tracer technique for Meister and conventional fertilizers.

A meta-analysis based on 27 observations (Linquist et al., 2013) inferred that when controlled- and slow-release fertilizers were applied 12–15 days before the rice fields

**Table 3** Comparison of nitrogen use efficiency‡ using Meister (a polyolefin-coated urea) and conventional urea on wetland rice in Northeast Japan

| | Nitrogen use efficiency (%) | | |
|---|---|---|---|
| Rice culture | Meister† | Urea† | Experimental site |
| Transplanted | 83 | 33 | Akita Agricultural Research Center |
| Direct seeding | 80 | 30 | Yamagata University |
| Direct seeding | 83 | 41 | Aichi Agricultural Research Center |

†Single basal dose application for Meister and 2/3 split doses application for urea.
‡Nitrogen use efficiency was determined by the tracer method by using $^{15}N$-labelled fertilizer sources.
Source: Tachibana (2007).

were flooded (Carreres et al., 2003; Slaton et al., 2009), the yields increased by up to 14–76%. In addition, the largest benefits of controlled- and slow-release fertilizers in terms of increasing yields were in high pH soils. In low pH soils (pH 5.7), a reduction in yields was recorded (Slaton et al., 2009).

Controlled- and slow-release fertilizers have been shown to be more efficacious than conventional fertilizers, and their consumption grew at an average annual rate of about 30% during 2009–14 (CEH, 2015). But their use is relatively limited as these fertilizers are very expensive (2–10 times) than the conventional fertilizers. China is the largest producer and consumer of controlled- and slow-release fertilizers. Consumption in China has been increasing significantly in recent years and is projected to grow at 12.8% annually during 2014–19. It also includes application on large areas under rice. Global demand for controlled- and slow-release fertilizers is expected to continue to increase at around 10% annually during 2014–19 (CEH, 2015) and is likely to be driven by increases in efficiency, sustained high yields and reduced adverse environmental impacts associated with nutrient loss, and also by increases in food demand by the growing population, especially among the third-world nations that are shifting to a more protein-based diet and away from traditional carbohydrate-based diets.

# 5 Urease and nitrification inhibitors

N is lost from rice system via two major pathways – $NH_3$ volatilization and nitrification–denitrification. Volatilization as $NH_3$ is of primary concern when N is applied as urea during non-flooded periods. Urea readily gets hydrolysed by urease enzymes to $NH_3$ and $CO_2$ resulting in an increase in soil pH and $NH_4^+$ around the fertilizer granule (Francis et al., 2008). Fertilizer N is generally applied to rice during non-flooded periods such as in dry-seeded, delayed flooded systems commonly practised in the Southern United States (Street and Bollich, 2003) and losses through $NH_3$ volatilization in such systems have been recorded from 24 to 32% of applied fertilizer N (Griggs et al., 2007; Norman et al., 2009). In flooded rice systems, $NH_4^+$-N originating from hydrolysed urea accumulates in floodwater and is prone to $NH_3$ volatilization due to elevated pH of floodwater during daylight hours (due to photosynthetic activity by aquatic biomass) and increased temperatures (Mikkelsen et al.,

1978; Fillery and Vlek, 1986). In these systems, N losses via $NH_3$ volatilization have been recorded in the range of 20–56% of applied fertilizer N (Mikkelsen et al., 1978; Fillery et al., 1984; Fillery and De Datta, 1986; De Datta et al., 1989). The use of urease inhibitors to reduce $NH_3$ volatilization from urea hydrolysis has emerged as an effective strategy to increase N use efficiency of urea-based N products in rice. More than 14 000 compounds or mixtures of compounds with a wide range of characteristics have been tested (Kiss and Simihaian, 2002) and many patented as urease inhibitors, but the compound N-(*n*-butyl) thiophosphoric triamide (NBPT) has been reported to significantly minimize $NH_3$ volatilization losses from urea (Buresh et al., 1988; Bremner and Chai, 1989; Clay et al., 1990; Al-Kanani et al., 1994; Rawluk et al., 2001; Norman et al., 2009). Besides hydroquinone (HQ) in China and very limited regional use of neem extracts in India, NBPT is at present the only urease inhibitor of commercial and practical importance in agriculture (Trenkel, 2010).

In rice systems, a portion of $NH_4^+$ can get nitrified and subsequently denitrified leading to fertilizer N losses. Nitrification is the biological conversion of $NH_4^+$ to $NO_3^-$ and requires free $O_2$, while denitrification is the reduction of $NO_3^-$ in the absence of $O_2$ to nitrous oxide and N gas. Losses via this mechanism can be substantial when an aerobic period, during which nitrification occurs, is followed by an anaerobic period when denitrification proceeds rapidly. Such conditions in rice systems are encountered in a drying and wetting cycle as in permeable soils or in intermittent wet and dry rice systems (Belder et al., 2004). In flooded rice fields, there exist adjoining aerobic zones where nitrification can occur and anaerobic zones where denitrification occurs. The transport of $NO_3^-$ between aerobic and anaerobic zones couples nitrification with denitrification (Buresh et al., 2008; Reddy and Patrick, 1986). Denitrification losses are affected by soil type and N fertilizer management and have been recorded in the 12–33% range (Buresh et al., 1993a; Aulakh et al., 2001). Nitrification inhibitors when added to N fertilizers and applied to soil delay the transformation of $NH_4^+$ to $NO_2^-$ by inhibiting or at least by slowing the action of *Nitrosomonas* spp. bacteria. Many compounds that can inhibit nitrification have been identified (Trenkel, 2010), but three products have come out on a commercial basis. These are (i) 2-chloro-6-(trichloromethyl) pyridine (Nitrapyrin) with the trade name 'N Serve', (ii) dicyandiamide (DCD, $H_4C_2N_4$) and (iii) 3,4-dimethylpyrazole phosphate (DMPP).

Urease inhibitor NBPT has been reported to significantly minimize $NH_3$ volatilization loss of urea (Bremner and Chai, 1989; Al-Kanani et al., 1994; Lee et al., 1999; Zhu, 2000; Rawluk et al., 2001; Norman et al., 2009; Dillon et al., 2012; Qi et al., 2012). In several studies, urea applied along with NBPT into the floodwater of transplanted, lowland rice exhibited reduced $NH_3$ volatilization loss and increased N uptake of rice (Freney et al., 1995; Chaiwanakupt et al., 1996; Aly et al., 2001). Several researchers recorded significant increases in grain yield of rice due to application of NBPT and urea over application of urea alone (Chaiwanakupt et al., 1996; Lee et al., 1999; Norman et al., 2009; Pang and Peng, 2010; Dillon et al., 2012; Marchesan et al., 2013; Liu et al., 2014; Rogers et al., 2015), while others measured no significant yield increase (Buresh et al., 1988; Freney et al., 1995; Aly et al., 2001). In China, HQ is being extensively used as a urease inhibitor in rice because of its lower price (Yeomans and Bremner, 1986). It is recommended to apply it in combination with DCD (Xu et al., 2000). Malla et al. (2005) and Xu et al. (2005) observed that urea amended with HQ can improve crop growth of rice. Khanif and Husin (1992), however, did not find significant effects of urea amended with HQ on yield, N uptake and

N use efficiency. HQ is photosensitive, and this has to be taken into account when urea is treated with HQ.

Among the three commercially available nitrification inhibitors – Nitrapyrin, DCD and DMPP – DCD has received the maximum attention of researchers. In recent years, some studies have been conducted with DMPP as well. Although continuous submergence in rice fields naturally suppresses the nitrification process, during drainage periods in rice fields or in soils experiencing alternate aerobic–anaerobic cycles, DCD has been found to reduce $N_2O$ emissions significantly when applied with urea (Kumar et al., 2000; Xu et al., 2002; Boeckx et al., 2005; Malla et al., 2005). Wilson et al. (1990) reported that DCD-amended urea successfully inhibited nitrification for up to 28 days on a silt loam soil. Experiments conducted in the rice belt throughout the Southern United States revealed that rice grain yields were 8% greater when DCD was applied with urea incorporated either pre-plant or pre-flood than with urea alone when applied pre-plant. However, yields were 20% greater when urea was applied pre-flood compared to pre-plant (Wells et al., 1989).

In a meta-analysis of yield and N uptake by rice based on eight studies in which urea was used as the primary N source (and served as the control treatment), Linquist et al. (2013) observed that DCD resulted in an overall increase in yield of 16.5% [95% confidence interval (CI) = 8.6–24.8%]. In most studies, DCD was applied at a rate that supplied 10–15% of the total N rate; DCD contains 67% N. As not all the studies included DCD-N as part of the total N budget, some of the yield and N uptake for DCD may be due to N supplied by DCD itself. In a separate analysis based on studies where DCD-N was accounted for as part of the N rate (Banerjee et al., 2002; Ghosh et al., 2003; Malla et al., 2005; Majumdar, 2005), yield benefit of DCD decreased to 6.4% and its effect on yield became non-significant (95% CI = −1.3 to 14.7%) (Linquist et al., 2013). In some studies, no significant effect of DCD has been observed on yield of rice (Majumdar et al., 2000; Dillon et al., 2012). In some studies based on $^{15}N$-labelled fertilizers, it has been observed that N uptake efficiency in rice is increased by applying DCD along with urea (Norman et al., 1989; Wilson et al., 1990), whereas Humphreys et al. (1992) found that DCD had no effect. One reason for the limited effectiveness of DCD in rice systems is that it degrades with increasing temperature; at 25°C, the half-life of DCD is reduced to 20 days (Kelliher et al., 2008). Since rice is typically grown in the tropics or warmer temperate regions, it is possible that high DCD degradation rates render it less effective. In China, combined use of DCD and urease inhibitor HQ in rice is currently attracting increasing attention of researchers. It has been observed that DCD in combination with HQ could substantially reduce $N_2O$ emissions during rice growth season (Xu et al., 2000, 2002) and effectively regulate the behaviour of applied urea–N in the soil–plant system (Zu et al., 2002; Ghosh et al., 2003; Boeckx et al., 2005). Li et al. (2009b) observed that application of HQ and DCD together with basal fertilizer, tillering fertilizer and panicle initiation fertilizer doses to lowland rice decreased the total $N_2O$ emission by 24, 56 and 17%, and increased grain yield by 10, 18 and 6%, respectively. Li et al. (2009c) recorded similar trend in rice yield when HQ and DCD were applied together with urea. Recently, Sun et al. (2015) found that application of N-Serve along with urea to rice significantly increased the grain yield of rice; yield at 180 kg N $ha^{-1}$ with N-Serve was equal to that recorded with 240 kg N $ha^{-1}$ without nitrification inhibitor, thus saving 60 kg N $ha^{-1}$ of fertilizer without any yield loss. Similarly, Li et al. (2009a) found that application of DMPP nitrification inhibitor with urea increased the rice grain yield by 6–18%.

Urease and nitrification inhibitors cannot completely control $NH_3$ and $N_2O$ losses when urea is surface applied to soils because the inhibitory effect depends on soil physical and chemical characteristics and also on environmental conditions. The urease inhibitors available so far can prevent urea hydrolysis for at most 1 or 2 weeks, during which time the fertilizer should ideally be incorporated into the soil by water (rain or irrigation) or mechanical methods (Chien et al., 2009). Similarly, application of nitrification inhibitors needs to be coordinated with drainage management of floodwater and/or permeability of soils.

# 6 Deep placement of nitrogen fertilizers

Fertilizer best management practices based on nutrient stewardship principles essentially consist of the right source of nutrients for application at the right rate, at the right time and in the right place. Placement of fertilizer nutrients, particularly deep placement in the soil, is crucial to increase the yield of rice because it influences potential N losses. It is an established fact that incorporation of N into the soil minimizes N losses (Mikkelsen and Finfrock, 1957), whereas broadcasting urea onto soil or into floodwater can increase N losses – particularly $NH_3$ volatilization losses (Mikkelsen et al., 1978; De Datta et al., 1989). Mikkelsen et al. (1978) could observe that incorporating N fertilizer at 10–12 cm depth reduced $NH_3$ volatilization losses to just 1% compared to losses of 20% when N fertilizer was surface applied. Management of floodwater depth, particularly during rainy season, in lowland rice can eliminate runoff, but has little effect on $NH_3$ volatilization (Li et al., 2008). If N can be placed at a depth in the soil, it can ensure adequate distance from the ponded water (Cao et al., 1984; Kapoor et al., 2008) and consequently drive $NH_3$ volatilization to an insignificant level.

Placement of fertilizer at a depth in the soil under rice is not simple task, particularly when rice seedlings are transplanted in a puddled soil (Mohanty et al., 1999). Two techniques have been developed to place fertilizer N in the soil: (i) urea supergranules (USG) hand or mechanical placement (Dupuy et al., 1990; Savant and Stangel, 1990) and (ii) urea band mechanical injection (Schnier et al., 1993; Bautista et al., 2001). USG or briquettes are made by compressing prilled or granular urea in small machines with indented pocket rollers to produce individual briquettes varying in weight from 0.9 to 2.7 g. Within a week after transplanting rice, the briquettes are inserted into the puddled soil by hand, being placed to a depth of 7–10 cm in the middle of alternating squares of four hills of rice. Recently, mechanical applicators have been developed for USG application at adjustable row spacing (Hoque et al., 2013), and these are being perfected to avoid the labour-intensive practice of placing USG with hand (Ahamed et al., 2014). In the urea band mechanical injection technique, urea solution is delivered by discharge tubes placed behind furrowers. Pneumatic injection of urea prills was also tested for point positioning (Scholten, 1992).

Several studies have shown yield increases, reduced losses of N and increased fertilizer N use efficiency from deep placement of USG (Obcemea et al., 1984, De Datta, 1987, Savant and Stangel, 1990; Bandaogo et al., 2015; Huda et al., 2016). Similar results have also been obtained with band placement of liquid urea (Schnier et al., 1988, 1993). Deep placement of USG and NPK briquettes leads to reduced nutrient losses, particularly from $NH_3$ volatilization, surface runoff of N and P (Sommer et al., 2004; Rochette et al., 2013), nitrous oxide and nitric oxide emissions (Gaihre et al., 2015) and nitrification and denitrification (Chien et al., 2009). Deep placement of nutrients as fertilizer remain in a reduced soil layer for a longer time and move very little to soil surface/floodwater. As a result of reduced nutrient concentration in floodwater, any water runoff from rice paddies reduces nutrient loss and the potential eutrophication problem (Singh et al., 1995; Kapoor et al., 2008; Chien et al., 2009). As evident from a large number of research reports showing better performance of USG than split application of broadcast urea, USG is increasingly being used by farmers in Asia, particularly in Bangladesh (Hasan et al., 2002; Kabir et al., 2009; Hasanuzzaman et al., 2009; Islam et al., 2011; Miah et al., 2012; Das et al., 2013; Mamun et al., 2013; Sikder and Xiaoying, 2014; Rahman and Barmon, 2015; Huda et al., 2016), Cambodia (Bhattarai et al., 2010), India (Daftardar et al., 1997; Kapoor et al., 2008), Vietnam (IFDC, 2007; CODESPA, 2011), China (Xiang et al., 2013) and Africa (Liverpool-Tasie et al., 2015). Rice is grown throughout the year in Bangladesh, with two principal cropping seasons: boro (irrigated dry season) and aman (wet season). Bowen et al. (2005) conducted 531 on-farm trials to measure rice yields obtained in side-by-side comparisons of USG versus the farmers' practice of applying split applications of broadcast urea. Average urea–N applied under farmers' practice was 149 kg N ha$^{-1}$ for boro and 95 kg N ha$^{-1}$ for aman rice. In contrast, the average amount of N applied for USG treatment was 79 kg N ha$^{-1}$ for boro and 59 kg N ha$^{-1}$ for aman rice. Figure 2 shows the relationship between yields obtained by farmers' practice and USG. Points to the left of the 1:1 line indicate USG dominance and it is very clear from the side-by-side comparisons that grain yields were consistently greater with deep placement of USG than broadcasting prilled urea in split doses. The yield increases with USG were obtained by applying much less urea than

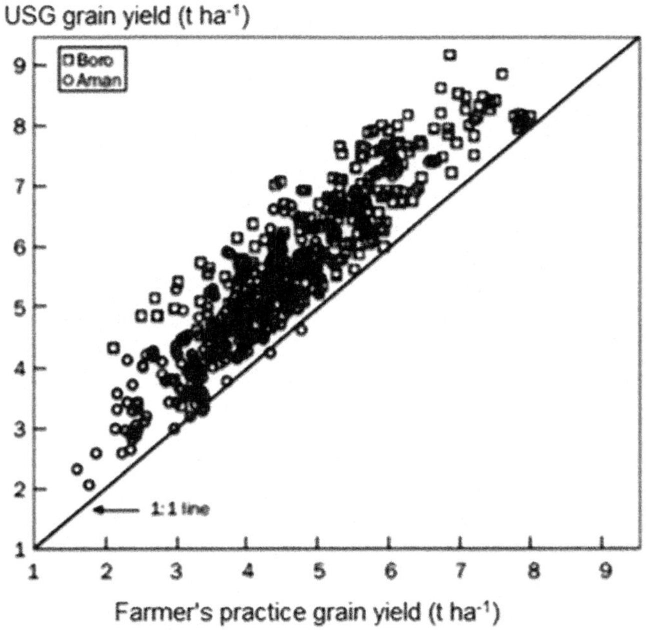


**Figure 2** Rice yields obtained by applying USG plotted against yields obtained by farmers' practice using broadcast urea in side-by-side comparisons at 531 on-farm locations for irrigated dry season (boro) and wet season (aman) rice in Bangladesh during 2000–4.

Source: Bowen et al. (2005).

with farmers' practice. Only situation where USG does not work is when rice is grown in coarse-textured highly percolating soils. In these soils, urea contained in USG is lost via leaching and does not remain available to rice, thereby resulting in very low yield levels (Katyal et al., 1985; Bijay-Singh and Katyal, 1987).

# 7 Phosphorus and potassium

Reduced soil conditions normally increase the P availability to lowland rice. Therefore, in many soils, P availability is not a yield-limiting factor for rice and significant response of modern rice varieties to fertilizer P may be observed after several years of intensive cropping (De Datta et al., 1988). Unbalanced P input/output can lead to either depletion or excessive enrichment of soil P in intensive irrigated rice systems. As a consequence, P management must focus on the buildup and maintenance of adequate available P levels in the soil, so that P supply does not limit crop growth. Since P fertilizer applications exhibit residual effects that can last several years, maintenance of soil P supply requires long-term strategies tailored to site-specific conditions (Fairhurst et al., 2007).

Sustainable P management requires the replenishment of soil P reserves, especially at high yield levels in double and triple rice-cropping systems, even if a direct yield response to P application is not expected or observed. Classical empirical approach for making fertilizer P recommendations requires a large number of site-specific field calibration studies and does not take into account crop P requirements based on a target yield. In recent decades, strategies based on estimates of the potential soil P supply and crop P uptake are being used to work out fertilizer P recommendations for rice (Fageria and Gheyi, 1999). Potential P supply can be estimated as P uptake by a rice crop from indigenous soil resources measured under field conditions by ensuring that other nutrients are amply supplied (Janssen et al., 1990). Fairhurst et al. (2007) described a practical version of this strategy for calculating P rates for lowland rice. Although blanket recommendations for large regions are still being widely used for applying P to rice in many developing countries in Asia, strategies for increasing the precision for fertilizer P recommendation for rice are being introduced because management-induced variations between farmers are much larger than differences among soil types.

In the vast Indo-Gangetic plains in South Asia, where rice is grown in annual rotation with upland wheat, P is managed in cropping system rather than in individual crops. General recommendation is that P should be applied to wheat and rice can use the residual P because the availability of soil and residual fertilizer P increases under submergence and high temperatures prevailing during rice season. For the rice–wheat system, when 26 kg P $ha^{-1}$ was applied to wheat, rice did not respond to P, but in a 7-year study, Yadvinder-Singh et al. (2000) concluded that P should also be applied to rice at rates of >15 kg P $ha^{-1}$ if rice yields greater than 6 t $ha^{-1}$ are targeted.

General strategy for K management for rice in soils with low soil K supply is to apply 25 kg K $ha^{-1}$ for each tonne of target grain yield increase over the yield of rice in the plots receiving no fertilizer K (Fairhurst et al., 2007). As more than 80% of K taken up by rice remains in the straw, it should be considered an important source of K when fertilizer K requirement of rice is worked out. In addition, because of high seasonal K inputs (7–60 kg K $ha^{-1}$ $year^{-1}$) via irrigation water and release of non-exchangeable K (Forno et al., 1975), significant responses of rice to fertilizer K application are not observed at many locations. Although the standard approach for the identification of K-deficient soils or plant K deficiency revolves around rapid chemical tests with empirical critical threshold ranges, it is inadequate for intensive irrigated rice systems in the tropics and subtropics. These systems with two or three rice crops grown in a year in submerged soil with soil drying in fallow periods or rice grown in rotation with an upland crop like wheat are extremely K demanding. Under alternating aerobic–anaerobic soil conditions, extractable soil K levels can fluctuate enormously. Extractable soil $K^+$ is still considered the most important indicator of available K in soils under rice, but its suitability as a measure of plant-available K remains controversial, particularly when soils with different textures and clay mineralogy are considered. Mixed-bed exchange resins incubated for 2 weeks under flooded soil conditions were found to be superior to K extracted by 1N ammonium acetate for prediction of K uptake by rice (Dobermann et al., 1996). Fertilizer K is applied to rice by broadcast method immediately before or after transplanting and in multiple split doses during the crop growth period. In general, a major portion, and sometimes all, of the K fertilizer should be applied at or near the time of seeding/transplanting of rice (Fageria et al., 2003; Bijay-Singh et al., 2004). A small portion of the total K fertilizer requirement should be top-dressed on soils where leaching losses of K are of concern. In the humid tropical soils with low cation exchange capacity and clay content, fertilizer K is commonly broadcast applied as a top-dressing.

Rice is grown in an annual rotation with upland wheat in more than 25 M ha area in China and Indo-Gangetic plains in South Asia. In long-term experiments on rice–wheat in the Indo-Gangetic plain, average grain yield response to application of 33 kg K $ha^{-1}$ to rice ranged from 0 to 0.5 t $ha^{-1}$. Low response to fertilizer K in the alluvial soils in the Indo-Gangetic plains is due to release of K from illitic minerals as it could meet the K needs of these crops (Bijay-Singh et al., 2004). In a long-term experiment in Hubei province in China, direct response of wheat to K application was larger than that of rice, while the residual response of rice was larger than that of wheat (Chen, 1997). Thus, when fertilizer K is not available in sufficient quantity, it should be preferably applied to wheat rather than to rice.

P and K fertilizers are applied to rice as broadcast, but with USG technology picking up, these can also be placed deep along with fertilizer N in the form of NPK briquettes. Kapoor et al. (2008) reported that deep placement of NPK briquettes provided significantly higher rice grain yield, total P and K uptake, although closer spacing (20 cm × 10 cm) led to better utilization of P and K, particularly in soils with low available P. Bulbule et al. (2008) also reported significantly higher grain yield of rice when the crop was fertilized through briquettes (56–30–30 kg NPK $ha^{-1}$) as compared to application of conventional fertilizers (100–50–50 kg NPK $ha^{-1}$). Farmers in Vietnam and Cambodia obtained 25% higher yields with deep placement of NPK briquettes over the broadcasting of fertilizer. In Bangladesh, yield of rice was increased by 15–25%, while expenditure on fertilizer was decreased by 24–32%, as less amount of fertilizer was required in the form of NPK briquettes (IFDC, 2007). Sarker et al. (2015) reported that deep placement of NPK briquettes resulted in 4–10% higher rice yield and nutrient savings of 20–35% N, 18% P and 17–24% K over the recommended practice of NPK incorporation. In China, deep application of fertilizer along with precision hill-drilling machine in super rice is emerging as a new mechanized mode for achieving high efficiency and labour-saving advantages. Using a precision rice hill-drop drilling machine developed by Luo et al. (2008) that can synchronize with deep application of fertilizer (NPK) could save fertilizer by over 30% and improve rice yield by 10% than manual fertilizing (Wang et al., 2010). Similar results have been reported by Baimba et al. (2014). Obviously, one-time deeply placed fertilizer application synchronous with sowing by precision hill-drilling machine will also lead to improved resource and energy management in rice.

# 8 Micronutrients

Zinc (Zn) deficiency in rice is a widespread phenomenon limiting productivity under lowland conditions (Quijano-Guerta et al., 2002). Zn nutrition of rice is best managed by application of suitable Zn fertilizers at the proper rates based on soil testing and at appropriate crop growth stages. In case Zn deficiency symptoms are observed in the field, but soil test for Zn is not readily available, broadcasting 20–40 kg $ZnSO_4{\cdot}7H_2O$ $ha^{-1}$ over the soil surface is recommended. For emergency treatment of Zn deficiency in rice crop, application of 0.5–1.5 kg Zn $ha^{-1}$ as a foliar spray (a 0.5% $ZnSO_4$ solution at about 200 L water $ha^{-1}$) immediately at the onset of symptoms is recommended (Fairhurst et al., 2007). Zn can be applied to rice as Zn sulphates, oxides, oxysulphates, lignosulphonates and a number of organic-chelated materials such as Zn- ethylenediaminetetraacetic acid (Zn-EDTA) and Zn-*N*-(2-hydroxyethyl)-ethylenediaminetriacetic acid (Zn-HEDTA), but highly water-soluble $ZnSO_4$ remains the most commonly used fertilizer. Liscano et al. (2001) suggested that 40–50% Zn in the fertilizer should be water soluble to optimize Zn uptake by rice seedlings. Since the recommended rates of soil applied Zn are about 20 times higher than the total crop uptake of Zn, a single Zn fertilizer application should provide adequate Zn for several years before additional Zn fertilizer is needed. Fairhurst et al. (2007) recommended dipping of rice seedlings or pre-soak seeds in a 2–4% ZnO suspension.

Iron (Fe) deficiency occurs commonly in rain-fed upland rice, rain-fed dry nurseries or when rice is grown on neutral, calcareous and alkaline upland soils, alkaline and calcareous lowland soils with low organic matter content, lowland soils irrigated with alkaline irrigation water and coarse-textured soils derived from granite. Fe deficiency in rice occurs due to low concentration of soluble $Fe^{2+}$ in upland soils, insufficient soil reduction under submerged conditions (low organic matter status of soils), high pH of alkaline or calcareous soils following submergence (decreased solubility and reduced uptake of Fe because of large bicarbonate concentrations) and wide P:Fe ratio in the soil (Fairhurst et al., 2007). In flooded rice, Fe deficiency does not commonly occur due to the increase in Fe availability associated with the anaerobic soil conditions. Because solubility of Fe increases during organic matter decomposition in flooded soils, Fe deficiency may occur when organic matter decomposition is insufficient. Soil analysis is not an effective means of identifying Fe-deficient soils. Fe deficiency can best be treated by applying solid $FeSO_4$ (about 30 kg Fe $ha^{-1}$) next to rice rows, or broadcast (larger application rate required) along with organic matter through crop residues, green manures (GM) or animal manures (Fairhurst et al., 2007). Foliar applications of $FeSO_4$ (2–3% solution) or Fe chelates can also cure Fe deficiency under emergent situations. Use of acidifying fertilizers such as $(NH_4)_2SO_4$ instead of urea on high pH soils can also be helpful.

Manganese (Mn) deficiency is more often observed in upland rice, alkaline and calcareous soils with low organic matter status and small amounts of reducible Mn, degraded rice soils high in Fe content, acid upland, leached old acid sulphate soils with low base content, leached sandy soils containing small amounts of Mn or in excessively limed acid soils. Deficiency of Mn can be corrected by foliar application of Mn or by banding Mn with an acidifying starter fertilizer. Mn sulphate or finely ground MnO (5–20 kg Mn $ha^{-1}$) can be applied in bands along rice rows. For rapid treatment of Mn deficiency,

$MnSO_4$ solution (1–5 kg Mn $ha^{-1}$ in 200 L water $ha^{-1}$) can be foliar sprayed. Application of farmyard manure (FYM) and acid-forming fertilizer such as $(NH_4)_2SO_4$ can prevent Mn deficiency in rice (Fairhurst et al., 2007).

Boron (B) deficiency in rice may lead to reduced grain yield from floret sterility. B deficiency in rice can be corrected by applying B in soluble forms as borax (0.5–3 kg B $ha^{-1}$) (Fairhurst et al., 2007). Borax should be broadcast and incorporated before planting, top-dressed, or as foliar spray during vegetative rice growth.

# 9 Integrated plant nutrient management based on organic resources and mineral fertilizers

Use of organic manures as source of nutrients and for general soil benefits dates back to the beginning of settled agriculture. Following the widespread use of mineral fertilizers that supplied adequate amounts of N, P and K as plant nutrients, organic manures were considered a secondary source of nutrients. But with increasing awareness about soil health and sustainability in agriculture, organic manures and many diverse organic materials have gained importance as components of IPNM strategies. In fact, return to managing soil fertility through the combination of organic resources and mineral nutrient inputs is one of the key features of sustainable soil management (Pieri, 1992), and the rediscovery of the benefits of the concept of IPNM has become the mainstay of soil fertility management practices at the turn of the twentieth century (Mokwunye and Hammond, 1992; Palm et al., 1997). Management of organic inputs has been able to draw on the knowledge gained from ecological studies of decomposition processes, nutrient cycles and nutrient balances (Myers et al., 1994; Cadisch and Giller, 1997; Smaling, 1998; Palm et al., 2001). The IPNM aims to manipulate judiciously the nutrient stocks and flows, in order to arrive at a satisfactory and sustainable level of agricultural production. The basic concept underlying IPNM is the maintenance and possible improvement of fertility and health of the soil for sustained crop productivity on a long-term basis, and use fertilizer as a supplement to nutrients supplied by different organic sources available at the farm to meet the nutrient requirement of the crops. As maintenance and/or improvement of the organic status of the soil is central to the philosophy of IPNM, major focus in sustainable agricultural systems is on the management of soil organic matter and plant nutrients through integrated use of mineral fertilizers with organic inputs such as FYM, animal manures, biological N fixation, crop residues, GM, sewage sludge and food industry waste.

As plant root growth is greatly affected by soil environment, application of organic manures into soil can positively influence root growth by improving physical and chemical environments of rhizosphere soil. Mandal et al. (2003) observed that integrated use of mineral fertilizers and FYM or GM could markedly improve crop root length density, root volume and root dry weight, as well as the depth of root penetration. Yang et al. (2004) studied the effects of different nutrient and water regimes on root growth of rice and observed that incorporation of organic sources into soil under rice improved root morphological characteristics and root activity of rice plants in terms of increased root density, active adsorption area, root oxidation ability of $\alpha$-naphthylamine and root surface phosphatase activity. However, continuous flooding of rice markedly decreased the improvement of rice root growth due to the application of organic manure compared to the alternately submerged regime.

Deficiency of soil organic matter is widespread in tropical soils and particularly in those under the influence of arid, semiarid and subhumid climates due to the controlling influence of climatic factors on primary productivity and biomass decomposition. Among rice-growing regions in Asia, the lowest soil organic C in agricultural soils has been reported in South Asia (Kyuma, 1988), particularly in the Indo-Gangetic plains spread over 13 M ha. In virgin or previously uncultivated soils, soil organic matter levels are environment specific and the highest for that particular environment. In well-managed cultivated soils, soil organic C fluctuates between a low steady-state value observed in the heavily cultivated soil and the highest value observed in the uncultivated soil (Buyanovsky and Wagner, 1998). Cultivation tends to lower the equilibrium soil C levels, but the addition of organic manures with fertilizers reduced the extent of soil organic matter decline with cultivation. Katyal et al. (2001) documented such changes with arable cropping from long-term field experiments in South Asia. Soil organic matter in a virgin soil remained stable for 10 years with mineral fertilizer application, but subsequently fell to about 40% of the initial value during the next 3 years. However, when manures and fertilizer were applied, the soil organic matter level was stable for 25 years, thus illustrating the value of integrated use of organic and inorganic nutrient sources in stabilizing and maintaining soil organic matter in cropping systems and ensuring sustainability regardless of the cropping system. Under irrigated conditions and regular application of recommended rates of NPK fertilizers, productivity stagnated or declined after initially increasing for 5–6 years. It was the combined application of fertilizers and FYM that unfailingly sustained productivity. This conclusion was valid, irrespective of the location and for all the cropping systems, including rice-based cropping systems. In continuous rice systems with two or three crops each year, the dominant land use systems in tropical lowlands of East and Southeast Asia where irrigation water is available, the soil organic C levels are relatively stable even if manure or straw incorporation is not practised (Cheng, 1984; Cassman et al., 1985; Nambiar, 1994). Apparently raising two or three lowland rice crops without an upland crop phase grown in aerated soil slows down the C decomposition. Thus, while IPNM in regions like South Asia results in improved soil health in terms of increased soil organic C as well as improved nutrient management, in regions with continuous rice systems, it is practised for judiciously manipulating the nutrient stocks and flows to rice crops.

Nambiar (1994) reported that soil organic matter in treatments receiving only mineral fertilizers declined in some rice–wheat long-term experiments in India, and that application of FYM to rice along with nutrients proved effective in building up soil organic matter and boosting crop yields. In recent years, a large number of long-term experiments in South Asia have shown that application of organic amendments such as FYM to rice tends to build up soil organic matter in different rice-based cropping systems (Yadav et al., 2000; Yadvinder-Singh et al., 2004; Yaduvanshi and Swarup, 2005; Manna et al., 2006; Singh et al., 2007; Tirol-Padre et al., 2007; Kumar et al., 2008; Majumder et al., 2008a,b; Benbi and Senapati, 2010; Brar et al., 2013). Pathak et al. (2011) analysed 12 rice-based long-term experiments from different locations in India and concluded that the NPK + FYM treatment has good potential for C sequestration in the soil, and it increased yield and net return in majority of the experiments. GM with relatively high N and C content contributes less to soil organic matter build up than does straw or FYM (Yadvinder-Singh et al., 1994). Bronson and Hobbs (1998) and Bronson et al. (1998) argued that soil organic matter and N-supplying capacity of soils under rice–wheat system in the Indo-Gangetic plains in South Asia is reduced due to intensive cropping and minimal return of crop organic matter in the form of crop residues.

Since 1980s, a large number of studies have been conducted in South Asia that show positive impact on yield of rice through integrated management of different organic materials and mineral fertilizers (Subbaiah et al., 1983; Khan et al., 1986, 2007; Maskina et al., 1988; Budhar et al., 1991; Rajput and Warsi, 1991; Sharma and Sharma, 1994; Jayakrishna Kumar et al., 1994; Prasad and Prasad, 1994; Kumar and Yadav, 1995; Yadvinder-Singh et al., 1995, 2004; Rabeya Khanam et al., 1997; Mondal and Chettri, 1998; Singh et al., 1999, 2000, 2007; Pandey et al., 1999; Patra et al., 2000; Duxbury, 2001; Hussain et al., 2001; Satyanarayana et al., 2002; Bajpai et al., 2006; Rahman and Parkinson, 2007; Mahajan et al., 2008; Hossaen et al., 2011; Arif et al., 2014). Incorporation of GM legumes and residues of grain legumes into soil before transplanting of rice in rice–wheat system in South Asia (Datt and Bhardwaj, 1995; Kumar and Yadav, 1995; Mahapatra and Sharma, 1995; Hegde, 1998a,b; Saxena and Yadav, 1998; Tiwari et al., 1998; Prasad et al., 1999; Sharma and Prasad, 1999; Tiwana et al., 1999) has shown yield benefits to rice during the first season ranging from 16 to 115%. In rain-fed lowland rice environments such as in Laos and many Southeast Asian countries, drought and fluctuating soil–water conditions (from aerobic to anaerobic states) can limit productivity and the efficient use of applied nutrients. Linquist et al. (2007) observed that additive or synergistic benefits from the combined application of organic and chemical fertilizers were more likely in seasons where soil–water conditions were less favourable (alternate flooding and drying). Zhang et al. (2009) analysed the results of long-term experiments conducted across different agro-ecological regions of China and observed that compared to NPK, the combined application of NPK and FYM resulted in increased rice yields during the initial years in the rice–wheat systems. In the Philippines, Javier et al. (2002) evaluated IPNM based on different organic materials and mineral fertilizers in six continuous rice seasons and concluded that it could increase and sustain rice growth and grain yield.

Ladha et al. (2003) analysed 12 rice–wheat long-term experiments from India in which treatments consisted of recommended dose of NPK, NPK + FYM and NPK + GM. The two IPNM treatments consisted of application of 50% NPK + 50% N through FYM and 40- to 45-day-old *Sesbania sesban* GM to rice (Yadav et al., 2000). The initial and final yields of rice, and yield trends over time as these are influenced by the three nutrient management treatments are shown in Table 4. The treatments showed no significant effect on the final yield, although the initial yield was significantly higher with NPK than with NPK + FYM (Table 4). The two-way analysis of variance (ANOVA) and pairwise multiple comparison indicated that the annual rate of yield change in rice was significantly ($p < 0.05$) higher with the addition of organic manures when compared with the NPK treatment. In four rice–wheat long-term experiments initiated in the 1980s in the Indo-Gangetic plains in India, Nayak et al. (2012) calculated sustainability yield index of different IPNM treatments as (Y-SD)/$Y_{max}$, where Y is the average yield of rice over years and SD is the standard deviation and $Y_{max}$ is the observed maximum yield in the experiment over the years of cultivation. It was observed that NPK + FYM and NPK + GM treatments sustained the rice yield more than NPK treatment at all the four locations. In a 28-year-old (1984–85 to 2011–12) long-term experiment, Das et al. (2014) observed rate of change of rice yields as 0.0193 and 0.0589 t $ha^{-1}$ $year^{-1}$ for NPK and NPK + FYM treatments, respectively; corresponding sustainability yield indices for rice worked out to be 0.82 and 0.95. It suggests that supplying a portion of total amount of nutrient requirement of rice through organic manures (such as FYM and GM) can contribute to maintaining and enhancing the productivity of rice-based cropping systems in the long run.

**Table 4** Comparison of yield and yield trends of rice among treatments with full NPK and NPK supplemented with farmyard manure (FYM) and green manure (GM) in 12 rice–wheat long-term experiments in India

| Treatment | Initial yield†, t $ha^{-1}$ | Final yield†, t $ha^{-1}$ | Rate of yield change, t $ha^{-1}$ $year^{-1}$ |
|---|---|---|---|
| NPK | 4.3a‡ | 4.3a | 0.019b |
| NPK + FYM | 3.9b | 4.3a | 0.053a |
| NPK + GM | 4.1ab | 4.5a | 0.050a |

†Initial yield is the average of the first 3 years and final yield is the average of the last 3 years.
‡Values followed by a common letter in a column are not significantly different from each other $p < 0.05$ by the Tukey's test.
Source: Adapted from Ladha et al. (2003).

# 10 Summary and future trends

Nutrient management in rice is continuously evolving since the introduction of mineral fertilizers in Asia during the Green Revolution era. Though blanket fertilizer recommendations for rice were formulated and gradually improved to apply nutrients using as much as possible the right source at the right rate, right time and right place, yet these could not achieve the desired levels of nutrient use efficiency. So as to increase the production of rice to meet the requirements of the burgeoning global population along with minimal deterioration of the environment due to leakage of nutrients from the soil–plant system, a number of new nutrient management strategies are being developed and introduced to the rice farmers. Of course, driving force behind these developments is also the economics of rice production because the cost of production of fertilizers is rapidly increasing all over the world. Recent advancements in nutrient management strategies have become possible due to both technological improvements and the urgency to keep the soil resources healthy for future generations.

Need-based and real-time management of fertilizer N in rice is possibly one of the most spectacular advancements in enhancing fertilizer N use efficiency in rice. It has been accomplished through the use of simple and inexpensive gadgets like LCCs and expensive gadgets like chlorophyll meter and optical sensors. Spectral properties of rice leaves tell the real-time need of N when the crop is standing in the field. Temporal variabilities in the rice-growing season are also taken care of through these strategies. The real-time N management in rice is integrated with the SSNM approach developed by IRRI and promoted through expanded partnerships with research and extension organizations, non-governmental organizations and the private sector in Asia. The key feature of the SSNM approach is the dynamic adjustments in fertilizer N, P and K management to accommodate field- and season-specific conditions. SSNM is now a matured technology for improving both fertilizer use efficiency and grain yield of rice crop. The major challenge ahead will be to retain the success of the approach while further simplifying the procedure of SSNM as it is disseminated to farmers (Peng et al., 2010). While not compromising with the nature of the SSNM approach, simplified versions should combine decisions that are made on a field-specific basis as well as that are valid for somewhat larger recommendation domains with similar soil and climatic conditions.

When applied to rice, slow- and controlled-release fertilizers supply N at a rate that closely matches the N demand by the crop. Thus, application of fertilizer as a single dose has the potential to increase not only N use efficiency but also the labour and handling costs. However, use of slow- or controlled-release fertilizers in rice is still limited by their cost compared to conventional fertilizers. It may change in the future. The fertilizer industry is very actively searching for less expensive controlled-release N products not only to help rice farmers, but also to address the global concerns about the environmental degradation due to use of soluble N fertilizers. The development of cost-effective and environment-friendly slow-release products will require coordinated effort of agronomists, modelling specialists, polymer chemists and materials scientists.

The interest in urease and nitrification inhibitors in rice systems is well justified because urea is the most important conventional N fertilizer for rice and risks of N losses via $NH_3$ volatilization and nitrification–denitrification significantly contribute to low fertilizer use efficiency. However, the short-lived effect of the inhibitor is a limitation when urea is surface applied to flooded rice. Nevertheless, urease and nitrification inhibitors represent an alternative that cannot be disregarded due to increasing production and availability of urea as N source for rice. Further research is needed to improve our understanding of the conditions under which the inhibitors are economically viable.

USG or NPK briquettes when placed at a depth in typical rice soils can very effectively curb losses via $NH_3$ volatilization and runoff. The number of farmers using these products is gradually increasing in several Asian countries, particularly in Bangladesh and Vietnam. Farmers with small holdings can apply USG with hand. As briquette applicators are perfected and become available at economical prices, even big farmers will opt for these products.

Increasing awareness about soil health and sustainability in agriculture are the driving forces for renewed interest in IPNM in rice based on organic nutrient sources and mineral fertilizers. Since 1980s, there is increasing research interest in IPNM in South Asian countries where soil organic matter is very low. Outcome of several long-term experiments has already prompted formulation of integrated nutrient management strategies in rice in India, Pakistan and Bangladesh. Owing to the consistent increase in the cost of mineral fertilizers and to sustain soil health for the future generations, the adoption of IPNM by rice farmers in South Asia is likely to increase in the future. IPNM is going to be the best way to utilize organic waste originating on the farm in a beneficial manner.

# 11 Where to look for further information

There is no single publication that provides a standard introduction to the subject. The reviews by Chien et al. (2009), Shaviv (2001), Varinderpal-Singh et al. (2010) and Trenkel (2010) provide good material to start with.

The best source for information on SSNM in rice is the IRRI. One should find useful material at: http://www.knowledgebank.irri.org/ericeproduction/IV.4_SSNM.htm

Further reading about controlled- and slow-release fertilizers can be Trenkel (2010) and Shaviv (2001). For urease and nitrification inhibitors, recent review by Timilsena et al. (2015) and Linquist et al. (2013) provide very useful information.

For SSNM, besides the IRRI website, the book edited by A. Dobermann, C. Witt and D. Dawe, (Dobermann et al., 2004) contains a wealth of information.

Pieri (1992), though making a case for low fertility African soils, provides a basis for revival of IPNM strategies. Katyal et al. (2001) contain relevant information about IPNM in South Asia, where it is most relevant due to low soil organic matter content.

# 12 References


Aarnio, T. and Martikainen, P. J. (1995), Mineralization of C and N and nitrification in Scots pine forest soil treated with nitrogen fertilizers containing different proportions of urea and its slow-releasing derivative, urea-formaldehyde, *Soil Biol. Biochem.*, 27, 1325–31.

Abdulrachman, S., Susanti, Z., Pahim, Djatiharti, A., Dobermann, A. and Witt, C. (2004), Site-specific nutrient management in intensive irrigated rice systems of West Java, Indonesia, in Doberman, A., Witt, C. and Dawe, D. (eds), *Increasing Productivity of Intensive Rice Systems Through Site-specific Nutrient Management*, pp. 171–92, Science Publishers, Inc., Enfield, NH and International Rice Research Institute, Los Baños, Philippines.

Acquaye, S. and Inubushi, K. (2004), Comparative effects of application of coated and non-coated urea in clayey and sandy paddy soil microcosms examined by $^{15}N$ tracer technique, *Plant Nutr.*, 50, 205–13.

Adhikari, C., Bronson, K. F., Panaullah, G. M., Regmi, A. P., Saha, P. K., Dobermann, A., Olk, D. C., Hobbs, P. R. and Pasuquin, E. (1999), On-farm N supply and N nutrition in the rice-wheat system of Nepal and Bangladesh, *Field Crops Res.*, 64, 273–86.

Agarwal, S. K., Shanker, H., Agarwal, M. M. and Azad, C. S. (1990), Effect of slow-release nitrogen and nitrification inhibitors on rice-wheat sequence, *Indian J. Agron.*, 35, 337–40.

Ahamed, M. S., Ziauddin, A. T. M. and Sarker, R. I. (2014), Design of improved urea super granule applicator, *Int. J. App. Sci. Engg. Res.*, 3, 98–104.

Alam, M. M., Ladha, J. K., Buresh, R. J., Rashid, H. U. and Khan, A. (2006a), Nitrogen management in lowland rice by the use of leaf color chart through farmer participatory approach in Bangladesh. In *Proceedings of the 18th World Congress of Soil* Science [CD-ROM, Abs. #155-96], 9–15 July 2006. Philadelphia, PA.

Alam, M. M., Ladha, J. K., Foyjunnessa, Rahman, Z., Khan, S. R., Rashid, H. U., Khan, A. H. and Buresh, R. J. (2006b), Nutrient management for increased productivity of rice-wheat cropping system in Bangladesh, *Field Crops Res.*, 96, 374–86.

Alam, M. M., Ladha, J. K., Khan, S. R., Foyjunnessa, Rashid, H. U., Khan, A. H. and Buresh, R. J. (2005), Leaf color chart for managing nitrogen fertilizer in lowland rice in Bangladesh, *Agron. J.*, 97, 949–59.

Ali, A. M., Thind, H. S., Sharma, S. and Varinderpal-Singh (2014), Prediction of dry direct-seeded rice yields using chlorophyll meter, leaf color chart and GreenSeeker optical sensor in northwestern India, *Field Crops Res.*, 161, 11–15.

Al-Kanani, T., MacKenzie, A. F., Fyles, J. W., Ghazala, S. and O'Halloran, I. P. (1994), Ammonia volatilization from urea amended with lignosulfonate and phosphoroamide, *Soil Sci. Soc. Am. J.*, 58, 244–8.

Aly, S. S. M., Awad, E., El-Sherbieny, A. E., Soliman, S. and Ismail, M. M. (2001), The efficiency of modified urea labelled with N-15 by rice plant, *Egyptian J. Soil Sci.*, 41, 15–26.

Aparicio, N., Villegas, D., Araus, J. L., Casadesús, J. and Royo, C. (2002), Relationship between growth traits and spectral vegetation indices in durum wheat, *Crop Sci.*, 42, 1547–55.

Arif, M., Tasneem, M., Bashir, F., Yaseen, G. and Iqbal, R. M. (2014), Effect of integrated use of organic manures and inorganic fertilizers on yield and yield components of rice, *J. Agric. Res.*, 52, 197–206.

Aulakh, M. S., Khera, T. S., Doran, J. W. and Bronson, K. F. (2001), Denitrification, $N_2O$ and $CO_2$ fluxes in rice-wheat cropping system as affected by crop residues, fertilizer N and legume green manure, *Biol. Fertil. Soils*, 34, 375–89.
Azeem, B., KuShaari, K., Man, Z. B., Basit, A. and Thanh, T. H. (2014), Review on materials & methods to produce controlled release coated urea fertilizer, *J. Controlled Release*, 181, 11–21.
Báez-González, A. D., Chen, P., Tiscareño-López, M. and Srinivasan, R. (2002), Using satellite and field data with crop growth modelling to monitor and estimate maize yield in Mexico, *Crop Sci.*, 42, 1943–9.
Baimba, M. K., Shenggang, P., Zhaowen, M., Xiwen, L., Hua, T., Farouque, H. and Ru, T. X. (2014), Effect of deep fertilizer application with precision hill-drilling machine on yield formation and fertilizer use efficiency in rice (*Oryza sativa*), *Int. J. Agric Biol.*, 16, 1091–7.
Bajpai, R. K., Chitale, C., Upadhyay, S. K. and Urkurkar, J. S. (2006), Long-term studies on soil physico-chemical properties and productivity of rice-wheat system as influenced by integrated nutrient management in inceptisol of Chhattisgarh, *J. Indian Soc. Soil Sci.*, 54, 24–9.
Bajwa, S., Mishra, A. R. and Norman, R. J. (2010), Canopy reflectance response to plant nitrogen accumulation in rice, *Precision Agric.*, 11, 488–506.
Balasubramanian, V. (2002), Leaf colour chart for real-time N management in rice. Available via DIALOG. http://www.knowledgebank.irri.org/ (Accessed 27 January 2008).
Balasubramanian, V., Ladha, J. K., Gupta, R. K., Naresh, R. K., Mehla, R. S., Bijay-Singh and Yadvinder-Singh (2003), Technology options for rice in the rice-wheat system in South Asia, in Ladha, J. K., Hill, J. E., Duxbury, J. M., Gupta, R. J. and Buresh, R. J. (eds), *Improving the Productivity and Sustainability of Rice-Wheat Systems: Issues and Impacts*, ASA Spec. Publ. 65, pp. 115–47, ASA, CSSA, SSSA, Madison, WI.
Balasubramanian, V., Morales, A. C., Cruz, R. T. and Abdulrachman, S. (1999), On-farm adaptation of knowledge-intensive nitrogen management technologies for rice systems, *Nutr. Cycl. Agroecosyst.*, 53, 59–69.
Balasubramanian, V., Morales, A. C., Cruz, R. T., Thiyagarajan, T. M., Nagarajan, R., Babu, M., Abdulrahman, S. and Hai, I. H. (2000), Adaptation of the chlorophyll meter (SPAD) technology for real time N management in rice: A review, *Int. Rice Res. Notes*, 25(1), 4–8.
Bandaogo, A., Bidjokazo, F., Youl, S., Safo, E., Abaidoo, R. and Andrews, O. (2015), Effect of fertilizer deep placement with urea supergranule on nitrogen use efficiency in Sourou Valley (Burkina Faso), *Nutr. Cycl. Agroecosyst.*, 102, 79–89.
Banerjee, B., Pathak, H. and Aggarwal, P. K. (2002), Effects of dicyandiamide, farmyard manure and irrigation on crop yields and ammonia volatilization from an alluvial soil under a rice (*Oryza sativa* L.) – wheat (*Triticum aestivum* L.) cropping system, *Biol. Fertil. Soils*, 36, 207–14.
Bautista, E. U., Koike, M. and Suministrado, D. C. (2001), Mechanical deep placement of nitrogen in wetland rice, *J. Agric. Engg. Res.*, 78, 333–46.
Belder, P., Bouman, B. A. M., Cabangon, R., Guoan, L., Quilang, E. J. P., Yuanhua, L., Spiertz, J. H. J. and Tuong, T. P. (2004), Effect of water-saving irrigation on rice yield and water use in typical lowland conditions in Asia, *Agric. Water Manage*, 65, 193–210.
Benbi, D. K. and Senapati, N. (2010), Soil aggregation and carbon and nitrogen stabilization in relation to residue and manure application in rice–wheat systems in northwest India, *Nutr. Cycl. Agroecosyst.*, 87, 233–47.
Bhattarai, S. P., Palada, M. C., Midmore, D. J., Wu, D. and Salas, R. (2010), On-farm evaluation of fertilizer briquettes and low-cost drip irrigation for smallholder vegetable production in Cambodia, *Irrig. Drainage*, 60, 318–29.
Bijay-Singh (2008), Crop demand-driven site-specific nitrogen applications in rice (*Oryza sativa*) and wheat (*Triticum aestivum*): Some recent advances, *Indian J. Agron.*, 53, 157–66.
Bijay-Singh, Gupta, R. K., Yadvinder-Singh, Gupta, S. K., Jagmohan Singh, Bains, J. S. and Vashistha, M. (2006), Need-based nitrogen management using leaf color chart in wet direct-seeded rice in northwestern India, *J. New Seeds*, 8, 35–48.
Bijay-Singh and Katyal, J. C. (1987), Relative efficacy of some new urea-based nitrogen fertilizers for growing wetland rice on a permeable alluvial soil, *J. Agril. Sci., Camb.*, 109, 27–31.

Bijay-Singh, Varinderpal-Singh, Purba, J., Sharma, R. K., Jat, M. L., Yadvinder-Singh, Thind, H. S., Gupta, R. K., Choudhary, O. P., Chandna, P., Khurana, H. S., Kumar, A., Jagmohan-Singh, Uppal, H. S., Uppal, R. K., Vashistha, M. and Gupta, R. K. (2015), Site-specific nitrogen management in irrigated transplanted rice (*Oryza sativa*) using an optical sensor, *Precision Agric.*, 16, 455–75.

Bijay-Singh, Varinderpal-Singh, Yadvinder-Singh, Thind, H. S., Ajay-Kumar, Gupta, R. K., Kaul, A. and Vashistha, M. (2012), Fixed-time adjustable dose site-specific fertilizer nitrogen management in transplanted irrigated rice (*Oryza sativa* L.) in South Asia, *Field Crops Res.*, 126, 63–9.

Bijay-Singh, Yadvinder-Singh, Imas, P. and Xie, J. (2004), Potassium nutrition of the rice-wheat cropping system, *Adv. Agron.*, 81, 203–59.

Bijay-Singh, Yadvinder-Singh, Ladha, J. K., Bronson, K. F., Balasubramanian, V., Singh, J. and Khind, C. S. (2002), Chlorophyll meter- and leaf color chart-based nitrogen management for rice and wheat in northwestern India, *Agron. J.*, 94, 821–9.

Binford, G. D., Blackmer, A. M. and Meese, B. G. (1992), Optimal concentrations of nitrate in cornstalks at maturity, *Agron. J.*, 84, 881–7.

Blaise, D. and Prasad, R. (1996), Relative efficiency of modified urea fertilizers in wetland rice (*Oryza sativa*), *Indian J. Agron.*, 41, 373–8.

Boeckx, P., Xu, X. and van Cleemput, O. (2005), Mitigation of $N_2O$ and $CH_4$ emission from rice and wheat cropping systems using dicyandiamide and hydroquinone, *Nutr. Cycl. Agroecosyst.*, 72, 41–9.

Bowen, W. T., Diamond, R. B., Singh, U. and Thompson, T. P. (2005), Urea deep placement increases yield and saves nitrogen fertilizer in farmers' fields in Bangladesh, in Toriyama, K., Heong, K. L. and Hardy, B. (eds), *Rice is Life: Scientific Perspectives for the 21st Century*, Proceedings of the World Rice Research Conference held in Tokyo and Tsukuba, Japan, 4–7 November 2004, pp. 369–72, International Rice Research Institute, Los Baños, Philippines, and Japan International Research Center for Agricultural Sciences, Tsukuba, Japan.

Brar, B. S., Kamalbir-Singh, Dheri, G. S. and Balwinder-Kumar (2013), Carbon sequestration and soil carbon pools in a rice–wheat cropping system: Effect of long-term use of inorganic fertilizers and organic manure, *Soil Tillage Res.*, 128, 30–6.

Bremner, J. M. and Chai, H. S. (1989), Effects of phosphoroamides on ammonia volatilization and nitrite accumulation in soils treated with urea, *Biol. Fertil. Soils*, 8, 227–30.

Bronson, K. F. and Hobbs, P. R. (1998), The role of soil management in improving yields in the rice-wheat systems of South Asia, in Lal, R. (ed.), *Soil Quality and Agricultural Sustainability*, pp. 129–39, Ann Arbor Science, Chelsea, MI.

Bronson, K. F., Cassman, K. G., Wassmann, R., Olk, D. C., Noordwijk, M. and Garrity, D. P. (1998), Soil carbon dynamics in different cropping systems in principal ecoregions of Asia, in Lal, R., Follet, R. F. and Stewart, B. A. (eds), *Management of Carbon Sequestration in Soil*, pp. 35–57, CRC Press, Boca Raton, FL.

Budhar, M. N., Palaniappan, S. P. and Rangaswamy, A. (1991), Effect of farm wastes and green manures on low land rice, *Indian J. Agron.*, 36, 251–2.

Bulbule, A. V., Durgude, A. G., Patil, V. S. and Kulkarni, L. K. (2008), Fertilization of low land transplanted rice through briquettes, *Crop Res.*, 35, 1–5.

Buresh, R. J. (2004), Farmers earn more from their rice crop by scientifically optimizing fertilizer use, *Rice Today*, 32–33 October to December.

Buresh, R. J. (2007), Fertile progress, *Rice Today*, 32–33 July to September.

Buresh, R. J., Castillo, E. G. and De Datta, S. K. (1993), Nitrogen losses in puddled soils as affected by timing of water deficit and nitrogen fertilization, *Plant Soil*, 157, 197–206.

Buresh, R. J., De Datta, S. K., Padilla, J. L. and Samson, M. I. (1988), Field evaluation of two urease inhibitors with transplanted low-land rice, *Agron. J.*, 80, 763–8.

Buresh, R. J., Reddy, K. R. and van Kessel, C. (2008), Nitrogen transformations in submerged soils, in Schepers, J. S. and Raun, W. R. (eds), *Nitrogen in Agricultural Systems*, Agronomy Monograph 49, pp. 401–36, American Society of Agronomy, Madison, WI.

Buyanovsky, G. A. and Wagner, W. H. (1998), Changing role of cultivated land in the global carbon cycle, *Biol. Fert. Soils*, 27, 242–5.

Cadisch, G. and Giller, K. E. (1997), *Driven by Nature: Plant Litter Quality and Decomposition*, CAB International, Wallingford, UK.
Campbell, J. B. (2002), *Introduction to Remote Sensing* (3rd ed.), The Guilford Press, New York.
Cao, Z. H., De Datta, S. K. and Fillery, I. R. P. (1984), Effect of placement methods on floodwater properties and recovery of applied nitrogen ($^{15}$N-labelled urea) in wetland rice, *Soil Sci. Soc. Am. J.*, 48, 196–203.
Carreres, R., Sendra, J., Ballesteros, R., Valiente, E. F., Quesada, A., Carrasco, D., Legane's, F. and de La Cuadra, J. G. (2003), Assessment of slow release fertilizers and nitrification inhibitors in flooded rice, *Biol. Fertil. Soils*, 39, 80–7.
Cassman, K. G. and Pingali, P. L. (1995), Intensification of irrigated rice systems: Learning from the past to meet future challenges, *Geo J.*, 35, 299–305.
CEH (Chemical Economics Handbook) (2015), CEH: Controlled- and slow-release fertilizers, https://www.ihs.com/products/controlled-and-slow-release-chemical-economics-handbook.html
Chaiwanakupt, P., Freney, J. R., Keerthisinghe, D. G., Phongpan, S. and Blakeley, R. L. (1996), Use of urease, algal inhibitors and nitrification inhibitors to reduce nitrogen loss and increase the grain yield of flooded rice, *Biol. Fertil. Soils*, 22, 89–95.
Chalk, PM., Craswell, E. T., Polidoro, J. C. and Chen, D. (2015), Fate and efficiency of $^{15}$N-labelled slow- and controlled-release fertilizers, *Nutr. Cycl. Agroecosyst.*, 102, 167–78.
Chang, T. C. and Youngdahl, L. J. (1997), Response of rice plant to combination nitrogen fertilizer containing urea and slowly available nitrogen (SAN) compounds, *Zhongguo Nongye Huaxue Huizhi.*, 35, 495–502.
Chen, D., Suter, H., Islam, A., Edis, R., Freney, J. R. and Walker, C. N. (2008), Prospects of improving efficiency of fertiliser nitrogen in Australian agriculture: a review of enhanced efficiency fertilisers, *Soil Res.*, 46, 289–301.
Chen, F. (1997), The effect of long-term K application on the characteristics of fixation and release of soil K, PhD diss., Huazhong Agricultural University, Wuhan, Hubei, China.
Cheng, Y. (1984), Effects of drainage on the characteristics of paddy soils in China, in *Organic Matter and Rice*, pp. 417–30, International Rice Research Institute, Los Baños, Philippines.
Chien, S. H., Prochnow, L. I. and Cantarella, H. (2009), Recent developments of fertilizer production and use to improve nutrient efficiency and minimize environmental impacts, *Adv. Agron.*, 102, 267–320.
Chubachi, T., Asano, I. and Oikawa, T. (1986), The diagnosis of nitrogen nutrition of rice plants (Sasanishiki) using a chlorophyll meter, *Japan. J. Soil Sci. Plant Nutr.*, 57, 190–3.
Clay, D. E., Malzer, G. L. and Anderson, J. L. (1990), Ammonia volatilization from urea as influenced by soil temperature, soil water content, and nitrification and hydrolysis inhibitors, *Soil Sci. Soc. Am. J.*, 54, 263–6.
CODESPA (2011), *Root Solutions Fertilizer Deep Placement (FDP) for Poverty Reduction in Vietnam*, Fundación CODESPA, Spain
Daftardar, S. Y., Wagle, S. M. and Savant, N. K. (1997), Agronomic performance of urea briquettes containing diammonium phosphate in rainfed transplanted rice on farmers' fields, *J. Agric. Sci.*, 128, 291–7.
Das, D. K., Maiti, D. and Pathak, H. (2009), Site-specific nutrient management in rice in Eastern India, *Nut. Cycl. Agroecosyst.*, 83,85–94.
Das, A., Sharma, R. P., Chattopadhyay, N. and Rakshit, R. (2014), Yield trends and nutrient budgeting under a long-term (28 years) nutrient management in rice-wheat cropping system under subtropical climatic condition, *Plant Soil Environ.*, 60, 351–7.
Das, S., Chaki, A. K., Sarker, A. R., Mondal, S., Sarker, M. M. R. and Moniruzzaman, M. (2013), Effect of spacing of transplanting, rate and depth of placement of urea super granule (USG) on the yield of BORO rice, *Bull. Inst. Trop. Agr. Kyushu Univ.*, 36, 85–94.
Datt, N. and Bhardwaj, K. K. R. (1995), Nitrogen contribution and soil improvement by legume green manuring in rice-wheat cropping on an acid clay loam soil, *J. Indian Soc. Soil Sci.*, 43, 603–7.
De Datta, S. K. (1987), Nitrogen transformation processes in relation to improved cultural practices for lowland rice, *Plant Soil*, 100, 47–69.

De Datta, S. K., Gomez, K. A. and Descalsota, J. P. (1988), Changes in yield response to major nutrients and in soil fertility under intensive rice cropping, *Soil Sci.*, 146, 350–8.

De Datta, S. K., Trevitt, A. C. F., Freney, J. R., Obcemea, W. N., Real, J. G. and Simpson, J. R. (1989), Measuring nitrogen losses from lowland rice using bulk aerodynamic and nitrogen-15 balance methods, *Soil Sci. Soc. Am. J.*, 53, 1275–81.

Dillon, K. A., Walker, T. W., Harrell, D. L., Krutz, L. J., Varco, J. J., Koger, C. H. and Cox, M. S. (2012), Nitrogen sources and timing effects on nitrogen loss and uptake in delayed flood rice, *Agron. J.*, 104, 466–72.

Dobermann, A. and Cassman, K. G. (2004), Environmental dimensions of fertilizer nitrogen: What can be done to increase nitrogen use efficiency and ensure global food security? in Mosier, A. R., Syers, J. K. and Freney, J. R. (eds), *Agriculture and the Nitrogen Cycle: Assessing the Impacts of Fertilizer Use on Food Production and the Environment*, Scientific Committee on Problems of the Environment (SCOPE) Ser. vol. 65, pp. 261–78, Island Press, Washington DC, USA.

Dobermann, A., Cassman, K. G., Mamaril, C. P. and Sheehy, J. E. (1998), Management of phosphorus, potassium, and sulfur in intensive, irrigated lowland rice, *Field Crops Res.*, 56, 113–38.

Dobermann, A., Sta Cru, P. C. and Cassman, K. G. (1996), Fertilizer inputs, nutrient balance, and soil nutrient-supplying power in intensive, irrigated rice systems. I. Potassium uptake and K balance, *Nutr Cycl. Agroecosyst.*, 46, l–10.

Dobermann, A. and Witt, C. (2004), The evolution of site-specific nutrient management in irrigated rice systems of Asia, in Doberman, A., Witt, C. and Dawe, D. (eds), *Increasing Productivity of Intensive Rice Systems Through Site-specific Nutrient Management*, pp. 3–9, Science Publishers, Inc., Enfield, NH and International Rice Research Institute, Los Baños, Philippines.

Dobermann,A., Witt, C. and Dawe, D. (eds) (2004), *Increasing Productivity of Intensive Rice Systems Through Site-specific Nutrient Management*, Science Publishers, Inc., Enfield, NH and International Rice Research Institute, Los Baños, Philippines.

Dupuy, J., D'Onofrio, G. and Gaudin, R. (1990), Etude a` l'aide d'ure´e enrichie en 15N de la fertilisation azote´e du riz inonde´ a`Madagascar. 1. Comparaison entre supergranules d'ure´e, perlure´e et solution d'ure´e place's a` 1 cm ou 10 cm de profondeur, *L'Agronomie Tropicale*, 45, 3–10.

Duxbury, J. M. (2001), Long-term yield trends in the rice-wheat cropping system: Results from experiments and Northwest India, *J. Crop Prod.*, 3, 27–52.

Esfahani, M., Ali Abbasi, H. R., Rabiel, B. and Kavousi, M. (2008), Improvement of nitrogen management in rice paddy fields using chlorophyll meter (SPAD), *Paddy Water Environ.*, 6, 181–8.

Fageria, N. K. and Baligar, V. C. (2005), Enhancing nitrogen use efficiency in crop plants, *Adv. Agron.*, 88, 97–185.

Fageria, N. K. and Gheyi, H. R. (1999), *Efficient Crop Production*, Federal University of Paraiba, Campina Grande, Brazil.

Fageria, N. K., Slaton, N. A. and Baligar, V. C. (2003), Nutrient management for improving lowland rice productivity and sustainability, *Adv. Agron.*, 80, 63–152.

Fairhurst, T. H., Dobermann, A., Quijano-Guerta, C. and Balasubramanian, V. (2007), Mineral deficiencies and toxicities, in Fairhurst, T. H., Witt, C., Buresh, R. J. and Dobermann, A. (eds), *Rice: A Practical Guide to Nutrient Management* (2nd edition), pp. 46–86, International Rice Research Institute, Los Baños, Philippines, International Plant Nutrition Institute, Norcross (GA), USA and International Potash Institute, Berne, Switzerland.

FAOSTAT (2016), http://faostat3.fao.org/download/Q/QC/E (Accessed 5 January 2016).

Fashola, O. O., Hayashi, K. and Wakatsuki, T. (2002), Effect of water management and polyolefin-coated urea on growth and nitrogen uptake of Indica rice, *J. Plant Nutr.*, 25, 2173–90.

Fillery, I. R. P. and De Datta, S. K. (1986), Ammonia volatilization from nitrogen sources applied to rice fields: I. Methodology, ammonia fluxes, and nitrogen-15 loss, *Soil Sci. Soc. Am. J.*, 50, 80–6.

Fillery, I. R. P., Simpson, J. R. and De Datta, S. K. (1984), Influence of field environment and fertilizer management on ammonia loss from flooded rice, *Soil Sci. Soc. Am. J.*, 48, 914–20.

Fillery, I. R. P. and Vlek, P. L. G. (1986), Reappraisal of the significance of ammonia volatilization as an N loss mechanism in flooded rice fields, *Fert. Res.*, 9, 79–98.

Forno, D. A., Yoshida, S. and Asher, C. J. (1975), Zinc deficiency in rice. I. Soil factors associated with the deficiency, *Plant Soil*, 42, 537–50.

Fox, R. H., Roth, G. W., Iversen, K. V. and Piekielek, W. P. (1989), Soil and tissue nitrate tests compared for predicting soil nitrogen availability to corn, *Agron. J.*, 81, 971–4.

Francis, D. D. and Piekielek, W. P. (1999), *Assessing Crop Nitrogen Needs with Chlorophyll Meters*, Site-Specific Management Guidelines (SSMG) No. 12, Potash and Phosphate Institute, Norcross (GA), USA.

Francis, D. D., Vigil, M. F. and Moiser, A. R. (2008), Gaseous losses of nitrogen other than through denitrification, in Schepers, J. S. and Raun, W. R. (eds), *Nitrogen in Agricultural Systems*, Agronomy Monograph 49, pp. 255–79, American Society of Agronomy, Madison, WI, USA.

Freney, J. R., Keerthisinghe, D. G., Phongpan, S., Chaiwanakupt, P. and Harrington, K. G. (1995), Effect of urease, nitrification and algal inhibitors on ammonia loss and grain yield of flooded rice in Thailand, *Fert. Res.*, 40, 225–33.

Fujita, T. and Shoji, S. (1999), Kinds and properties of Meister fertilizers, in Shoji, S. (ed.), *Meister Controlled Release Fertilizer – Properties and Utilization*, pp. 13–34, Konno Printing Company Ltd., Sendai, Japan.

Furuya, S. (1987), Growth diagnosis of rice plants by means of leaf color, *Japan Agric. Res. Quart.*, 20, 147–53.

Gaihre, Y. K., Singh, U., Islam, S. M. M., Huda, A., Islam, M. R., Satter, M. A., Sanabria, J., Islam Md, R. and Shah, A. L. (2015), Impacts of urea deep placement on nitrous oxide and nitric oxide emissions from rice fields in Bangladesh, *Geoderma*, 259/260, 370–9.

Ghosh, S., Majumdar, D. and Jain, M. C. (2003), Methane and nitrous oxide emissions from an irrigated rice of North India, *Chemosphere*, 51, 181–95.

Gines, H. C., Redondo, G. O., Estigoy, A. P. and Dobermann, A. (2004), Site-specific nutrient management in irrigated rice systems of Central Luzon, Philippines, in Doberman, A., Witt, C. and Dawe, D. (eds), *Increasing Productivity of Intensive Rice Systems Through Site-specific Nutrient Management*, pp. 145–69, Science Publishers, Inc., Enfield, NH and International Rice Research Institute, Los Baños, Philippines.

Golden, B. R., Slaton, N. A., Norman Jr., R. J., Wilson, C. E. and DeLong, R. E. (2009), Evaluation of polymer-coated urea for direct-seeded, delayed-flood rice production, *Soil Sci.Soc. Am. J.*, 73, 375–83.

Gonzalez, M. E., Gonzalez, A., Toro, C. A., Cea, M., Sepúlveda, N., Diez, M. C. and Navia, R. (2012), Biochar as a renewable matrix for the development of encapsulated and immobilized novel added-value bioproducts, *J. Biobased Mat. Bio.*, 6, 237–48.

Griggs, B. R., Norman, R. J., Wilson Jr., C. E. and Slaton, N. A. (2007), Ammonia volatilization and nitrogen uptake for conventional and conservation tilled dry-seeded, delayed-flooded rice, *Soil Sci. Soc. Am. J.*, 71, 745–51.

Guertal, E. A. (2009), Slow-release nitrogen fertilizers in vegetable production: a review, *Hort. Technol.*, 19, 16–19.

Han, S., Hendrickson, L. L. and Ni, B. (2002), Comparison of satellite and aerial imagery for detecting leaf chlorophyll content in corn, *Trans. ASAE*, 45, 1229–36.

Haque, M. A., Husain, M. M., Jharna, D. E., Uddin, M. N. and Hussain, A. S. M. I. (2003), A comparative study of leaf color chart based and farmer's practice of nitrogen fertilizer application on rice under on-farm conditions of Bangladesh, *Pakistan J. Biol. Sci.*, 6, 1685–8.

Hasan, M. S., Hossain, S. M. A., Salim, M., Anwar, M. P. and Azad, A. K. M. (2002), Response of hybrid and inbred rice varieties to the application methods of urea supergranules and prilled urea, *Pakistan J. Biol. Sci.*, 5, 746–8.

Hasanuzzaman, M., Nahar, K., Alam, M. M., Hossain, M. Z. and Islam, M. R. (2009), Response of transplanted rice to different application methods of urea fertilizer, *Int. J. Sustainable Agric.*, 1, 1–5.

Hassan, Z. A., Young, S. D., Hepburn, C. and Arizal, R. (1992), Urea-rubber matrices as slow-release fertilizers. II. Efficacy under flooded rice conditions, *Fert. Res.*, 32, 83–90.
Heffer, P. (2013), *Assessment of Fertilizer Use by Crop at the Global Level, 2010–2010/11*, International Fertilizer Industry Association (IFA), Paris, France.
Hegde, D. M. (1998a), Effect of integrated nutrient supply on crop productivity and soil fertility in rice-wheat system in semi-arid and humid ecosystem, *Indian J. Agron.*, 43, 7–12.
Hegde, D. M. (1998b), Integrated nutrient management effect on rice-wheat system productivity in subhumid ecosystem, *Indian J. Agric. Sci.*, 68, 144–8.
Hong, S. D., Fox, R. H. and Piekielek, W. P. (1990), Field evaluation of several chemical indexes of soil nitrogen availability, *Plant Soil*, 123, 83–8.
Hoque, M. A., Wohab, M. A., Hossain, M. A., Saha, K. K. and Hassan, M. S. (2013), Improvement and evaluation of BARI USG applicator, *Agric. Engg. Int.*, 15, 87–94.
Hossaen, M. A., Shamsuddoha, A. T. M., Bhuiyan, M. S. I. and Zobaer, A. S. M. (2011), Efficacy of different organic manures and inorganic fertilizer on the yield and yield attributes of Boro rice, *The Agriculturists*, 9, 117–25.
Hu, R., Cao, J., Huang, J., Peng, S., Huang, J., Zhong, X., Zou, Y., Yang, J. and Buresh, R. J. (2007), Farmer participatory testing of standard and modified site-specific nitrogen management for irrigated rice in China, *Agric. Syst.*, 94, 331–40.
Huda, A., Gaihre, Y. K., Islam, M. R., Singh, U., Islam, Md. R., Sanabria, J., Abdus Satter, M., Afroz, H., Halder, A. and Jahiruddn, M. (2016), Floodwater ammonium, nitrogen use efficiency and rice yields with fertilizer deep placement and alternate wetting and drying under triple rice cropping systems, *Nutr. Cyc. Agroecosyst.*, 104, 53–66.
Humphreys, E., Chalk, P. M., Muirhead, W. A. and White, R. J. G. (1992), Nitrogen fertilization of dry-seeded rice in south-east Australia, *Fert. Res.*, 31, 221–34.
Hussain, F., Bronson, K. F., Yadvinder-Singh, Bijay-Singh and Peng, S. (2000), Use of chlorophyll meter sufficiency indices for nitrogen management of irrigated rice in Asia, *Agron. J.*, 92, 875–9.
Hussain, N., Hassan, G., Arshadullah, M. and Mujeeb, F. (2001), Evaluation of amendments for the improvement of physical properties of sodic soil, *Int. J. Agric. Bio.*, 3, 319–22.
IFDC (2007), *Mitigating Poverty and Environmental Degradation through Nutrient Management in South Asia*, IFDC Report, March 2007, International Fertilizer Development Centre, Muscle Shoals, AL.
Inada, K. (1963), Studies on a method for determining the deepness of green color and chlorophyll content of intact crop leaves and its practical applications. I. Principle for estimating the deepness of green color and chlorophyll content of whole leaves, *Proc. Crop Sci. Soc. Japan*, 32, 157–62.
IRRI (International Rice Research Institute) (1999), *Use of Leaf Color Chart (LCC) for N Management in Rice*, CREMNET Tech. Brief No. 2, International Rice Research Institute, Los Baños, Philippines.
IRRI (International Rice Research Institute) (2007), *Nutrients for Irrigated Rice: SSNM*, http://www.knowledgebank.irri.org/ericeproduction/IV.4_SSNM.htm (Accessed 2 February 2016).
Islam, M. S., Rahman, F. and Hossain, A. T. M. S. (2011), Effects of NPK briquettes on rice (*Oryza sativa*) in tidal flooded ecosystem, *The Agriculturists*, 9, 37–43.
Jackson, R. D. (1986), Remote sensing of biotic and abiotic plant stress, *Ann. Rev. Phytopath.*, 24, 265–87.
Janssen, B. H., Guiking, F. C. T., van der Eijk, D., Smaling, E. M. A., Wolf, J. and van Reuler, H. (1990), A system for quantitative evaluation of the fertility of tropical soils (QUEFTS), *Geoderma*, 46, 299–318.
Jat, R. A., Wani, S. P., Sahrawat, K. L., Singh, P., Dhaka, S. R. and Dhaka, B. L. (2012), Recent approaches in nitrogen management for sustainable agricultural production and eco-safety, *Arch. Agron. Soil Sci.*, 58, 1033–60.
Javier, E. F., Marquez, J. M., Grospe, F. S., Mamucod, H. F. and Tabien, R. E. (2002), Three-year effect of organic fertilizer use on paddy rice, *Philippine J. Crop Sci.*, 27(2), 11–15.

Jayakrishna Kumar, V., Sundresan Nair, C., Shahul Hameed, S. M., Toyuddin, E. and Ramachandran Nair, V. (1994), Influence of Integrated supply of nitrogen through organic and inorganic sources on grain yield of wet land rice, *Oryza*, 21, 40–2.

Justes, E., Meynard, J. M., Mary, B. and Plénet, D. (1997), Management of N nutrition: diagnosis using stem base extract: JUBIL method, in Lemaire, G. (ed.), *Diagnosis of the Nitrogen Status in Crops*, pp. 163–87, Springer-Verlag, Berlin, Germany.

Kabir, M. H., Sarkar, M. A. R. and Chowdhury, A. K. M. S. H. (2009), Effect of urea super granules, prilled urea and poultry manure on the yield of transplant *Aman* rice varieties, *J. Bangladesh Agril. Univ.*, 7, 259–63.

Kaneta, Y. (1995), Novel applications of controlled-release fertilizers. 2. The single application of fertilizer in a nursery box to no-till transplanted rice, *Nippon Dojo Hiryogaku Zasshi*, 66, 176–81.

Kapoor, V., Singh, U., Patil, S. K., Magre, H., Shrivastava, L. K., Mishra, V. N., Das, R. O., Samadhiya, Sanabria, J. and Diamond, R. (2008), Rice growth, grain yield, and flood water nutrient dynamics as affected by nutrient placement method and rate, *Agron. J.*, 100, 526–36.

Katyal, J. C., Bijay-Singh, Vlek, P. L. G. and Craswell, E. T. (1985), Fate and efficiency of nitrogen fertilizers applied to wetland rice II. Punjab, India, *Fert. Res.*, 6, 279–90.

Katyal, J. C., Rao, N. H. and Reddy, M. N. (2001), Critical aspects of organic matter management in the Tropics: the example of India, *Nutr. Cyc. Agroecosyst.*, 61, 77–88.

Kelliher, F. M., Clough, T. J., Clark, H., Rys, G. and Sedcole, J. R. (2008), The temperature dependence of dicyandiamide (DCD) degradation in soils: a data synthesis, *Soil Biol. Biochem.*, 40, 1878–82.

Khan, S. K., Mohanty, S. K. and Chalam, A. B. (1986), Integrated Management of organic manure and fertilizer nitrogen for rice, *J. Indian Soc. Soil Sci.*, 24, 505–9.

Khan, U. M., Qasim, M. and Khan, U. I. (2007), Effect of integrated nutrient management on crop yields in rice-wheat cropping system, *Sarhad J. Agr.*, 23, 1019–25.

Khanif, Y. M. and Husin, A. (1992), Improvement of fertilizer N recovery of rice by dicyandiamide and hydroquinone. In *Proceedings International Symposium on Paddy Soils,* Academia Sinica, Beijing, China.

Khurana, H. S. (2005), *Site-specific Nutrient Management (SSNM) for Improving Productivity of Rice-Wheat Cropping System in Punjab*, PhD diss., Punjab Agricultural University, Ludhiana, Punjab, India.

Kiran, J. K., Khanif, Y. M., Amminuddin, H. and Anuar, A. R. (2010), Effects of controlled release urea on the yield and nitrogen nutrition of flooded rice, *Commun. Soil Sci. Plant Anal.*, 41, 811–19.

Kiss, S. and Simihaian, M. (2002*), Improving Efficiency of Urea Fertilizers by Inhibition of Soil Urease Activity*, Kluwer Academic Publishers, Dordrecht, p. 417.

Kitamura, H. and Imai, K. (1995), Innovation of fertilizer application by using controlled-release fertilizers. 1. The method of non-split application of fertilizer in transplanting of rice, *Nippon Dojo Hiryogaku Zasshi*, 66, 71–9.

Kondo, M., Singh, C. V., Agbisit, R. and Murty, M. V. R. (2005), Yield response to urea and controlled-release urea as affected by water supply in tropical upland rice, *J. Plant Nutr.*, 28, 201–19

Kumar, A. and Yadav, D. S. (1995), Use of organic manure and fertilizer in rice (*Oryza sativa*) –wheat (*Triticum aestivum*) cropping system for sustainability, *Indian J. Agril. Sci.*, 65, 703–7.

Kumar, B., Gupta, R. K. and Bhandari, A. L. (2008), Soil fertility changes after long-term application of organic manures and crop residues under rice-wheat system, *J. Indian Soc. Soil Sci.*, 56, 80–5.

Kumar, U., Jain, M. C., Pathak, H., Kumar, S. and Majumdar, D. (2000), Nitrous oxide emissions from different fertilizers and its mitigation by nitrification inhibitors in irrigated rice, *Biol. Fertil. Soils*, 32, 474–8.

Kyuma, K. (1988), Paddy soils of Japan in comparison with those in Tropical Asia. In *Proceedings, First International Symposium on Paddy Soil Fertility*, Chiangmai, Thailand, pp. 5–19.

Ladha, J. K., Pathak, H., Padre, A. T., Dawe, D. and Gupta, R. K. (2003), Productivity trends in intensive rice-wheat cropping systems in Asia, in Ladha, J. K., Hill, J. E., Duxbury, J. M., Gupta, R. J.,

Buresh, R. J. (eds), *Improving the Productivity and Sustainability of Rice-Wheat Systems: Issues and Impacts*, ASA Spec. Publ. 65, pp. 45–76, ASA, CSSA, SSSA, Madison, WI.

Ladha, J. K., Pathak, H., Krupnik, T. J., Six, J. and van Kessel, C. (2005), Efficiency of fertilizer nitrogen in cereal production: Retrospects and prospects, *Adv. Agron.*, 87, 85–156.

Lee, J. H., Lee, H. J. and Lee, B. W. (1999), Effects of urease inhibitor, nitrification inhibitor, and slow-release fertilizer on nitrogen fertilizer loss in direct-seeding rice, *Korean J. Crop Sci.*, 44, 230–5.

Li H, Liang, X., Chen, Y., Tian, G. and Zhang, Z. (2008), Ammonia volatilization from urea in rice fields with zero-drainage water management, *Agril. Water Manag.*, 95, 887–94.

Li, H., Chen, Y. X., Liang, X. Q., Lian, Y. F. and Li, W. H. (2009a), Mineral-nitrogen leaching and ammonia volatilization from a rice-rapeseed system as affected by 3,4-dimethylpyrazole phosphate, *J. Environ. Qual.*, 38, 2131–7.

Li, X. L., Zhang, G. B., Xu, H., Cai, Z. C. and Yagi, K. (2009b), Effect of timing of joint application of hydroquinone and dicyandiamide on nitrous oxide emission from irrigated lowland rice paddy field, *Chemosphere*, 75, 1417–22.

Li, X. L., Zhang, G. B., Xu, H., Cai, Z. C. and Yagi, K. (2009c), Methane and nitrous oxide emissions from rice paddy soil as influenced by timing of application of hydroquinone and dicyandiamide, *Nutr. Cyc. Agroecosyst.*, 85, 31–40.

Linquist, B. A., Liu, L., van Kessel, C. and van Groenigen, K. J. (2013), Enhanced efficiency nitrogen fertilizers for rice systems: Meta-analysis of yield and nitrogen uptake, *Field Crops Res.*, 154, 246–54.

Linquist, B. A., Phengsouvanna, V. and Sengxue, P. (2007), Benefits of organic residues and chemical fertilizer to productivity of rain-fed lowland rice and to soil nutrient balances, *Nutr. Cycl. Agroecosyst.*, 79, 59–72.

Liscano, J. F., Wilson Jr., C. E., Norman, R. J. and Slaton, N. A. (2001), Zinc availability to rice from seven granular fertilizers, Ark Agric Exp Stn Res Bull No 963. Arkansas Agricultural Research Station, Fayetteville (AR), USA.

Liu, K., Li, Y. and Hu, H. (2014), Estimating the effect of urease inhibitor on rice yield based on NDVI at key growth stages, *Front. Agr. Sci. Engg.*, 1, 150–7.

Liverpool-Tasie, L. S. O., Adjognon, S. and Kuku-Shittu, O. (2015), Productivity effects of sustainable intensification: The case of urea deep placement for rice production in Niger State, Nigeria, *African J. Agril. Resource Econ.*, 10, 51–63.

Lobell, D. B. (2007), The cost of uncertainty for nitrogen fertilizer management: A sensitivity analysis, *Field Crops Res.*, 100, 210–17.

Luo, X. W., Jiang, E. H., Wang, Z. M., Tang, X. R., Li J. H. and Chen, W. T. (2008), Precision rice hill drop drilling machine, *Trans. Chin. Soc. Agric. Engg.*, 24, 52–6.

Ma, B. L., Morrison, M. J. and Dwyer, L. (1996), Canopy light reflectance and field greenness to assess nitrogen fertilization and yield of maize, *Agron. J.*, 88, 915–20.

Magdoff, F. R., Jokela, W. E., Fox, R. H. and Griffin, G. F. (1990), A soil test for nitrogen availability in the northeastern United States, *Commun. Soil Sci. Plant Anal.*, 21, 1103–15.

Mahajan, A., Bhagat, R. M. and Gupta, R. D. (2008), Integrated nutrient management in sustainable rice-wheat cropping system for food security in India, *SAARC J. Agric.*, 6, 29–32.

Mahapatra, B. S. and Sharma, G. L. (1995), Effect of summer legumes on growth and yield of lowland rice and its residual effect on succeeding wheat in rice-wheat system, *Indian J. Agril. Sci.*, 65, 557–61.

Maiti, D., Das, D. K., Karak, T. and Banerjee, M. (2004), Management of nitrogen through the use of leaf color chart (LCC) and soil plant analysis development (SPAD) or chlorophyll meter in rice under irrigated ecosystem, *The Sci. World J.*, 4, 838–46.

Majumdar, D. (2005), Crop yield and soil nitrogen dynamics in an intermittently flooded rice field affected by nitrification inhibitors, *Arch. Agron. Soil Sci.*, 5, 645–53.

Majumdar, D., Kumar, S., Pathak, H., Jain, M. C. and Kumar, U. (2000), Reducing nitrous oxide emission from an irrigated rice field of North India with nitrification inhibitors, *Agric. Ecosyst. Environ.*, 81, 163–9.

Majumder, B., Mandal, B. and Bandyopadhyay, P. (2008a), Soil organic carbon pools and productivity in relation to nutrient management in a 20-year-old rice–berseem agroecosystem, *Biol. Fertil. Soils*, 44, 451–61.

Majumder, B., Mandal, B., Bandyopadhyay, P. K., Gangopadhyay, A., Mani, P. K., Kundu, A. L. and Mazumder, D. (2008b), Organic amendments influence soil organic carbon pools and rice–wheat productivity, *Soil Sci. Soc. Am. J.*, 72, 775–85.

Malla, G., Bhatia, A., Pathak, H., Prasad, S., Jain, N. and Singh, J. (2005), Mitigating nitrous oxide and methane emissions from soil in rice-wheat system of the Indo-Gangetic plain with nitrification and urease inhibitors, *Chemosphere*, 58, 141–7.

Mamun, M. A. A., Islam, M. R. and Shahidullah, S. M. (2013), Effect of nitrogen source and water management on rice yield and nitrogen use efficiency, *Agril. Adv.*, 2, 292–8.

Mandal, U. K., Singh, G., Victor, U. S. and Sharma, K. L. (2003), Green manuring: its effect on soil properties and crop growth under rice–wheat cropping system, *Eur. J. Agron.*, 19, 225–37.

Manna, M. C., Swarup, A., Wanjari, R. H., Singh, Y. V., Ghosh, P. K., Singh, K. N., Tripathi, A. K. and Saha, M. N. (2006), Soil organic matter in a West Bengal Inceptisol after 30 years of multiple cropping and fertilization, *Soil Sci. Soc. Am. J.*, 70, 121–9.

Maskina, M. S., Bijay-Singh, Yadvinder-Singh, Baddesha, H. S. and Meelu, O. P. (1988), Fertilizer requirement of rice-wheat and maize-wheat rotations on coarse textured soils amended with farmyard manure, *Fert. Res.*, 17, 153–64.

Marchesan, E., Grohs, M., Walter, M., da Silva, L. S. and Formentini, T. C. (2013), Agronomic performance of rice to the use of urease inhibitor in two cropping systems, *Rev. Ciênc. Agron.*, 44, 594–603.

Miah, I., Chowdhury, M. A. H., Sultana, R., Ahmed, I. and Saha, B. K. (2012), Effects of prilled urea and urea super granule on growth, yield and quality of BRRI dhan28, *J. Agroforest. Environ.*, 6, 57–62.

Mikkelsen, D. S. and Finfrock, D. C. (1957), Availability of ammoniacal nitrogen to lowland rice as influenced by fertilizer placement, *Agron. J.*, 49, 296–300.

Mikkelsen, D. S., De Datta, S. K. and Obcemea, W. N. (1978), Ammonia volatilization losses from flooded rice soils, *Soil Sci. Soc. Am. J.*, 42, 725–30.

Mohanty, S. K., Singh, U., Balasubramanian, V. and Jha, K. P (1999), Nitrogen deep-placement technologies for productivity, profitability, and environmental quality of rainfed lowland rice systems, *Nutr. Cycl. Agroecosyst.*, 53, 43–57.

Mokwunye, U. and Hammond, L. L. (1992), Myths and science of fertilizer use in the tropics, in Lal, R. and Sanchez, P. A. (eds), *Myths and Science of Soils of the Tropics*, Soil Science Society of America, Madison, WI, pp. 121–34.

Mondal, S. S. and Chettri, M. (1998), Integrated nutrient management for sustaining productivity and fertility under rice (*Oryza sativa*)-based cropping system, *Indian J. Agril. Sci.*, 68, 337–40.

Myers, R. J. K., Palm, C. A., Cuevas, E., Gunatilleke, I. U. N. and Brossard, M. (1994), The synchronisation of nutrient mineralisation and plant nutrient demand, in Woomer, P. L. and Swift, M. J. (eds), *The Biological Management of Tropical Soil Fertility*, pp. 81–116, Wiley-Sayce Publication, London, UK.

Nagarajan, R., Ramanathan, S., Muthukrishnan, P., Stalin, P., Ravi, V., Babu, M., Selvam, S., Sivanatham, M., Dobermann, A. and Witt, C. (2004), Site-specific nutrient management in irrigated rice systems of Tamil Nadu, India, in Doberman, A., Witt, C. and Dawe, D. (eds), *Increasing Productivity of Intensive Rice Systems Through Site-specific Nutrient Management*, pp. 101–23, Science Publishers, Inc., Enfield, NH and International Rice Research Institute, Los Baños, Philippines.

Nambiar, K. K. M. (1994), *Soil Fertility and Crop Productivity under Long-Term Fertilizer Use in India*. Indian Council of Agricultural Research, New Delhi.

Nayak, A. K., Gangwar, B., Shukla, A. K., Mazumdar, S. P., Kumar, A., Raja, R., Kumar, A., Rai, P. K. and Mohan, U. (2012), Long-term effect of different integrated nutrient management on soil organic carbon and its fractions and sustainability of rice–wheat system in Indo Gangetic Plains of India, *Field Crops Res.*, 127, 129–39.

Nguyen, H. T., Kim, J. H., Nguyen, A. T., Nguyen, L. T., Shin, J. C. and Lee, B. W. (2006), Using canopy reflectance and partial least squares regression to calculate within-field statistical variation in crop growth and nitrogen status of rice, *Precision Agri.*, 7, 249–64.

Norman, R. J., Wells, B. R. and Moldenhauer, K. A. K. (1989) Effect of application method and dicyandiamide on urea-nitrogen-15 recovery in rice, *Soil Sci. Soc. Am. J.*, 53, 1269–74.

Norman, R. J., Wilson Jr., C. E., Slaton, N. A., Griggs, B. R., Bushong, J. T. and Gbur, E. E. (2009), Nitrogen fertilizer sources and timing before flooding dry-seeded, delayed-flood rice, *Soil Sci. Soc. Am. J.*, 73, 2184–90.

Obcemea, W. N., De Datta, S. K. and Broadbent, F. E. (1984), Movement and distribution of fertilizer nitrogen as affected by depth of placement in wetland rice, *Fert. Res.*, 5, 125–48.

Palm, C. A., Myers, R. J. K. and Nandwa, S. M. (1997), Combined use of organic and inorganic nutrient sources for soil fertility maintenance and replenishment, in Buresh, R. J., Sanchez, P. A. and Calhoun, F. (eds), *Replenishing Soil Fertility in Africa*, pp. 193–217, American Society of Agronomy and Soil Science Society of America, Madison, WI.

Palm, C. A., Giller, K. E., Mafongoya, P. L. and Swift, M. J. (2001), Management of organic matter in the tropics: translating theory into practice, *Nutr. Cycl. Agroecosyst.*, 61, 63–75.

Pampolino, M. F., Manguiat, I. J., Ramanathan, S., Gines, H. C., Tan, P. S., Chi, T. T. N., Rajendran, R. and Buresh, R. J. (2007), Environmental impact and economic benefits of site-specific nutrient management (SSNM) in irrigated rice systems, *Agril. Systems*, 93, 1–24.

Pandey, J. and Singh, R. P. (1987), Effect of urea, slow release fertilizers, and nitrification inhibitors on the yield of rainfed low land rice, *Indian J. Agron.*, 32, 303–5.

Pandey, N., Sarawgi, A. K., Rastogi, N. K. and Tripathi, R. S. (1999), Effect of farmyard manure and chemical N fertilizer on grain yield and quality of scented rice (*Oryza sativa*) varieties, *Indian J. Agril. Sci.*, 69, 621–3.

Pang, G. B. and Peng, S. Z. (2010), Research advances of nitrogen fertilizer application technologies in Chinese rice field, *Soils*, 42, 329–35. (in Chinese).

Pasuquin, J. M. C. A., Witt, C. and Mutters, R. (2004), A new, standardized leaf color chart for efficient N management in irrigated rice, *Philipp. J. Crop Sci.*, 29, 46–7.

Patil, M. D., Das, B. S., Barak, E., Bhadoria, P. B. S. and Polak, A. (2010), Performance of polymer-coated urea in transplanted rice: effect of mixing ratio and water input on nitrogen use efficiency, *Paddy Water*, 8, 189–98.

Pathak, H., Byjesh, K., Chakrabarti, B. and Aggarwal, P. K. (2011), Potential and cost of carbon sequestration in Indian agriculture: Estimates from long-term field experiments, *Field Crops Res.*, 120, 102–11.

Patra, A. K., Nayak, B. C. and Mishra, M. M. (2000), Integrated nutrient management in rice (*Oryza sativa*)-wheat (*Triticum aestivum*) cropping system, *Indian J. Agron.*, 45, 453–57.

Peng S., Buresh R. J., Huang J., Yang J., Zou Y., Zhong X., Wang G. and Zhang F. (2006), Strategies for overcoming low agronomic nitrogen use efficiency in irrigated rice systems in China, *Field Crops Res.*, 96, 37–47.

Peng, S., Buresh, R. J., Huang, J., Zhong, X., Zou, Y., Yang, J., Wang, G., Liu, Y., Hu, R., Tang, Q., Cui, K., Zhang, F. and Dobermann, A. (2010), Improving nitrogen fertilization in rice by site-specific N management – A review, *Agron. Sustainable Dev.*, 30, 649–56.

Peng, S. and Cassman, K. G. (1998), Upper thresholds of nitrogen uptake rates and associated nitrogen fertilizer efficiencies in irrigated rice, *Agron. J.*, 90, 178–85.

Peng, S., Garcia, F. V., Laza, R. C. and Cassman, K. G. (1993), Adjustment for specific leaf weight improves chlorophyll meter's estimate of rice leaf nitrogen concentration, *Agron. J.*, 85, 987–90.

Peng, S., Garcia, F. V., Laza, R. C., Sanico, A. L., Visperas, R. M. and Cassman, K. G. (1996), Increased N-use efficiency using a chlorophyll meter on high-yielding irrigated rice, *Field Crops Res.*, 47, 243–52.

Peñuelas, J., Gamon, J. A., Fredeen, A. L., Merino, J. and Field, C. B. (1994), Reflectance indices associated with physiological changes in nitrogen- and water-limited sunflower leaves, *Remote Sensing Environ.*, 48, 135–46.

Pieri, C. (1992), *Fertility of Soils: A Future for Farming in the West African Savannah*, Springer Series in Physical Environment, Springer-Verlag, Berlin.

Prasad, B. and Prasad, J. (1994), Integrated nutrients management for specific yield of rice (*Oryza sativa* L) based on targeted yield concept and soil test values in old alluvial soils, *Oryza*, 31, 140–3.

Prasad, R., Gangaiah, B. and Aipe, K. C. (1999), Effect of crop residue management in a rice-wheat cropping system on growth and yield of crops and on soil fertility, *Exp. Agric.*, 35, 427–35.

Qi, X., Wu, W., Peng, S., Shah, F., Huang, J., Cui, K., Liu, H. and Nie, L. (2012), Improvement of early seedling growth of dry direct-seeded rice by urease inhibitors application, *Aust. J. Crop Sci.*, 6, 525–31.

Quijano-Guerta, C., Kirk, G. J. D., Portugal, A. M., Bartolome, V. I. and McLaren, G. C. (2002),Tolerance of rice germplasm to zinc deficiency, *Field Crops Res.*, 76, 123–30.

Rabeya Khanam, Sahu, S. K. and Mitra, G. N. (1997), Yield maximization of rice through integrated Nutrient Management on Aeric Ustochrept, *J. Indian Soc. Soil Sci.*, 45, 396–7.

Rahman, S. and Barmon, B. K. (2015), Productivity and efficiency impacts of urea deep placement technology in modern rice production: An empirical analysis from Bangladesh, *J. Developing Areas*, 49, 119–34.

Rahman, S. and Parkinson, R. J. (2007), Productivity and soil fertility relationships in rice production systems, Bangladesh, *Agril. Syst.*, 92, 318–33.

Rajput, A. L and Warsi, A. S. (1991), Contribution of organic materials to nitrogen economy in rice (*Oryza sativa*) production, *Indian J. Agron.*, 36, 455–6.

Ramesh, K. and Reddy, D. D. (2011), Zeolites and their potential uses in agriculture, *Adv Agron.*, 113, 215–36.

Raun, W. R., Johnson, G. V., Stone, M. L., Solie, J. B., Lukina, E. V., Thomason, W. E. and Schepers, J. S. (2001), In-season prediction of potential grain yield in winter wheat using canopy reflectance, *Agron. J.*, 93, 131–8.

Rawluk, C. D. L., Grant, C. A. and Racz, G. J. (2001), Ammonia volatilization from soils fertilized with urea and varying rates of urease inhibitors NBPT, *Canadian J. Soil Sci.*, 81, 239–46.

Reddy, K. R. and Patrick Jr., W. H. (1986), Denitrification losses in flooded rice fields, *Fert. Res.*, 9, 99–116.

Richardson, A. D., Duigan, S. P. and Berlyn, G. P. (2002), An evaluation of noninvasive methods to estimate foliar chlorophyll content, *New Phytol.*, 153,185–94.

Rochette, P., Angers, D. A., Chantigny, M. H., Gasser, M. O., MacDonald, J. D., Pelster, D. E. and Bertrand, N. (2013), Ammonia volatilization and nitrogen retention: how deep to incorporate urea? *J. Environ. Qual.*, 42, 1635–42.

Rogers, C. W., Norman, R. C., Brye, K. R., Slaton, N. A. and Smartt, A. D. (2015), Comparison of urease inhibitors for use in rice production on a silt-loam soil, *Crop, Forage and Turfgrass Management*, 1(1),1–6.

Sarker, B. S., Ali, M. G., Mridha, M. A. J. and Mazid Miah, M. A. (2015), Effect of deep placement of NPK briquette for rice yield maximization during Boro, T Aus and T Aman seasons at different locations in Bangladesh. In *Building Productive, Diverse and Sustainable Landscapes*, Proceedings of the 17th ASA Conference, 20–24 September 2015, Hobart, Australia, www.agronomy2015.com.au.

Satawathananont, S., Chatuporn, S., Niyomvit, L., Kongchum, M., Sookthongsa, J. and Dobermann, A. (2004), Site-specific nutrient management in irrigated rice systems of Central Thailand, in Doberman, A., Witt, C. and Dawe, D. (eds), *Increasing Productivity of Intensive Rice Systems Through Site-specific Nutrient Management*, pp. 125–243, Science Publishers, Inc., Enfield, NH and International Rice Research Institute, Los Baños, Philippines.

Sato, T., Shibuya, K., Shaigusa, M. and Abe, T. (1993), Single basal application of total nitrogen-fertilizer with controlled-release coated urea on non-tilled rice culture, *Japan J. Crop Sci.*, 62, 408–13.

Satyanarayana, V., Vara Prasad, P. V., Murthy, V. R. K. and Boote, K. J. (2002), Influence of integrated use of farmyard manure and inorganic fertilizers on yield and yield components of irrigated lowland rice, *J. Plant Nutr.*, 25, 2081–90.

Savant, N. K. and Stangel, P. J. (1990), Deep placement of urea supergranules in transplanted rice: principles and practices, *Fert. Res.*, 25, 1–83.
Saxena, A. and Yadav, D. S. (1998), Effect of summer green gram on rice-wheat system, *Indian J. Agril. Sci.*, 68, 583–6.
Schnier, H., De Datta, S. K., Fagi, A. M., Eaqub, M., Ahmed Faruque, Tejasarwana, R. and Mazid, A. (1993), Yield response of wetland to band placement solution of urea in various soils in the tropics, *Fert. Res.*, 36, 221–7
Schnier, H. F., De Datta, S. K., Mengel, K., Marqueses, E. P. and Faronilo, J. E. (1988), Nitrogen use efficiency, floodwater properties and nitrogen-15 balance in transplanted lowland rice as affected by liquid band placement, *Fert. Res.*, 16, 241–55.
Scholten, J. H. M. (1992), Increasing urea-N efficiency for transplanted lowland rice by pneumatic injection: Yield and economics at the farm level, *Fert. Res.*, 33, 107–14.
Sharma, G. D. and Sharma, H. L. (1994), Utilization of weed plants as organic manure under different methods of rice establishment, *Indian J. Agril. Sci.*, 64, 184–6.
Sharma, S. N. and Prasad, R. (1999), Effects of sesbania green manuring and mungbean residue incorporation of productivity and nitrogen uptake of a rice-wheat cropping system, *Bioresource Technol.*, 67, 171–5.
Shaviv, A. (2001), Advances in controlled-release fertilizers, *Adv. Agron.*, 71, 1–49.
Shoji, S. and Kanno, H. (1994), Use of polyolefin-coated fertilizer for increasing fertilizer efficiency and reducing nitrate and nitrous oxide emissions, *Fert. Res.*, 39, 147–52
Shoji, S. and Kanno, H. (1995). Innovation of new agrotechnology using controlled release fertilizers for minimizing the environmental deterioration, in Hagin Y. (ed.), *Proceedings of the Dahlia Greidinger Memorial International Workshop on Controlled/Slow Release Fertilizers*, March 1993, Technion, Haifa, Israel.
Shukla, A. K., Ladha, J. K., Singh, V. K., Dwivedi, B. S., Balasubramanian, V., Gupta, R. K., Sharma, S. K., Singh, Y., Pathak, H., Pandey, P. S., Padre, A. T. and Yadav, R. L. (2004), Calibrating the leaf color chart for nitrogen management in different genotypes of rice and wheat in a systems perspective, *Agron. J.*, 96, 1606–21.
Sikder, R. and Xiaoying, J. (2014), Urea super granule (USG) as key conductor in agricultural productivity development in Bangladesh, *Developing Country Studies*, 4, 132–9.
Sims, J. T., Vasilas, B. L., Gartley, K. L., Milliken, B. and Green, V. (1995), Evaluation of soil and plant nitrogen tests for maize on manured soils of the Atlantic coast plain, *Agron. J.*, 87, 213–22.
Singh, A. K., Amgain, L. P. and Sharma, S. K. (2000), Root characteristics, soil physical properties and yield of rice (*Oryza sativa*) as influenced by integrated nutrient management in rice–wheat (*Triticum aestivum*) system, *Indian J. Agron.*, 45, 217–22.
Singh, G., Jalota, S. K. and Yadvinder-Singh (2007), Manuring and residue management effects on physical properties of a soil under the rice–wheat system in Punjab, India, *Soil Tillage Res.*, 94, 229–38.
Singh, N. P., Sachan, R. S., Pandey, P. C. and Bisht, R. S. (1999), Effects of decade long fertilizer and manure application on soil fertility and productivity of rice-wheat system in a Mollisol, *J. Indian Soc. Soil Sci.*, 47, 72–80.
Singh, U., Cassman, K. G., Ladha, J. K. and Bronson, K. F. (1995), Innovative nitrogen management strategies for lowland rice systems, In *Fragile Lives in Fragile Ecosystems*, pp. 229–54, International Rice Research Institute, Manila, Philippines.
Singh, Y., Gupta, R. K., Singh, B. and Gupta, S. (2007), Efficient management of fertilizer nitrogen in wet direct-seeded rice (*Oryza sativa*) in northwest India, *Indian J. Agril. Sci.*, 77, 561–4
Slaton, N. A., Golden, B. R. and Norman, R. J. (2009), Rice response to urea and two polymer-coated urea fertilizers, *AAES Res. Ser.*, 581, 211–19.
Smaling, E. M. A. (1998), Nutrient balances as indicators of productivity and sustainability in sub-Saharan African agriculture, *Agric. Ecosyst. Environ.*, 71, 1–346.
Sommer, S. G., Schjoerring, J. K. and Denmead, O. T. (2004), Ammonia emission from mineral fertilizers and fertilized crops, *Adv. Agron.*, 82, 557–622.

Son, T. T., Chien, N. Y., Thoa, V. T. K., Dobermann, A. and Witt, C. (2004), Site-specific nutrient management in irrigated rice systems of the Red River Delta of Vietnam, in Doberman, A., Witt, C. and Dawe, D. (eds), *Increasing Productivity of Intensive Rice Systems Through Site-specific Nutrient Management*, pp. 217–42, Science Publishers, Inc., Enfield, NH and International Rice Research Institute, Los Baños, Philippines.

Stevens, G. and Hefner, S. (1999), *Use of a Portable Chlorophyll Meter to Manage Crop Nitrogen in Rice*, Agric. Pub. No. MP729, 15 March 1999.

Street, J. E. and Bollich, P. K. (2003), Rice production, in Smith, C. W. and Dilday, R. H. (eds), *Rice: Origin, History, Technology, and Production*, pp. 271–96, John Wiley & Sons, Inc., New Jersey.

Subbaiah, S. V., Pillai, K. G. and Singh, R P. (1983), Effect of complementary use of organic and inorganic sources of N on the growth, N uptake and grain yield of rice variety 'Rasi', *Indian J. Agril. Sci.*, 53, 325–9.

Sun, H., Zhang, H., Powlson, D., Min, J. and Shi, W. (2015), Rice production, nitrous oxide emission and ammonia volatilization as impacted by the nitrification inhibitor 2-chloro-6-(trichloromethyl)-pyridine, *Field Crops Res.*, 173, 1–7.

Tachibana, M. (2007), *Annual Report*, Chisso Asahi Fertilizer Co., Ltd., Tokyo, Japan.

Tan, P. S., Tuyen, T. Q., Huan, T. T. N., Khuong, T. Q., Hoai, N. T., Diep, L. N., Dung, H. T., Phung, C.V., Lai, N. X. and Dobermann, A. (2004), Site-specific nutrient management in irrigated rice systems of the Mekong Delta of Vietnam, in Doberman, A., Witt, C. and Dawe, D. (eds), *Increasing Productivity of Intensive Rice Systems Through Site-specific Nutrient Management*, pp. 193–215, Science Publishers, Inc., Enfield, NH and International Rice Research Institute, Los Baños, Philippines.

Tang, S., Yang, S., Chen, J., Xu, P., Zhang, F., Ai, S. and Huang, X. (2007), Studies on the mechanism of single basal application of controlled release fertilizers for increasing yield of rice (*Oryza sativa* L.), *Agric. Sci. China*, 6, 586–96

Thenkabail, P. S., Smith, R. B. and DePauw, E. (2000), Hyperspectral vegetation indices and their relationships with agricultural crop characteristics, *Remote Sensing Environ.*, 71, 158–82.

Thind, H. S., Bijay-Singh, Pannu, R. P. S., Yadvinder-Singh, Varinderpal-Singh, Gupta, R. K., Vashistha, M., Jagmohan-Singh and Kumar, A. (2010), Relative performance of neem (*Azadirachta indica*) coated urea *vis-a-vis* ordinary urea applied to rice on the basis of soil test or following need based nitrogen management using leaf colour chart, *Nutr. Cycl. Agroecosyst.*, 87, 1–8.

Tilman, D., Cassman, K., Matson, P. A., Naylor, R. and Polasky, S. (2002), Agricultural sustainability and intensive production practices, *Nature*, 417, 671–7.

Tilman, D., Fargione, J., Wolff, B., D'Antonio, C., Dobson, A., Howarth, R., Schindler, D., Schlesinger, W. H., Simberloff, D. and Swackhamer, D. (2001), Forecasting agriculturally driven global environmental change, *Science*, 292, 281–4.

Timilsena, Y. P., Adhikari, R., Casey, P., Muster, T., Gilla, H. and Adhikari, B. (2015), Enhanced efficiency fertilisers: a review of formulation and nutrient release patterns, *J. Sci. Food Agric.*, 95, 1131–42.

Tirol-Padre, A., Ladha, J. K., Regmi, A. P., Bhandari, A. L. and Inubushi, K. (2007), Organic amendments affect soil parameters in two long-term rice-wheat experiments, *Soil Sci. Soc. Am. J.*, 71, 442–52.

Tiwana, U. S., Narang, R. S. and Gosal, K. S. (1999), Nutrient management for yield maximization of rice-wheat cropping system, *Indian J. Agron.*, 44, 1–7.

Tiwari, K. N., Dev, G., Sharma, D. N. and Singh, U. V. (1998), Maximizing yield of a rice-wheat sequence in recently reclaimed saline-sodic soils, *Better Crops Int.*, 12(2), 9–11.

Trenkel, M. E. (2010), *Improving Fertilizer Use Efficiency – Controlled-Release and Stabilized Fertilizers in Agriculture*, International Fertilizer Industry Association (IFA), Paris, France, p.151.

Tubaña, B. S., Harrell, D., Walker, T. and Phillips, S. (2011), Mid-season nitrogen fertilization rate decision tool for rice using remote sensing technology, *Better Crops*, 95(1), 22–4.

Turner, F. T. and Jund, M. F. (1991), Chlorophyll meter to predict nitrogen topdress requirement for semi-dwarf rice, *Agron. J.*, 83, 926–8.

Turner, F. T. and Jund, M. F. (1994), Assessing the nitrogen requirements of rice crops with a chlorophyll meter, *Australian J. Expt. Agric.*, 34, 1001–5.

Varinderpal-Singh, Bijay-Singh, Yadvinder-Singh, Thind, H. S. and Gupta, R. K. (2010), Need based nitrogen management using the chlorophyll meter and leaf colour chart in rice and wheat in South Asia: a review, *Nutr. Cycl. Agroecosyst.*, 88, 361–80.

Varinderpal-Singh, Yadvinder-Singh, Bijay-Singh, Baldev-Singh, Gupta, R. K., Jagmohan-Singh, Ladha, J. K. and Balasubramanian, V. (2007), Performance of site-specific nitrogen management for irrigated transplanted rice in northwestern India, *Arch. Agron. Soil Sci.*, 53, 567–79.

Vitousek, P. M., Mooney, H. A., Lubchenco, J. and Melillo, J. M. (1997), Human domination of earth's ecosystems, *Science*, 277, 494–9.

Wakimoto, K. (2004), Utilization advantages of controlled release nitrogen fertilizer on paddy rice cultivation, *Japan Agril. Res. Quart.*, 38, 15–20.

Wang, G., Dobermann, A., Witt, C., Sun, Q. and Fu, R. (2001), Performance of site-specific nutrient management for irrigated rice in Southeast China, *Agron. J.*, 93, 869–78.

Wang, G., Zhang, Q. C., Witt, C. and Buresh, R. J. (2007), Opportunities for yield increases and environmental benefits through site-specific nutrient management in rice systems of Zhejiang province, China, *Agril. Syst.*, 94, 801–6.

Wang, Z. M., Luo, X. W., Tang, X. R., Ma, G. H., Zhang, G. Z. and Zeng, S. (2010), Precision rice hill-direct-seeding technology and machine based on the combination of agricultural machinery and agronomic technology, *J. South China Agric. Univ.*, 31, 91–5.

Wang, S., Zhao, X., Xing, G., Yang, Y., Zhang, M. and Chen H. (2015), Improving grain yield and reducing N loss using polymer-coated urea in southeast China, *Agron. Sustainable Dev.*, 35, 1103–15.

Wells, B. R., Bollich, P. K., Ebelhar, W., Mikkelsen, D. S., Norman, R. J., Brandon, D. M., Helms, R. S., Turner, F. T. and Westcott, M. P. (1989), Dicyandiamide (DCD) as a nitrification inhibitor for rice culture in the United States, *Commun. Soil Sci. Plant Anal.*, 20, 2023–47.

Wells, B. R., Norman, R. J. and Helms, R. S. (1993), *Use of Plant Area Measurements to Estimate Mid-season Nitrogen Rates for Rice*, Res. Ser. No. 431, pp. 137–41, Rice Res. Studies 1992, University of Arkansas, Fayetteville, AR.

Wilson, C. E., Jr., Norman, R. J. and Wells, B. R. (1990), Dicyandiamide influence on uptake of preplant-applied fertilizer nitrogen by rice, *Soil Sci. Soc. Am. J.*, 54, 1157–61.

Witt, C., Buresh, R. J., Peng, S., Balasubramanian, V. and Dobermann, A. (2007), Nutrient management, in Fairhurst, T. H., Witt, C., Buresh, R. J. and Dobermann, A. (eds), *Rice: A Practical Guide to Nutrient Management* (2nd edition), International Rice Research Institute, Los Baños, International Plant Nutrition Institute, Norcross (GA) and International Potash Institute, Berne, Switzerland.

Witt, C., Dobermann, A., Abdulrachman, S., Gines, H. C., Wang, G. H., Nagarajan, R., Satawathananont, S., Son, T. T., Tan, P. S., Tiem, L. V., Simbahan, G. C. and Olk, D. C. (1999), Internal nutrient efficiencies of irrigated lowland rice in tropical and subtropical Asia, *Field Crops Res.*, 63, 113–38.

Witt, C., Fairhurst, T. H., Sheehy, J. E., Dobermann, A. and Gfroerer-Kerstan, A. (2005a), A nutrient decision support system software for irrigated rice, *Better Crops*, 4, 26–8.

Witt, C., Pasuquin, J. M. C. A., Mutters, R. and Buresh, R. J. (2005b), New leaf color chart for effective nitrogen management in rice, *Better Crops*, 89 (1), 36–9.

Xiang, J., Haden, V. R., Peng, S., Bouman, B. A. M., Hunag, J. Cui, K., Visperas, M., Zhu, D., Zhang, Y. and Chen, H. (2013), Effect of deep placement of nitrogen fertilizer on growth, yield, and nitrogen uptake of aerobic rice, *Australian J. Crop Sci.*, 7, 870–7.

Xu, X., Boeckx, P., van Cleemput, O. and Kazuyuki, I. (2005), Mineral nitrogen in a rhizosphere soil and in standing water during rice (*Oryza sativa* L.) growth: effect of hydroquinone and dicyandiamide, *Agric. Ecosyst. Environ.*, 109, 107–17.

Xu, X. K., Boeckx, P., Van Cleemput, O. and Zhou, L. K. (2002), Urease and nitrification inhibitors to reduce emissions of $CH_4$ and $N_2O$ in rice production, *Nutr. Cycl. Agroecosyst.*, 64, 203–11.

Xu, X. K., Wang, Y. S., Zheng, X. H., Wang, M. X., Wang, Z., Zhou, L. K. and van Cleemput, O. (2000), Methane emission from a simulated rice field ecosystem as influenced by hydroquinone and dicyandiamide, *Sci. Total Environ.*, 263, 243–53.
Xue, L. H., Cao, W. X., Luo, W. H. and Dai, T. B. (2004), Monitoring leaf nitrogen status in rice with canopy spectral reflectance, *Agron. J.*, 96, 135–42.
Xue, L., Li, G., Qin, X., Yang, L. and Zhang, H. (2014), Topdressing nitrogen recommendation for early rice with an active sensor in South China, *Precision Agric.*, 15, 95–110.
Yadav, R. L., Dwivedi, B. S., Prasad, K., Tomar, O. K., Shurpali, N. J. and Pandey, P. S. (2000), Yield trends, and changes in soil organic-C and available NPK in a long-term rice–wheat system under integrated use of manures and fertilisers, *Field Crops Res.*, 68, 219–46.
Yaduvanshi, N. P. S. and Swarup, A. (2005), Effect of continuous use of sodic irrigation water with and without gypsum, farmyard manure, press mud and fertilizer on soil properties and yields of rice and wheat in a long term experiment, *Nutr. Cycl. Agroecosyst.*, 73, 111–18.
Yadvinder-Singh, Bijay-Singh, Ladha, J. K., Bains, J. S., Gupta, R. K., Jagmohan-Singh and Balasubramanian, V. (2007), On-farm evaluation of leaf color chart for need-based nitrogen management in irrigated transplanted rice in northwestern India, *Nutr. Cycl. Agroecosyst.*, 78, 167–76.
Yadvinder-Singh, Bijay-Singh, Ladha, J. K., Khind, C. S., Gupta, R. K., Meelu, O. P. and Pasuquin, E. (2004), Long-term effects of organic inputs on yield and soil fertility in the rice–wheat rotation, *Soil Sci. Soc. Am. J.*, 68, 845–53.
Yadvinder-Singh., Bijay-Singh., Maskina, M. S. and Meelu, O. P. (1995), Response of wetland rice to nitrogen from cattle manure and urea in a rice-wheat rotation, *Trop. Agric.*, 72, 91–6.
Yadvinder-Singh, Dobermann, A., Bijay-Singh, Bronson, K. F. and Khind, C. S. (2000), Optimal phosphorus management strategies for wheat-rice cropping on a loamy sand, *Soil Sci. Soc. Am. J.*, 64, 1413–22.
Yadvinder-Singh, Ladha, J. K., Bijay-Singh and Khind, C. S. (1994), Management of nutrient yields in green manure systems, in Ladha, J. K. and Garrity, D. P. (eds), *Green Manure Production Systems for Asian Ricelands*, pp. 125–53, International Rice Research Institute, Philippines.
Yan, X., Jin, J., He, P. and Liang, M. (2008), Recent advances on the technologies to increase fertilizer use efficiency, *Agric. Sci. China*, 7, 469–79.
Yang, C., Yang, L., Yang, Y. and Ouyang, Z. (2004), Rice root growth and nutrient uptake as influenced by organic manure in continuously and alternately flooded paddy soils, *Agril. Water Manag.*, 70, 67–81.
Yang, W. H., Peng, S., Huang, J., Sanico, A. L., Buresh, R. J. and Witt, C. (2003), Using leaf color charts to estimate leaf nitrogen status of rice, *Agron. J.*, 95, 212–17.
Yang, Y. C., Zhang, M., Li, Y. C., Fan, X. H. and Geng, Y. Q. (2012a), Improving the quality of polymer-coated urea with recycled plastic, proper additives, and large tablets, *J. Agric. Food Chem.*, 60, 11229–37.
Yang, Y. C., Zhang, M., Li, Y. C., Fan, X. H. and Geng, Y. Q. (2012b), Controlled release urea improved nitrogen use efficiency, activities of leaf enzymes, and rice yield, *Soil Sci. Soc. Am. J.*, 76, 2307–17.
Yang, Y. C., Zhang, M., Zheng, L., Cheng, D. D., Liu, M. and Geng, Y. Q. (2013), Controlled release urea for rice production and its environmental implications, *J. Plant Nutri.*, 36, 781–94.
Ye, Y., Liang, X., Chen, Y., Liu, J., Gu, J., Gu, R. and Li, L. (2013), Alternate wetting and drying irrigation and controlled-release nitrogen fertilizer in late-season rice. Effects on dry matter accumulation, yield, water and nitrogen use, *Field Crops Res.*, 144, 212–24.
Yeomans, J. C. and Bremner, J. M. (1986), Effects of urease inhibitors on denitrification in soil, *Commun. Soil Sci. Plant Anal.*, 17, 63–73.
Zhang, A. P., Liu, R. L., Yang, S. Q., Zhang, Q. W., Li, Y. H. and Yang, Z. L. (2012), Effect of side bar fertilization technology based on slow-release fertilizer on rice yield and nitrogen losses, *J. Agro-Environ. Sci.*, 31, 555–62.
Zhang, H., Xu, M. and Zhang, F. (2009), Long-term effects of manure application on grain yield under different cropping systems and ecological conditions in China, *J. Agric. Sci.*, 14, 31–42.

Zhang, M., Yang, Y. C. and Wan, L. B. (2006), *Production Method of Controlled Release Fertilizer by Use of Recycled Thermoplastic resin*, Chinese Patent ZL.2004100357833. Shandong Agriculture University and Kingenta Ecological Engineering Co. Ltd.

Zhu, Z. L. (2000), Loss of fertilizer N from plants-soil system and the strategies and techniques for its reduction, *Soil Environ. Sci.*, 9, 1–6

Zu, X., Boeckx, P., Wang, Y. S., Huang, Y., Zheng, X. H., Hu, F. and van Cleemput, O. (2002), Nitrous oxide and methane emissions during rice growth and through rice plants: effect of dicyandiamide and hydroquinone, *Biol Fertil Soils*, 36, 53–8.

Chapter 3

# Sustainable rice cultivation in coastal saline soils: a case study

*Sukanta K. Sarangi and Buddheswar Maji, ICAR-Central Soil Salinity Research Institute, India*

1 Introduction
2 The Sundarbans region of India
3 Management of coastal acid saline soils
4 Salt- and flood-tolerant rice varieties for coastal saline soils
5 Nursery management for wet season rice
6 Rice field management after transplanting
7 Management practices for dry season rice
8 Managing toxic gases from coastal rice fields
9 Removing excess soluble salts from the root zone
10 Integrated nutrient management (INM) for coastal saline soils
11 Reducing salinity build-up through mulching/cover crops
12 Management practices for increasing income from rainfed lowlands
13 Contingency plan for cyclone affected area
14 Recommended practices for cultivation of *kharif* and *rabi* rice
15 Alternative livelihood options
16 Future trends and conclusion
17 Where to look for further information
18 References

## 1 Introduction

Rice is the staple food for more than half of the world's population and it is grown in more than 100 countries. It is predicted that human population might reach up to 9.4 billion by 2050, and to meet this population's need, it is estimated that the world needs 8–10 million tonnes of rice each year (Patra, 2013). As the productivity gains from favourable irrigated rice environments are almost saturated, there is a great potential to harness the productivity

http://dx.doi.org/10.19103/AS.2016.0003.07

gain from unfavourable rice-growing environments such as drought-prone, submergence-prone, salinity affected and upland ecosystems. In India, about 3 million hectare (mha) rice area is affected by waterlogging, mostly in the coastal lowland regions. The risk of frequent submergence due to flash floods and waterlogging makes the ecosystem less productive. In these areas, stagnant flooding of 25–50 cm water depth occurs for a substantial period of the crop-growing season (Sarangi et al., 2016). Waterlogging at the early stages of crop growth suppresses tillering and increases plant mortality. Poor drainage in waterlogged areas leads to accumulation of toxic substances (iron toxicity, sulphide injury, etc.) and increase in incidence of pests and diseases.

Soil salinity in coastal deltas of south and south-east Asia is one of the major challenges of rice cultivation (Haque, 2006; Mu et al., 2015; Nhan et al., 2011; www.fao.org). In India, 20 mha of land (potential rice cultivation area) is affected by varying degrees of salinity or sodicity or both, of which 7 mha is seriously affected by salinity (Sadana, 2002). In coastal areas of India, about 1.26 mha soil is affected by soil salinity (see Table 1). More than 1 mha of coastal saline soils are found in three states: Gujarat (0.46 mha), West Bengal (0.44 mha) and Odisha (0.15 mha), which constitute 83% of the total coastal saline soils found in India. A common definition of these saline soils is that their saturated soil paste extract has an electrical conductivity (ECe) of more than 4 m mhos $cm^{-1}$ (=4 dS $m^{-1}$), pH of less than 8.5 and exchangeable sodium percentage (ESP) of less than 15% (U.S. Salinity Laboratory, 1954). The dominant salts present in these saline soils include chlorides and sulphates of sodium, calcium and magnesium. Besides these, the saline soils also comprise smaller amounts of potassium salts and bicarbonates and nitrates (Das, 1993).

Though rice is considered to be moderately sensitive to salinity (IRRI, 1997), it is one of the crops ideally suited for cultivation under coastal saline environments (Amanullah et al., 2007). Mishra (1995) observed that, for salt-affected soils, rice is the most preferred and adapted crop than any other crop because of the inherent genetic mechanism for salinity tolerance. Further, salt is managed by addition of water, and rice can be grown or has the capability to grow under flooded conditions. Also, rice is the staple food for majority of population in these environments; therefore, cultivation of this crop is a bare necessity for the food and livelihood security of the farming communities in coastal regions.

Salinity can affect all the growth and development stages of rice (Shereen et al., 2005); however, it is comparatively tolerant of salt stress during germination, active tillering, and towards maturity and is sensitive during early seedling and reproductive stages (Pandey et al., 2010). The symptoms of salt injury in rice are stunted growth, rolling of leaves, white tips, drying of older leaves and grain sterility (www.fao.org). Soil salinity limits the rice plant's growth and development, resulting in yield losses of more than 50% (Zeng and Shanon, 2000). Mass (1998) classified plants according to their salt tolerance and reported that rice tolerates up to an ECe of 3 dS $m^{-1}$ without yield loss and 50% of yield was reduced when EC exceeded 7.2 dS $m^{-1}$. Guidelines available in literature (Mass and Grattan, 1999; Hanson et al., 1999) revealed that rice yields are reduced by 12% for every unit of salinity (dS $m^{-1}$).

The main obstacle to intensification of crop production in the coastal areas is seasonally high content of salts in the root zone of the soil, which enter inland through rivers and channels, especially during the latter part of the dry season, when the downstream flow of fresh water becomes very low. During this period, the salinity of the river water increases. During flooding with saline river water or seepage from the rivers, the salts accumulate and become concentrated through evaporation in the soil surface layers, which may also result in an increase in salinity of the groundwater and make it unsuitable for irrigation (Haque,

**Table 1** Extent and distribution of coastal saline soils in India (rounded off to three decimal places)

| State/Union territory/Island | Coastal saline soil (mha) |
|---|---|
| Andhra Pradesh | 0.078 |
| Andaman and Nicobar Islands | 0.077 |
| Goa | 0.015 |
| Gujarat | 0.462 |
| Karnataka | 0.001 |
| Kerala | 0.020 |
| Maharashtra | 0.007 |
| Odisha | 0.147 |
| Puducherry | 0.001 |
| Tamil Nadu | 0.013 |
| West Bengal | 0.441 |
| TOTAL | 1.262 |

2006). With the onset of monsoon, salinity decreases gradually due to washing out of salts beyond root zone, making rice cultivation possible in the wet season with another challenge of flooding/submergence. Coastal rice growing soils are acidic to neutral in reaction, low in organic carbon and available nitrogen (N), medium in available phosphorus (P) and potassium (K) content is high (Sarangi et al., 2012). Prevalence of acid saline soils with twin problems of soil acidity (as low as pH 4.0) and salinity due to excess salts in various parts of coastal deltas also poses another challenge to rice cultivation. Effective nutrient and soil health management in saline soils is a necessity for improving productivity.

In the coastal deltas majority of farmers grow local rice varieties, which are photoperiod sensitive, and have long duration and low yields (Sarangi et al., 2016; SeinnSeinn et al., 2015). Rice varieties respond to salinity differently; the salinity sensitivity of a rice variety is not only determined by the salinity level but also by the salinity stress time. The major effect of salinity on lowering yield is due to lesser filled grains per panicle and grain weight in sensitive varieties than tolerant ones (Grattan et al., 2002; Motamed et al., 2008; Clermont-Dauphin et al., 2010). On an average, under saline conditions rice varieties suffered a yield loss of around 50%, 30% and 10% of their potential yields for varieties sensitive, moderately tolerant and tolerant to salinity, respectively. Under freshwater conditions, these rice varieties yielded the same. Therefore, if farmers grow salinity-tolerant rice varieties, they would gain 1–2 tonnes rice per ha per crop, compared to growing salinity-sensitive rice varieties (Nhan et al., 2011).

To increase rice production and productivity in these areas, a holistic approach combining better management practices with stress-tolerant varieties is the alternative. These varieties should have suitable features for different seasons and ecology, such as for low-lying areas during the wet season tall plants (155–70 cm), stiff culm, erect leaves, drought and salinity tolerance at seedling stage, submergence tolerance with long duration and strong seed dormancy are desirable. For dry season varieties should be tolerant to salinity, medium to short duration, high yielding, having long and slender grain quality and so on. Achieving higher productivity from the underutilized high potential areas is thus an immediate need of India, particularly in the eastern region. Government of India is now focusing on these potential areas for achieving the second green revolution. One such ecosystem is the Sundarbans in India, where rice is the predominant crop grown under salinity and submergence stress situations, as a result productivity is very low. However, with adoption of improved technologies, there is scope to increase the rice productivity and livelihood security in this region.

# 2 The Sundarbans region of India

The Indian part of Sundarbans (Bengal Basin) represents the estuarine phase of the Ganges and the Brahmaputra river systems and lies between 21°30'N–22°40' 48"N latitude and 88°01' 48"E–89°04' 48"E longitude at an altitude of 3m (above mean sea level). It is in the lower Gangetic delta, which is one of the largest coastal deltas in the world, which is located in the eastern part of India and southern part of Bangladesh. Indian Sundarbans comprises coastal regions of the districts of North 24 Parganas and South 24 Parganas of West Bengal, India (see Fig. 1).

## 2.1 Crops and cropping pattern

The existing cropping pattern in Sundarbans region is primarily mono-cropped with paddy cultivation during the wet season, which accounts for almost 98% of cultivable area and the rest used for vegetables. Major cropping patterns are rice–fallow, rice–rice with very negligible share of vegetables cultivation. The wet season (*kharif/aman*) covers the period from mid-June to December and dry season (*rabi/boro*) spreads from mid-December to May (see Fig. 2). Cropping intensity in the area is quite low (120–125%) as most of the cultivated areas remain fallow during the dry season. Growing crops other than rice is very difficult during the wet season due to deep waterlogging/flooding of low-lying agricultural fields by ample monsoonal rain (see Fig. 3). This is also a constraint for successful rice cultivation, therefore the productivity of rice during the wet season is very low (2 $t ha^{-1}$) due to the complex nature of coastal agro-ecosystem. Transplanting is the prevalent method of rice cultivation by farmers during *kharif* and *rabi* seasons. However, this method of rice crop establishment is adversely affected by flood and salinity during wet and dry seasons, respectively. Therefore, alternative rice crop establishment methods and efficient management practices are of prime importance for sustainable rice cultivation in these unfavourable ecologies.

During the initial period of wet season, the rice crop suffers from soil salinity and submergence due to flood, whereas during the dry season availability of irrigation water is the major constraint, and there is stress of soil and water salinity. Due to these constraints, most of the land (70–80%) remain fallow during dry (*rabi*) season; however about 15–20% of agricultural land is covered by rice crop (*boro* rice), and in about 5–10% land vegetables are grown. Although the area under *boro* rice is limited, rice productivity is higher during this season due to availability of congenial weather conditions such as ample sunshine, control over irrigation water and higher efficiency of applied inputs such as fertilizers, plant protection chemicals and so on (Sarangi et al., 2014).

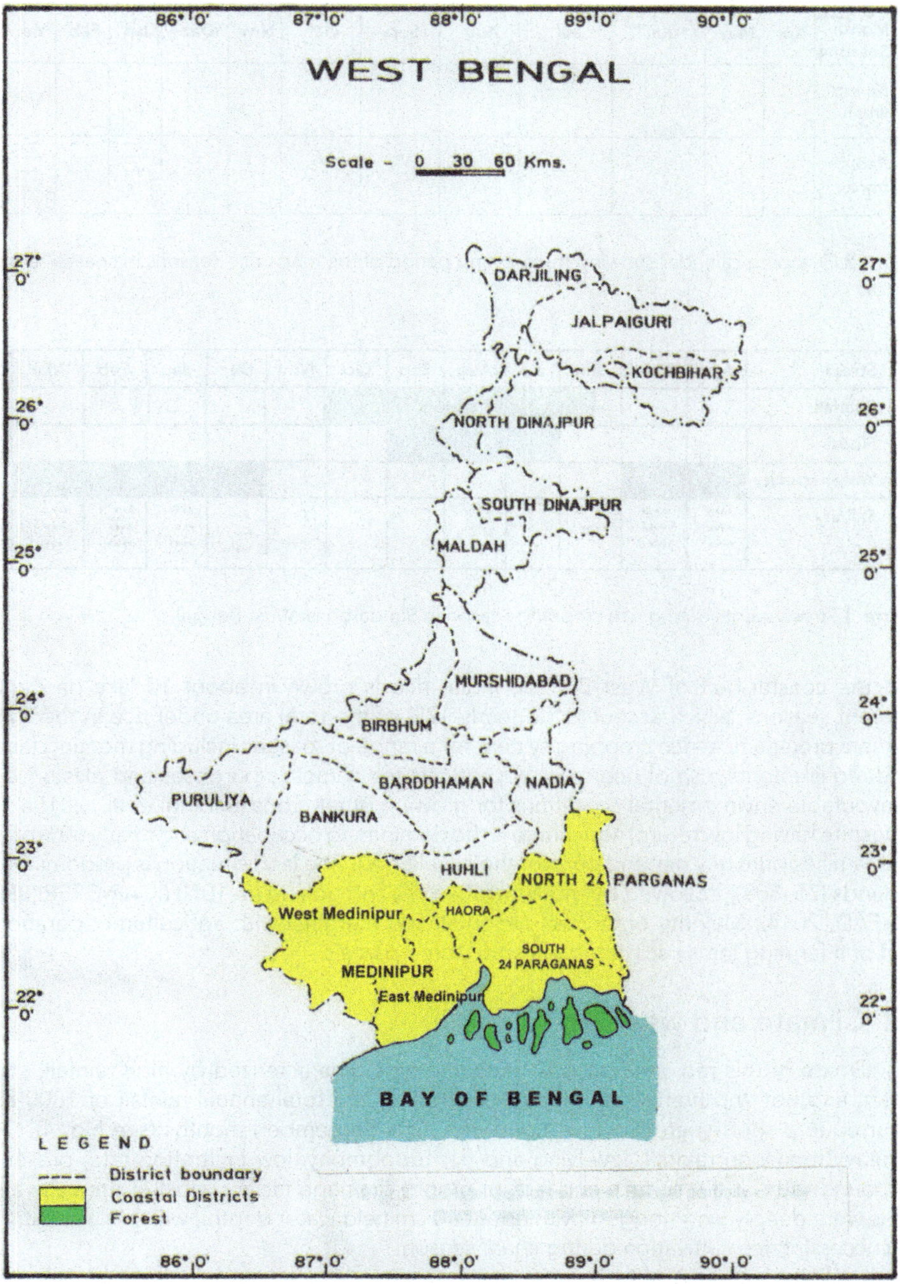


**Figure 1** Coastal districts of West Bengal, India (coastal region of North and South 24 Parganas districts form the Indian Sundarbans, Bandyopadhyay et al., 2003).

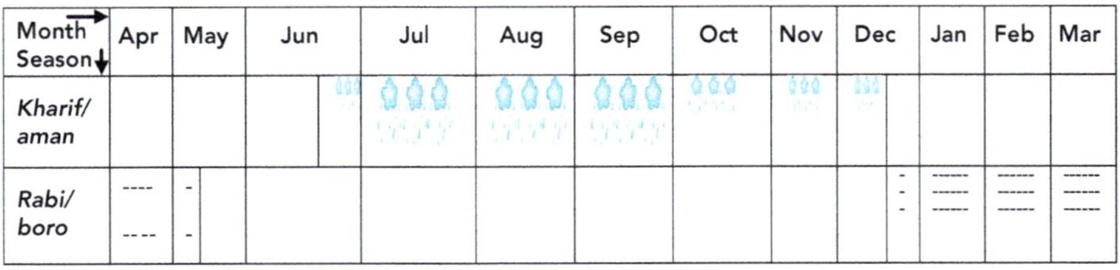


**Figure 2** Cropping calendar showing the growing period of the major rice seasons in coastal regions of India.

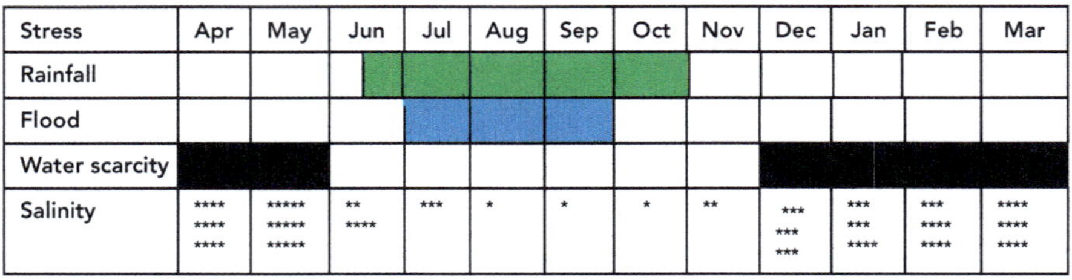


**Figure 3** Stress events during rice cropping season in Sundarbans, West Bengal.

In the coastal belt of West Bengal, India, rice is grown in about 10 lakh ha during different seasons, which accounts for nearly 17% of the total area under rice in the state. Farmers prefer a rice–rice cropping system for a range of reasons including meeting family food requirements, use of rice crop by-products for domestic purposes and existence of unfavourable environmental conditions for growing other crops (Sarangi et al., 2015a).

Despite having low return, agriculture is the dominating occupation and nearly 80% of the farm families primarily depend on it for their livelihood. The land situation is predominantly lowlands (75–80%), followed by medium (10–15%) and upland (5–10%) (ICAR-CSSRI, IRRI-EC-IFAD, 2014). Majority of farmers are small and marginal with agricultural operational land of a farming family scattered in two to three parcels.

## 2.2 Climate and weather

The climate of this region is tropical monsoon type, characterized by high rainfall, short winter, and hot and humid summer. About 80% of the total annual rainfall of 1800 mm occurred in a short span of a few monsoon (June–September) months (see Fig. 4). Due to heavy monsoon rainfall, low-lying and flat topography, low infiltration rate, presence of groundwater at the surface and lack of proper drainage facility, most of the cultivated fields were deeply waterlogged (with about 40 cm field water depth), which is a constraint for successful rice cultivation during *kharif* season.

Soil salinity increases after the cessation of monsoon rain and the highest salinity is reached in April and May (see Fig. 4); during this period both soil and groundwater reach the highest salinity level. However, waterlogging in the paddy fields is highest in August and September when the monsoon is at its peak. May is the hottest month of the year with mean maximum and minimum temperatures of 35°C and 26°C, respectively (see Fig. 5).

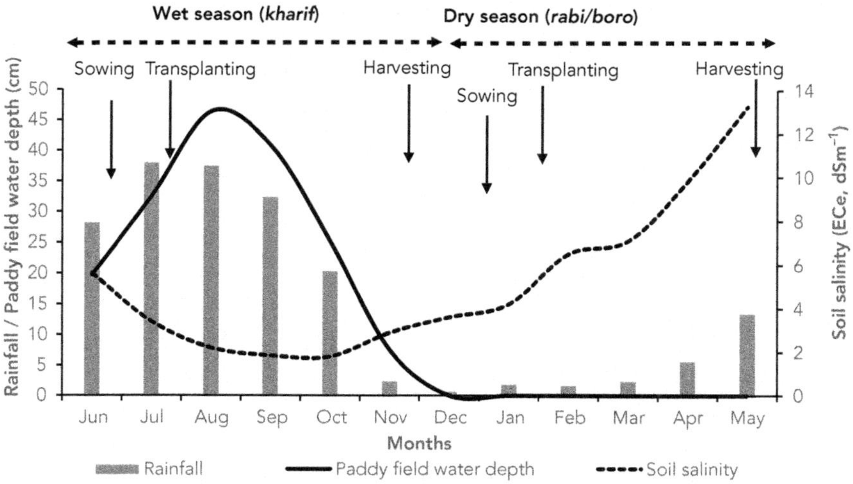


**Figure 4** Monthly and seasonal variation of rainfall (mean of 2001–2 to 2013–14), paddy field water depth (mean of 2011–12 to 2015–16) and soil salinity (mean of 2011–12 to 2015–16) at Canning Town (Latitude: 22°15′N, Longitude: 88°40′E, Altitude: 3 m), West Bengal, India.

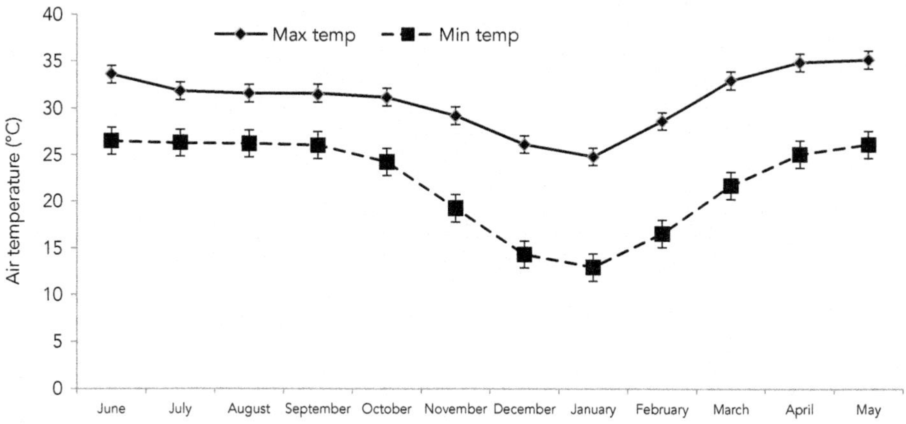


**Figure 5** Mean (2001–2 to 2013–14) monthly maximum and minimum temperature at Canning Town, West Bengal, India.

Temperature gradually decreases from May and reaches lowest in January (maximum and minimum of 24.7°C and 12.8°C, respectively).

## 2.3 Soil types

The coastal area of West Bengal is spread over an area of about 1.29 mha, which include 0.86 mha under cultivation/habitation and about 0.43 mha under mangrove forest. About

**Table 2** Morphological, physical and physico-chemical characteristics of a soil profile of Sundarbans (Bandyopadhyay et al., 2003)

| Horizon | Depth (cm) | Colour (moist) Matrix | Colour (moist) Mottles | Texture* | Clay (%) | pH (1:2) | EC ($dSm^{-1}$) | SAR | ESP | OC (%) |
|---|---|---|---|---|---|---|---|---|---|---|
| Ap | 0–18 | 2.5Y 5/2 | 10 YR 6/6, c1d | sic | 46 | 7.4 | 3.4 | 9.0 | 6.3 | 0.33 |
| A1 | 18–67 | 2.5Y 5/0 | – | sic | 46 | 7.6 | 4.0 | 9.7 | 5.8 | 0.36 |
| Bwg | 67–167 | 5Y 5/1 | 10 YR 3/3, m1d | sic | 50 | 7.6 | 10.2 | 8.9 | 10.0 | 0.75 |
| Cg1 | 167–228 | 2.5Y 4/0 | 10 YR 3/3, f1d | sicl | 38 | 7.9 | 12.9 | 10.0 | 11.3 | 0.84 |
| Cg2 | 228–288 | 2.5Y 4/0 | – | sil | 26 | 4.0 | 10.8 | 6.3 | 11.7 | 1.96 |

*sic: silty clay; sicl: silty clay loam; sil: silty loam

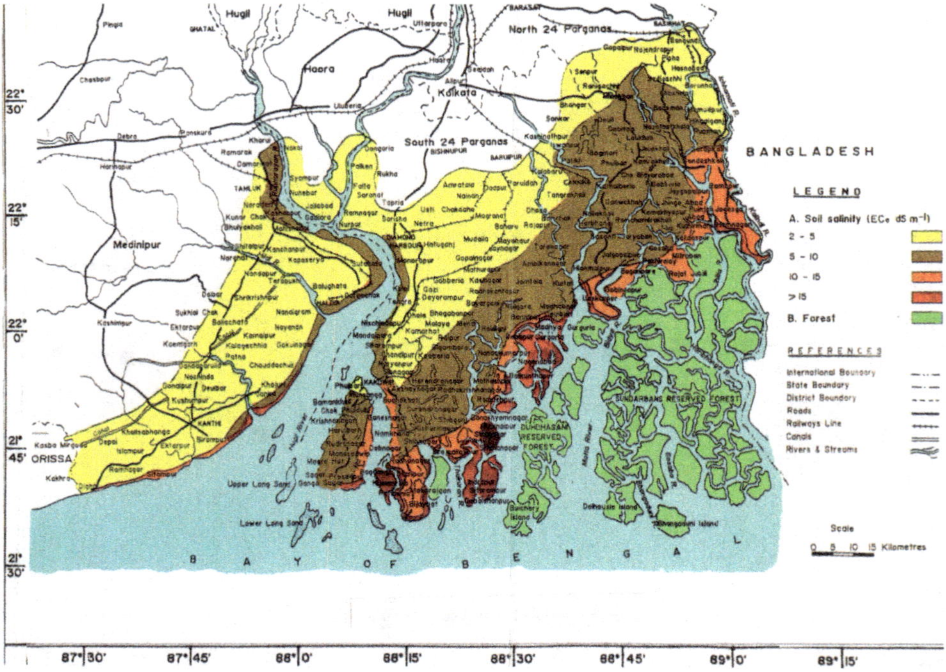


**Figure 6** Extent and distribution of saline soil in coastal region of West Bengal (Bandyopadhyay et al., 2003).

0.44 mha cultivated area is affected by salt, which has *mixed* mineralogy, *hyperthermic* temperature and *aquic* moisture regimes. The majority of agricultural soil is heavy textured with >40% clay in surface soil (see Table 2); presence of Jarosite mottles [$KFe_3(SO_4)_2(OH)_6$] is also observed in some horizons. The soils show negligible variation in hue ranging from 2.5Y to 5Y with matrix chroma of around 2.0, many a times associated with yellowish red mottles in the sub-surface horizons indicating the prevalence of *aquic* moisture regime in the pedons.

The coastal soils of West Bengal are formed on alluvial deposits carried by major rivers in the coastal region. Depending on the nature of deposits, hydrology, proximity to the existing creeks, river and seas, nature of forest cover and so on, different soils are found in the coastal region. Based on the pH of surface soil the coastal soils are grouped under four pH categories, viz. pH < 4.5, 4.5–6, 6–8 and >8. The approximate areas under different pH categories are 0.04, 0.01, 0.70 and 0.02 mha, respectively (Maji and Bandyopadhyay, 1995). It is observed that the pH of the soils is highly variable in the region and also varies widely within a small distance. Highly acidic saline soils are occasionally found at a number of places in Sundarbans region of coastal West Bengal. Studies have shown that the acidity of the soils was due to oxidation of sulphidic materials present in the soil. The yield of crops on such soils is extremely poor. Many a time crops on such soils show even the visual toxicity and deficiency symptoms for several elements (Bandyopadhyay and Maji, 1995).

The soils show various degrees of salinity and the salinity gradient of the area increases eastward from Matla River (see Fig. 6). Based on the degree of salinity, the soils have been subdivided into four groups, viz. ECe 2–5, 5–10, 10–15 and >15 $dSm^{-1}$, with approximate corresponding areas in each group as 0.50, 0.28, 0.07 and 0.01 mha, respectively. Salinity builds up due to gradual drying up of lands after monsoon season and attains maximum in April and May (see Fig. 4). It builds up in the soil due to upward capillary movement of saline groundwater present at shallow depths and occasional inundation with saline water from the river (Das and Maji, 2001). It is observed that the salinity of the soils is highly variable both spatially and temporarily, and there may be a wide variation of salinity within a small distance. Based on field morphology and other physico-chemical characteristics, the coastal soils of West Bengal have been classified, namely: *Typic Endoaquepts, Fluventic Endoaquepts, Aeric Endoaquepts, Sulphic Endoaquepts, Fluvaquentic Endoaquepts* and *Vertic Endoaquepts*. The dominant entisols are *Typic Ustifluvents, Typic Fluvaquents* and *Sulphic Endoaquents*.

Based on the physiographic situations, agro-ecosystem, climatic conditions and key challenges for growing rice in coastal saline soils of Sundarbans, the potential management strategies include sustainable management of acid saline soils, growing salt- and flood-tolerant varieties during the wet season and salt-tolerant short-duration varieties during the dry season, adoption of suitable establishment methods (producing healthy seedlings during the wet season for transplanting and drum seeding of pre-germinated seeds for dry season), efficient agronomic practices after crop establishment, washing out of salts from root zone, integrated nutrient management (INM), efficient management practices (mulching/cover crops) to curtail upward movement of salts through capillary pores and so on.

## 3 Management of coastal acid saline soils

The soils (pH < 4.0) are rich in exchangeable Al and Fe and highly deficient in available P. These soils show very poor crop yields. The soils are very low in available P with high P fixation capacity, and the available P in soil increases slightly due to liming and application of P fertilizers. Over-liming decreases crop yield. The available P content of soil also increased due to application of rock phosphate, whose benefit may not be immediately reflected; its effect may be observed after a few initial cropping seasons. The excess availability of Fe and Al could be controlled by proper levels of liming. Lime dose determined by KCl extraction method with a lime factor of 4.0 (or lime dose determined by Shoemaker's method with a lime factor of 0.5) was found to be most suitable for coastal acid soils of Sundarbans. Liming of the soil @ 50% of recommended dose ($L_{0.5}$ =1.5 tha$^{-1}$) and application of high doses of P fertilizer, particularly, water-soluble phosphate fertilizers would improve the crop yield. Re-liming may be necessary after a few years. Good yield can be obtained from such soils by applying lime (@$L_{0.5}$) and higher P fertilizer (double or more of recommended dose) along with green manure or other organic manures. Oyster shell can be used as locally available cheap source of lime. Rock phosphate can also be used as an alternative cheap source of P, whose availability (from rock phosphate) is favourable under low soil pH (acid soil), low exchangeable calcium and low concentration of P in soil solution.

In Kerala, India, the management of acid saline soils is done by indigenous technology called *Pokkali* cultivation by growing salt- and acid-tolerant traditional landrace *Pokkali*. *Pokkali* or *Kari* soil is a local name for acid sulphate soil, covering about 50,000 ha in Kerala; this typically peaty saline soil is highly acidic (pH around 3.5), deep black in colour, high in organic matter, poorly drained, remains submerged in water during the monsoon and is put to rice cultivation as soon as the water recedes (Biswas and Mukherjee, 1994). According to the traditional practice, *Pokkali* fields are prepared for paddy cultivation from mid-April every year. The land had to be dried under the hot sun for almost a month, by which time the salt content left over from prawn farming would be ready to be washed off by the first monsoon rains. Earthen mounds of 1 m base and 50 cm height are made all across the field. In June, after the southwest monsoon brings in a few good showers, germinated *Pokkali* seeds are sown on the flattened tops of the mounds. In a month, the mounds are dismantled and the seedlings in clefts are dispersed around the flattened mounds (www.thehindu.com). Some areas near the mouth of the rivers and close to the sea are prone to flooding and salinity. The nature of the soil here is also highly acidic and has elemental toxic content, making the areas uncongenial for paddy cultivation. But this traditional method has been proved by scientists as the best method to prepare the fields for *Pokkali* because of the larger surface area, and it also enhances the integration of the toxic contents of the soil with rainwater. The *Pokkali* paddy, which to an extent is saline, flood-resistant and acid-resistant, is the wonder crop cultivated in these regions (Joseph, 2008).

# 4 Salt- and flood-tolerant rice varieties for coastal saline soils

Landraces are ecotypes cultivated for a long time in pristine habitats. The traits that characterize such landraces are expressed only in their sites of adaptation and may not express fully if grown in other areas. Therefore, landraces are defined as site-specific (Arunachalam et al., 2006). The varieties grown by many farmers in the Sundarbans region were either landraces or old varieties (landraces that were introduced from other areas to Sundarbans), although these landraces or old varieties are the important germplasm source for either pure-line selections or development of high-yielding varieties (HYVs).

Salt-tolerant selections made from landraces or locally adapted varieties are given in Table 3. The research station at Canning Town was established in 1959 as Central Saline Rice Research Station under the Central Rice Research Institute (CRRI), Cuttack, Odisha, India. The salt-tolerant rice germplasm was conserved at this station. With the establishment of Central Soil Salinity Research Institute (CSSRI) in 1969 at Karnal, Haryana, India, the Canning station was transferred to CSSRI in February 1970.

During the 1970s, SR 26B (pure-line selection from landrace '*Kalambanka*' from Odisha, India), a tall indica rice variety of 150 days duration, was considered as salt tolerant and was the standard recommended salt-resistant variety of India. However, the local landraces *Damodar*, *Dasal* and *Getu* were superior to SR 26B not only in tolerance but also in grain yield, duration and height. Experiment under salt- and flood-resistant paddy scheme at Rice Research Station, Chinsurah, West Bengal, India, for two consecutive years (1968–69 and 1969–70) showed that the local landraces *Nona Bokra*, *Getu* and *Benisail* are suitable for cultivation in the highly saline tracts as their salt tolerance capacities are higher (up to 12.5 $dSm^{-1}$), while Patnai-23, *Rupsail*, OC 1393, *Kumargore* and SR 26B are suitable for medium saline areas (up to 9.5 $dSm^{-1}$), as their yields were very high, but the HYVs such as IR-8 and TN-1 may be suitable for low saline (4.5 $dSm^{-1}$) tracts (Datta, 1972). Due to higher yield potential of HYVs such as IR-8, Jaya and TN-1, these were used as parent for crossing with *Damodar*, *Dasal* and *Getu*. Chemical mutagens and X-ray were also used to introduce dwarfness, high yield and earliness in the existing salt-tolerant landrace *Damodar* and salt tolerance capacity in IR-8 (Annual Report, CSSRI, 1972). Yields of IR-8 mutants (by using chemicals such as dimethyl sulphoxide and triethanol amine and X-ray) were found to be higher compared to *Damodar* and IR-8. Crossings involving *Damodar*, SR 26B as salt-tolerant parent and IR-20, and IR-24 as high-yielding parent were also carried out and evaluated at salinity 8–9 $dSm^{-1}$ during 1978. Promising crosses even produced grain yields of 5.8 $tha^{-1}$ during the dry season and up to 3.6 $tha^{-1}$ in wet season (Annual Report, CSSRI, 1978). During 1970s, CSR-1, CSR-2 and CSR-3 were identified as salt-tolerant selections from local landraces *Damodar*, *Dasal* and *Getu*, respectively. CSR-4, commonly known as Mohan, was a mutant of IR-8. Subsequently, the salt-tolerant landraces *Nona Bokra*, *Nonasail*, *Pokkali*, *Vytilla* and *Anakodan* were also used in crossing programme in collaboration with International Rice Research Institute (IRRI) and under All India Co-ordinated Rice Improvement Project. CSR-6 was selected from local landrace *Nonasail* and it was used as a check alongwith *Pokkali*, *Getu* and Jaya during 1980s.

Twenty-four salt-tolerant landraces were collected in April–May 1981 from the coastal belt of Orissa (presently Odisha), India, and were evaluated in 1982 at Canning Town with soil salinity level of 9.0–10.4 $dSm^{-1}$ at transplanting and 6.7–9.3 $dSm^{-1}$ at flowering. Out of the 24 landraces, about 11 were promising and yielded at par with the control variety CSR-6. Some of these landraces such as *Patnai* were widely cultivated in the Sundarbans region and even cultivated at present under different names such as *Sabita*, *Patnai 23* and so on. This may be due to their salinity as well as flood tolerance.

In 1984, under the Saline Alkali Tolerant Varietal (SAVT) trial, the designation CST 7-1 (IET No. 9341), which is a cross between CSR-1 and IR 24, produced grain yields of 3.31 $tha^{-1}$ with flowering duration of 106 days under soil salinity (ECe) of 6.7–9.4 $dSm^{-1}$ at transplanting and 5.2–6.1 $dSm^{-1}$ at flowering. This was later used as a national control for evaluation of salt-tolerant genotypes under coastal saline soils. In 1987, the selection CAC 615 from landrace *Velki* (collected from coastal belt of Orissa in 1981) produced grain yields of 3.69 $tha^{-1}$, which was significantly higher than CST 7-1, local check (CSR-1) and landrace *Velki*. Canning 7 and Canning 8 were salt-tolerant rice varieties for dry/*boro* season developed in 1989 as secondary selections of BG 35-2 and IR 32307-107-3-2-2, respectively. Canning 7 with flowering duration of 101 days and average grain yield of 4.0 $tha^{-1}$ is also presently preferred by some farmers in this region due to its salt tolerance, short duration and flexibility to time of sowing.

During the 1990s (1994–8) four of the most significant salt-tolerant rice varieties, viz. CSRC 2-1-7 (IET No. 13428), CSRC(S) 11-5-0-2 (IET No. 13422), CSRC(S) 5-2-2-5 (IET 12855) and CSRC(S) 7-1-4 (IET No. 14199) were developed; these were released in the names SUMATI, UTPALA, BHUTNATH and AMAL-MANA in 2003, 2003, 2005 and 2008, respectively (see Table 5). Salt-tolerant rice varieties were also developed from other coastal states of India. Panvel 1 and Panvel 2 were released by Maharashtra during 1987 and 1988, respectively. Panvel 1 was of 125 days duration with long bold grains, tolerant to salt and blast disease. Panvel 2 was of 129 days duration with long slender grains and recommended for irrigated condition. Vytilla 2, Vytilla 3 and Vytilla 4 were released by Kerala, India, in 1980, 1987 and 1993, respectively. Vytilla 2 matures in 125 days having long bold grains, red rice and tall, recommended for saline areas in Ernakulam and Alleppey districts. Vytilla 3 matures in 115 days, having medium bold grain, tall, red kernel, and suitable for *Pokkali* (acid saline) soils. Vytilla 4 matures in 125 days having medium bold grains, tolerant to salinity and flooding. In Odisha, India, Lunisree (CRM 30) was recommended for coastal saline soils, which was developed from *Nonasail* mutant

**Table 3** Salt-tolerant rice selections from landraces or locally adapted varieties

| Landrace | Origin | Salt-tolerant selection |
|---|---|---|
| *Arya* | Maharashtra | Arya 33 |
| *Benisail* | West Bengal | Matla |
| *Bhura Ratta* | Maharashtra | BR 4-10 |
| *Budda Molagolukulu* | Andhra Pradesh | MCM 2 |
| *Chattivirippu* | Kerala | Mo 1 |
| *Choottupokkali* | Kerala | Vytilla 1 |
| *Damodar* | West Bengal | CSR 1 |
| *Dasal* | West Bengal | CSR 2 |
| *Getu* | West Bengal | CSR 3 |
| *Jhona* | Punjab, Haryana, Western UP | Jhona 341 |
| *Kalambanka* | Odisha | SR 26B |
| *Kala Ratta* | Maharashtra | KR 1-24 |
| *Kalladachampavu* | Kerala | Mo 2 |
| *Kujathikkara* | Kerala | Mo 3 |
| *Nona Bokra* | West Bengal | Hamilton |
| *Nonasail* | West Bengal | CSR 6 |
| *Patnai* | Odisha | Patnai 23 |
| *Velki* | Odisha | CAC 615 |

**Table 4** Characteristics of the promising rice entries for cultivation in Sundarbans region during the *boro* (dry) season

| Entry | Salinity tolerance ($dSm^{-1}$) | Plant height (cm) | Duration (days) | Grain yield ($tha^{-1}$) | Grain type[1] | Year released |
|---|---|---|---|---|---|---|
| Canning 7 | 6–8 | 95–105 | 130 | 4.0–4.5 | LB | 1989 |
| CSR4 | 6–8 | 95–100 | 135 | 3.5–4.0 | LB | 1985 |
| WGL 20471 | 6–8 | 95–105 | 130 | 5.0–5.5 | LS | 1991 |
| CST 7-1 | 6–8 | 90–95 | 130 | 4.5–5.0 | MB | 1992 |
| IET 4786 | 4–6 | 80–95 | 135 | 3.5–5.6 | LS | 2000[3] |
| Annada | 4–6 | 95–105 | 135 | 3.5–4.5 | SB | 1987 |
| Boby | 4–6 | 100–105 | 145 | 4.0–4.5 | LB | |
| Lalat | 2–4 | 102 | 135 | 4.0–4.5 | LS | 1988 |
| BRRI dhan47[2] | 4–6 | 105 | 152 | 6.0 | MB | 2007 |
| Binadhan-8[2] | 10 | 95 | 145 | 6.0 | MB | 2010 |
| BRRI dhan55[2] | 4–6 | 100 | 145 | 7.3 | LS | 2011 |
| Bidhan 2 | 6–8 | 105–110 | 135 | 4.0–5.0 | LS | |
| Parijat | 4–6 | 85–95 | 135 | 3.5–4.0 | MS | 1976 |
| N. Sankar | 2–4 | 95–100 | 140 | 3.0–4.0 | LS | TV[4] |
| S. Sankar | 2–4 | 95–100 | 140 | 3.0–4.0 | LS | TV |

[1]LB: long bold, MB: medium bold, SB: short bold, LS: long slender, MS: medium slender.
[2]Varieties developed and released in Bangladesh; characteristics are for conditions in Bangladesh.
[3]Year of notification.
[4]TV: traditional variety.

(Gamma irradiation) in 1992 with duration 145 days, long slender grain, tall plants, tolerant to salinity and waterlogging (50 cm).

After 2008, large-scale evaluation of rice varieties was conducted for salt tolerance and flood tolerance at regional research station, Canning, as well as in the farmer's fields in the Sundarbans region during dry and wet seasons, respectively. The significant salt-tolerant varieties identified were WGL 20471, Bidhan 2, Binadhan-8, BRRI dhan47, Annada, Boby, Lalat and Canning 7 for dry season (see Table 4). The promising flood-tolerant rice varieties were Amal-Mana, Swarna-Sub1, CSRC(D) 7-0-4 and CSRC(D) 12-8-12 (see Table 5) for cultivation in the flood-prone areas of Sundarbans region during the wet season. During 2010–13 four salt-tolerant HYVs of rice (Luna Suvarna, Luna Sampad, Luna Barial and Luna Shankhi) were released by CRRI, Cuttack, Odisha, India. These varieties can tolerate soil salinity of 5.0–8.0 $dSm^{-1}$ and are late in maturity (150 days) except Luna Shankhi, which is an early maturing variety (110 days). Luna Suvarna is a tall (135 cm) HYV with an average productivity of 3.5–4.0 $tha^{-1}$, medium slender grains and can withstand up to 45 cm water stagnation. Luna Sampad is also tall (130 cm) with medium bold grains and having average productivity of 3.6–4.2 $tha^{-1}$. Luna Barial has short bold grains with an average productivity

**Table 5** Characteristics of the promising rice entries for cultivation in Sundarbans region during the *aman* (wet) season

| Variety | Salinity tolerance ($dSm^{-1}$) | Water tolerance | Plant height (cm) | Duration (days) | Grain yield ($tha^{-1}$) | Grain type[3] | Year released |
|---|---|---|---|---|---|---|---|
| SR 26B | 4–6 | SF[1] 30–45 cm | 130–145 | 165 | 3.5–4.0 | MS | TV[4] |
| Sabita | 4–6 | -do- | 180–190 | 210 | 3.5–4.0 | LS | 1986 |
| Geetanjalli | 4–6 | -do- | 160–170 | 165 | 4.0–4.5 | LS | TV |
| Sumati | 6–8 | 15–30 cm | 90–100 | 145 | 3.5–4.0 | LS | 2003 |
| Utpala | 6–8 | -do- | 95–105 | 140 | 4.0–4.5 | LS | 2003 |
| Bhutnath | 6–8 | -do- | 100–115 | 135 | 4.0–5.0 | LS | 2005 |
| Amal-Mana | 6–8 | SF 30–45 cm | 160–170 | 165 | 5.0–5.5 | LS | 2008 |
| Patnai 23 | 4–6 | -do- | 160–170 | 165 | 3.0–3.5 | MS | TV |
| NC 678 | 4–6 | SF >45 cm | 165–170 | 160 | 2.0–2.5 | MS | TV |
| CSRC (D) 7-0-4 | 4–6 | -do- | 160–175 | 170 | 4.0–4.5 | MS | 2008 |
| CSRC (D) 13-16-9 | 4–6 | -do- | 160–175 | 170 | 4.0–4.5 | MS | 2008 |
| CSRC (D) 12-8-12 | 4–6 | -do- | 160–175 | 170 | 4.0–4.5 | MS | 2008 |
| CSRC (D) 2-17-5 | 4–6 | -do- | 155–175 | 165 | 4.0–4.5 | MS | 2008 |
| Swarna-Sub1 | – | FF[2] 2 weeks | 95–105 | 140 | 4.5–5.0 | MS | 2009 |

[1]SF: stagnant flooding.
[2]FF: flash flooding.
[3]LB: long bold, MB: medium bold, SB: short bold, LS: long slender, MS: medium slender.
[4]TV: traditional variety.

of 3.9 $tha^{-1}$, whereas Luna Shankhi has medium slender grains with an average yield capacity of 4.6 $tha^{-1}$ and suitable for dry season cultivation (Anonymous, 2016).

The on-station and on-farm experiments on varietal evaluation in Sundarbans region suggested that there is a considerable potential for increasing rice production in the coastal zone of West Bengal through replacement of locally grown wet and dry season varieties with improved, stress-tolerant varieties (Sarangi et al., 2015a, 2016). With adoption of promising stress-tolerant HYVs and matching agronomic management practices, the attainable grain yields of 4.0–5.0 $tha^{-1}$ can be achieved in these areas (Sarangi et al., 2015a).

# 5 Nursery management for wet season rice

The rainfed lowland environment is risky for crop production due to uncontrolled drought, floods, and heavy pests and disease infestations. Rice cultivation in such fragile ecosystems needs stress-tolerant variety and good management strategies for further enhancing the tolerance level of existing varieties. One such strategy is raising healthy seedlings through proper nursery management such as optimal seeding density, proper seedling age and balanced nutrients application to the nursery. Management of rice crop in the nursery for getting healthy and robust seedlings is equally important as that in the main field to improve the productivity of transplanted rice under coastal salinity stress.

Farmers of the coastal region often practice suboptimal nursery management, which leads to reduced plant population due to seedling mortality and hence, poor yields (1.5–2 $tha^{-1}$), below the national average. However, by adopting the improved nursery management practices developed at Canning Town, farmers could raise the productivity significantly (4.0–5.0 $tha^{-1}$). Nursery areas should be chosen close to a water source, with good drainage facility and free from shade. Seed bed should be 1m wide and of a convenient length (maximum 10 m). The wet nursery bed is prepared like a puddle rice field but with extra care. Application of organic material like farmyard manure (FYM) is essential for a healthy nursery bed. FYM is the well-decomposed mixture of dung with litter and leftover material from roughages and fodder fed to cattle. The N, P and K concentrations of FYM are 0.50, 0.25 and 0.40%, respectively. At least 500 kg FYM in a nursery area of 1000 $m^2$ should be applied. The use of FYM significantly increase soil organic matter, soil N and P status, hydraulic conductivity, water holding capacity and decreased soil bulk density, thereby creating a good environment for growth and development of rice and also micronutrients like zinc supplied for growth improvement. After puddling and levelling, beds are made to a height of 25–30 cm and the top is properly levelled to avoid submergence of seedlings due to heavy rain after emergence.

Seedlings raised from 25 kg pure seeds with minimum 90% germination are sufficient for transplanting 1 ha of area. While sowing the pre-germinated seeds, throw them uniformly with a mild force onto the raised bed to fix them in the puddle soil. This would help seeds to anchor quickly and not to be carried away by the irrigation/flood water. Protect seeds from damage caused by birds, squirrels, other animals and pests. It is preferable to raise nursery during the wet season in high lands with good water control. Irrigation water application should be well managed according to the length of seedlings, and excess water should be drained out. Excess water may lead to the formation of taller and weaker seedlings, and many a times, prolonged flooding also causes seedling mortality. It is recommended to maintain a water level of 2–3 cm during initial 10 days. When there is a sudden heavy rain causing submergence of seedbed, do not completely drain it suddenly at one go, as the entire shoot along with leaves touch the muddy soil and seedlings are damaged due to contact with muddy soil. In this case, drain the water slowly as the leaves float over water, then allow the seedlings to put forth new leaves above the water level, get stronger and then drain the excess water gradually. When the water level recedes and there is no further rain, progressively raise the water level until seedlings are pulled for transplantation. In saline coastal zones of Sundarbans, high salinity stress retards growth of seedlings in the nursery. A lower seed density of 25 g $m^{-2}$ with application of 50-30-15 kg $N\text{-}P_2O_5\text{-}K_2O$ + 5 t FYM $ha^{-1}$ produce healthy and robust seedlings with high seedling vigour (see Fig. 7). The full dose of P and K and 50% of N were applied as basal at sowing. The rest of nitrogen was applied in two equal splits at 15d of sowing and 7d before uprooting. Combining inorganic and organic fertilizers in the seedbed enhanced seedling growth (dry weight, shoot and root length, number of green leaves, leaf area) and leaf nitrogen concentration. In some areas, soils deficient in zinc (Zn) affect the growth of seedlings. To manage this, two sprays of zinc sulphate [0.5% $ZnSO_4$ + 0.25% $Ca(OH)_2$ mixture], that is, 50 g $ZnSO_4$ + 25 g $Ca(OH)_2$ mixed in 10 litres of water for 100 $m^2$ area is applied after 10 days interval from

**Figure 7** Improved nursery management practices include optimum (a) seed density, (b) nutrient application and (c) age for transplanting.

sowing. Adding organic sources such as FYM supplies not only macronutrients but also essential micronutrients such as Fe, Mn, Cu and Zn.

Use of balanced nutrients and lower seeding rate in the nursery produced healthier and more vigorous seedlings that withstood floods and salinity better than seedlings produced through traditional farmers' practices. Good seedling vigour in the nursery ensured faster recovery and growth and higher yield in the transplanted rice systems when abiotic stresses are anticipated. Application of small amount of nutrients at 5-3-1.5 kg N-$P_2O_5$-$K_2O$ + 500 kg FYM in a nursery area of 1000 $m^2$, enough to produce seedlings for 1 ha, can increase grain yields by up to 30% over current farmers' practice. These cost-effective nursery management options have great potential to enhance yield in coastal areas affected by salt stress and could help to cope up with stagnant as well as flash flood incidences.

Under high salinity stress there is mortality of germinating seeds as well as problem of availability of proper nutrients to the seedlings during growth period. Seedlings raised conventionally in seedbeds are less vigourous and have poor establishment upon transplanting, especially when water is relatively deep. This problem can be overcome by raising healthy seedlings through proper nursery management, taking into account the prevailing land situation and suitable varieties. Among the different components of nursery management, optimum seed density and adequate nutrition in terms of organic and inorganic sources are important. Age of seedlings at transplanting is also important under salinity and waterlogging stress situations since younger seedlings are vulnerable to death. Appropriate seeding density, nursery management and seedling age are obviously dependent on many factors and might vary in different production systems. In the Sundarbans, transplanting of 40-day-old seedlings produced significantly higher yields (4.27 $tha^{-1}$) than 25-day-old seedlings (3.50 $tha^{-1}$) (Sarangi et al., 2015). Transplanting older seedlings is important in systems without good water control, un-levelled fields, and soils affected by salt stress; taller seedlings help avoid losses caused by floods.

Nursery bed should be kept weed free by regular monitoring and removing the unwanted plants, off types and so on. Insects like stem borers, rice leaf folder and hispa, grasshopper and white-backed planthopper may affect the nursery. An integrated approach involving good nursery management practices and use of insecticides if necessary may be followed to manage these insect pests before they reach the threshold level. Insecticides like monocrotophos 36 EC @ 0.05% or chloropyriphos 20 EC @ 0.015–0.02% may be sprayed to control the above insect pests. If water in the nursery bed is controlled, granular insecticides like Phorate (Thimet 10 G) may be applied @ 1.25 kg per 1000 $m^2$ nursery area. To control diseases like blast, spray Hinosan 50 EC @ 2 ml per litre of water or

Bavistin 50 WP @ 2.5 grams per litre of water. If any insecticides are not applied in the nursery bed, during the time of transplanting dip the seedlings in chloropyriphos solution @ 1 ml per litre of water for 10–12 h.

# 6 Rice field management after transplanting

Improved main field management practices include improved stress-tolerant variety combined with fertilizer dose of 50-20-10 kg N-$P_2O_5$-$K_2O$ + 5 t FYM $ha^{-1}$, two seedlings per hill and a hill spacing of 15 × 15 cm. The grain yields of rice under stagnant flooding increased from 2.6 $tha^{-1}$ under farmer's method of management with farmer's variety to 4.5 $tha^{-1}$ with improved management and variety (Amal-Mana). The variability of grain yields with different stagnant flooding water depths in the improved management with improved variety (Amal-Mana) is significantly less compared to farmer's management with farmer's varieties like Pankaj, Sadamota, Pratikshya, Geetanjali and Talmugur. Grain yield decline with increasing water depth was highest under farmers' management – farmers' varieties combination, but stable when improved variety and management were combined (see Fig. 8). The slope increased from − 0.07 $tha^{-1}$ $cm^{-1}$ to−0.04 $tha^{-1}$ $cm^{-1}$ with the introduction of improved management practices, and further to−0.02 $tha^{-1}$ $cm^{-1}$ with the introduction of the improved variety farmers' management practices. The yield loss is negligible (slope of−0.002 $tha^{-1}$ $cm^{-1}$) for the improved management combined with improved variety (see Fig. 8). Association of grain yield with water depth was strongest under farmer's management with farmer's variety ($R^2$ = 0.81**) and weakest when improved management was combined with improved variety ($R^2$ = 0.015).

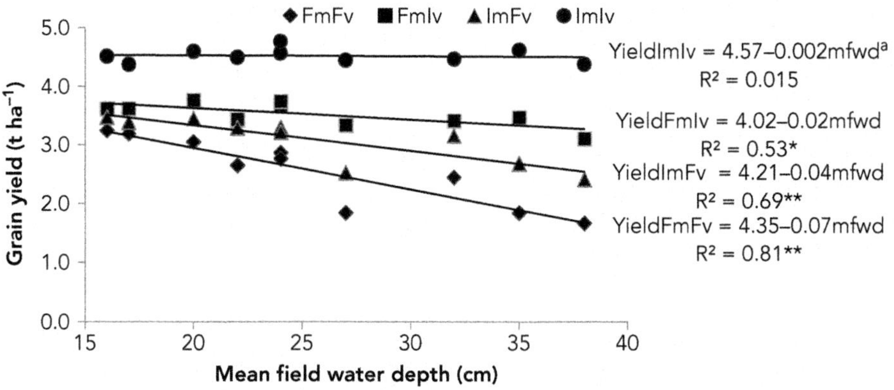


**Figure 8** Responses of combinations of varieties and management options with mean field water depth (from transplanting to harvest) in farmers' fields. FmFv: Farmer's management with farmer's variety, FmIv: Farmer's management with improved variety, ImFv: Improved management with farmer's variety, ImIv: Improved management with improved variety; [a]mfwd: mean field water depth in cm; *, ** significant at $P < 0.05$ and 0.01, respectively.

This data indicate that grain yield is stable when using a tolerant variety and improved management, but is highly dependent on water depth under the current farmers' system using their traditional varieties and management. Introduction of a stagnant flood-tolerant variety alone also resulted in relatively higher and more stable yields (3.51 $tha^{-1}$) than under improved management but using traditional varieties (3.08 $tha^{-1}$).

# 7 Management practices for dry season rice

Farmers in the coastal areas also prefer to grow rice during the dry season though there are constraints of salinity and irrigation water scarcity. As salinity can be managed better with standing water, rice stands out to be the most suitable crop under such situation. In addition, productivity potential of *boro* rice is much higher than *kharif* season rice, especially when improved salt-tolerant varieties are used. Higher productivity of *boro* rice is mainly due to higher solar radiation, favourable temperature, control over irrigation water and comparatively more efficient use of resources. Farmers use more inputs such as fertilizers and pesticides, during the dry season, to get higher yield. However, cultivation of rice during the dry season in coastal saline soils is challenging and requires careful choice of a suitable rice variety and good management practices.

The cost of rice cultivation is higher during the dry season due to the high cost of irrigation water. In coastal areas, and especially in the dry season, soil salinity is a common problem mainly due to capillary rise of saline groundwater and deposition of salt on the soil surface. In this situation, use of salt-tolerant rice varieties is essential to obtain a good rice crop, otherwise the chance of crop failure due to salt injury is high. Varieties like Canning 7, WGL 20471 and Bidhan 2 are found to be tolerant to salinity and can be grown with improved management practices. Traditionally, rice is grown by manual transplanting of 25–30-day-old seedlings after puddling. Puddling reduces percolation losses, controls weeds and enables easy transplanting. However, due to labour scarcity for transplanting and excessive water requirements for puddling, and negative effect on soil physical properties, alternative methods of crop establishment such as drum seeding have been found to solve such constraints in coastal salt-affected soils for their suitability for *boro* rice cultivation.

Another newly developed establishment option is the system of rice intensification, recommending 8–12-day-old seedlings and a wide spacing. However, such very young seedlings may be adversely affected by higher salinity stress in the *boro* season of coastal salt-affected areas. Modified system of rice intensification includes use of salt-tolerant variety, nursery in raised beds with irrigation facility to wash out salts, optimum FYM application, use of relatively older seedlings (15–18 days instead of 8–12 days), gap filling in case of mortality, application of extra irrigation water for washing out salts from root zone, cultivation of *Azolla* along with rice and use of *Azospirillum*, phosphate solubilizing bacteria and *Trichoderma* as recommended (Sarangi et al., 2011 and Sarangi, 2013).

The drum seeder (see Fig. 9) is one of the revolutionary equipments that can solve the problem of rice establishment during the dry season in coastal areas. It is developed for direct sowing of pre-germinated rice seeds under puddled conditions. Paddy drum seeder has eliminated the need of transplantation and hours of manual work which literally break the back of the farmers particularly women farmers involved in transplanting paddy seedlings to the field. Drum seeder at one stretch with single operator's effort

**Figure 9** Drum seeder machine for efficient establishment of dry season rice in salt-affected areas of Sundarbans.

**Figure 10** Properly germinated seeds for use in the drum seeder machine.

covers eight rows of sowing at 20 cm row-to-row spacing. This equipment is made up of plastic except that the shaft and handle used in the equipment is steel, as a result it is light in weight and easy to handle. The light weight plastic wheels are having floating characteristics and not meshed up in the puddle unlike iron wheels.

The eight-row drum seeder is 1.8 m wide, and the diameter and length of each drum are 0.18 m and 0.25 m, respectively. Each drum has a capacity of 2 kg of pre-germinated seeds; however, the drums should not be filled completely and about one-third of each drum should be kept empty for easy flow of seeds through the perforated holes. The seeds are soaked in water for a day and then incubated for another 24 h for sprouting.

**Figure 11** (a) Drum seeding of pre-germinated rice seeds on the puddle soil and (b) field after emergence of seedlings.

**Figure 12** Simultaneous cultivation of rice and *Azolla*: an efficient method for soil health management in coastal saline region.

To prevent sprouts intermingling with each other, sprout length should not be more than 7–8 mm. If the radicles are allowed to grow further, there is intermingling with each other, which creates problem in proper dropping from drum. Properly sprouted seeds (see Fig. 10) that are not too long are dried in the shade for about half an hour for smooth and easy dropping of seeds through the holes of the drum. Before 7–10 days of sowing the land should be soaked with water and ploughed 2–3 times followed by levelling; proper levelling is required for smooth running of the machine. There should not be weeds, clods and depressions in the field. The soil should not be too soft or too hard. Water from the puddled field should be drained out at least 24 h before sowing to form only a thin layer of water on the puddled surface. Drum seeder is drawn manually over the puddle soil with triangle marks on the drums pointed towards the pulling direction.

The machine is pulled at a normal walking speed (1 km $h^{-1}$) in backward direction (see Fig. 11). The wheel impression in the first pass will serve as a marker. In the second pass the wheel should pass on the same wheel impression of the previous pass to maintain the row-to-row spacing of 20 cm. The sowing is continued and occasionally the dropping of the seeds through the holes of the seeder is watched. Refill the drums when they reach one-fourth capacities. One person can sow the seeds in one hectare within 7–8 h. The field is not irrigated for 2–3 days after sowing to allow roots to anchor and then the depth of water is increased gradually as the seedlings grow.

Proper nutrient management is another important component of crop management, especially for high-yielding dry season rice crop. The target is to maintain or even improve soil fertility, to limit the costs of fertilizer and to obtain high yields. This can be achieved by optimum use of chemical fertilizer in integration with the locally available organic sources of plant nutrients. Since most of the farmers in coastal areas of West Bengal possess cattle as an integral component of their farming system, FYM as a cheap and easily available source of organic nutrients can be one of the important components of INM. *Azolla*, another possible component of INM, is a free floating freshwater fern (see Fig. 12). Its incorporation improves soil nutrients primarily due to nitrogen fixed by the *Anabaena* species of blue-green algae (BGA) present in the lobes of *Azolla* leaves. Farmers can produce their own *Azolla* inoculums in ponds and ditches that are common in coastal areas. The mean temperature during the *boro* season in most the coastal areas of West Bengal falls within the range (20–30°C) required for *Azolla* cultivation, making climatic conditions mostly congenial for its good growth.

Before the application of *Azolla* into the main field, the basic stock of viable material is maintained in a nursery with a water depth of 3–6 cm. *Azolla* stocks were put into beds at 1 kg $m^{-2}$, and are allowed to grow for a month. Fresh *Azolla* at 2 $tha^{-1}$are inoculated in the main field about 4 weeks after drum sowing. The *Azolla* mat was incorporated into the soil 4 weeks after inoculation. Compared with RDF treatment, the organic carbon increased by 20.4 and 38.6% in *Azolla* and FYM treatments, respectively. The application of organic nutrient sources had also a positive effect on the availability of major nutrients.

Weed management is one of the important challenges of drum seeding technology during the dry season. An integrated management strategy involving mechanical, manual and chemical methods should be followed for efficient and cost-effective weed management. Initial weed control is achieved through the use of selective herbicides. Pre-emergence application of pretilachlor @ 0.75 kg $ha^{-1}$ on 8 days of sowing or pretilachlor + safener (Sofit) @ 0.45 kg $ha^{-1}$ on 3–4 days of sowing provide effective weed control for about one month. Use of cono weeders help in removing the weeds between rows (see Fig. 13); manual weeding is done to remove weeds within the rows. Cono weeder consists of two inversely mounted rotors, attached to a main frame and handle assembly. The height of the handle is adjustable according to the suitability of the operator. It is run in back-and-forth movement between rows to uproot the weeds and bury in the soil. Cono weeder is used on 30 days of sowing, second at 45 days of sowing and subsequent ones depending

**Figure 13** (a) Cono weeder and (b) its use for effective weed management in drum-seeded rice plots.

**Figure 14** Drum-seeded rice matures early and produces higher yield in saline soils of Sundarbans region.

on the weed population. This implement is gender friendly, women farmers can operate easily due to less drudgery. About 0.15 ha area can be covered by a person in a day.

Drum seeding technology in Sundarbans region saved about 10–18% variable costs compared with transplanting, mainly because the latter had higher requirements of labour and inputs to raise seedlings in a nursery, pull up seedlings, transport and transplant them. Benefit–cost ratio (BCR) was significantly higher in drum seeding (1.94) in comparison with transplanting (1.27) during the dry season.

Use of drum seeder also facilitates early establishment of dry season rice crop immediately after harvesting of wet season rice, thereby efficient utilization of carry over soil moisture, less development of salinity and earlier maturity. During the initial period of dry season, the underground water is less saline and the soil profile is saturated with rainwater; therefore, the soil salinity is less and amount of irrigation water required to prepare the land and remove the salts is lesser. Early seeding helps in reducing irrigation water input by about 17% than with late seeding, mainly due to lower irrigation water requirements for land preparation, reduced loss of irrigation water as evaporation as the crop matures in congenial weather conditions (see Fig. 14). Irrigation water productivity with early seeding (41–5 kg grain $ha^{-1}$ $cm^{-1}$) was 30% higher than with late seeding (31–35 kg grain $ha^{-1}$ $cm^{-1}$) at Canning Town (Sarangi et al., 2015b).

# 8 Managing toxic gases from coastal rice fields

A considerable amount of $H_2S$ gas is evolved from rice field of coastal saline soil due to reduction of sulphate salts to sulphides. A portion of sulphides formed in soil is released as $H_2S$ gas in the atmosphere cause environmental pollution. The gas gives strong smell of $H_2S$ in coastal rice fields. The other portion of the $H_2S$ formed reacts with soluble metallic cations in soil forming mostly insoluble metallic sulphides, of which iron sulphide is the primary one. The insoluble metallic sulphides form coating on the surface of rice roots and interfere with the nutrient absorption capacity of the roots. Free $H_2S$ has high toxicity to plants and animals. Application of ferrous sulphate to soil reduces the concentration of $H_2S$ in soil solution by precipitating it as FeS. The precipitated sulphides give a dark colouration to the rice soil as commonly encountered in the coastal rice fields. A considerable amount of methane gas is also produced in waterlogged rice fields, causing environmental pollution. The amount of methane produced increases with the application of nitrogenous fertilizers but the effect could be reduced with management practices such as direct sowing, addition of composted or partially composted organic matter (low C:N ratio), alternate wetting and drying, incorporation or deep placement of urea super granules and wherever possible growing of short-duration rice varieties with low emission potential.

# 9 Removing excess soluble salts from the root zone

Reclamation of saline soils requires removal or leaching of excess soluble salts from the root zone of crops. Good-quality water is needed for leaching salts from the soil and subsequent removal of leached water beyond the desired depth for effective growing of roots. Leaching requirement is the fraction of irrigation water that should pass through the plant root zone in order to maintain the soil salinity to a specified or desired level. It is the ratio of depth of drainage water to the depth of applied water and expressed as per cent. Under ideal conditions, the LR is the inverse ratio of corresponding electrical conductivities (EC), that is, the ratio of EC of the applied water to the EC of drainage water, or LR = ECaw/ECdw, where ECaw and ECdw indicate EC of applied water and drainage water, respectively. Thus, if the EC of the drainage water or soil solution at the root zone is to be maintained below 4 $dSm^{-1}$, and EC of the applied water is 1 $dSm^{-1}$, LR will be 25%, that is, 25% of applied water must pass through the root zone (Biswas and Mukherjee, 1994).

Leaching process in the Sundarbans region can be done with the initial monsoon phase by accumulating the rainwater for some time and then draining the water with the salts. This process can be repeated for several times depending upon the degree of salinity of the soil. However, for growing dry season crops, during which there is non-availability of rainwater, the scarce irrigation water mostly stored in farm ponds or limited amount of good-quality groundwater should be used judiciously for leaching the salts. The soil salinity starts increasing after the monsoon season or beginning of the dry season due to capillary rise of salty water from underground brackish water table present at shallow depths. Therefore, early land preparation requires less irrigation water for leaching the excess salts than that of late. There was saving of 16 cm irrigation water for land preparation of *boro* rice due to the early seeding than that for late seeding (Sarangi et al., 2015b). Further, leaching of salts should be carried out under unsaturated condition through intermittent ponding or sprinkler irrigation instead of continuous ponding. The former process facilitates removal of salts through fine pores as well as large pores instead of passing only through the large pores. Therefore, the quantity of water required for the desired leaching of salts may be lowered in case of intermittent irrigation as much as two to five times of that under ponding (Bandyopadhyay et al., 2001).

# 10 Integrated nutrient management (INM) for coastal saline soils

INM or integrated plant nutrient supply (IPNS) system involves monitoring of all the pathways of plant nutrient supply in crops and cropping systems and calls for a judicious combination of fertilizers, organic manures and biofertilizers (Prasad, 2008). The balance fertilization through integrated use of manure, fertilizer and biofertilizer along with micronutrients has been found useful in rice crops (Yadav et al., 2013). Though the cost of cultivation is marginally increased when the nutrients are applied through the combination of different sources, due to higher grain and straw yields, the net income and BCR are also higher under INM. The highest net income of C 12840 $ha^{-1}$ and BCR (1.69) was obtained with application of 75% recommended dose of fertilizers along with 5 t FYM $ha^{-1}$ + BGA + phosphate solubilizing bacteria + 5 kg Zn $ha^{-1}$ (Sharma et al., 2015). Further, the soil health was improved in the long run due to combined application of organic and inorganic sources of nutrients (Sarangi et al., 2014).

## 10.1 Nitrogen management in rice fields

Coastal soils are extensively used for paddy cultivation during *kharif* under submerged conditions. Nitrogen is the major nutrient required for salt-affected soils, but its efficiency is very poor under saline soil condition. Major form of nitrogen losses from soil is through volatilization, which increases with increase in soil salinity and accounts for 30–50% losses of applied inorganic nitrogen fertilizer to soil. Volatilization loss from soil can be substantially reduced through placement of N fertilizer at shallow soil depth (5 cm). Volatilization loss from ammonical-N fertilizers was more than that from urea. Volatilization loss can be reduced by increasing the number of split of N application. The loss of N through denitrification was comparatively low (<5%) compared to volatilization loss. Leaching losses of N was also relatively low.

Important components of INM include inorganic or chemical fertilizers, organic manures such as FYM, green manuring with *Sesbania aculeata* and biofertilizers such as *Azolla*.

**Table 6** Most common inorganic fertilizers and organic manures used in the Sundarbans region

| | | Nutrient content (%) | | | |
|---|---|---|---|---|---|
| S. no | Name of the fertilizer/manure | N | $P_2O_5$ | $K_2O$ | S |
| 1 | Urea | 46.0 | – | – | – |
| 2 | Ammonium sulphate | 20.6 | – | – | 24.2 |
| 3 | Calcium ammonium nitrate | 25.0 | – | – | – |
| 4 | Single super phosphate | – | 16.0 | – | – |
| 5 | Diammonium phosphate | 18.0 | 46.0 | – | – |
| 6 | Muriate of potash | – | – | 60.0 | – |
| 7 | Suphala | 10.0 | 26.0 | 26.0 | – |
| 8 | Suphala | 12.0 | 32.0 | 16.0 | – |
| 9 | Zinc sulphate (Zn: 33–36%) | – | – | – | 17.8 |
| 10 | Farmyard manure (FYM) | 0.50 | 0.30 | 0.50 | – |
| 11 | Green manure *(Sesbania aculeata)* | 3.97 | 0.37 | 4.80 | – |

Best yield of crops under rice-based cropping system in coastal region of West Bengal was obtained when city compost was used as the organic N component in the combined sources (organic + inorganic) of nitrogen. Chemical fertilizers are important component of INM (see Table 6); however, these should be economically used to reduce the deleterious effects on environment including soil health.

Nitrogen content and biometric characteristics of plants were better when combined sources of nitrogen were applied. The fertility status of soil was improved due to application of organic manures while application of inorganic sources of N did not show any residual improvement in soil nitrogen status. Application of biofertilizers (*Azotobacter*, *Azospirillum* and BGA) produced significant improvement in yield of crops. Application of different organic manures to soil in combination with inorganic nitrogen fertilizer improved the microbial activities in soil, thus leading to better soil health. Green manuring with *S. aculeata* (locally called *dhaincha*) during pre-*kharif* season improves rice growing soil health. It is able to withstand flood, drought, saline and alkaline conditions (Barik, 2004). Use of green manure in rainfed lowland rice, especially *Sesbania rostrata*, which nodulates in roots as well as in stem, gives better performance and maintains the nutrients supply. It is considered as the fastest $N_2$ fixing plant known, of the order of 60–270 kg N $ha^{-1}$ in 45–55 days. It has been found to fix $N_2$ even when the soil N status is higher. It can withstand flood, drought and salinity, and is therefore considered as a good green manure crop for coastal saline soils.

## 10.2 Management of P and K fertilizers for coastal soils

Study on short-term and long-term effects of omission and addition of P and K fertilizers in the coastal soils of Sundarbans, West Bengal, showed initially that the soils showed response of crops to N fertilizers only but not to P and K fertilizers (Bandyopadhyay and Maji, 1993). After few years the soils showed gradual decline in the available P status when no P fertilizer is applied and in the long run the soil become deficient in P. Application of higher doses of P fertilizer increases the available P status of soil, particularly when P was applied without N. Application or omission of N fertilizer (urea) did not show any effect on available N content in soil. Regular application of N is essential for higher yield of crops as well as sustainable productivity and maintenance of soil health. Application of P @ 11 kg $ha^{-1}$ for each crop of rice and 5.5 kg $ha^{-1}$ for each crop of maize is necessary for maintaining P fertility status of coastal soils of Sundarbans. However, for hybrid maize the P and K doses are 34.9 and 33.2 kg $ha^{-1}$ respectively. Application of K or its omission had no effect either on yield of crops or on available K status of soil. However, when tall indica rice varieties are grown, to prevent lodging of crops a maintenance dose @ 10 kg $K_2O$ $ha^{-1}$ is recommended.

## 10.3 Management of micronutrients

The coastal salt-affected soils are rich in most or all the micronutrients except Zn (Maji and Bandyopadhyay, 1991). Coastal soils show poor response of micronutrients. Most of the soils have high content of active Fe, which is sometimes present in toxic concentrations, particularly, when the soil is highly acidic (acid sulphate soils) or under submerged condition. Zinc and Cu availability in soil is found to be reduced under submerged condition. Hence, rice is most likely to be affected with Zn deficiency in soil compared to upland crops (Maji and Bandyopadhyay, 1992). Due to Zn deficiency, a disease called '*Khaira*' is caused in rice crop. This disease generally appears during seedling stage in patches in the nursery, and also after 10–15 days of transplanting. The symptoms are stunted growth; chlorosis between the veins of new leaves, where brown spots are formed; on lower leaves a large number of small, brown to bronze spots appear which coalesce to form bigger spots; and ultimately the entire leaf turns bronze coloured and dries. Root growth is also restricted and usually the main roots turn brown. The finer roots are destroyed. In severe cases, tillering decreases, plants fail to grow further and time to crop maturity may be increased, but sometimes there is natural recovery after 6 weeks of age (Singh, 1983; Dobermann and Fairhurst, 2000). When natural recovery did not happen, to overcome this disease, apply 0.5–1.5 kg Zn $ha^{-1}$ as a foliar spray (e.g. 0.5% solution at about 200 litres of water $ha^{-1}$) for emergency treatment of Zn deficiency in growing plants. If Zn deficiency is prevalent in the soil, apply 10–25 kg $ha^{-1}$ $ZnSO_4.7H_2O$ to the soil. Uptake of $ZnSO_4$ is more efficient when broadcast over the soil surface (compared with incorporated) particularly in direct-sown rice. To facilitate more homogeneous application, mix the Zn sulphate with sand in 1:3 ratio (Dobermann and Fairhurst, 2000).

## 10.4 Management of algal weeds in rainfed lowland rice fields

The rainfed lowland ecosystem in Sundarbans provides ideal conditions for growth of cyanobacteria and algae as well as rice. Species of $N_2$-fixing cyanobacteria are desirable for N fertilization, but excessive growth of some of these algae become undesirable, as they compete with and dominate rice crop as weed. These algal weeds cause a significant loss of grain yield due to restricted tillering and growth of rice plants. During initial 30 days of transplanting, algal mats entangle the transplanted hills tightly, and did not allow for tillering of seedlings. Sometimes due to flooding, these algae form a thick mat on the plant surface and suffocate the plants. Therefore, effective management is needed to control the growth of these weeds. Copper

**Table 7** Rice yield under different methods for control of algae at Canning Town

| Treatments | Grain yield ($tha^{-1}$) |
|---|---|
| Control (no measure) | 2.80 |
| $CuSO_4$@10 kg $ha^{-1}$ as basal | 2.93 |
| $CuSO_4$@10 kg $ha^{-1}$ as top dressed | 3.20 |
| $CuSO_4$@10 kg $ha^{-1}$ as basal + 5 kg $ha^{-1}$ as top dressed | 3.27 |
| $CuSO_4$@20 kg $ha^{-1}$ as basal | 3.20 |
| $CuSO_4$@20 kg $ha^{-1}$ as top dressed | 3.33 |
| $CuSO_4$@10 kg $ha^{-1}$ as basal + 10 kg $ha^{-1}$ as top dressed | 3.70 |
| Hand weeding | 4.08 |
| Mahua cake @ 120 kg $ha^{-1}$ | 4.30 |
| CD (P = 0.05) | 0.62 |

Source: Biswas, 1987

sulphate ($CuSO_4$) is generally used as an algaecide to control this weed; however, use of locally available materials like mahua (*Madhuca indica*) cake can be effective for this purpose (see Table 7).

Management approaches involving use of copper sulphate have raised environmental concerns (Hill et al., 2006) and are becoming less economical to apply. Recent research in California reveal that altering phosphorus fertilizer applications by incorporating fertilizer P or delaying P application by up to 30 days reduces rice field algae and cyanobacteria abundance (Spener and Linquist, 2014).

## 11 Reducing salinity build-up through mulching/cover crops

The salinity build-up in coastal soils of Sundarbans can be controlled by growing crops with large foliage cover (see Fig.15) rather than leaving fallow. Mulching of soil during the dry season by using rice husk and straw, organic farm waste or any other suitable organic materials can reduce salinity development and add organic matter to soil. When mulching material is not available, the soil is to be kept ploughed instead of leaving unploughed. Due to mulching or ploughing, the capillary pores are broken down, evaporation of water from soil is checked and thereby the build-up of salt deposit is prevented. Ploughing of soil prior to pre-monsoon shower and sowing of green manuring crop, for example, *Sesbania*, will reduce salinity and add nitrogen through symbiotic N-fixation.

**Figure 15** Cover crop of *Sesbania* for management of salinity build-up in Sundarbans region of West Bengal, India.

# 12 Management practices for increasing income from rainfed lowlands

During *kharif* season, for increasing income from rainfed lowlands, freshwater fishes can also be cultivated along with rice within the confined field. Vegetable crops such as brinjal/eggplant (*Solanum melongena*), lady's finger (*Abelmoschus esculentus*), Indian spinach (*Basella alba/rubra*), broad beans (*Vicia faba*) and so on can be grown on the bunds around the field. Thus, harvests of rice, freshwater fishes and vegetables are possible from the same field. Income from a particular field increases due to increased harvest index, and land and water use. During the dry season brackish water fish can be grown. However, soil salinity increases due to brackish water aquaculture during summer season, but it may be brought down to a safe limit during *kharif* season through leaching and run-off processes with the help of excess rainfall (more than 1500 mm). It is, however, advisable to create drainage facilities for washing out of excess soluble salts from soil profile; in addition, entry of saline water into the field should be stopped at the beginning of rainy season. For the purpose of drainage and leaching of salts, canals of 1 m width and 1 m depth may be dug out at certain intervals (depending on the size of the field) inside the field and also along the border. Soils from the canals may be utilized to raise a strip of land at least 1.5 m width around the field plot for growing vegetables.

Under paddy-cum-fish cultivation technology, trenches (3m top width x 1.5 m bottom width x 1.0 m depth) are dug around the farm land periphery, leaving about 3.5m wide as outer boundary, and the dugout soil is used for making dikes (about 1.5 m top width x 1.0 m height x 3m bottom width). The dikes are used for growing vegetables and/or green manuring crops/fruit crops/multipurpose tree species throughout the year. Dikes protect free flow of water from the field and trenches are used for harvesting rainwater. Deep trenches in the field provide better drainage condition in the field during the non-monsoon months. A small ditch is dug out at one corner of the field for sheltering of fish when water is drained out from trenches. Remaining portion of the farm land including the trenches is used for profitable paddy + fish cultivation during *kharif*. Utilizing the rainwater harvested during monsoon in deep trenches, the farm land containing non-trench and non-dike area is used for cultivation of crops requiring less water during dry (*rabi*/summer) season. Rice–fish farming system with higher water and land productivity and employment opportunities can ensure food, nutrition and livelihood security for the farming communities, particularly the largest groups of small and marginal farmers (Mahapatra et al., 2008).

# 13 Contingency plan for cyclone affected area

A large area of the Sundarbans in West Bengal is frequently affected by cyclones, causing loss of large area of rice crop, number of livestock, animals and human lives. Due to this, standing crops are submerged and the majority of agricultural lands are flooded with saline water. Submergence of agricultural lands with saline tidal water results in great damage to the soil and water quality of that area. Under these situations, hardly any crop other than salt-tolerant rice is possible even during the *kharif* season. To manage such situations and reduce the effect of natural calamity the following steps need to be followed for washing away of salts, to the maximum extent, with the use of monsoon rains.

- i Repairing of broken embankment to prevent further entry of saline water into agricultural land.
- ii Renovating the surface drainage system of the land/mini-watershed area to drain out the standing saline water through sluice gate.
- iii Pumping out of standing saline water, if needed, from the low-lying area/ditches when the natural drains fail to drain out the water.
- iv Raising of earthen bunds' heights around the field to retain rainwater for efficient washing away of salts.
- v Stirring the soil and standing rainwater (manually or with power tiller, tractor, etc.) in the field and allowing the muddy water to settle overnight and then drain out the standing water.
- vi Repeating the above step for several times within a span of 15 days or so, depending on the salinity of the land. This will remove majority of salt from the soil surface.
- vii Sowing of dhaincha (*Sesbania* spp.) seeds @30–35 kg $ha^{-1}$ during April-May and after 8-10 weeks of sowing incorporation of the biomass into the soil before transplanting will reduce the harmful effect of soil salinity to some extent and will increase soil fertility.
- viii Growing rice varieties tolerant to salinity. Some of the varieties named below may be selected depending on the land situation and soil salinity:

    - Upland (Salinity ECe 6–9 $dSm^{-1}$, waterlogging up to 0–20 cm): Canning 7, CSR-4, CSR-36, Bidhan-2, CST 7-1, WGL 20471, Boby, Annada, Lalat
    - Medium-land (Salinity 6–8 $dSm^{-1}$, waterlogging up to 0–30 cm): Namita-Dipti, Sumati, Utpala, Bhutnath, CSR 1, CSR 2, CSR 3, CSR 6, Talmugur, Nona Bokra, IR-72046
    - Low-land (Salinity ECe 5–7 $dSm^{-1}$, stagnant waterlogging up to 30–40 cm): Amal-Mana, SR 26B, Sabita, Patnai 23, Geetanjali
    - Semi-deep water land (Salinity ECe 4-6 $dSm^{-1}$, stagnant waterlogging up to 40–60 cm): CSRC(D) 7-0-4, CSRC(D) 12-8-12, Malabati, Kalamota, Sadamota (Selection), Tilak Kachari, FR 13A, NC-678, Asfal, Najani, Kumragour
    - Flash flood affected areas: Swarna –Sub 1 and CR 1009-Sub 1

- ix Raising of nursery on a good-quality soil at higher elevation with addition of organic manure. The seeds should be soaked overnight in 0.1% salt (NaCl) solution to increase the salt tolerance of the crop.
- x Transplanting of aged seedling (40 days), recommended number of seedlings per hill (2 nos.) and closer spacing (15 cm × 15 cm).
- xi Applying nitrogen (N) fertilizer as a basal dose should be avoided. Instead, neem-coated urea may be applied one week after transplanting when there is clear weather and no possibility of heavy rain within a couple of days. The N fertilizer may be placed into the soil (3.5–5.0 cm depth) or may be applied in the form of mudball for better efficiency. Combined sources (organic + inorganic) of N will produce better yields.

xii Adopting paddy-cum-fish culture wherever possible. For paddy-cum-fish culture the rice field should be linked with contiguous pond/ditch/channels.
xiii Cultivating brackish water fish on a community basis or individually by protecting the land area with nylon net or suitable bamboo structure, if the river embankment fails to prevent the entry of tidal water into the fields.
xiv Growing of salt-tolerant grasses such as Para grass [*Brachiaria mutica* (Forsk) Stapf.] and Karnal grass [*Leptochloa fusca* (Linn.)] (may be used as fodder for cattle in integrated farming system) on unutilized land.
xv Harvesting and conserving excess monsoon rain in farm ponds and furrows/trenches during the wet season prevents the ingress of saline water/build-up of salinity during the dry season. Harvested freshwater is used conjunctively with groundwater to reduce the salinity build-up. Limited quantity of groundwater should be used efficiently by installing the pump at appropriate depth and location. Dry season crop is sown early to use residual soil moisture. Short-duration and salt-tolerant rice varieties should be adopted to increase water use efficiency and yield.

# 14 Recommended practices for cultivation of *kharif* and *rabi* rice

## 14.1 *Kharif* rice

- Grow green manuring crops, viz. dhaincha (*S. aculeata*) before *kharif* rice to save about 50% of chemical fertilizer.
- In case of difficulties in growing of green manuring crops apply green leaves of *Gliricidia* or other suitable plant species @ 5–10 $tha^{-1}$.
- Use optimum seed rate depending upon the establishment method.
- Treat seeds before sowing with amisan (0.5%) + streptocycline (0.1%).
- Land should be levelled with clear bunds (50–60 cm high).
- Ensure field level drainage.
- Transplant two seedlings per hill at 15 × 15 cm spacing.
- Maintain shallow but continuous water regimes in the field.
- Apply only neem-coated urea as N fertilizer and P and K on soil test basis.
- Apply first dose of N one week after transplanting.
- Avoid ammonium sulphate on saline soil.
- Use 5 t FYM and fertilizer dose of 50-20-10 kg $N\text{-}P_2O_5\text{-}K_2O$ $ha^{-1}$.
- Apply 60–70% of total N by tillering and the rest at boot leaf stage.
- Protect crops from pest and diseases on regular schedule but on economic basis.
- Use mahua cake @ 120 kg $ha^{-1}$ or $CuSO_4$ @ 20 kg $ha^{-1}$ for algal weed control in lowland situation.

## 14.2 *Rabi*/summer/*boro* rice

- Dig pond in the field (5:1 ratio) for water harvesting and use the stored water for *rabi*/ summer cropping.

- Avoid summer fallowing of lands, adopt rice straw mulching or grow cover crops.
- Select short-duration salt-tolerant varieties, viz. Canning-7, WGL 20471, Bidhan 2, IET 4786 and so on.
- Use drum seeder to sow pre-germinated seeds treated with seed treating chemicals and sowing should be done early immediately after harvest of *kharif* rice to utilize the residual soil moisture.
- Early sowing needs less irrigation water and results in higher yield.
- Adopt INM involving organic (FYM or *Azolla*) and inorganic sources.
- Wherever possible follow conjunctive use of underground irrigation water with harvested rainwater in farm ponds.

# 15 Alternative livelihood options

In many areas of Sundarbans, saline water ingresses into the agricultural fields, because of which the levels of soil and water salinity are increased, resulting in large amount of cultivable area becoming unsuitable for growing any of the *kharif* crops. Under these circumstances alternative livelihood option may be promoted for sustaining the farm families. Options may be – promotion of duckery, piggery, mushroom cultivation, nursery raising for trees and fish seed, wherever possible. Creation of employment through rural non-farm activities such as repairing of damaged embankment, plantation of mangroves, and repairing of damaged rural roads, buildings and other infrastructure may be taken up in a large scale through utilizing the existing government schemes. Strategies such as production/processing of good-quality seeds of salt-tolerant and high-yielding rice varieties through community participation or promotion of seed village concept by active engagement of government machinery may be encouraged in areas affected by salinity.

# 16 Future trends and conclusion

Salinity and submergence are the major constraints for sustainable rice cultivation in coastal region. These abiotic stresses are likely to increase in the near future due to the effects of climate change, intrusion/ingress of seawater, over exploitation of groundwater and flooding due to heavy rain. Therefore, to make the rice cultivation sustainable, improved agronomic practices suitable to different rice lands, soil types and season must be adopted. Using salt-, flood- and lodging-tolerant rice varieties during wet season and short-duration, salt-tolerant varieties with good grain quality during dry season result in higher yield and profit. Appropriate establishment methods, INM, and cultural and reclamation measures are essential coping strategies under these unfavourable environments. Prevention of intrusion of seawater through construction/renovation of dikes, leaching of excess salts from root zone, mulching/cover crops during dry season, rainwater harvesting during wet season for conjunctive use with groundwater during dry season and efficient management of acid saline soils through liming and application of rock phosphate are the key aspects for a climate-resilient rice crop production in coastal saline region.

# 17 Where to look for further information

Relevant information on management of coastal saline soils is available at the website of Indian Council of Agricultural Research-Central Soil Salinity Research Institute (cssri.org), IRRI research projects on Stress Tolerant rice for Africa and South Asia (STRASA)/EC-IFAD and so on. Partners and important publications of STRASA project can be seen at http://strasa.irri.org. The section titled 'Rice-based system intensification and diversification' in the conference proceedings on 'Revitalizing the Ganges Coastal Zone: Turning Science into Policy and Practices' also deals with recent advancements on rice cultivation in Sundarbans region. This conference was organized by the CGIAR Challenge Program on Water and Food (CPWF), particularly focussing on Ganges Basin Development Challenges. CPWF was a partner of the CGIAR Research Program on Water, Land and Ecosystems. More on this could be seen at https://waterandfood.org/river-basins/ganges-2 and wle.cgiar.org.

# 18 References

Amanullah, M. M., Natarajan, S., Vanathi, D., Ramasamy, S. and Sathyamoorthi, K. (2007), 'Lowland rice in coastal saline soils – A review', *Agricultural Reviews*, 28(4), 245–53.

Annual Report. (1972), Central Soil Salinity Research Institute, Karnal (Haryana), India – 132 001.

Annual Report. (1978), Central Soil Salinity Research Institute, Karnal (Haryana), India – 132 001.

Anon. (2016), 'Popular and recent CRRI Rice Varieties for Different Ecologies' crri.icar.gov.in/popular_var.pdf(accessed 23 September 2016).

Arunachalam, V., Chaudhury, S. S., Sarangi, S. K., Ray, T., Mohanty, B. P., Nambi, V. A. and Mishra, S. (2006), Rising on Rice: The Story of Jeypore, M. S. Swaminathan Research Foundation, Chennai, India.

Bandyopadhyay, B. K. and Maji, B. (1993), 'Long term effect of NPK fertilisers in coastal saline soil under rice-fallow cropping system', *Journal of Indian Society of Soil Science*, 41(2), 374–5.

Bandyopadhyay, B. K. and Maji, B. (1995), 'Nature of acid soils of Sundarbans delta and suitability of classifying them as acid sulphate or potential acid sulphate soils', *Journal of Indian Society of Soil Science*, 43 (2), 251–5.

Bandyopadhyay, B. K., Maji, B., Sen, H. S. and Tyagi, N. K. (2003), *Coastal Soils of West Bengal – their Nature, Distribution and Characteristics*, Bulletin No. 1/2003, Central Soil Salinity Research Institute, Regional Research Station, Canning Town, West Bengal, India.

Bandyopadhyay, B. K., Sen, H. S., Maji, B. and Yadav, J. S. P. (2001), *Saline and alkali soils and their management*, ISCAR Monograph Series 1, Indian Society of Coastal Agricultural Research (ISCAR), Canning Town, South 24 Parganas, West Bengal, India.

Barik, T. (2004), *Fundamentals of field crops (Part – I)*, Saraswat publishers, Orissa, India.

Biswas, C. R. (1987), *Agronomic Management of Crops in Coastal Salt Affected Soils*, Annual Report, Central Soil Salinity Research Institute, Karnal, India.

Biswas, T. D. and Mukherjee, S. K. (1994), *Text Book of Soil Science*, Second edition, Tata McGraw-Hill Publishing Company Limited, New Delhi, India.

Clermont-Dauphin, C., Suwannang, N., Grunerger, O., Hammecker, C. and Maeght, J. L. (2010), 'Yield of rice under water and soil salinity risks in farmers' fields in northeast Thailand', *Field Crops Research*, 118, 289–96.

Das, M. and Maji, B. (2001), 'Seasonal fluctuation in salinization of soil and ground water and its spatial heterogeneity with time', *Journal of Indian Society of Soil Science*, 49(4), 773–6.

Das, P. C. (1993), *Manures and Fertilisers*, Kalyani Publishers, Ludhiana, India.

Datta, S. K. (1972), 'A study of salt tolerance of twelve varieties of rice', *Current Science*, 41(12), 456–7.

Dobermann, A. and Fairhurst, T. H. (2000), *Rice: Nutrient Disorders and Nutrient Management*, Potash & Phosphate Institute (PPI), Potash & Phosphate Institute of Canada (PPIC) and International Rice Research Institute (IRRI).

Grattan, S. R., Zeng, L., Shannon, M. C. and Roberts, S. R. (2002), 'Rice is more sensitive to salinity than previously thought', *California Agriculture*, 56(6), 189–95.

Hanson, B., Grattan, S. R. and Fulton, A. (1999), *Agricultural Salinity And Drainage*, Oakland, CA, USA, University of California, Division of Agriculture and Natural Resources Publication 3375.

Haque, S. A. (2006), 'Salinity problems and crop production in coastal regions of Bangladesh', *Pakistan Journal of Botany*, 38(5), 1359–65.

Hill, J. E., Williams, J. F., Mutters, R. G. and Greer, C. A. (2006), 'The California rice cropping system: agronomic and natural resource issues for long-term sustainability', *Paddy Water Environment*, 4, 13–19.

ICAR-CSSRI, IRRI-EC-IFAD. (2014), Final report of IRRI-EC-IFAD project on: Improved rice crop management for raising productivity in submergence-prone and salt-affected rainfed lowlands in South Asia, in Maji, B., Sarangi, S. K., Sharma, D. K. and Singh, S. (eds), ICAR-Central Soil Salinity Research Institute (ICAR-CSSRI), Regional Research Station (RRS), West Bengal, India.

IRRI. (1997), *Screening Rice For Salinity Tolerance*, G. B. Gregorio, D. Senadhira and R. D. Mendoza, IRRI Discussion Paper series No. 22.

Ismail, A. M., Johnson, D. E., Ella, E. S., Vergara, G. V. and Baltazar, A. M. (2012), 'Adaptation to flooding during emergence and seedling growth in rice and weeds, and implications for crop establishement', *AoB PLANTS* 2012, pls019, doi:10.1093/aobpla/pls019.

Joseph, R. (2008), 'Kerala Owns Pokkali',http://www.panap.net/en/r/post/rice-awards-aprja-2008-09/630.

Kumar, V. and Ladha, J. K. (2011), 'Direct seeding of rice: recent developments and future research needs', *Advances in Agronomy*, 111, 297–413.

Mahapatra, I. C., Sinhababu, D. P., Roy, J. K. and Behera, U. K. (2008), 'Rice-based farming systems for livelihood improvement of Indian farmers', in Singh, D. P., Dani, R. C., Rao, K. S., Rao, G. J. N., Nayak, S. K., Panda, D., Dash, R. N., Misra, A. K. and Ghosh, A. (eds), *Rice Research Priorities And Strategies For Second Green Revolution. Proceedings of the National Symposium On "Research Priorities And Strategies In Rice Production System For Second Green Revolution"*, 20–22 November 2007, Central Rice Research Institute, Cuttack (Orissa), India.

Maji, B. and Bandyopadhyay, B. K. (1991), 'Micronutrient research in coastal salt affected soils', *Journal of Indian Society of Coastal Agricultural Research*, 9(1 and 2), 219–23.

Maji, B. and Bandyopadhyay, B. K. (1992), 'Effect of submergence on Fe, Mn and Cu contents in some coastal rice soils of West Bengal', *Journal of Indian Society of Soil Science*, 40(2), 390–2.

Maji, B. and Bandyopadhyay, B. K. (1995), 'Characterization and classification of coastal soils of various pH groups in Sundarbans, West Bengal', *Journal of Indian Society of Soil Science*, 43(1), 103–7.

Mass, E. V. (1998), *Agricultural Salinity Assessment And Management Manual*. ASCE Publication.

Mass, E. V. and Grattan, S. R. (1999), Crop yields as affected by salinity, In Skaggs, R. W. and van Schilfgaarde, J. (eds), *Agricultural Drainage*, pp. 55–108. Agronomy Monograph 38, ASA, CSSA, SSA, Madison, WI.

Mishra, B. (1995), Annual Report, Central Soil Salinity Research Institute, Karnal, India.

Motamed, M. K., Asadi, R., Rezael, M. and Amiri, E. (2008), 'Response of high yielding rice varieties to NaCl salinity in green house circumstances', *African Journal of Biotechnology*, 7(2), 3866–73.

Nhan, D. K., Phap, V. A., Phuc, T. H. and Trung, N. H. (2011), Rice production response and technological measures to adapt to salinity intrusion in the coastal Mekong delta, http://www.mpowernetwork.org/Knowledge_Bank/Key_Reports/PDF/Research_Reports/Rice_response_to_salinity.pdf.

Pandey, M. P., Verulkar, S. B. and Sharma, D. (2010), 'Rice research: past achievements, present scenario and future thrust', *Indian Journal of Agricultural Sciences*, 80(6), 447–69.

Patra, B. C. (2013), Sustainable use of rice germplasm for livelihood security. In: Souvenir International Symposium on Sustainable Rice Production and Livelihood Security: Challenges and Opportunity, Association of Rice Research Workers, Cuttack, Odisha, India.

Prasad, R. (2008), 'Integrated plant nutrient supply (IPNS) for sustainable agriculture', *Indian Journal of Fertilisers,* 4(12), 71–90.

Sadana, U. S. (2002), *Problem Soils And Their Amelioration For Crop Production.* 21 February–13 March 2002. Center for advanced studies. Punjab Agricultural University, India.

Sarangi, S. K., Mahanta, K. K., Mandal, S. and Maji, B. (2012), 'Characterization of soils and cropping pattern of coastal areas of Haldia, Paradeep and Visakhapatnam ports', *Journal of Indian Society of Coastal Agricultural Research,* 30(2),12–18.

Sarangi, S. K., Mandal, S., Mahanta, K. K. and Maji, B. (2011), 'Assessment of system of rice intensification (SRI) during *Rabi* season in coastal salt-affected soils', *Journal of Indian Society of Coastal Agricultural Research,* 29(1), 54–8.

Sarangi, S. K. (2013), 'Modifying SRI to suit coastal salt affected soils of West Bengal', *LEISA INDIA,* 15(1), 36.

Sarangi, S. K., Maji, B., Singh, S., Sharma, D. K., Burman, D., Mandal, S., Ismail, A. M. and Haefele, S. M. (2014), 'Crop establishment and nutrient management for dry season (boro) rice in coastal areas', *Agronomy Journal,* 106(6), 2013–23, doi:10.2134/agronj14.0182.

Sarangi, S. K., Maji, B., Singh, S., Burman, D., Mandal, S., Sharma, D. K., Singh, U. S., Ismail, A. M. and Haefele, S. M. (2015), 'Improved nursery management further enhances the productivity of stress-tolerant rice varieties in coastal rainfed lowlands', *Field Crops Research,* 174, 61–70, http://dx.doi.org/10.1016/j.fcr.2015.01.011.

Sarangi, S. K., Burman, D., Mandal, S., Maji, B., Humphreys, E., Tuong, T. P., Bandyopadhyay, B. K. and Sharma, D. K. (2015a), Promising rice genotypes for the wet and dry seasons in coastal West Bengal. In Humphreys, E., Tuong, T. P., Buisson, M. C., Pukinskis, I., Phillips, M. (eds), *Revitalizing the Ganges Coastal Zone: Turning Science into Policy and Practices: Conference Proceedings CGIAR Challenge Program on Water and Food (CPWF),* Colombo, Sri Lanka.

Sarangi, S. K., Burman, D., Mandal, S., Maji, B., Tuong, T. P., Humphreys, E., Bandyopadhyay, B. K. and Sharma, D. K. (2015b), Reducing irrigation water requirement of dry season rice (boro) in coastal areas using timely seeding and short duration varieties. In Humphreys, E., Tuong, T. P., Buisson, M. C., Pukinskis, I., Phillips, M. (eds), *Revitalizing the Ganges Coastal Zone: Turning Science into Policy and Practices: Conference Proceedings CGIAR Challenge Program on Water and Food (CPWF),* Colombo, Sri Lanka.

Sarangi, S. K., Maji, B., Singh, S., Sharma, D. K., Burman, D., Mandal, S., Singh, U. S., Ismail, A. M. and Haefele, S. M. (2016), 'Using improved variety and management enhances rice productivity in stagnant flood-affected tropical coastal zones', *Field Crops Research,* 190, 70–81, http://dx.doi.org/10.1016/j.fcr.2015.10.024.

SeinnSeinn, M. U., Ahmad, M. M., Thapa, G. B. and Shreshta, R. P. (2015), Farmers' adaptation to rainfall variability and salinity through agronomic practices in lower Ayeyarwady Delta, Myanmar', *Journal of Earth Science and Climate Change,* 6, 258, doi: 10.4172/2157-7617.1000258.

Sharma, G. D., Thakur, R., Chouhan, N. and Keram, K. S. (2015), 'Effect of integrated nutrient management on yield, nutrient uptake, protein content, soil fertility and economic performance of rice (*Oryza sativa* L) in a Vertisol', *Journal of the Indian Society of Soil Science,* 63(3), 320–6.

Shereen, A., Mumtaz, S., Reza, S., Khan, M. A. and Solangi, S. (2005), 'Salinity effects on seedling growth and yield components of different inbred rice lines', *Pakistan Journal of Botany,* 37(1), 131–9.

Singh, C. (1983), *Modern techniques of raising field crops,* Oxford & IBH Publishing Co. Pvt. Ltd., New Delhi, India.

Spencer, D. F. and Linquist, B. A. (2014), 'Reducing rice field algae and cyanobacteria abundance by altering phosphorus fertilizer applications', *Paddy Water Environment,* 12, 147–54, doi: 10.1007/s10333-013-0370-6.

U.S. Salinity Laboratory. (1954), Diagnosis and improvement of saline and alkali soils, Handbook 60, U.S. Government Printing Office, Washington DC.

www.thehindu.com/todays-paper/tp-features/tp-sundaymagazine/the-shrinking-fields-of-pokkali/article2275531.ece. Accessed on 24 September 2016.

www.fao.org/3/a-i3084e/i3084e18.pdf. Accessed on 15 September 2016.
Yadav, B., Khamparia, R. S. and Kumar, R. (2013), 'Effect of zinc and organic application on various zinc fractions under direct- seeded rice in Vertisol', *Journal of the Indian Society of Soil Science*, 61, 128–34.
Zeng, L. and Shannon, M. C. (2000), 'Salinity effects on the seedling growth and yield components of rice', *Crop Science*, 40, 996–1003.

Chapter 4

# Dry-seeded and aerobic rice cultivation

*T. Parthasarathi, Ben-Gurion University of the Negev, Israel; M. Kokila, D. Selvakumar, V. Meenakshi and A. Kowsalya, Tamil Nadu Agricultural University, India; K. Vanitha, Tamil Nadu Rice Research Institute, Tamil Nadu Agricultural University, India; A. Tariq, University of Copenhagen, Denmark; A. Surendran, Rutgers University, USA; and Eli Vered, Netafim Irrigation Ltd, Israel*

1 Introduction
2 Dry direct seeded rice cultivation
3 Good agricultural practices (GAP) in dry direct seeding
4 Aerobic rice cultivation
5 Good agricultural practices (GAP) in aerobic rice cultivation
6 Conclusion and future trends
7 Where to look for further information
8 References

## 1 Introduction

Rice is one of the most important food crops worldwide and is cultivated over 100 countries. Rice has traditionally required large volumes of water for its cultivation. It has been estimated that it is responsible for up to 80% of total water used in agriculture (Tripathi, 1990). An increasing population and depletion of water resources are major challenges. Finding an efficient irrigation strategy that would minimize water consumption in rice cultivation is a key objective in dealing with the potential future competition for and scarcity of water as a resource. Several water-saving techniques have been developed, including dry-seeded rice and aerobic rice cultivation. This chapter explores these techniques and good agricultural practices (GAP) for optimizing their use.

## 2 Dry direct seeded rice cultivation

Direct seeding (sometimes known by the acronym DSR (direct seeded rice)) is an alternative to more traditional transplanting techniques in rice cultivation. It is particularly suited to rainfed regions. There are two basic methods of direct seeding or DSR, wet or dry. In dry direct seeding, the soil is not puddled or flooded at the time of planting. The dry rice

http://dx.doi.org/10.19103/AS.2016.0003.15

**Figure 1** View of dry-seeded rice at different growth phases.

seeds are sown on a dry or moist non-puddled soil surface, and then incorporated by ploughing or harrowing (Fig. 1). Dry direct seeding has become an important method of rice cultivation technique in rainfed upland, lowland and flood-prone areas in countries such as India, the Philippines, Vietnam, Thailand, Korea, America, Japan and the sub-Saharan Africa, and is gaining importance in irrigated areas where water is becoming scarce (Kumar and Ladha, 2011). The main advantages of dry direct seeding are potential savings in water and labour costs of up to 30 and 40%, respectively (Pandey et al., 2012; NB Kamboj et al. (2012) quote savings of 20–25%). About 10% of the world's rice cultivation is by dry seed upland rice cultivation. In Asia, West Africa and Latin America, upland rice is an important rice growing system.

The potential advantages of the technique mean there has been significant research into its application. Kumar and Ladha (2011) have reviewed the range of direct seeding techniques, including dry direct seeding. Detailed guidance on practical application is provided by Gupta et al. (2006) and Gopal et al. (2010). A recent review of direct seeded rice systems in Asia is by Kamboj et al. (2012). There are many potential variations in application methods such as:

- Planting: by broadcasting, dibbling or drilling
- Tillage: conventional or zero-till

Kumar and Ladha (2011) have identified a number of key issues in successful dry direct seeding:

- Effective land levelling (e.g. to ensure effective irrigation)
- Good crop establishment (to ensure optimum plant density e.g. via optimizing planting date, seed treatment, seed rate and planting depth)
- Precise water management (especially in the early stages of growth to ensure good rooting and seedling growth)
- Effective weed management (to avoid competition with young rice plants).

Kamboj et al. (2012) also identified land levelling and weed management as critical to success. Land levelling is reviewed by Jat et al. (2006, 2009). Irrigation is discussed by Bouman et al. (2007). Weed management is assessed by Rao et al. (2007). Other factors that can limit yields in dry direct seed cultivation identified by Kamboj et al. (2012) and Gopal et al. (2010) include poor timing and depth of sowing, inappropriate seeding rates, insufficient fertilizer application and weed competition. These can be addressed by establishing GAP in dry direct seed cultivation. These are discussed in the next section.

# 3 Good agricultural practices (GAP) in dry direct seeding

The GAP guidelines are developed for Cauvery Delta Zone of Tamil Nadu, India, by the scientists from Tamil Nadu Rice Research Institute and IRRI (Sudhir Yadav et al., 2014). However, these guidelines may be applicable for all rice-planting areas of the world except the sowing date and season.

The following guidelines set out suggested best practice in dry-seeded cultivation:

i Tillage: Dry-seeded rice can be sown into dry tilled soil ('conventional tillage') or into non-tilled soil ('zero tillage').

ii Land levelling: The fields must be accurately levelled. Good land levelling helps ensure high yield and reliable production of dry-seeded rice. This is best achieved using laser-assisted land levelling.

iii Cultivar: Many of the inbred varieties and hybrids bred for puddled transplanted rice have also been found to be suitable for dry-seeded rice. Seeding with strip tillage using a seeder with tiller powered by a 2-wheel tractor can be used for sowing. A seed rate of 25–35 kg/ha (using good-quality seeds with more than 95% germination) is optimal for a row spacing of 20 cm sown with a multi-crop planter.

iv Seed hardening of paddy seeds with 1% KCl is recommended for dry-seeded rice to empower the tender young seedlings to withstand drought in the early phase. The combination treatment will protect the seed from both soil-borne fungi and insects. The use of imidacloprid, alone or in combination with tebuconazole, is also suitable for treating dry seeds. Mix the chemicals in 15 mL of water/kg seed.

v Sowing: Sowing date for Kuruvai: 20 May to 25 June. The sowing date for Samba: 10 August to 10 September (Specific for Cauvery Delta Zone of Tamil Nadu).

vi Spraying of herbicides: Use clean water and a plastic container to make spray solution as herbicides bind with suspended soil particles and metal surfaces (e.g. iron bucket). Use a multiple-nozzle boom fitted with flat-fan nozzles for full coverage. Use protective clothing when applying herbicides. Ensure that the soil surface is moist before applying.

vii Pre-emergence herbicide: Use a high volume of water when pre-emergence and herbicide is applied to moist soil. Apply irrigation 24 h after post-emergence herbicide application if the soil is not moist.

viii Fertilizer: Avoid basal dose of urea at sowing. Use a minimum of three splits of N fertilizer.

ix Irrigation: Keep the soil in the seed/root zone moist during establishment. Keep the soil close to saturation from the start of heading to the start of grain filling. Dry-seeded rice needs an assured water supply for the first 3 weeks after sowing for good establishment.

The dry-seeded rice is becoming an attractive option for farmers due to the elimination of the labour requirement for nursery preparation and maintenance, pulling out and transport of seedlings, and transplanting. Because the soil is not puddled, it also has a lower water requirement for crop establishment. Furthermore, the total crop cycle is shorter by 10–15 days because of the absence of transplanting shock. These features are of major importance for the increasing scarcity of water for irrigation in the area. Small farmers, as well as large farmers, can readily adopt direct seeded rice, provided that the required machinery is locally available.

## 3.1 Cultivation season

May/June to September/October and August/September to January are the suitable seasons for dry-seeded rice cultivation. Seeding after the first rains but just before the main onset of the rainy season (e.g. the monsoon) is best to avoid problems in land preparation and seeding in very wet conditions.

## 3.2 Soil and field preparation

Seeds establish best in sandy loam to heavy clay soil. Plough the land after the first rain followed by harrowing (e.g. with a two disc harrow). For precision sowing, uniform irrigation, uniform crop stand, improved weed control and fertilizer use efficiency, land levelling is critical in dry-seeded rice cultivation. This can best be achieved by laser-assisted land levelling (Fig. 2).

## 3.3 Cultivars

To achieve an early harvest and reduce water requirements, short duration cultivars and hybrids are preferred for dry seed rice cultivation. The short duration varieties, ADT 43, ADT 47, ADT 36, ADT 37, CO 47, CO51 and the hybrid CORH 3 are mostly used in Tami Nadu. Other varieties are listed in Gopal et al. (2010) and Kamboj et al. (2012).

## 3.4 Seed treatment

The following seed-priming treatments are recommended:

- **Dormancy breaking:** Before sowing, dormancy of freshly harvested rice seed should be broken by heat treatment or drying the seed under direct sunlight for 1 or 2 days.

**Figure 2** Land leveling using laser-assisted land leveler.

- **Seed hardening:** Seed hardening with 1% KCl enhances drought resistance in young seedlings.
- **Seed treatment for fungicides and insecticides:** Treat the seed with Imidacloprid (Gaucho 350 FS®- 3 ml/Kg) or combination of Tebuconazole (Raxil Easy®- 3ml/Kg) and Imidacloprid (Gaucho 350 FS®- 3ml/Kg)

## 3.5 Seed rate and sowing depth

- **Seed rate**: 25–35 Kg/ha.
- **Sowing depth:** 1–2 cm.

## 3.6 Sowing

Farmers have practised the following three types of sowing method (Fig. 3):

### Broadcasting on to dry soils:

- Uniform broadcasting of 60–80 kg of rice seed for one hectare of land

**Broad casting** (www. Thehindu.com) | **Drilling** (www.un-csam.org) | **Dibbling** (www. knowledgebank.irri.org)

**Figure 3** Sowing operation in direct seeded rice cultivation.

**Table 1** Direct seeding of rice: a comparison

| Details | Hand sowing | Sowing with country plough | Sowing by seeding drill |
|---|---|---|---|
| Time | 1 hour | 5 hours | 45 minutes |
| Seed rate | 35 kg | 35 kg | 15 kg |
| Persons required | 1 | 2 | 1 |
| Cost (INR) | 1000 | 1200 | 600 |

- After broadcasting, the seeds were covered with soil by using spike-tooth harrow
- Depth of seeding: surface or 0–3 cm

## Drilling:

- Drill 80–100 kg of seeds in one hectare
- The seeds were drilled manually or mechanically at less than 10–15 cm depth in both dry and moist soil with simultaneous application of fertilizers
- Depth of seeding: 1–3 cm

## Dibbling:

- The best sowing method for mountain slopes and hill area where ploughing and harrowing are difficult.
- The seeds were dropped into the hole opened by long wood or bamboo stick.
- Depth of seeding: 2–3 cm

## 3.7 Irrigation

The irrigation management is to minimize the use of irrigation water while maintaining yield because of increasing water scarcity and/or the high cost of pumping groundwater. Rice does not need to be continuously flooded for good growth and yield. It can be grown with periodic irrigation, allowing the soil surface to dry for a few days between irrigations (alternate wetting and drying). This reduces the amount of irrigation water needed to grow the crop, and is recommended where there is water scarcity, and to reduce irrigation costs. However, if the soil becomes too dry too often between irrigations, the rice crop will suffer and a loss of yield will occur. Therefore, irrigation needs to be managed carefully.

The irrigation requirement for dry deeded rice depends very much on the weather and the soil type. The lighter (less clayey, more sandy) the soil, the more frequently it will need irrigation in the absence of adequate rain. Water supply is must during the first 3 weeks after sowing for good crop establishment. When dry-seeded rice is established in hot and dry conditions, irrigation is done on the day of sowing or within 1 day after sowing. The next one to two irrigations are required at intervals of 3–5 days to keep the soil moist in the root zone. During the active tillering phase, and the heading to grain-filling stage, the topsoil (0–15 cm) should be kept close to saturation, with irrigation applied as needed. At other stages, the topsoil can be allowed to become drier, but never to the degree that the leaves show signs of rolling (no longer flat) in the early morning. For clayey soils, the appearance of hairline cracks on the soil surface is a general indication of the need to irrigate.

## 3.8 Fertilizer use

The fertilizer requirement will vary according to the soil nutritional status. The basal application of nitrogen should be avoided. The recommended fertilizer for dry-seeded upland rice is $PO_5$ at 50 kg/ha, $K_2O$ at 50 kg/ha, $ZnSO_4$ at 25 kg/ha and 100–120 kg N/ha. The nitrogen requirement can be determined by a leaf colour chart. Table 2 describes ways of dealing with mineral deficiencies.

## 3.9 Thinning

To avoid overcrowding, seedlings should be thinned to 2–4 seedling per hill in dibbed seed and only one seedling per hill in drilled seed.

## 3.10 Weed management

Weeds are a major challenge in dry seed rice cultivation. If rice is seeded in rows in dry soil, hand tools or a tractor-driven rototiller can be used to bury weeds.

**Table 2** Deficiency and remedy measures of dry-seeded rice

| Symptom | Deficiency | Remedy |
|---|---|---|
| Yellowing of seedling at 1–2 weeks after emergence | Iron | • Spray 1% ferrous sulphate with 0.2% citric acid<br>• In between rows, apply 25 kg of iron chelates and urea per hectare |
| At 3–4 weeks after emergence scattered light yellow spots on the older leaves followed by leaf drying | Zinc | • Soil application of zinc sulphate @ 25 kg/ha<br>• Foliar spraying of 1% zinc sulphate with 0.5% lime |

**Figure 4** Manual weeding in dry-seeded rice cultivation.

The first weeding should be at 2–3 weeks after seedling emergence followed by second weeding at 6–7 weeks after seedling emergence either by manual weeding or chemical treatment (Fig. 4).

Alongside manual methods, cultural techniques for weed control include:

- Stale seed beds
- Well-spaced seedlings (to encourage vigorous early growth)
- Surface mulch (e.g. using residues of previous crops) and cover crops (e.g. sesbania which also adds nitrogen to the soil and can be used for fodder).

The appropriate use of pre-plant, pre-emergence and post-emergence herbicides is summarized in Gopal et al. (2010) and Kamboj et al. (2012).

## 3.11 Roguing (removal of off types)

To improve grain purity, from tillering to harvest identify and remove off types based on morphological characters. Off types can be identified by:

- Plant height
- Leaf colour
- Tillering nature
- Panicle length
- Awn
- Grain colour
- Grain size
- Grain length
- Crop duration

## 3.12 Harvesting and threshing

Plants are ready to harvest 30–45 days after flowering. The indication for paddy harvest is yellowing of paddy straw and grains. Either a manual or mechanical method can be used for paddy harvesting. To avoid losses, thresh immediately after harvesting.

# 4 Aerobic rice cultivation

Dry direct seeding is one form of aerobic rice cultivation. Aerobic cultivation is a water-saving production system, in which rice is grown in well-drained, non-puddled and non-saturated soil conditions. Aerobic rice production systems can be rainfed or irrigated. Irrigation techniques include furrow irrigation using raised beds or micro irrigation. This production system typically includes intermittent irrigation during the crop-growing period. Aerobic cultivation has traditionally been used in upland areas, providing stable, though low yields. The following areas are suitable for aerobic rice cultivation:

- **Favourable uplands:** Flat type lands where monsoon with or without supplemental irrigation is sufficient to bring soil moisture content close to field capacity and where farmers have access to external inputs such as fertilizers.
- **Rainfed lowlands:** Fields on upper slopes or terraces in undulating lands which support rainfed conditions during the growing season.
- **Irrigated lowlands:** Areas that do not have access to sufficient water to keep rice fields flooded for a substantial period of time.

The main advantages of aerobic cultivation include (provide reference):

- Reduced water requirement (of up to 50%)
- Lower labour requirement and costs
- Need for fewer seeds
- Improved efficiency in fertilizer use
- Retention of soil texture and quality

The main challenges facing aerobic rice cultivation include:

- Lower yields compared to traditional transplanting (Peng et al., 2006).
- Competition from weeds (Sunyob et al., 2012).
- Difficulties in marketing aerobic varieties (Bouman et al., 2007).

These can be addressed by appropriate GAP.

# 5 Good agricultural practices (GAP) in aerobic rice cultivation

The different growth stages of aerobic rice (shown in Fig. 5) are:

1. Seedling phase (20 days after sowing – DAS)
2. Active tillering phase (40–55 DAS)
3. Flowering phase (80 DAS)
4. Grain-filling phase (80–110 DAS)

**Figure 5** Field view of aerobic rice at different phenophases (Image courtesy: 1&4. Parthasarathi et al. (2013); 2. ugent.be; 3. agriculturewire.

## 5.1 Variety selection

Suitable aerobic rice varieties selected for better yield and quality are shown in Table 3. Aerobic rice varieties are adapted to aerobic soils and have higher yield potential than traditional upland varieties, achieving yields of up to 5–6 t $ha^{-1}$ (Nie et al., 2014).

## 5.2 Seeds and seed treatment

Before sowing, seeds are treated with bio fungicide (*Pseudomonas fluorescens*) at the rate of 200 g 10 $kg^{-1}$ of seeds. Incubated seeds are dry-sown by hand dibbling at 3-cm depth in rows at a seeding rate of 30 kg $ha^{-1}$. Seeds are sown directly in the field at a spacing of 20 × 10 cm. Maintenance of optimum plant population in the field is 50 hills per $m^2$.

## 5.3 Land preparation

Fields are dry ploughed and harrowed using tractor-drawn cultivator. To achieve fine soil tilth, a rotavator can be used. Levelling of field can be done using a laser land leveller. Ridges and furrows or raised flat beds are prepared manually/using ridge or bed formers. Intercropping with sesbania can improve the soil fertility in aerobic rice.

## 5.4 Weed management

Aerobic rice cultivation leads to alternate wetting and drying of soil that enhances weed growth. Increased weed populations in aerobic rice fields compete for soil moisture and

**Table 3** Rice varieties suitable for aerobic rice cultivation

| S.No. | Location | Varieties | Reference |
|---|---|---|---|
| *Sub-tropical environment* | | | |
| 1. | India | PMK (R) 3,<br>ADT(R) 45,<br>Apo,<br>Pyari,<br>Sahabhagi dhan, Annada,<br>Mas-26, Puspa,<br>Pusa Hybrid 10,<br>Proagro 6111 (Hybrid),<br>Pusa 834 and<br>IR55423-01 (Apo1) | Vanitha and Mohandass (2013)<br>Parthasarathi et al. (2013)<br>Govindarasu et al. (2015)<br>Jana et al. (2013), Sunil and Shankaralingappa (2014),<br>Bouman et al. (2007) |
| 2. | The Philippines, Laos | B6144F-MR-6-0-0, Apo (PSBRc9), UPLRi5 and PSBRc80 | Bouman et al. (2007) |
| *Temperate environment* | | | |
| 1. | China | HD277, HD297 and HD502 | Bouman et al. (2007) |

nutrients with the crop. The main weeds in aerobic rice cultivation are *Echinochloa colona, Leptochloa chinensis, Digitaria adscendens, Cyperus iria* and *Eleusine indica.* Techniques to allow rice crops to outcompete weed, such as closer spacing of rice to encourage a denser canopy to inhibit weed growth, are important tools in integrated weed management programmes for aerobic rice (Sunyob et al., 2012). Application of pre-emergence herbicide pendimethalin at 0.75 kg $ha^{-1}$ followed by hand weeding twice or mechanical weeding on 25 and 45 DAS can control weeds effectively. An alternative is pendimethalin application at 1.0 kg $ha^{-1}$ along with single type sweep weeding on 45 DAS.

## 5.5 Pest control: nematodes

Continuous mono-cropping of aerobic rice is favourable to an increase in root knot nematode populations (Nie et al., 2014). An increase in nematode infestation in aerobic rice induces root galls which lead to yellowing of leaf, stunted growth and ultimately reduction in grain yield (Fig. 6.).

Nematodes can be dealt with by:

- Seed treatment with *Pseudomonas* powder formulation @ 10g per Kg of rice seed.
- Application of *P. fluorescens* in the field with FYM. Pestigation of *Pseudomonas* liquid formulation @ 1 ml per 10 litres of water.
- In the case of severe nematode infestation, application of neem cake @ 300 kg per acre.

## 5.6 Fertilizer management

A leaf colour chart (LCC) can be used to assess fertilizer requirements. Fertilizer dose: 150: 50: 50 kg NPK/ha. N in four stages:

**Figure 6** Appearance of nematode-infested root and field in aerobic rice (Image courtesy: Parthasarathi et al., 2012).

**Figure 7** Leaf colour chart' before (LCC); Image source: IRRI.

- 20% at 15 DAS,
- 30% at tillering
- 30% at Panicle Initiation
- 20% at flowering

Nitrogen management at LCC cv. of 4 can be followed (Fig. 7). Basal application of $ZnSO_4$ at 25 kg/ha and $FeSO_4$ at 50 kg/ha is desirable wherever zinc and iron deficiency were noted or apply TNAU Rainfed rice MN mixture @12.5 kg/ha as EFYM at 1:10 ratio incubated for 30 days at friable moisture. Need-based foliar application of 0.5% ZnSO4 and 1% FeSO4 may be taken up at tillering and PI stages.

For the optimal use and management of fertilizers, site-specific nutrient management (SSNM; http://irri.org/ssnm/) can be used. In the absence of knowledge on SSNM, 70–90 kg N $ha^{-1}$ could be a useful starting point to obtain a yield of 4–6 t $ha^{-1}$. The first split can best be given at 10–12 days after emergence, the second at active tillering, and the third at panicle initiation stages.

## 5.7 Irrigation

Irrigation can be achieved through flooding or flash flooding to bring the soil water content to field capacity. Aerobic rice cultivation requires an irrigation scheduling of 1.0 IW/CPE

**Table 4** Comparison of irrigation methods on water use and yield of rice (CPG, 2012)

| Methods of irrigation | Water requirement (cm) | % watersaving | Yield kg ha$^{-1}$ | % increase in yield | Water use efficiency (kg ha mm$^{-1}$) |
|---|---|---|---|---|---|
| Conventional aerobic rice | 74.30 | 38.10 | 4747 | – | 6.39 |
| Surface drip | 61.90 | 48.40 | 5940 | 14.20 | 9.60 |
| Sub-surface drip | 61.90 | 48.40 | 6227 | 19.80 | 9.74 |
| Transplanted | 120.00 | – | 5200 | – | 4.33 |

ratio to 3.0-cm depth throughout crop growth. Initial irrigation is to be given immediately after sowing, followed by further irrigation 3 DAS and subsequent irrigation given based on the pan evaporation values from USWB Open Pan Evaporimeter (OPE). The optimum soil water condition would be maintained around field capacity (30–40 kPa or 0.3–04 bar soil moisture potential) across the growth stages. The total water requirement of aerobic rice is 650 mm.

Table 4 compares different irrigation techniques. Drip irrigation techniques can be very effective in aerobic rice cultivation. Advantages include:

- Precise water application with 25% increase in water saving over conventional irrigation
- Improvement in nutrient uptake and saving of fertilizers by scheduled fertigation
- Increase in grain yield
- Less manpower required
- Lower incidence of pests (and resulting diseases) because moisture is more precisely targeted at and taken up by plants
- Application of nematicides and biopesticides through drip systems.

For aerobic rice condition, adoption of surface drip irrigation laid out at 0.8 m lateral spacing with 0.3 m dripper distance scheduled at 125% pan evaporation is recommended for clay soil condition. Evaluation of drip irrigation studies revealed that 0.8 m lateral spacing

**Surface Drip Irrigation** **Sub-surface Drip Irrigation**

Image courtesy: Parthasarathi *et al.* (2013)

**Figure 8** Comparison between Surface and Sub-surface Drip Irrigation.

**Figure 9** Sub-surface drip irrigated rice (Image source: Parthasarathi et al., 2013).

along with 1.0 lph low discharge rate drippers improved the grain yield (Parthasarathi et al., 2015) in clay loamy soil (Fig. 8). In the case of sandy soil conditions, 150% pan evaporation is recommended (Fig. 9).

Irrigation can also be combined with fertilizer application (known as fertigation). NPK fertigation (STCR based) + Biofertigation of Azophosmet 500 mL $ha^{-1}$ during panicle initiation and flag leaf stages are recommended. In the case of sandy soil condition 150% pan evaporation has been recommended (CPG 2012, TNAU).

# 6 Conclusion and future trends

Dry seeding is a planting method and water is managed which can range from flooding to alternate wetting and drying to rainfed. Before dry seeding, the soil may have undergone dry land preparation by conventional tillage or through zero-till dry-seeded rice. In dry seeding, the soil is not puddled or flooded at the time of seeding.

In the irrigated areas of India where dry-seeded rice is being introduced, the goal is to achieve similar yield to that of puddled transplanted rice, while minimizing the irrigation input to the rice. Therefore, the dry-seeded rice is not continuously flooded but is irrigated frequently to avoid yield loss. This means keeping soil water content of the root zone (~0–20 cm) between saturation and field capacity much of the time.

Aerobic rice is grown under conditions where the root zone is maintained in a non-flooded or aerobic condition for most of the time. The soil water content in a significant part of the root zone is allowed to dry down to less than field capacity. This comes with a yield penalty due to water loss stress. To reduce the penalty, special cultivars are being developed with greater drought tolerance, but at the same time with the ability to respond to inputs, especially nutrients.

The System of Aerobic Rice involves growing these specially developed cultivars under aerobic conditions. At present, the yield potential of aerobic cultivars (under well-water conditions) is less than the yield potential of high yielding cultivars bred for puddled transplanted systems. Aerobic rice is usually dry-seeded into dry or moist tilled soil like in dry-seeded rice. Aerobic conditions can also be introduced into fields after puddling and transplanting.

# 7 Where to look for further information

## Websites

http://agriculturewire.com/aerobic-rice-cultivation-reduces-water-usage/
http://irri.org/our-work/research/better-rice-varieties/aerobic-rice
http://irri.org/ssnm/
http://www.knowledgebank.irri.org/step-by-step-production/growth/water-management/aerobic-rice
http://www.molecularbiotechnology.ugent.be/group-research-4.html

# 8 References

Bouman, B. A. M., R. M. Lampayan and T. P. Tuong. 2007. Water management in irrigated rice: coping with water scarcity. Los Baños (Philippines): International Rice Research Institute. 54p.

CPG – Crop Production Guide. 2012. Aerobic rice. Tamil Nadu Agricultural University, p. 388.

Gopal, R., R. Jat, R. Malik, V. Kumar, M. Alam, M. Jat, M. Mazid, Y. Saharawat, A. McDonald and R. Gupta. 2010. 'Direct Dry Seeded Production Technology and Weed Management in Rice-based Systems', *Technical Bulletin*, International Maize and Wheat Improvement Center, New Delhi, India.

Govindarasu, R., K. Paramasivam, S. Nadaradjan, N. Shashidhara and M. Vengatesh. 2015. Aerobic rice: a production system for water scarceness. *AEIJST*, 3 (6): 2348–6732.

Gupta, R. K., J. K. Ladha, S. Singh, R. Singh, M. Jat, Y. Saharawat, V. Singh, S. Singh, G. Singh, G. Sah, M. Gathala and R. Sharma. 2006. 'Production Technology for Direct Seeded Rice', *Technical Bulletin Series 8*, Rice-Wheat Consortium for the Indo-Gangetic Plains, New Delhi, India.

Jana, K., G. K. Mallick and S. Ghosh. 2013. Yield of aerobic rice affected by high temperature stress during summer season – A study from red and laterite zone of West Bengal, India. *J. Appl. Nat. Sci.* 5 (2): 394–6.

Jat, M., R. ChandGupta, S. Sharma and M. Gill. 2006. 'Laser land levelling: a precursor technology for resource conservation', Rice-Wheat Consortium Technical Bulletin 7, New Delhi, India.

Jat, M., M. Gathala, J. Ladha, Y. Saharawat, A. Jat, K. Vipin, S. Sharma, V. Kumar and R. Gupta. 2009. Evaluation of precision land levelling and double zero-till systems in the rice-wheat rotation, *Soil Till. Res.* 105: 112–21.

Kamboj, B. R., A. Kumar, D. Bishnoi, K. Singla, V. Kumar, M. Jat, N. Chaudhary, H. Jat, D. Gosain, A. Khippal, R. Garg, O. Largwal, S. Goyal, N. Goyal, A. Yadiv, D. Malik, A. Mishra and R. Bhatia. 2012. 'Direct Seeded Rice Technology in Western Indo-Gangetic Plains of India: CSISA Experiences', *CSISA Technical Bulletin* (CSISA, IRRI and CIMMYT), New Delhi, India.

Kumar, V., and J. K. Ladha. 2011. Direct seeded rice: recent developments and future research needs, *Adv. Agron.* 111: 297–413.

Nie, L., S. Fahad, S. Peng, J. Huang and K. Cui. 2014. Causes of soil sickness associated with aerobic rice continuous monocropping. *Int. J. Agric. Biol.* 16: 431–4.

Nie, L., S. Peng, M. Chen, F. Shah, J. Huang, K. Cui and J. Xiang. 2012. Aerobic rice for water-saving agriculture. *A review. Agronomy for Sustainable Development, Springer Verlag/EDP Sciences/INRA*, 32 (2): 411–18. DOI:10.1007/s13593-011-0055-8.

Pandey, S., N. Suphanchaimat, Ma. L.Velasco. 2012. The patterns of spread and economics of a Labor-saving innovation in rice production: the case of direct seeding in Northeast Thailand. *Quart. J. Inter. Agric.* 4: 333–56.

Parthasarathi, T., K. Vanitha, P. Lakshamanakumar and D. Kalaiyarasi. 2012. Aerobic rice-mitigating water stress for the future climate change. *Int. J. Agron. Plant Prod.* 3 (7): 241–54.

Parthasarathi, T., K. Vanitha, S. Mohandass, S. Senthilvel and E. Vered. 2015. Effects of impulse drip irrigation systems on physiology of aerobic rice. *Ind J Plant Physiol.* DOI 10.1007/s40502-014-0131-6.

Parthasarathi, T., S. Mohandass, S. Senthilvel and E. Vered. 2013. Effect of various micro irrigation treatments on growth and yield response of aerobic rice. *Int. Agric. Eng. J.* 22 (4): 49–62.

Rao, A., D. Johnson, B. Sivaprasad, J. Ladha and A. Mortimer. 2007. Weed management in direct-seeded rice, *Adv. Agron.* 93: 153–255.

Sudhir Y., R. Ganeshamoorthy, E. Humphreys, R. Rajendran, V. Ravi, F. Mussgnug, V. Kumar, B. S. Chauhan, T. Ramesh, B. R. Kamboj, M. Gathala, R. K. Malik, M. L. Jat and A. J. McDonald. 2014. Guidelines for Dry Seeded Rice (DSR) in the Cauvery Delta Zone, Tamil Nadu, India.

Sunil, C. M., and B. C. Shankaralingappa. 2014. Rooting behaviour of Aerobic rice under integrated package of Agrotechniques. *Asian J. of Agric. Res.* 8 (2): 105–13.

Sunyob, N. B., A. S. Juraimi, M. M. Rahman, M. P. Anwar, A. Man and A. Selamat. 2012. Planting geometry and spacing influence weed competitiveness of aerobic rice. *J. Food Agric. Environ.* 10 (2): 330–6.

Surajit K. De Datta. 1981. *Principles and Practices of Rice Production. The International Rice Research Institute*, Los Baños: The Philippines.

Tripathi, R. P. 1990. Water requirement in rice-wheat system. RiceWheat Workshop, Modipuram.

Vanitha, K., and S. Mohan dass. 2013. Influence of drip fertigation on root characters and grain yield of aerobic rice (*Oryza sativa* L.). *J. Appl. Biosci.* 39 (2): 115–20.

Chapter 5

# Processing rice straw and husks as co-products

*Nguyen Van Hung, Carlito Balingbing, James Quilty, Bjoern Ole Sander, Matty Demont and Martin Gummert, International Rice Research Institute (IRRI), The Philippines*

1 Introduction: rice by-products
2 Availability of rice straw and husks
3 In-field rice residue processing and management
4 Off-field rice residue processing and management
5 Using rice residues for energy production
6 Non-energy uses of rice residues
7 Summary
8 Future trends
9 Where to look for further information
10 References

## 1 Introduction: rice by-products

Each kilogram of milled rice produced results in roughly 1.4 kg of rice straw, 0.28 kg of husk and 0.15 kg of rice bran as by-products from rice production during harvesting and milling. Rice straw is separated from the grains when the plant is threshed either manually or mechanically using stationary threshers or combine harvester, while rice husk is a by-product produced during milling. Rice bran, also a by-product of milling, which has multiple established uses as animal feedstock and in the food and cosmetics industries, is not discussed further in the chapter. One of the most important differences between the by-products, rice straw and rice husk, is the location of their production within the rice value chain. Rice husk is produced at a centralized location (the rice mill) and is relatively uniform in moisture content, composition and particle size. Rice straw is produced at the point of threshing, which, where manual threshing or stationary threshers are used, is a central point within the field, or where combine harvesters operate the straw is distributed across the harvested field. In addition to being an uncentralized source of biomass, the moisture content (MC), composition, and particle size of rice straw vary greatly depending on a range of environmental and management variables. Open-field burning has always

http://dx.doi.org/10.19103/AS.2016.0003.18

been an option for rice farmers, but was not practised much in traditional rice production systems.

In traditional rice-based cropping systems, the rice plants are cut and are then carried to a central threshing location, where the grains are separated from the straw in the threshing process. The straw can subsequently be used off-field for animal fodder or bedding; as a substrate for mushroom production; as a construction material; as a raw material for handicrafts; or as biofuel, when mixed with animal manure. It is also sometimes brought back to the field either unprocessed or as compost and then incorporated.

Rice cropping system intensification with a shorter turnaround time between crops and higher yields means that large amounts of straw can be produced in a relatively short period of time. The rapid introduction of combine harvesters is a game changer in these systems, as these machines leave the straw spread over the field. Manual collection is unprofitable because of high labour costs. Moreover, incorporation into the soil is also undesirable in intensive systems with two to three crops per year. The limited period available for decomposition means that incorporation of rice straw can have negative impacts on the subsequent crop, and will increase greenhouse gas emissions (GHGE). As a result of combine harvesters leaving straw in the field after harvest, open-field burning of straw has increased dramatically over the last decade.

All stakeholders involved – from farmers to policymakers – have only limited knowledge on (i) alternative management options and (ii) the negative effects of burning on health, climate and soil nutrition. Existing legislation prohibiting field burning is usually not enforced by government agencies because of the lack of political will or power to implement unpopular measures.

The detrimental impact of open straw burning on human health has long been a concern for policymakers and scientists. The smoke comprises a wide range of pollutant gases and black carbon that collectively impair air quality and visibility. Over the Indian subcontinent, the vast amount of straw burning contributes to the so-called 'Asian Brown Cloud', a layer of polluted air appearing after each monsoon season. In Southeast Asia (SEA), the haze caused by straw burning has often been reported as a threat to health and sometimes even to aircraft operations, especially during the dry El Niño years, when straw burning is often aggravated by fires in forests and peatlands.

The impacts of straw burning on GHGE are much less documented and quantified than the effects of air pollution. The complete combustion of straw to $CO_2$ does not constitute a net GHGE because this process only completes an annual carbon cycle that has started with photosynthesis a few months earlier. Strictly speaking, the net GHGE is independent of the nature of the process of $CO_2$ formation, which could be either biological (aerobic decomposition) or chemical (combustion). However, field burning of straw is by no means complete and entails in most instances a large proportion of non-$CO_2$ gases. In the first instance, this applies to carbon monoxide, which is a highly toxic gas and an important indicator of air quality, but has a low global warming potential (GWP = 1.9) due to its short atmospheric lifetime of only two months. The most important GHGE from straw burning is methane ($CH_4$), which has a relatively high GWP of 25. The burning also emits nitrous oxide ($N_2O$), but the amount produced is small due to the low nitrogen content of straw (Gadde et al., 2009; Sander et al., 2014).

Another reason why farmers burn the straw is that they cannot capture value-adding opportunities from selling straw or straw products, despite the emerging market potential of this by-product. For example, in Vietnam the price for rice straw increased from US$20 per ton in 2012 to US$80 per ton in 2014 as an indirect effect of increased combine harvester use. Other residues are often marketed in the form of pellets, such as rice husk pellets which sell for US$80 per ton, but farmers do not have access to similar technologies for collecting and processing straw. They also have limited knowledge on the value chains for straw products.

Rice husks were long considered a waste in the rice milling process and were often dumped and/or burned. However, because it can be easily collected and is cheap, rice husk has been used as an energy source for small applications, such as for brick production, steam engines and gasifiers to power rice mills in Myanmar and for generating heat for rice dryers. The high silica content of rice husk ash makes it a good additive for the steel and concrete industries. To a lesser degree, rice husk ash is used as a soil conditioner, as an insulator, and in the manufacturing of activated carbon. More recently, the generation of electrical power on a small to medium scale has been piloted throughout Asia, for example 200 kW in Cambodia, 5–10 kW in Myanmar (Gummert and Haefele, 2012) and between 1 and 22 MW in Thailand (Carbon Market Watch, 2009). These rice husk power pilots have illustrated promising approaches, but have also demonstrated some limitations. The limitations are mostly due to feedstock supply problems that have cropped up ever since the rice husk waste, formerly free of charge, became a traded commodity, and also due to logistical problems and high cost of transportation. In particular, in Thailand problems arose as a result of inappropriate and inefficient technology causing excess release of silica and ash, which descended on farmer's fields, and led to widespread demonstrations in more than 20 Thai provinces (Carbon Market Watch, 2009).

Improved by-product management and technologies that can help reduce the carbon footprint and GWP of rice production and increase revenues from rice production and processing are therefore important for sustainability of rice production systems. This chapter outlines the different existing and promising technologies and management options for rice straw and rice husk utilization in the context of sustainable rice production.

## 2 Availability of rice straw and husks

The cultivation of rice results in the production of straw and husk, both of which are considered residues or by-products of the system. Rice husk, the main by-product from rice milling, constitutes a ratio of about 0.20–0.22 by paddy weight, whereas straw to paddy ratio ranges from 1.0 to 4.3 (Zafar, 2015). The technology for rice utilization is well established worldwide, whereas the handling of its by-products (e.g. straw and husk) has some degree of difficulty. Their availability is also limited at harvest time. Presently, these residues have a significant economic and environmental impact. The amount of rice straw harvested depends on various factors, such as harvesting methods, operational conditions and lodging with different harvesting methods in different regions and, more importantly, the varieties of rice. A case study at the International Rice Research Institute (IRRI) Experiment Station (Nguyen et al., 2015) recommended that the straw-to-grain yield ratios of RC238 variety for manual and combine harvested crop be 0.79 and 0.74, respectively. Along with the progress in the rice production system, approximately 731 million tons of rice straw per year is produced globally (Africa: 20.9 million tons, Asia: 667.6 million tons, Europe: 3.9 million tons, America: 37.2 million tons and Oceania: 1.7 million tons) (Gummert, 2013). The amount of rice husk and rice straw produced in Asia in 2013 is shown in Fig. 1. The rice straw amount is calculated based on an assumed 1:1

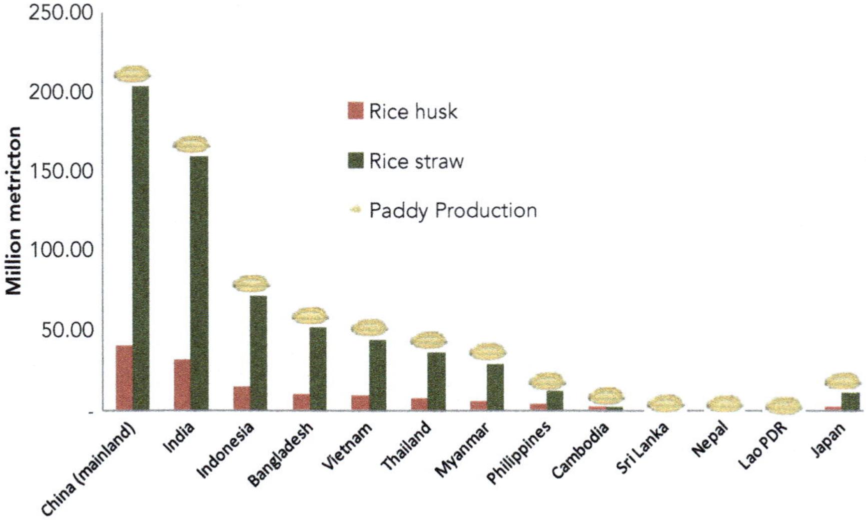


* Source: Calculated data from leading rice producers in Asia in 2013 obtained from *http://www.statista.com/statistics/255937/leading-rice-producers-worldwide/*

**Figure 1** Availability of rice husk and straw in the leading rice-producing countries in Asia in 2013.

ratio of straw and paddy yield (Zafar, 2015 and Nguyen et al., 2015), whereas the rice husk output is assumed to be 20% of paddy yield (Gummert, 2013; Zafar, 2015). The biggest amount of residues came from mainland China, which contributed to around 203 million metric tons. The lowest contribution of paddy by-products and straw biomass came from Lao People's Democratic Republic (PDR), which has comparatively a smaller area for rice cultivation. Large quantities of rice and wheat straw were produced in multiple cropping systems as a result of the Green Revolution that spread through South Asia and SEA from the late 1960s onwards.

The advent of mechanized harvesting (e.g. combine harvesters, reapers) has prompted farmers to burn in-situ large quantities of crop residues. Surplus crop residue (after removing some for other purposes) is usually burned because it interferes with tillage operations for the next season, especially in intensive cropping systems. In India, for example, farmers often burn rice straw as it is seen to be the cheapest, fastest and easiest way to remove large volumes of residue produced by the rice crop. This quick and cheap removal of residual biomass enables the farmers to establish the next crop in the crop rotation, often a wheat crop, in a timely manner and without interference from large rice straw loads in the field (Singh and Sidhu, 2014). The amount of surplus crop residue available in India is estimated to be approximately 82 million tons (Mt) and nearly 70 Mt (44.5 Mt rice straws and 24.5 Mt wheat straws) is burned annually (Singh and Sidhu, 2014). It was reported by Yadvinder-Singh et al. (2005) that more than 80% of total rice straw produced annually is being burned by farmers in India in a span of 3–4 weeks during October–November. In Thailand, Kanokkanjana and Garivait (2013) reported that, in 2010, the amount of straw burned in the field after harvesting amounted to 10.41 Mt. Burning

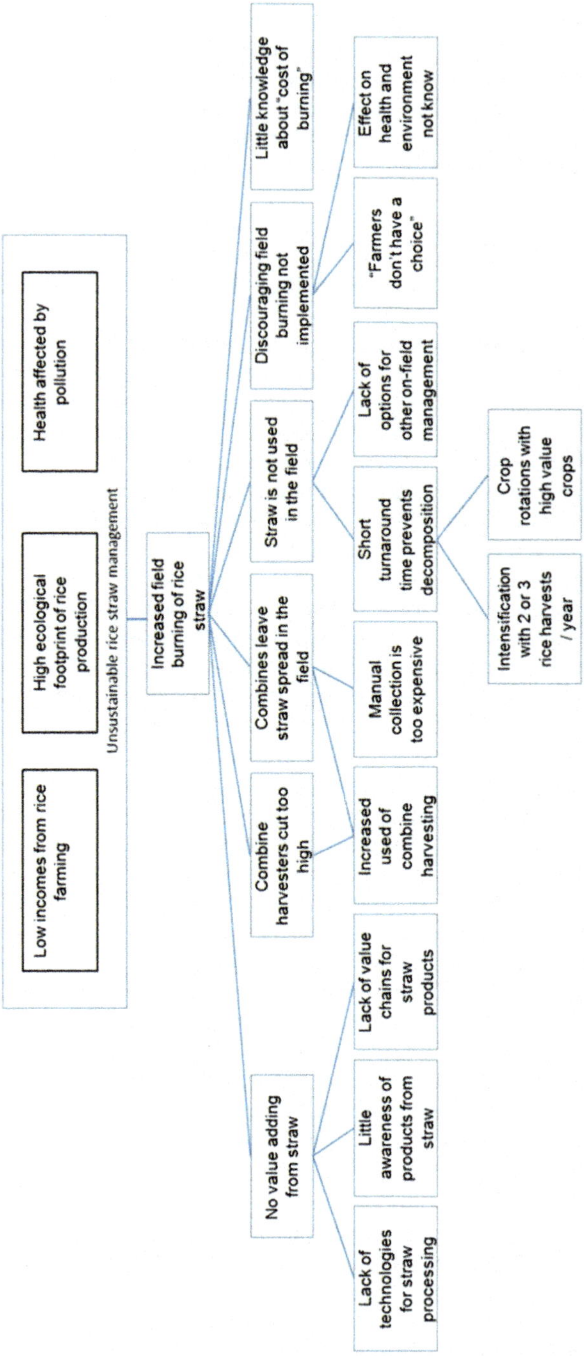


**Figure 2** Problem tree of rice straw management in the field.

Source: Adapted from Gummert, 2015.

has been the most convenient way to eliminate rice straw, especially in irrigated paddy fields with intensive cropping systems due to short turnaround periods (around 1–2 weeks) in between cropping seasons. The option of soil incorporation is not feasible due to the very short fallow period in intensive cropping systems. The lack of knowledge on the negative effects of field burning and the lack of political will to implement environmental laws are challenges for sustainable management of straw in the field. These challenges on sustainable rice straw management are outlined in a problem tree by Gummert (2015), and is shown in Fig. 2.

# 3 In-field rice residue processing and management

Traditionally, in South Asia and SEA countries, rice straw and husk are considered as wastes, either dumped into rivers or burned in the field, causing GHGE, contamination and pollution. With the development of recent technologies, rice residues can be processed and managed with better practices. Scalable processing and management options of rice residues can be classified as in-field and off-field options.

In small farmers' fields, where draught animal power and two-wheel tractors operate, large amounts of biomass can be difficult to effectively incorporate into the soil. Large amounts of biomass can also be an impediment to agronomic operations, reducing the efficiency of work and slowing the process of land preparation. This is a more significant problem in intensive double- and triple-irrigated rice cropping systems than it is in low-intensity, single-cropping systems. In the intensive systems with short fallow periods, biomass has limited time to decompose between successive cultivation of crops. The in-field management choices of rice residues can have both positive and negative impacts on the efficiency and sustainability of rice production. Management of rice residues influences the energy and nutrient balance, efficiency of fertilizer inputs, emissions of $CH_4$ and atmospheric pollutants and the pest and disease risks within rice agro-ecosystems. The changing social, climatic, environmental and economic circumstances, particularly in Asia, are driving transformations in rice agro-ecosystems that will undoubtedly result in changes in the in-field residue management strategies of rice farmers.

Although rice crops are still harvested manually in many regions across Asia, the adoption of combine harvesters is becoming increasingly common in Asian rice production. With manual harvesting, rice straw is cut below the panicle and the cut straw and the panicle are then taken to a location, usually within or beside the field, where threshing is done. The process of manual harvesting, therefore, generally results in a concentration of rice straw at points within or at the edge of fields. In contrast, after mechanical harvesting with combine harvesters, which both cut and thresh rice in the harvesting process, rice straw is either piled in windrows or spread across the field.

## 3.1 Open-field burning

In many regions, rice straw has little to no value and is an impediment to the efficiency of rice cropping operations. Historically, rice farmers would burn rice straw in the field as a quick, affordable and simple method of reducing biomass quantities that must otherwise be incorporated into the soil during land preparation for the subsequent crop. It is estimated that approximately 60% and 80% of rice straw is burned in the field in Vietnam and Philippines, respectively (Jamieson et al., 2014). However, open-field burning of rice straw poses many problems and is a constraint to improving the sustainability of intensive rice systems in Asia.

There is a large amount of evidence in scientific literature concluding that incorporation of rice straw residues provides benefits to soil fertility (Dobermann and Farihurst, 2002; Ponnamperuma, 1984; and Yadvinder-Singh et al., 2004). Research such as that conducted by Prasad et al. (1999) and Surekha et al. (2003) demonstrates that burning has a negative impact on soil carbon stocks and the fertility of soil. A range of environmental factors will influence the magnitude of the impact that open-field burning has on soil condition. The heat and duration of fire; soil moisture, both at the time of burning and during tillage; the time elapsed and climatic condition between burning and tillage; and the chemical, physical and biological properties of the soil, all will influence the change in soil properties resulting from the burning of rice straw. The impact of straw burning on soil fertility accumulates over time. Therefore, as was explained by Ponnamperuma (1984), the length of the experimental work must also be considered when making conclusions about the sustainability of this method of straw management.

Burning of rice straw does effectively remove large volumes of biomass and also helps control weeds and a variety of pests and diseases (Ponnamperuma, 1984). However, research indicates that the advantages of burning are offset by the disadvantages, including nutrient loss, depletion of soil organic matter (SOM) and reduction in the presence of beneficial soil biota (Mandal et al., 2004). Rice straw burned in the field also causes GHGE, including 0.7–4.1 g of $CH_4$ and 0.019–0.057 g of $N_2O$ per kg of dry rice straw (Table 1) and emission of other gaseous pollutants such as $SO_2$, NOx, HCl and, to some extent, dioxins and furans (Oanh et al., 2011 and Jenkins et al., 2003). Rice straw burning activities are also an important source of aerosol particles such as coarse dust particles ($PM_{10}$) and fine particles ($PM_{2.5}$) (Chang et al., 2013), affecting regional air quality and the radiation budget of the earth (Engling et al., 2009). For these reasons, and for the unaffordable health risks

**Table 1** Emission factors from rice straw burning

| | Emission factor (g kg-1 DM) | | |
|---|---|---|---|
| Moisture content | $CH_4$ | $N_2O$ | Source |
| 15 | 1.20 | – | a |
| 10.6 | 2.10 | | b |
| 14.2 | 4.10 | | b |
| 10.6 | 0.70 | 0.033 | c |
| 20 | 2.50 | 0.057 | c |
| 10 | 1.04 | 0.018 | d |
| 22 | 1.36 | 0.027 | d |
| 30 | 1.77 | 0.019 | d |

Sources: a: EPA (1995); b: Miura and Kanno (1997); c: Hayashi et al. (2014); d: Romasanta and Sander (2016).

that rice straw burning poses to societies, open-field burning of straw has been banned or is very strictly regulated in agricultural regions across the world.

## 3.2 Incorporation

As farmers stop burning rice straw because of health and environmental risks it poses, they are likely to increasingly turn to incorporating rice straw into the soil during land preparation. In circumstances where manual harvesting is done, only a proportion of the total crop residue is often incorporated into the soil. During manual harvesting, a significant proportion of the biomass is removed from the field and threshed. However, where harvesting is conducted using a combine harvester, unless straw is removed from the field after harvesting, the total residual straw biomass of the rice crop will be incorporated into the soil during the land preparation process. With careful management of rice straw incorporation, soil fertility can be maintained and enhanced, and the sustainability of rice production can be improved (Li et al., 2014). However, ineffective management of straw incorporation can result in a decrease in production efficiency (Mandal et al., 2004; Yadvinder-Singh et al., 2005; Dobermann and Fairhurst, 2002) and an increase in GHGE (Sander et al., 2014).

Rice straw can constitute a major challenge in seedbed preparation and crop establishment, but its incorporation into the soil can play an important role in maintaining the nutrient balance in rice production. Nutrients in straw must be mineralized through decomposition prior to becoming available to plants. The speed with which rice straw degrades in soil is influenced by litter quality, biomass composition and structure, as well as external factors, such as temperature, moisture, method of incorporation, time, surface area of straw, and the concentration and composition of gases in the soil (Li et al., 2014). It should be noted that slow decomposition is often the reason why farmers do not want to incorporate crop residues. Some research has been undertaken on speeding up this process, for example, by using fungal inoculums (Goyal and Sindhu, 2011; Ngo et al., 2012); however, research in this area of irrigated rice systems is limited.

Removal of rice residues from the field, which is practised in some systems where straw has alternative uses and value, does not necessarily deplete SOM in intensive irrigated rice production (Bijay-Singh et al., 2008). This is predominantly due to the saturated soil conditions in irrigated rice production that creates a reduced environment in which oxidative processes are restricted and the root biomass is sufficient to sustain carbon stocks. While the removal of rice straw does not necessarily have an impact on SOM, it can have a negative impact on the availability of soil nutrients (Dobermann and Fairhurst, 2002). However, where a nutrient balance approach is used to determine fertilizer application rates for rice, such as Site-Specific Nutrient Management (SSNM) (Witt and Dobermann, 2002), yields can be sustained in systems where straw is removed.

When calculating a nutrient balance for rice where the straw is removed from the field after harvest, consideration of K removal is of great importance. Rice straw is approximately 1.5% K. Hence, when large quantities of straw are regularly removed from a field at, or after, harvest, K deficiencies can potentially occur, unless the K in the soil is replenished. In contrast, the large amount of K in rice straw means that straw incorporation can have a significant and positive influence on soil K availability (Witt et al., 2002). If farmers are removing rice straw from their fields, allowing a period of time for K to leach from the straw can help reduce the risk of the occurrence of K deficiency. This is because the K in rice straw is largely soluble (Yadvinder-Singh et al., 2005). The period of time necessary to allow for sufficient leaching will vary depending on climatic conditions.

If not carefully managed, the efficiency of N fertilizer use can be negatively influenced by the incorporation of rice straw (Thuy et al., 2008; Yoneyama and Yoshida, 1977). Inorganic nitrogen can become immobilized as organic matter decomposes in the soil. Nitrogen can become bound in both the microbial biomass and the humic fractions of the SOM. The risk of immobilization, often referred to as N tie-up, tends to be greater in aerobic soils than in anaerobic conditions. However, rice straw, which has a high C/N, can cause tie-up in irrigated rice soils, particularly when time is not enough to achieve sufficient straw decomposition between incorporation and transplanting. Dobermann and Fairhurst (2002) suggested a minimum of two to three weeks between incorporation of rice straw and the transplanting of rice seedlings. Composting rice straw before it is added back to the paddy field reduces the risk of N immobilization, recycles most nutrients in the field and produces less $CH_4$ than incorporating non-composted rice straw directly into saturated soil (Corton et al., 2000; Silalertruksa and Gheewala, 2013).

Although careful management of rice straw incorporation can help sustain and improve soil fertility in rice production, increasing the organic matter in irrigated rice soils can result in increased GHG emissions. The incorporation of rice straw into saturated soil conditions of a puddled rice field leads to an increase in the production of $CH_4$ (Peng et al., 2008; Yuan and Conrad, 2014). Methane is the GHG that contributes the most to the GWP of irrigated rice production. The adoption of combine harvesters raises some concern about the possibility of increased GWP of rice production as the total straw biomass remains in the field where it is generally removed during manual harvesting (Alberto et al., 2015). However, irrigation scheduling, also known as alternate wetting and drying, which introduces periods of non-saturated soil conditions during each cropping, can significantly reduce $CH_4$ production in irrigated rice, including where straw is incorporated (Johnson-Beebout et al., 2009).

# 4 Off-field rice residue processing and management

Rice residues, straw and husk, play a critical role in terms of economic and environmental sustainability of rice production. Some features of rice husk and rice straw are summarized in Table 2.

**Table 2** Main features of rice residues

| Husks | Straws |
|---|---|
| Uniform in size | Bulky |
| Dry | Dry or wet depending on harvesting seasons |
| Accumulated at factory level | Field-based resource |
| Market access, traded | Only local market |
| Price structure available | High variation in prices |
| Direct use for energy (power plant, heat) possible | Needs further processing for efficient energy use |
| High ash content | High ash content |

Source: Adapted from Siemers, 2011.

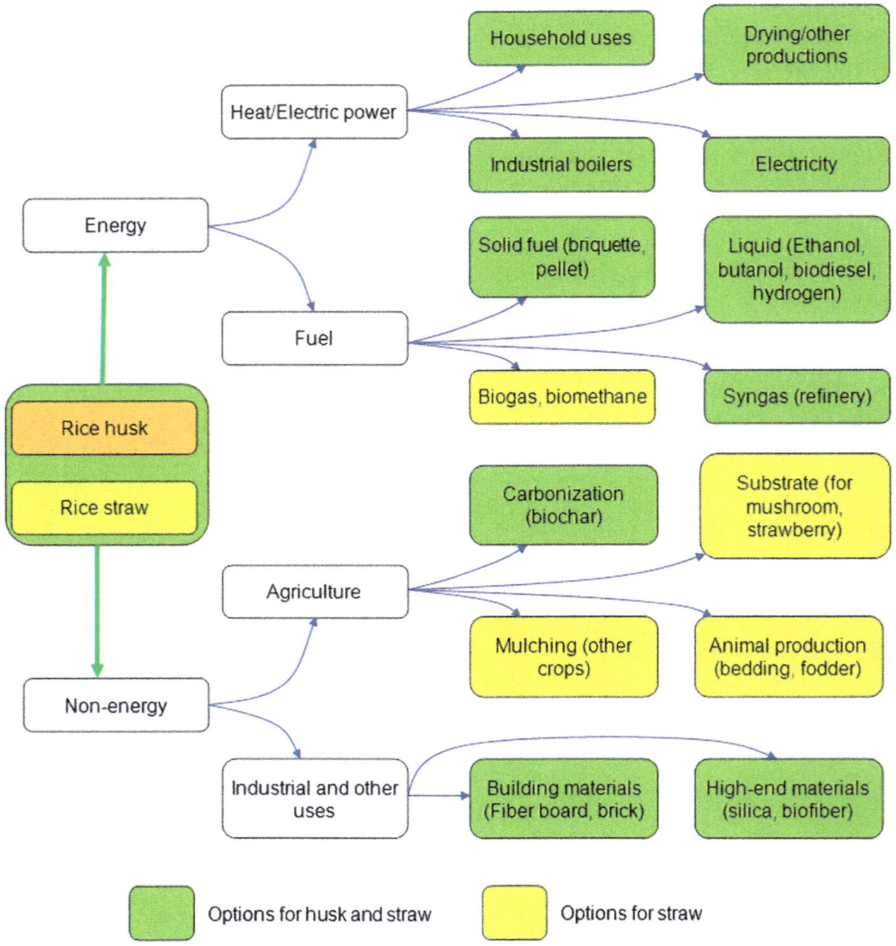


**Figure 3** Off-field options for rice husk and straw.

There is no processing required for burning rice husk; thus, husk is the by-product mostly used for energy production, especially in processing solid fuels, such as briquettes and pellets, to provide feedstock to generate heat in drying and industrial boilers. A case study in the Mekong Delta (MD) of Vietnam (Nguyen et al., 2012) has shown that about 10%, 10% and 60% of husk is used for paddy drying, processing briquettes and pellets and heating in brick production, respectively. Rice straw, on the other hand, does not seem to be the best material for major biofuel development because it is bulky, wet and has high silica content, which acts as a barrier to further processing. Recently, with the development of mechanized collection and processing technologies, rice straw and rice husk can be used for various options in agriculture, energy and industry as shown in Fig. 3.

## 4.1 Collection of rice residues

Rice husk is collected at the rice mill after milling. The pneumatic conveying technology used for collecting and transporting rice husk is already a developed technology in many parts of the world. This section therefore focuses on updated information on rice straw collection in South Asia and SEA countries.

Collection of rice straw is still a major challenge in the supply chain of rice straw. Straw needs to be gathered from the field and compressed into bales to make it compact and easy to transport. A case study in Vietnam (Nguyen et al., 2013) cited that

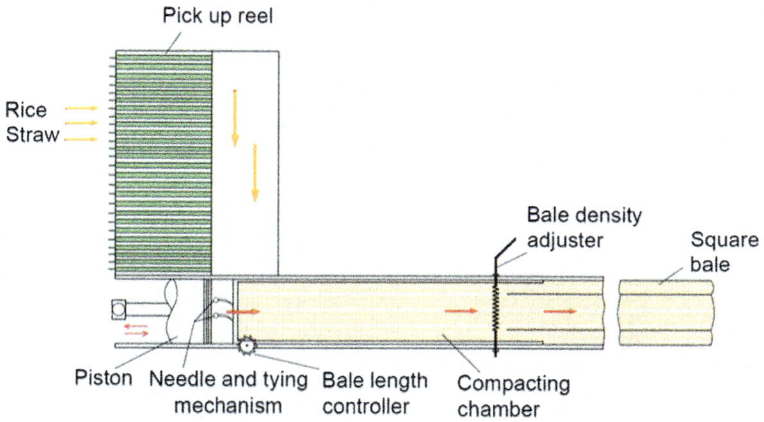


**Figure 4a** Piston type rice straw baler.

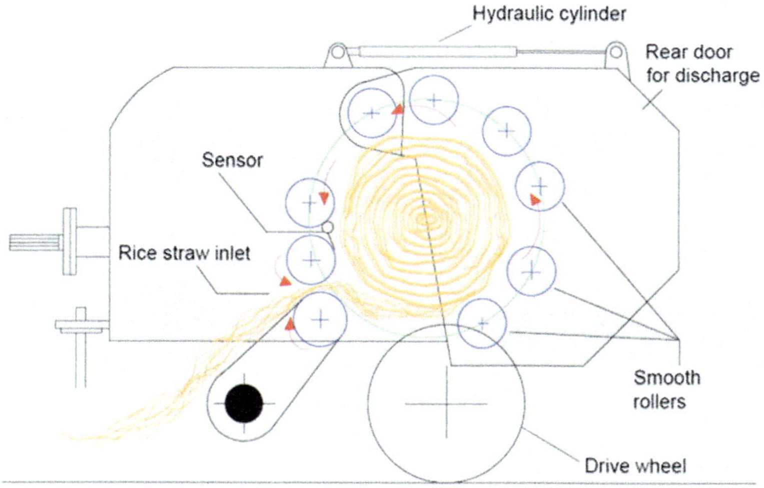


**Figure 4b** Principle of a roller type rice straw baler.

rice straw is mostly gathered manually, despite the current research and development of rice straw baling machinery. With the introduction of combine harvesters, the collection of rice straw has become harder and more costly – such an example is the increase in the price of rice straw from US$20 to US$80 per ton within three years, from 2012 to 2014, in Vietnam. This affected other businesses that use rice straw; for example, mushroom production declined by two-thirds (or 67%) in the MD of Vietnam from 2009 to 2012.

There are two basic concepts of crop residue collection in the field – the simultaneous collection of crop and residue, known as the total harvest concept, and the collection of residue after the crop is harvested, known as the post-harvest concept. The concept of total harvest was discussed by Horsfield et al. (1977) and Dobie et al. (1973). This system removes the straw and the grain in a single operation and hauls them to a designated location at the edge of the field, to the farmstead or to the grain terminal for separation. The total harvest system that collects the straw and the grain simultaneously will avoid the problem of reentering wet and rutted fields, a problem that occurs in post-harvest systems. However, the total harvest system is not suitable for working in small fields and wet and soft soils, which hamper the entry of heavy vehicles. For post-harvest collection

a. Roller baler Z755 in Vietnam, 2013.

b. Baler CLAAS Markant 55 at IRRI, 2015.

c. Roller baler CLAAS Rollant R250 at IRRI, 2014.

d. Self-propelled baler Phan-Tan, operating in Vietnam 2015.

**Figure 5** Four typical balers used for collecting rice straw in SEA countries.

of rice straw, the collection of residue after harvest could be accomplished by the use of either a small or medium baling system or a large baling system.

Collection involves three main operations: picking up of rice straw materials, compacting them into a bale and hauling them to a storage or processing area. In these activities, a collecting machine, called a baler, is the main compacting unit (baling) and sometimes includes the other two operations – picking up of straw and hauling of baled material – which depend on operational conditions and purposes, such as the field, harvesting, cost, etc. A stationary baler with only a compaction unit can be used properly in collecting straw disposed from paddy threshers. On the other hand, a mobile baler (either self-propelled or hauled by a tractor) is suitable for collecting rice straw spread in the field by a combine harvester. There are two main classifications of a baler, which are defined by the principle of operation of the compacting unit. The piston type is used in making square bales (Fig. 4a), whereas the roller type is applied to make round bales (Fig. 4b).

Four typical balers (Fig. 5) used for collecting rice straw in SEA countries are described in this section based on the assessment of some balers conducted by the authors recently. The smaller balers – such as the roller baler Z755 made in Vietnam (Fig. 5a) and the square-baler Markant 55 made by CLAAS Germany (Fig. 5b), with capacities from 0.5 to 1.5 ton rice straw per hour – are seen more suitable for the small and wet fields in SEA. The bigger baler, Rollant 250 made by CLAAS Germany (Fig. 5c), has a higher capacity of about 3 tons per hour, but its implement is not easy to manoeuvre in wet and soft soil. The self-propelled roller baler Phan-Tan (Fig. 5d), developed recently in Vietnam, is the best rice straw collection vehicle for working in the wet fields among these four balers. This self-propelled baler uses a rubber chain moving system and operates properly in wet and soft soil. It is matched with a bin to bring about 30 straw bales weighing about 20 kg per bale.

# 5 Using rice residues for energy production

Rice residues contain about 30–50% of organic carbon and have a high heat value (HHV) of 13–16 MJ per kg of straw (Jenkins, 1998; Goyal and Sindhu, 2011; Munder, 2013; Nguyen et al., 2015). These residues can be used to generate fuel, heat or electricity through thermal, chemical or bioprocesses. The energy conversion processes of rice residues are presented in a diagram in Fig. 6.

Rice husk is collected after rice milling, with MC of about 14–15%, which fits the requirement of further pretreatment or processing. On the other hand, rice straw is collected after cutting the crop from the field. MC is about 15–25% during the dry season and 26–40% in the wet season. For most energy conversion processes, rice straw needs to be cut into small pieces of 2–10 mm length and the MC should be lower than 20%. Thermal processes include combustion, gasification and pyrolysis, while anaerobic digestion (AD) and fermentation are classified as bioprocesses. Energy products from rice husk are heat, electricity and biofuel. Heat could be used for house heating and cooking, industrial boilers, drying and generating electricity. The biofuels generated are biogas, bio-methane and refined syngas. Fig. 7 shows a scheme of combined system for processing biomass, including mechanical pretreatment, pyrolysis, gasification and gas refinery. This system has been observed in the UK recently and known as the newest

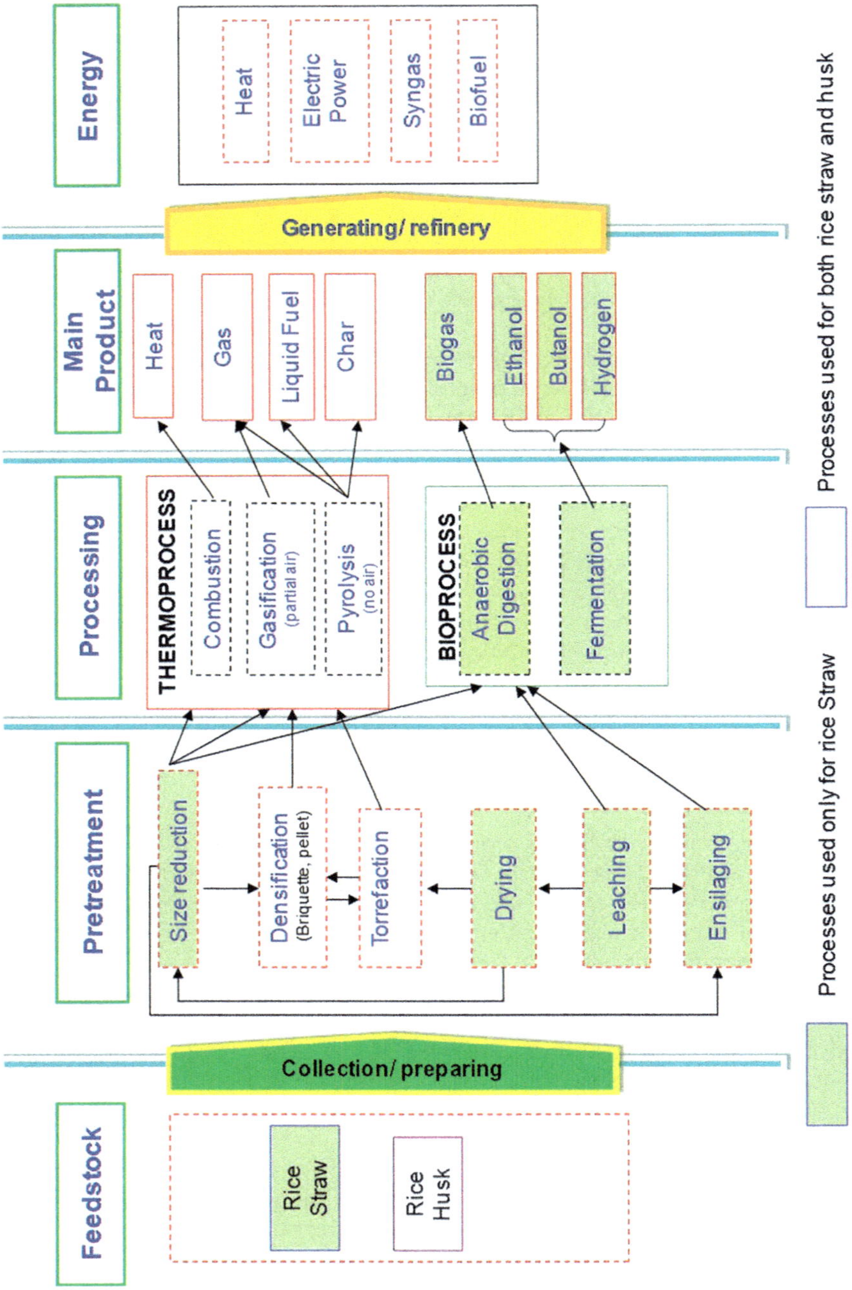


**Figure 6** Potential energy conversion processes of rice residues.

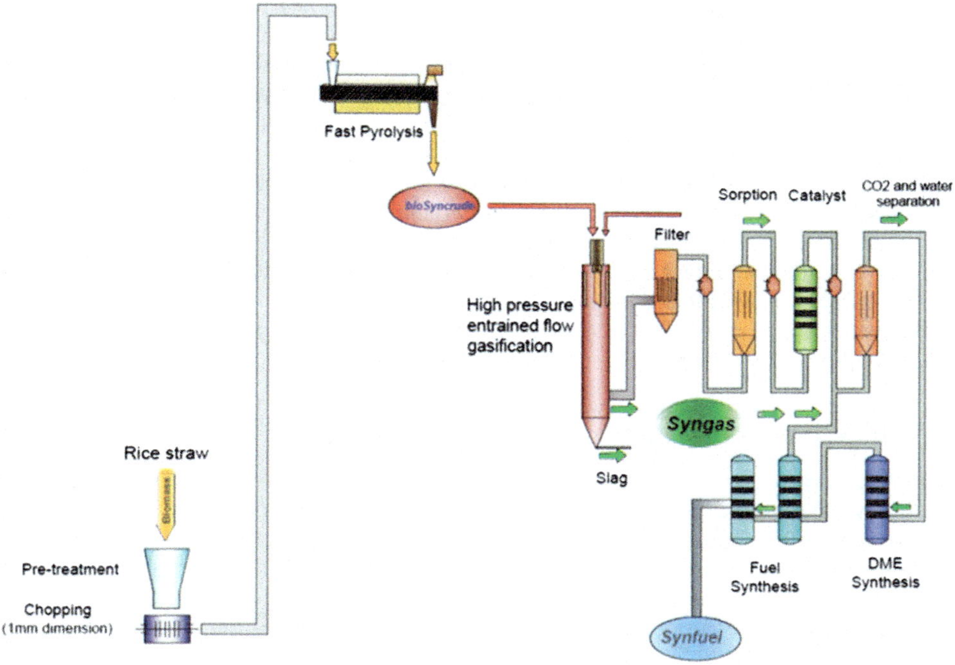


**Figure 7** Biofuel process route of biomass.

technology for biofuel conversion from biomass. However, it is currently being used for wheat straw and has not yet been evaluated in terms of economic benefit and energy balance.

Rice husk can be more appropriately used for a thermo-process set-up, while rice straw is better suited for bioprocesses. This idea is based on the observation of an efficient wet processing system using wet material, and dry material is best for use in dry processing. Rice husk can be used in loose, ground, torrefied or densified form, whereas straw requires additional pretreatments such as chopping, leaching and ensilaging prior to bioprocessing. The characteristics of rice residues compared with other solid fuels in term of energy use can be summarized as follows:

- High silica content causes wearing down of the components in processing machines such as the chopper or grinder. The content of volatile matter in rice straw is higher than that of wood and much higher than that of coal, whereas fixed carbon is much lower than that in coal. Ash content in rice residues is much higher than in wood and coal, which acts as a barrier to energy conversion (Jenkins et al., 1996; Jenkins, 1998).
- High content of ash, alkali and potassium causes agglomeration, fouling and melting in the components of combustors or boilers (Baker, 2000; Lim et al., 2012).
- Lignin wall of rice straw hampers digestion, which reduces the efficiency of bioenergy conversion processes (Klass, 1998).

One solution to resolve barriers in energy conversion from rice residues is pretreatment. Rice residues could be pretreated through mechanical, physical, chemical, bioprocess or a combination of these processes to increase efficiency in further energy conversion. These pretreatment technologies can be referred to in Klass (1998), Baker (2000), Kargbo et al. (2009), Sadaka and Negi (2009), Munder (2013) and Liu et al. (2013). Other solutions for ash-related problems in rice residues are to combine rice straw with other fuels that are lower in alkali and chlorine, to reduce temperatures in combustion systems and to use other types of conversion systems such as Organic Rankin Cycle (ORC) and two-stage combustion. Case studies on the use of rice residues for the energy sector in South and SEA are described in recent publications, such as Ghosh and Bhattacharyya (1999), Tillman (2000), Jung et al. (2008) and Lim et al. (2012).

Below are some case studies on energy conversion technologies of rice residues that were conducted by our group in some collaborative projects.

## 5.1 Rice husk combustion for paddy drying

Mechanical drying is an important process to reduce post-harvest losses. Paddy rice is usually harvested at the grain MC of between 24 and 26% (wet basis). Drying is an energy-intensive operation, which entails high energy cost. Using rice husk in loose form is one cost-effective solution to heat the air for drying. The cost of energy from rice husk is lower than that of kerosene and coal. In the Philippines, for example, energy cost (e.g. peso per MJ) from rice husk is about 20% of the cost of using coal and kerosene.

Several rice husk furnace technologies have been researched and developed in recent years, such as the inclined-grate manual-fed and the automatic non-grate furnaces. The type of rice husk furnace selected for a given dryer is based on efficiency, quality of flue gas

**Table 3** Types of rice husk furnace used for paddy dryers

| Furnace types | Scope of using | Drying air efficiency % | Investment cost USD $(MJ\ h^{-1})^{-1}$ | Advantages | Disadvantages |
|---|---|---|---|---|---|
| Inclined grate – cylindrical combustion chamber | Small-scale flatbed dryers (4–20 tons batch$^{-1}$) | 60 | 4–6 | Simple, low cost maintenance | High required labour |
| Spraying-fed cyclone furnace | Columnar and fluidized bed dryer with higher capacity | 50–55 | 14 | Automatic and continuous operating | Lower efficiency High investment cost |
| NLU-IRRI non-grate furnace | Columnar and fluidized bed dryer with higher capacity | 50–55 | 4–7 | Automatic and continuous operating | Lower efficiency |
| IRRI dRHF | Paddy dryers (4–10 tons batch$^{-1}$) | 60–80 | 5–7 | Semi-automatic operating | Not developed for the higher capacity yet |

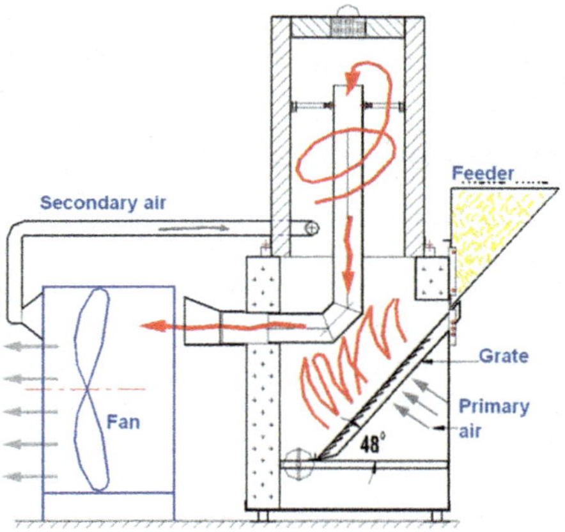


**Figure 8a** Principle of rice husk furnace with inclined grate and cylindrical combustion chamber.

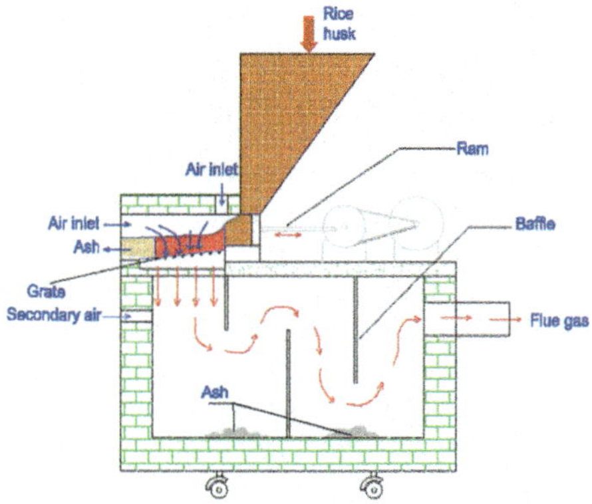


**Figure 8b** Principle of downdraught rice husk furnace.

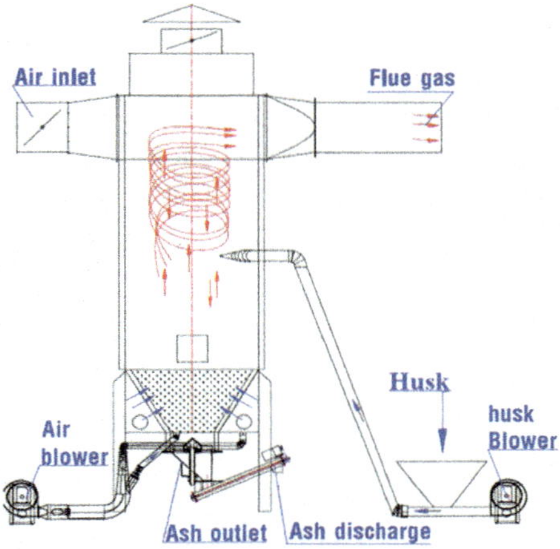


**Figure 9a** Continuously fed furnace with cyclone combustion chamber.

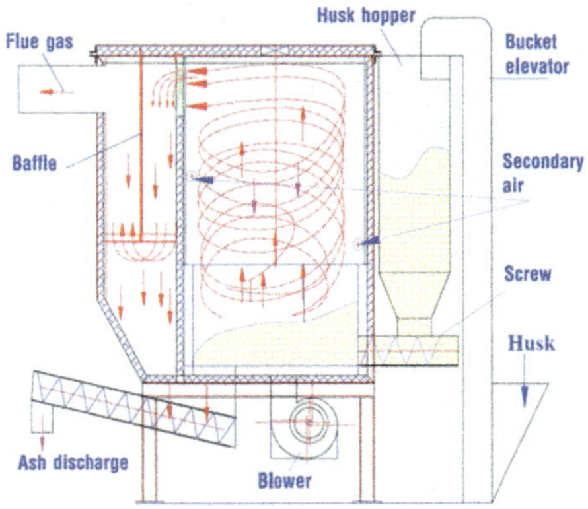


**Figure 9b** IRRI-NLU non-grate furnace.

(clean), heat capacity, nature of operation (manual or automatic), investment cost and energy cost. There are two types of small-scale rice husk furnaces that are popularly used in SEA: the inclined grate with cylindrical combustion chamber furnace, shown in Fig. 8a (Nguyen et al., 1996), and the downdraught rice husk furnace, shown in Fig. 8b (Gummert et al., 2007). In large-scale applications, continuously fed furnaces are preferred. There are two typical examples of this type of furnace: cyclone combustion furnace as shown in Fig. 9a (Nguyen et al., 2012) and the IRRI-NLU (Nong Lam University) non-grate furnace developed by the

ADB-IRRI post-harvest project (IRRI, 2014) shown in Fig. 9b (Tran et al., 2013). A comparison of these furnaces is summarized in Table 3 based on recent assessments.

## 5.2 Case study of rice straw combustion power plant

A feasibility study on a 1-MW rice straw power plant was conducted by IRRI and ENERTIME Company (Guillemot et al., 2014). The system (Fig. 10) uses an advanced technology for biomass combustion, which is the ORC. An ORC turbo-machine generates electricity using thermal energy from thermal oil (or saturated steam) generated from combustion of biomass in a boiler. This thermal exchanger transfers energy from the oil to the organic working fluid, heating the fluid and causing it to vaporize. The oil (or steam) is condensed and cooled and circulates back to the boiler. The vaporized working fluid fuels the turbine, which then sets the asynchronous generator in motion.

## 5.3 Case study on energy balance analysis of rice straw for anaerobic digestion

A potential small-scale energy conversion technology for rice straw is AD, which is used to generate biogas fuel that is used for cooking, heating air for drying or generating electricity. Rice straw is mixed with manure from livestock at a rate of 50–80%. Biogas yield is about 600 litres per kg of straw and the $CH_4$ content is about 50–55%. Fig. 11a shows a single-stage AD system for rice straw and pig dung and Fig. 11b shows a pilot of two-stage AD system for rice straw and cow dung. The study evaluated the energy balance of the rice straw supply chain and energy conversion through AD. Input energy was accounted for as direct and indirect energy. Direct energy included fuel consumption and manpower. Indirect energy was calculated based on the energy for the manufacture, lubrication and maintenance of

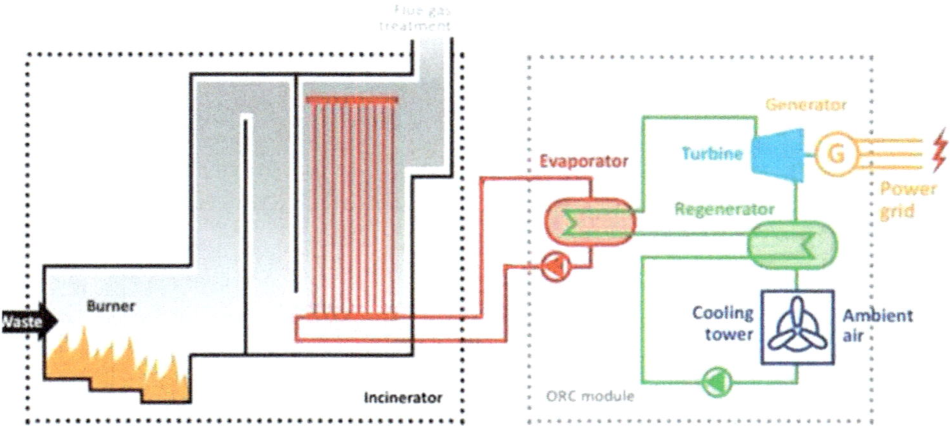


**Figure 10** Schematic diagram of 1-MW power plant with ORC.
Source: Guillemot et al., 2014.

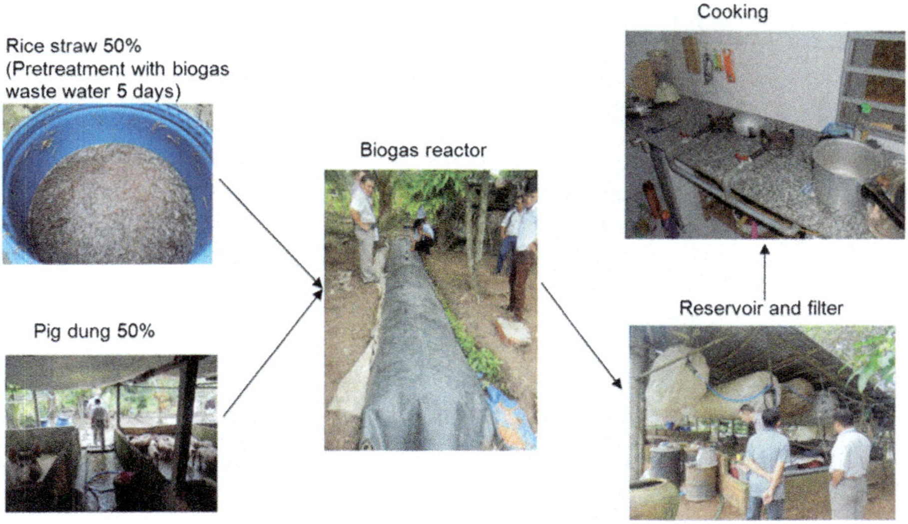


**Figure 11a** Small-scale, single-stage AD system for rice straw and pig dung in Can Tho City, Vietnam.

**Figure 11b** A pilot of a small-scale, two-stage AD system for rice straw and cow dung at IRRI.

machines and equipment. The total energy input of rice straw from harvesting to storage ranges from 500 to 900 MJ per ton of straw, depending on the nature of operation (i.e. manual or mechanized). The input energy for AD of rice straw with a 6-$m^3$ digester made of high-density polyethylene is about 3400 MJ per ton of straw. The net energy of the rice straw supply chain for biogas generation through AD is 3500 MJ per ton of straw. This rice straw management option can provide 70% net output energy benefit.

# 6 Non-energy uses of rice residues

Rice residues contain carbon, nitrogen, potassium and alkali, which make them suitable materials to produce biochar, fertilizer, substrates for mushroom production or animal bedding and fodder. Aside from these, the high content of lignin and silica of rice residues are best for making industrial materials such as medium density fibre board, brick, composite and organic plastic. These are also used for high-end materials such as silica used in low-roll resistance tyres, paint and cosmetic industries (Subbukrishna et al., 2015); bio-fibres used in the textile industry (Reddy and Yang, 2004); and as a resource of cellulose nanofibrils (Lu and Hsieh, 2012). This section provides an overview of three popular uses of rice straw in the agriculture sector, particularly as biochar for mushroom and animal fodder production.

## 6.1 Biochar production and utilization

Biochar, a carbon-rich product, is produced by thermal decomposition of organic material or biomass under a limited supply of oxygen ($O_2$) and at relatively low temperatures (less than 700°C). The process is known as carbonization. The carbonization process for organic material at 1 MPa can be represented by the following equation (eq. 18.1) introduced by Antal and Gronli (2003).

$$C_6H_{10}O_5 \rightarrow 3.74C + 2.65H_2O + 1.17CO_2 + 1.08CH_4 \quad (18.1)$$

Based on this research, biochar yield decreases as the peak temperature increases, whereas increasing peak temperature leads to an increase in fixed carbon yield. In another study by Azhar Uddin and Phuong (2009) on carbonization of rice husk and rice straw, carbon contents increased from 48 to 51% and from 53 to 58%, with increasing temperature from 350 to 550°C, respectively and a heating rate of 10°C per minute. Biochar yield and quality are also affected by pressure and heating rate (Manya et al., 2013).

Biochar is used as soil amendment to improve soil productivity, carbon storage and filtration of percolating soil water (Lehmann and Joseph, 2009). Libra et al. (2011) have found beneficial effects in increasing water holding capacity, aeration and hydraulic conductivity – reducing the tensile strength of hard-setting soils and stimulating growth activity after biochar amendment. It was found by Wua et al. (2012) that biochar generated at 400°C might be a better option for improving soil quality. A study of Haefele et al. (2011) showed that the application of untreated and carbonized rice husks increased total organic carbon, total soil N, the C:N ratio and available P and K, whereas the high C:N ratio of carbonized rice husk (biochar) limits N availability, thereby slightly reducing grain yield. On poor soil – which is identified as having low organic carbon, nitrogen content, base saturation, K availability and cation exchange capacity, and dominated by sand – soil chemical and physical improvements increased yields by 16–35% with an amendment of 4.13 kg carbonized rice husk per square metre cropping. On the other hand, biochar processing needs to be considered with respect to energy balance and production and application cost. Using biochar as soil amendment instead of burning the fuel more completely reduces the energy efficiency of bioenergy production. However, the carbon sequestration process in biochar production helps mitigate climate change (Woolf et al., 2010), and the emission reduction of using biochar as a soil amendment is higher than the fossil fuel offset in its use as fuel (Gaunt and Lehmann, 2008).

An advanced carbonization technology has recently been developed and is called the hydrothermal carbonization (HTC). The HTC of lignocellulosic biomass is a process that completely breaks down the plant cell wall and allows rapid conversion of biomass into a carbon-rich and lignite-like product (hydrochar). The carbonized product is composed of water soluble compounds and solid residues. The aqueous phase contains sugar and lignin-derived compounds, which may be a desirable feedstock for biodiesel and chemical production. The resulting solid products consist mainly of lignin, with a high degree of aromatization and a large amount of oxygen-containing groups. Moreover, hydrochar has a considerably higher heating value than the raw material (Xiao et al., 2012). Compared with biochar, hydrochar is not yet applied as a soil amendment. This process requires high amounts of water (Libra et al., 2011) when used as soil amendment.

GHG emission changes after the incorporation of biochar in flooded rice soils are not well understood. Although there have been many studies conducted on the utility of biochar as a mitigation option in rice production, the results are inconsistent. As an example in studies conducted with biochar incorporation rate ≤ 10t/ha, Shen et al. (2014) and Zhang et al. (2012) found no significant changes in $CH_4$ emissions, whereas an increase in $CH_4$ emissions of up to 30% was reported by Pandey et al. (2014). Similarly, with rates of biochar incorporation of 40 t/ha Zhang et al. (2012) observed an increase of 38–51% in $CH_4$ emissions, while Shen et al. (2014) found that $CH_4$ emissions decreased at this application rate. However, it must be noted that $CH_4$ emissions after biochar incorporation are consistently lower as compared to treatments with fresh straw incorporation at comparable rates of application.

Biochar is produced through incomplete combustion, that is, a combustion process with limited oxygen supply. It can be expected that during this process, considerable amounts of $CH_4$ are being produced. However, results of emission measurements from this process are extremely limited in the available literature.

## 6.2 Using rice straw for mushroom production

Cultivation of mushroom on rice and wheat straw may offer economic incentives for farmers and agri-businesses. The oyster mushroom, *Pleurotus spp.* (Fig. 12a), can effectively degrade lignin as well as cellulose, and its cultivation offers an on-farm technology for the bioconversion of poor-quality straw into nutritious food products. Mushroom cultivation is

**Figure 12a** Oyster mushroom.

**Figure 12b** Paddy straw mushroom.

considered a profitable agri-business that could produce food from rice and wheat straw, as well as help dispose them in an environmentally friendly manner.

The paddy straw mushroom, *Volvariella volvacea* (Fig. 12b), is considered to be one of the easiest mushrooms to cultivate because of its short incubation period of 14 days (Reyes, 2000). It is a tropical species that thrives best at 30–35°C for mycelia development and 28–30°C for fruiting body production. Being cellulytic, the paddy straw mushroom can be grown on a number of cellulosic substrates such as cotton wastes, sugarcane bagasse, rice straw and stubbles (Reyes, 2000). Using rice straw for mushroom production can yield about 5–10% of mushroom products (50–100 kg of mushroom per ton of dried straw) (Zhang et al., 2002; Ngo, 2011).

The cultivation of mushroom includes three salient features as described by Reyes (2000) – composting, pasteurization and cultivation – inside a mushroom house using stacks of open wooden crates. Composting allows microbial decomposers to reduce the tensile strength of rice straw. This also prepares the straw for easy colonization by the mycelium of the paddy straw mushroom. Pasteurization, on the other hand, is a critical process that eliminates undesirable microorganisms that may compete with the paddy straw mushroom during production. The general procedure for indoor cultivation involves eight major processes before harvesting (Reyes, 2000), as shown in Fig. 13.

Developing the mushroom culture as an industry will also encourage rice straw trading. In Vietnam for example, straw mushroom is popular with many dishes and serves as a means to augment farmers' income. Besides being food, mushroom is considered as

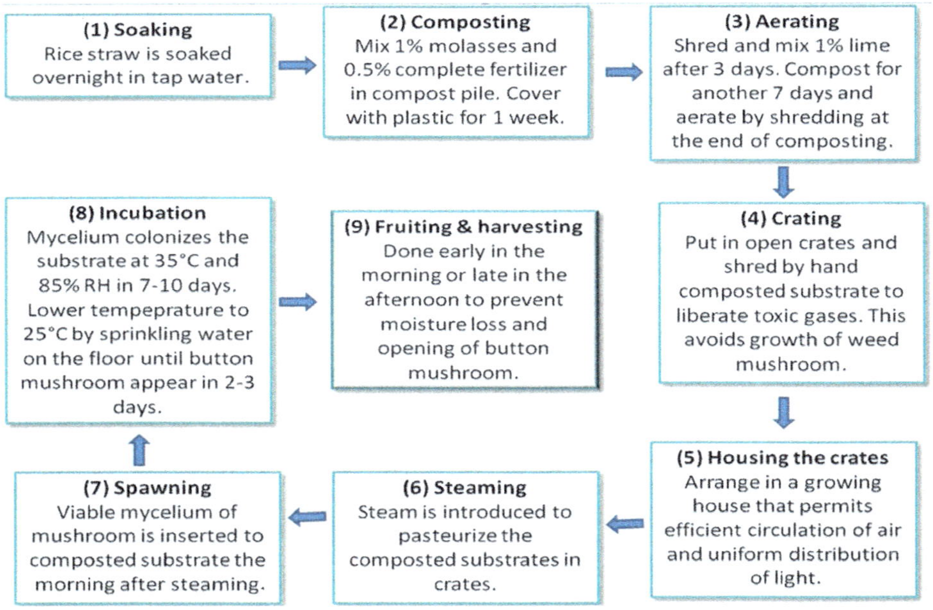


**Figure 13** Process of mushroom cultivation from paddy straw.
Source: Adapted from Reyes, 2000.

organic matter and medicine (Ngo, 2011). Mushroom cultivation seems to present non-negligible strengths as it is a straightforward process with a low-cost set-up, high price margin and quick returns (Chang et al., 2014). According to Marketsandmakers.com (2015), mushroom, which is consumed either fresh or processed, is gaining rapid demand in the market. The global market for mushrooms was valued at US$29 427 million in 2013 and is projected to grow at a compound annual growth rate of 9.5% from 2014 to reach $50 034 million by 2019. Europe dominated the market in 2013 and is expected to be the fastest-growing market for mushrooms between 2014 and 2019, followed by the Asia Pacific region.

Mushroom production emits around 73 g $CH_4$/t of straw (dry weight) according to Ngo (2011). This amount is small compared to the increase of $CH_4$ emissions after incorporating fresh straw into the field. In addition, another product is grown, which further brings down the emissions on a yield basis.

## 6.3 Using rice straw for livestock feed

Livestock ownership of cattle or buffalo is common in rice-based farming communities and has been a way of augmenting on-farm income. In traditional systems, these animals are also used as sources of power during land preparation and as means of transportation. Approximately 80% of the world's rice is grown by small-scale farmers in many developing countries, including SEA, and it is common to use rice straw as animal feed (Sarnklong, 2010). Rice straw is the principal crop residue fed to more than 90% of the ruminant livestock in this area (Devendra and Thomas, 2002). Rice straw is especially important during periods when other feeds are inadequate. In general, the daily maximum intake of rice straw by ruminants is about 1.0–1.2 kg per 100 kg live weight (Devendra and Thomas, 2002).

On the other hand, unsupplemented rice straw, according to Clawson and Garrett (1970), is too low in digestible energy, crude protein, calcium and phosphorus for it to be used as the source of these nutrients for cattle. Rice straw is also apparently low in cobalt, copper, magnesium and sulfur, with the possibility of bordering deficiencies of these minerals. Rice straw is low in protein (2–5% DM), high in fibre and lignin content (NDF > 50%) and low in DM digestibility (< 65%), thus resulting in low voluntary feed intake (1.5–2.0%) (Wanapat et al., 2009). Rice straw is considered as a poor feed for animals due to its high silica content (Singh and Sidhu, 2014); however, their stems, which have lower silica content than the leaves, are more digestible (Drake et al., 2002). Therefore, the rice crop should be cut as close to the ground as possible if the straw is to be fed to livestock. Rice straw has a slight pubescence (small hairs) that cows may take some time to adapt to. It is also high in oxalates, which decrease the absorption of calcium (Drake et al., 2002).

Rice straw can be treated with urea, which releases ammonia after dissolving in water (Schiere et al., 1989). Urea is safer for practical use by farmers than using anhydrous or aqueous ammonia; it also provides a source of nitrogen (crude protein), of which straw is deficient. The feeding value of rice straw is significantly improved when it is treated with sodium hydroxide (NaOH) or ammonia ($NH_3$), both of which improve the digestibility of cellulose, which comprises 35–40% of the straw (Hanafi et al., 2012). Feeding trials demonstrated that treated rice straw is consumed more readily by cattle and sheep than untreated straw. It is also a better source of carbohydrates and produces more weight gain.

# 7 Summary

A concept of integrated utilization of rice straw and rice husk is shown in Fig. 14. Rice husk is used to generate heat for drying. Rice straw is used for AD, processing livestock fodder or mushroom production. Ash/biochar from rice husk combustion, digestate/waste from AD, livestock and mushroom production is used in composting and producing organic fertilizer. The growth of mushrooms on rice straw helps decompose its structural carbohydrates and reduce the indigestible cell wall components, thus increasing the cell wall digestibility of the straw (Kholif, 2015). Fungal lignocellulolytic enzymes break the polysaccharide–lignin complex, which results in higher degradation of the straw.

Table 4 summarizes a number of characteristics of rice straw processing and management options, including barriers to adoption, outcomes and consequences of particular activities and GHGE. Open-field burning is the simplest and lowest cost management option for rice straw. Hence, about 60–80% of rice straw is burned in SEA. However, this practice results in the loss of nutrients from the system, releases hazardous pollutants into the

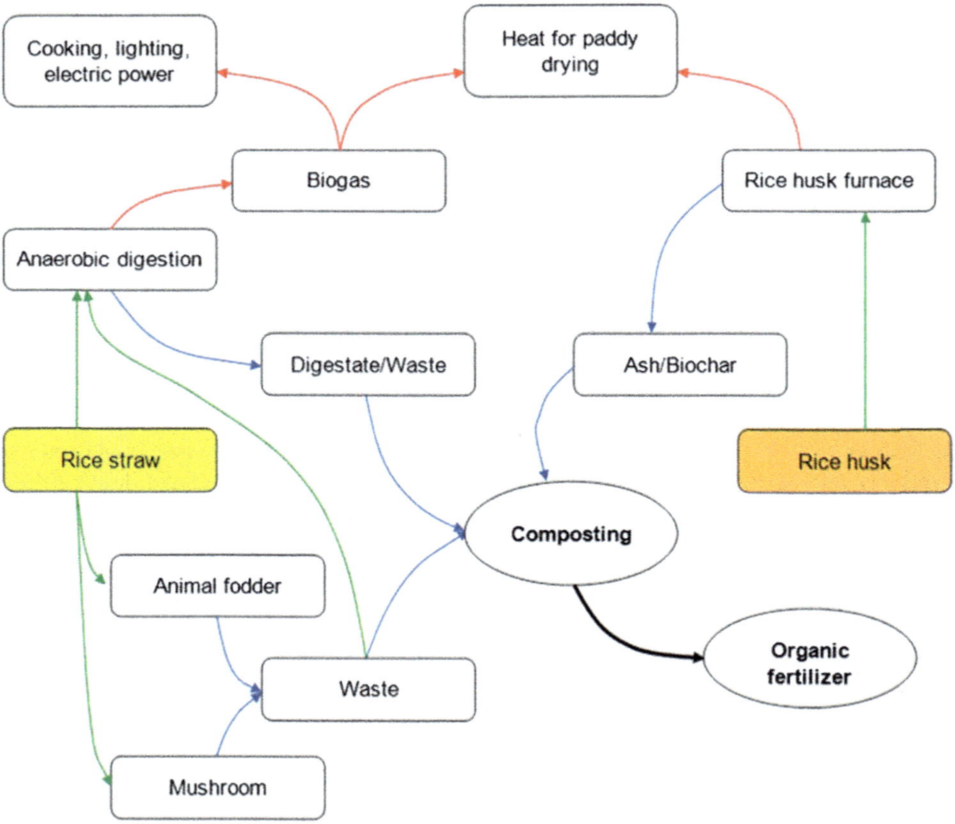


**Figure 14** Integrated utilization of rice straw and husk.

**Table 4** Characteristics of different rice straw processing and management options

| Options | Barriers | Consequences | GHGE |
|---|---|---|---|
| *On-field* | | | |
| Open-field burning | Illegal activity in most countries | 100% of N, 25% of P and 20% of K present in straw are lost<br>Hazardous air pollution | 0.7–4.1 g $CH_4$ and 0.019–0.057 g of $N_2O$ per kg rice straw burned |
| Incorporation | Time for decomposition (> 3 weeks) | Maintaining the nutrient balance | Strongly increases field $CH_4$ emissions in the following season |
| *Off-field* | | | |
| Collection (for rice straw) | Bulky (70–80 kg $m^{-3}$)<br>Cost of collection | Feedstock for further processing | Some emissions from diesel consumption |
| Chopping | Silica content of straw causes wearing down of machinery components | Materials for further processing | Some emissions from energy consumption |
| Combustion to generate energy | Melting and fouling on the grate (ash-related problem) | Rice husk: use for small-scale paddy dryers; efficiency 50–80%<br>Rice straw: 12–30% efficiency | Very low, if managed properly |
| Anaerobic digestion (rice straw) | Silica and lignin hamper the digestibility. | Efficiency about 30% | Very low, if digester is not leaking |
| Biochar production | Related to energy balance, production and application cost | Can increase yields, particularly in poor soils | Lower than fresh straw incorporation |
| Mushroom production | High cost of feed stock (collection and transportation) | Yield: 5–10% of rice straw (50–100 kg mushroom per 1 ton of dried straw) | 35–110 g $CH_4$ (or 0.9–2.8 kg $CO_2$-eq) per ton of straw, but generates another product |
| Livestock feed | High cost of feedstock; silica and lignin hamper digestibility | Intake of rice straw by ruminants is about 1.0–1.2 kg per day for 100 kg live weight | Emissions might vary compared to other feed sources |

atmosphere and emits more than 100 kg $CO_2$ equivalent (converted from $CH_4$ and $N_2O$ emissions) and also tract amount of human toxicity per ton of straw.

Incorporation of rice straw can help maintain soil fertility, but at least three weeks are required between incorporation and crop establishment to allow adequate decomposition. With a short turnaround time, this practice can lead to the formation of organic compounds in the rhizosphere that are toxic to rice. There are several existing technologies to speed up the decomposition time, such as using inoculums of *Trichoderma* sp. However, the availability of these products is limited and their cost tends to be high and are hence not being widely adopted in rice production in SEA.

Off-field uses of rice straw for further processing, such as bioenergy production, composting, mushroom production and livestock fodder are less common. The collection of rice straw is costly, but the development and dissemination of rice straw baling systems can help overcome this barrier. Combustion of rice husk for paddy drying is highly efficient and has a low cost, as rice husk is accrued in rice mills. A promising energy conversion technology for rice straw is AD, which has a low energy cost and small environmental footprint. However, for rice straw, this technology is still in development.

# 8 Future trends

Value chain development often follows an upgrading trajectory, which begins with process upgrading, moves on to product upgrading and then on to functional and inter-sectoral upgrading (Gereffi, 1999; Kaplinsky and Morris, 2000). In the case of rice value chains, Latin America is already well advanced in this upgrading trajectory, followed by Asia, where these transformations are currently rapidly occurring and then by Africa, where this trend of upgrading is now also emerging. However, within continents, transformations of rice value chains are not occurring uniformly; rather, they occur in waves, which are correlated with initial levels of income and development and degree of openness and market liberalization (Reardon and Timmer, 2014).

Urbanization, rising income levels and diet change trigger the demand for rice with superior quality attributes such as aroma, purity, homogeneity, packaging, food safety, traceability, nutrition, health and convenience. In some markets, consumers increasingly care about certification of production practices (organic, fair trade, sustainably produced) and supporting cultural heritage of traditional varieties and practices. This currently induces investment in process upgrading (e.g. post-harvest and processing technology, good agricultural practices), product upgrading (of intrinsic quality and extrinsic quality cues such as packaging, branding and advertising) and functional upgrading (e.g. contract farming by midstream value chain actors to ensure quality in sourcing paddy from farmers more efficiently).

Rising income levels also trigger a demand for a diverse diet toward (i) non-grains (meat, fish, dairy, fruit, vegetables) and animal feed; (ii) processed products and (iii) prepared foods bought away from home (Reardon et al., 2014). Together with the development of the industrial and manufacturing sectors and their concomitant demand for energy and fibre, this provides tremendous opportunities for inter-sectoral upgrading of rice value chains.

In Asia, rice straw is a tradeable commodity in fodder markets and transported for hundreds of kilometres to deficit areas, a direct reflection of the strong demand in the market. Evidence from dairy producers in India suggests that the market is able to differentiate and reward differences in straw quality, with observed price premiums of 33% and higher for higher quality straw. This implies that rice breeders can strengthen rice straw value chains and increase the livelihoods of both rice and dairy farmers in India by taking into account the dual 'food-feed' purpose of rice in their breeding programmes. In Japan, the development of rice varieties for whole crop silage (WCS) is strongly promoted by the government because of the increasing dairy and meat consumption and decreasing rice per capita consumption. Rice varieties for WCS have better resistance to pests such as rice blast and stripe, with better lodging resistance, and yields have high total digestible nutrients as compared with ordinary rice varieties (Kato, 2008).

Both examples illustrate that rice breeders can achieve maximum impact if they manage to find the optimal mix between breeding for (i) grain quality traits, hence fostering product upgrading in rice value chains; (ii) agronomic and stress tolerance traits in order to improve the resilience of rice varieties and sustainability of the production process, hence facilitating process upgrading in rice value chains and (iii) by-product quality traits, which will enable tapping rice value chains in other markets (feed, mushroom, energy, fibre, etc.). More generally, the success of value chain development of rice residues will hinge on investments in inter-sectoral upgrading, triggered by the demand for food, feed, energy and fibre from a growing urban population and the growing food and non-food industries.

This chapter has shown that there are already many technical and management options for rice husk and rice straw utilization in the context of sustainable rice production. It also has elaborated on new research activities, particularly focusing on better use of rice straw. However, these have been initiated very recently and there are, so far, very few practical and commercialized solutions out there that could help farmers manage by-products more sustainably, adding value to their rice crop. Although value chain development of rice residues is still in the initial stage, it is expected to become more dominant in regions with a large surplus in production – those with increasing demand for food, feed, energy and fibre.

Further research and interventions are suggested in the following areas:

- Forecasting of developments and trends in shifting rice production practices, as well as the markets for rice by-products that will have an effect on their availability and market potential.
- Further quantification of GHGE from different rice straw management practices in a cropping system context since on-field straw management also affects GHGE during rice production. Incorporation, for example, significantly increases $CH_4$ emissions.
- Pilot and verify business models for farmers and service providers for rice straw collection.
- Adaptive research and development of component technologies that currently constitute bottlenecks for an effective by-product management value chain. For straw densification and chopping, for example, technologies usually designed for other cereals with less abrasive properties are used, but have very low durability on rice straw or husk.
- Matching of technologies and scale to prevailing conditions in terms of feedstock availability, demand for by-products and value-adding potential for different stakeholders.
- Capacity building of intermediaries on techniques for the quantification of GHGE.
- Life cycle and sustainability assessments for different by-product management options in different cropping systems and agro-ecological conditions.
- Advocacy and policy advice on the effects of open-field burning, on other factors that affect sustainability of rice production and on improved management options.
- Market research on the needs of consumers, rice farmers and value chain stakeholders in terms of products or technologies to determine the optimal mix between (i) breeding for grain quality, agronomic, stress tolerance and by-product traits and (ii) investment in process, product, functional and inter-sectoral upgrading of rice value chains.

# 9 Where to look for further information

Some resources that provide relevant information on rice by-products can be found in the references cited in this chapter and in the online platforms of research institutions such as International Rice Research Institute, International Livestock Research Institute, Cuu Long Rice Research Institute, University of Arkansas and University of California.

# 10 References


Alberto, M. C. R., Wassmann, R., Gummert, M., Buresh, R., Quilty, J., Correa, T., Centeno, C. A. R. and Oca, G. 2015. Straw incorporated after mechanized harvesting of irrigated rice affects net emissions of CH4 and CO2 based on eddy covariance measurements. *Field Crops Res.* 184: 162–75.

Antal, M. J. and Gronli, M. 2003. Science and technology of charcoal production. *Ind. Eng. Chemical Res.* 42: 1619–40.

Azhar Uddin, Md. and Phuong, H. T. 2009. *Development of Technologies for the Utilization of Agricultural and Forestry Wastes: Preparation of Biochar from Rice Residues.* Okayama University. 3-1-1 Tsushima Naka, Kita-ku, Okayama 700–8530, Japan.

Baker, R. 2000. Biomass fuel leaching for the control of fouling, slagging, and agglomeration in biomass power generation. PhD Dissertation, University of Califonia.

Bijay-Singh, Shan, Y. H., Johnson-Beebout, S. E., Yadvinder-Singh and Buresh, R. J. 2008. Crop residue management for lowland rice-based cropping systems in Asia. *Adv. Agron.* 98: 117–99.

Carbon Market Watch. 2009. Mitsubishi's rice husk projects in Thailand (Newsletter #5). http://carbonmarketwatch.org/mitsubishis-rice-husk-projects-in-thailand-newsletter-5/.

Chang, C.-H., Jeon, S., Cosadio, A., Icalina, C., Panganiban, R., Quirino, R. and Song, Y. 2014. Status and prospect of mushroom industry in the Philippines. *JPAIR Multidisc. Res.* 16(1). doi:http://dx.doi.org/10.7719/jpair.v16i1.268.

Chang, C.-H., Liu, C.-C. and Tseng, P.-Y. 2013. Emissions inventory for rice straw open burning in Taiwan based on burned area classification and mapping using Formosat-2 satellite imagery. *Aerosol Air Qual. Res.* 13: 474–87.

Clawson, W. J. and Garrett, W. N. 1970. Ricestraw utilization by livestock. *Calif. Agric.* 24(9): 13–14. September 1970.

Corton, T. M., Bajita, J. B., Grospe, F. S., Pamplona, R. R., Assis, Jr., C. A., Wassmann, R., Lantin, R. S. and Buendia, L. V. 2000. Methane emission from irrigated and intensively managed rice fields in central Luzon (Philippines). *Nutr. Cycl. Agroecosys.* 58: 37–53.

Devendra, C. and Thomas, D. 2002. Crop-animal interactions in mixed farming systems in Asia. *Agric. Syst.* 71: 27–40.

Dobermann, A. and Fairhurst, T. H. 2002. Rice straw management. *Better Crops Int.*, Vol. 16, Special Supplement, *May*. http://www.ipni.net/publication/bci.nsf/0/163087B956D0EFF485257BBA006531E8/$FILE/Better%20Crops%20International%202002-3%20p07.pdf.

Dobie, J. B., Parsons, P. S. and Curley, R. G. 1973. Systems for handling and utilizing rice straw. *Transact. ASAE* 16(3): 533–6.

Drake, D., Nader, G. and Forero, L. 2002. Feeding rice straw to cattle. Publication 8079. University of California. Division of Agriculture and Natural Resources. http://anrcatalog.ucdavis.edu.

Engling, G., Lee, J. and Tsai, Y. 2009. Size-resolved anhydrosugar composition in smoke aerosol from controlled field burning of rice straw. *Aerosol Sci. Tech.* 43(7): 662–72. doi:10.1080/02786820902825113.

EPA, U.S., 1995. AP 42: Compilation of Air Pollutant Emission Factors, *Volume 1*: Stationary Point and Area Sources.

Gadde, B., Bonnet, S., Menke, C. and Garivait, S. 2009. Air pollutant emissions from rice straw open field burning in India, Thailand and the Philippines. *Environ. Pollut.* [Online] 157(5): 1554–8. doi:10.1016/j.envpol.2009.01.004.

Gaunt, J. L. and Lehmann, J. 2008. Energy balance and emissions associated with biochar sequestration and pyrolysis bioenergy production. *Environ. Sci. Technol.* 42(11): 4152–8.

Gereffi, G. 1999. International trade and industrial upgrading in the apparel commodity chain. *J. Int. Econ.* 48(1): 37–70.

Ghosh, A. and Bhattacharyya, B. C. 1999. Biomethanation of white rotted and brown rotted rice straw. *Bioprocess Eng.* 20: 297–302.

Goyal, S. and Sindhu, S. S. 2011. Composting of rice straw using different inocula and analysis of compost quality. *Microbiol. J.* 4: 126–38.

Guillemot, A., Bruant, R., Pasquiou, V. and Boucher, E. 2014. Feasibility study for the implementation of two ORC power plants of 1 MWe each using rice straw as a fuel in the context of a public-private partnership with the institutions PhilRice and UPLB. Courbevoie, France: ENERTIME.

Gummert, M. 2013. Rice straw and husks. Our Science, IRRI, Accessed on October. http://irri.org/index.php?option=com_k2&view=item&id=9967:straw-and-hull&lang=en.

Gummert, M. 2015. Scalable straw management options for improved farmer livelihoods, sustainability, and low environmental footprint in rice based production systems. BMZ proposal, IRRI Ref. No.A-2015-48.

Gummert, M., Chandrasekar, V., Phan, H. H., Nguyen, T. N., Ban, L. V., Aquino, E. and Rickman, J. F. 2007. Rice hull furnace – IRRI's experience. Paper presented at the 5th National Grains Postproduction Conference 18–19 July 2007, Philippines.

Gummert, M. and Haefele, S. M. 2012. Rice-based systems and rice residues. International workshop on the innovative use of rice straw and rice husk. http://postharvestla.irri.org/irrigated-rice-research-consortium/uk-bioenergy/uses-of-rice-straw-and-husks/haefele%20and%20gummert.pdf?attredirects=0&d=1.

Haefele, S. M., Konboon, Y., Wongboon, W., Amarante, S., Maarifat, A. A., Pfeiffer, E. M. and Knoblauch, C. 2011. Effects and fate of biochar from rice residues in rice-based systems. *Field Crops Res.* 121(3): 430–40.

Hanafi, E. M., Khadawry, H. H., Ahmed, W. M. and Zaabal, N. M. 2012. Some observations on rice straw with emphasis on updates of its management. *World Appl. Sci. J.* 16(3): 354–61.

Hayashi, K., Ono, K., Kajiura, M., Sudo, S., Yonemura, S., Fushimi, A. and Tanabe, K. 2014. Trace gas and particle emissions from open burning of three cereal crop residues: increase in residue moistness enhances emissions of carbon monoxide, methane, and particulate organic carbon. *Atmos. Environ.* 95: 36–44.

Horsfield, B. C, Jenkins, B. M. and Becker, C. 1977. Agricultural residue as an alternative source of energy for the Pacific gas and electric company. Research report. University of California, Davis: Department of Agricultural Engineering.

IRRI. 2014. Addressing the Pre- and Postharvest Challenges. IRRI Ref. No.: A-2012-69. Final report.

Jamieson, C., Nguyen, V. H, Gummert, M., Thornly, P. and Shield, I. 2014. Availability of rice straw for energy in South and Southeast Asia. International Rice Congress – Bangkok.

Jenkins, B. M. 1998. Physical properties of biomass. In O. kitani and C. W. Hall (eds), *Biomass Handbook*, Chapter 5.2. Gordon and Breach, New York.

Jenkins, B. M., Bakker, R. R. and Wei, J. B. 1996. On the properties of washed straw. *Biomass Bioenergy* 10(4): 177–200.

Jenkins, B. M., Mehlschau, J. J., Williams, R. B., Solomon, C., Balmes, J., Kleinman, M. and Smith, N. 2003. Rice straw smoke generation system for controlled human inhalation exposures. *Aerosol Sci. Technol.* 37(5): 437–54. doi:10.1080/02786820300977.

Johnson-Beebout, S. E., Angeles, O. R., Alberto, M. C. R. and Buresh, R. J. 2009. Simultaneous minimization of nitrous oxide and methane emission from rice paddy soils is improbable due to redox potential changes with depth in a greenhouse experiment without plants. *Geoderma* 149(1): 45–53.

Jung, S.-H., Kang, B.-S. and Kim, J.-S. 2008. Production of bio-oil from rice straw and bamboo sawdust under various reaction conditions in a fast pyrolysis plant equipped with a fluidized bed and a char separation system. *J. Anal. Appl. Pyrolysis*. 82(2008): 240–7.

Kanokkanjana, K. and Garivait, S. April 2013. Alternative rice straw management practices to reduce field open burning in Thailand. *Int. J. Environ. Sci. Dev.* 4(2): 119–23.

Kaplinsky, R. and Morris, M. 2000. A Handbook for Value Chain Research. International Development Research Center (IDRC), Ottawa, Canada.

Kargbo, F. R., Xing, J. and Zhang, Y. 2009. Pretreatment for energy use of rice straw: a review. *Afr. J. Agric. Res.* 4(13): 1560–65.

Kato, H. 2008. Review development of rice varieties for Whole Crop Silage (WCS) in Japan. *JARQ* 42(4): 231–6.

Klass, L. 1998. Biomass for Renewable Energy and Fuels. Entech International, Inc. Barrington, Illinois, United States.

Kholif, A. E. 2015. *Utilization of Biodegraded Rice Straw in Lactating Animal Rations Utilization of Biodegraded Rice Straw in Lactating Animal Rations*. [Online]. https://www.lap-publishing.com/catalog/details//store/gb/book/978-3-659-49825-1/utilization-of-biodegraded-rice-straw-in-lactating-animal-rations.

Lehmann, J. and Joseph, S. 2009. Biochar for Environmental Management: An Introduction. *Book Publication: Biochar Environ. Manage*. Chapter 1: pp. 1–12.

Li, J., Lu, J., Li, X., Ren, T., Cong, R. and Zhou, L. 2014. Dynamics of potassium release and adsorption on rice straw residue. *PLoS ONE* 9(2): e90440. doi: 10.137/journal.pone.0090440.

Libra, J. A., Ro, K. S., Kammannn, C., Funke, A., Berge, N. D. and Neubauer, Y. 2011. Hydrothermal carbonization of biomass residuals: a comparative review of chemistry, processes and applications of wet and dry pyrolysis. *Biofuels* 2(1): 71–106.

Lim, J. S., Manan, Z. A., Wan Alwi, S. R. and Hashim, H. 2012. A review on utilisation of biomass from rice industry as a source of renewable Energy. Elsevier Renewable and Sustainable Energy Reviews 16(2012): 3084–94.

Liu, H., Zhang, L., Han, Z., Xie, B. and Wu, S. 2013. The effects of leaching methods on the combustion characteristics of rice straw. *Biomass Bioenergy* 49: 22–7.

Lu, Ping and Hsieh, Y.-L. 2012. Preparation and characterization of cellulose nanocrystals from rice straw. *Carbohydr. Polym.* 87(1): 564–73.

Mandal, K. G., Misra, A. K., Hati, K. M., Bandyopadhyay, K. K., Ghosh, P. K. and Mohanty, M. 2004. Rice residue management options and effects on soil properties and crop productivity. *Food Agric. Environ.* 2: 224–31.

Manya, J. J., Roca, F. X. and Perales, J. F. 2013. TGA study examing the effect of presuure and peak temperature on biochar yield during pyrolysis of two-phase olive mill waste. *J. Anal. Appl. Pyrolysis*. 103: 86–95.

Marketsandmakers.com, accessed 2015. Mushroom market by type and by Region – Global Trends and Forecast to 2019. http://www.marketsandmarkets.com/PressReleases/mushroom.asp.

Miura, Y. and Kanno, T. 1997. Emissions of trace gases ($CO_2$, CO, $CH_4$, and $N_2O$) resulting from rice straw burning. *Soil Sci. Plant Nutr.* 43(4): 849–54.

Munder, S. 2013. Improving thermal conversion properties of rice straw by briquetting. Master-Thesis. University of Hohenheim.

Mussoline, W., Esposito, G., Giordano, A. and Lens. P. 2013. The anaerobic digestion of rice straw: a review. *Crit. Rev. Env. Sci. Tec.* 43(9): 895–915.

Ngo, T. T. T.. 2011. Comparative assessment of using rice straw for rapid composting and straw mushroom production in mitigating greenhouse gas emissions in Mekong Delta, Vietnam and Central Luzon, Philippines. Unpublished dissertation for PhD in Environmental Science. SESAM, UPLB.

Ngo, T. T. T., Sumalde, Z. M., Espaldon, M. V. O., Pacardo, E. P., Rapera, C. L. and Palis, F. G. 2012. Farmers' awareness and factors affecting adoption of rapid composting in Mekong Delta, Vietnam and Central Luzon, Philippines. *J. Environ. Sci. Manag.* 15(2): 59–73.

Nguyen, V. H., Nguyen, TN., Tran, V.T., Nguyen, V.X., and Phan, H. H.. 2012. Rice husk uses in the Mekong Delta of Vietnam. International workshop on the innovative use of rice straw and rice husk, *IRRI*-2013. https://docs.google.com/r?a=v&pid=sites&srcid=aXJyaS5vcmd8cG9zdGhhdmVzdC11bml0fGd4OjcyOWMzZTg2MGUwZDZlZDU.

Nguyen, V. H., Topno, S., Balingbing, C., Roeder, M., Quilty, J., Jamieson, C., Thornley, P. and Gummert, M. 2015. Energy balance analysis of rice straw for anaerobic digestion. *EUBCE*. http://www.etaflorence.it/proceedings/?detail=11054&mode=topic&categories=0&items=--&searchstring=Energy%20balance%20analysis%20of%20rice%20straw%20for%20anaerobic%20digestion&limit=2015.

Nguyen, V. H., Truong, Q.T., Nguyen, T.N., and Phan, H. H. 2013. Final report of ADB RETA NO 6489, subcomponent 2 'Reducing postharvest losses and ncreasing farmers' income by producing better-quality rice' in Vietnam.

Nguyen, V. X., Truong, V. and Phan, H. H. 1996. Development of rice husk furnaces for grain drying. http://aciar.gov.au/files/node/316/grain_drying_in_asia_part_9_11398.pdf.

Oanh, N. T., Ly, B. T. and Tipayarom, D. 2011. Characterization of particulate matter emission from open burning of rice straw. *Atmos. Environ.* 45(2): 493–502. doi:10.1016/j.atmosenv.2010.09.023.

Pandey, A., Mai, V. T., Duong, Q. V., Bui, T. P. L., Mai, T. L. A., Jensen, L. S., de Neergaard, A. 2014. Organic matter and water management strategies to reduce methane and nitrous oxide emissions from rice paddies in Vietnam. *Agric. Ecosyst. Environ.* 196: 137–46.

Peng, J., Lü, Z., Rui, J. and Lu, Y. 2008. Dynamics of the methanogenic community during plant residue decomposition in an anoxic rice field soil. *Appl. Environ. Microb.* 74(9): 2894–901.

Ponnamperuma, F. N. 1984. Straw as a source of nutrients for wetland rice. Organic Matter and Rice, pp. 117–35. International Rice Research Institute, Los Baños, Philippines.

Prasad, R., Gangaiah, B. and Aipe, K. C. 1999. Effect of crop residue management in a rice-wheat cropping system on growth and yield of crops and soil fertility. *Exp. Agric.* 35: 427–35.

Reardon, T., Chen, K. Z., Minten, B., Adriano, L., Dao, T. A., Wang, J. and Das Gupta, S. 2014. The quiet revolution in Asia's rice value chains. *Ann. N. Y. Acad. Sci.* 1331: 106–18.

Reardon, T. and Timmer, C. P. 2014. Five inter-linked transformations in the Asian agrifood economy: Food security implications. *Global Food Security* 3(2): 108–17.

Reddy, N. and Yang, Y. 2004. Biofibers from agricultural byproducts for industrial applications. *Trends Biotechnol.* 23(1): 22–7, January 2005. doi: 10.1016/j.tibtech.2004.11.002.

Reyes, R. G. November 2000. Indoor cultivation of paddy straw mushroom, Volvariella volvacea, in crates. Center for Tropical Mushroom Research and Development, Dept. of Biological Sciences, College of Arts and Sciences, CLSU, Nueva Ecija, Philippines. Mycologist, Volume 14, Part 4.

Romasanta, R. and Sander, B. O.2016. Personal Communication About Ongoing Research at IRRI.

Sadaka, S. and Negi, S., 2009. Improvements of biomass physical and thermochemical characteristics via torrefaction process. *Environ. Prog. Sustainable Energ.* 28(3). http://www.scopus.com/inward/record.url?eid=2-s2.0.

Sander, B. O., Samson, M. and Buresh, R. J. 2014. Methane and nitrous oxide emissions from flooded rice fields as affected by water and straw management between rice crops. *Geoderma* 235–36 (2014): 355–62.

Sarnklong, C., Cone, J. W., Pellikaan, W. and Hendriks, W. H. May 2010. Utilization of rice straw and different treatments to improve its feed value for ruminants: a review. *Asia-Aust. J. Anim. Sci.* 23(5): 680–92.

Schiere, J. B., Ibrahim, N. M., Sewalt, V. J. H. and Zemmelink, G. 1989. Response of growing cattle given rice straw to lick blocks containing urea an molasses. *Anim. Feed Sci. Technol.* 26: 179–89.

Shen, J., Tang, H., Liu, J., Wang, C., Li, J., Ge, T., Jones, D. and Wu, J. 2014. Contrasting effects of straw and straw-derived biochar amendments on greenhouse gas emissions within double rice cropping systems. *Agric. Ecosyst. Environ.* 188(2014): 264–74

Siemers, D. W. 2011. Technical and Economic Prospects of Rice Residues for Energy in Asia. Sustainable Bioenergy Symposium at RE Asia Bangkok, 2 June 2011. http://www.fao.org/fileadmin/templates/rap/files/meetings/2011/110602_technical.pdf.

Silalertruksa, T. and Gheewala, S. H. 2013. A comparative LCA of rice straw utilization for fuels and fertilizer in Thailand. *Biores. Technol.* 150(2013): 412–19.

Singh, Y. and Sidhu, H. S. March 2014. Management of cereal crop residues for sustainable rice-wheat production system in the Indo-gangetic plains of India. *Proc. Indian Natn. Sci. Acad.* 80(1): 95–114.

Subbukrishna, D. N., Suresh, K. C., Paul, P. J., Dasapa, S. and Rajan, N. K. S. 2015. Precipated silica from rice husk ash by ipsit process. 15th European Biomass Conference and Exhibition. http://cgpl.iisc.ernet.in/site/Portals/0/Publications/NationalConf/Preciptedsilicafromricehusk.pdf.

Surekha, K., Padma Kumari, A. P., Narayana Reddy, M., Satyanarayana, K. and Sta Cruz, P. C. 2003. Crop residue management to sustain soil fertility and irrigated rice yields. *Nut. Cyc. Agroeco.* 67: 145–54.

Thuy, N. H., Shan, Y., Bijay-Singh, Wang, K., Cai, Z., Yadvinder-Singh and Buresh, R. J. 2008. Nitrogen supply in rice-based cropping systems as affected by crop residue management. *Soil Sci. Am. J.* 72: 514–23.

Tillman, D. A. 2000. Biomass co-firing: the technology, the experience, the combustion consequences. *Biomass Bioenergy* 19(2000): 365–84.

Tran, V. T., Nguyen, V. H., Nguyen, V. H., Nguyen, H. T., Le, Q. V., Truong, Q. T. and Phan, H. H. 2013. Design and development of the 10-ton/batch columnar paddy dryer. In Proceedings of International Workshop on Agricultural Engineering and Post-harvest Technology for Asia Sustainability. *Science and Technology Publishing House*, pp. 152–63.

Wanapat, M., Polyorach, S., Bonnop, K., Mapato, C. and Cherdthong, A. 2009. Effects of treating rice straw with urea or urea and calcium hydroxide upon intake, digestibility, rumen fermentation and milk yield of dairy cows. *Livest. Sci.* 125(2009): 238–43.

Witt, C., Buresh, R. J., Balasubramanian, V., Dawe, D. and Dobermann, A. 2002. Improving nutrient management strategies for delivery in irrigated rice in Asia. *Better Crops International* 16(2): 24–31.

Witt, C. and Dobermann, A. 2002. A site specific nutrient management approach for irrigated, lowland rice in Asia. *Better Crops International* 16(1): 20–4.

Woolf, D., Amonette, J. E., Street-Perrott, F. A., Lehmann, J. and Joseph, S. 2010. Sustainable biochar to mitigate global climate change. *Nat. Commun.* 1(5): 1–9. doi:10.1038/ncomms1053. ISSN 2041-1723.

Wua, W., Yang, M., Feng, Q., Kim, M., Wang, H., Lu, H. and Chen, Y. 2012. Chemical characterization of rice straw-derived biochar for soil amendment. *Biomass Bioenergy* 47(2012): 268–76.

Xiao, L.-P., Shi, Z.-J., Xu, F. and Sun, R.-C. 2012. Hydrothermal carbonization of lignocellulosic biomass. *Bioresour. Technol.* 118(2012): 619–23.

Yadvinder-Singh, Bijay-Singh, Ladha, J. K., Khind, C. S., Khera, T. S. and Bueno, C. S. 2004. Effects of residue decomposition on productivity and soil fertility in rice–wheat rotation. *Soil Sci. Am. J.* 68: 854–64

Yadvinder-Singh, Bijay-Singh and Timsina J., 2005. Crop residue management for nutrient cycling and improving soil productivity in rice-based cropping systems in the tropics. *Adv. Agron.* 85: 269–407.

Yoneyama, T. and Yoshida, T. 1977. Decomposition of rice residue in tropical soils III. Nitrogen mineralization and immobilization of rice residue during its decomposition in soil. *Soil Sci. Plant Nutr.* 23(2): 175–83.

Yuan, Q. and Conrad, R. 2014. Straw application in paddy soil enhances methane production also from other carbon sources. *Biogeosciences* 11: 237–46.

Zafar, S. 2015. Rice straw as bioenergy resource. Bioenergy Consult. 26 March 2015. http://www.bioenergyconsult.com/tag/rice-residues/.

Zhang, A., Bian, R., Pan, G., Cui, L., Hussain, Q., Li, L., Zheng, J., Zheng, J., Zhang, X., Han, X. and Yu, X. 2012. Effects of biochar amendment on soil quality, crop yield and greenhouse gas emission in a Chinese rice paddy: a field study of 2 consecutive rice growing cycles. *Field Crops Res.*, 127: 153–60.

Zhang, R., Li, X. and Fadel, J. G. 2002. Oyster mushroom cultivation with rice and wheat straw. *Bioresour. Technol.*, 82, 277–84.

# Part 2

# Overall management of rice cultivation

Chapter 6

# Yield gap analysis towards meeting future rice demand

*Kazuki Saito, Pepijn van Oort, Ibnou Dieng, Jean-Martial Johnson, Abibou Niang, Kokou Ahouanton and Amakoe Delali Alognon, Africa Rice Center, Benin; Atsuko Tanaka, JICA and Africa Rice Center, Benin; Kalimuthu Senthilkumar and Elke Vandamme, Africa Rice Center, Tanzania; Cyriaque Akakpo, INRAB, Benin; Zacharie Segda, CNRST/INERA, Burkina Faso; Ibrahim Bassoro and Delphine Mapiemfu Lamare, IRAD, Cameroon; Moundibaye D. Allarangaye, ITRAD, Chad; Henri Gbakatchetche, CNRA, Côte d'Ivoire; Belay A. Bayuh, EIAR, Ethiopia; Famara Jaiteh, NARI, The Gambia; Ralph K. Bam, CSIR-CRI; Wilson Dogbe, CSIR-SARI, Ghana; Keita Sékou, IRAG, Guinea; Raymond Rabeson, FOFIFA, Madagascar; Nianankoro Kamissoko, IER, Mali; Illiassou Maïga Mossi, INRAN, Niger; Oladele S. Bakare, NCRI, Nigeria; Fanny L. Mabone, INERA, DR Congo; Elie R. Gasore, RAB, Rwanda; Idriss Baggie, SLARI, Sierra Leone; Geophrey J. Kajiru, Ministry of Agriculture, Tanzania; Komlan A. Ablede, ITRA, Togo; and David Nanfumba, NARO, Uganda*

1 Introduction

2 Studies on quantification of yield gaps

3 Studies to identify causes of yield gaps

4 Future trends and conclusion

5 Where to look for further information

6 References

## 1 Introduction

The global demand for staple crop products is expected to grow by 60% between 2010 and 2050 mainly as a result of increased population, per capita income growth and use of biofuels (Alexandratos and Bruinsma, 2012). This is also the case for rice, which provides 20% of the world's food energy (Fischer et al., 2014). Global demand for rice will continue to increase from 479 million tons in 2014/15 to 551 million tons in 2029/30 (GRiSP, 2016). To meet this estimated future demand, agricultural production must be enhanced in an environmentally sustainable manner in the context of increasing competition for water, land and labour, as well as the constantly evolving threat from abiotic and biotic stresses, and under potentially more extreme weather conditions associated with climate change. Agriculture also needs to address problems of poor food and nutrition security for smallholder rice farmers, including women farmers, in Asia, Latin America and sub-Saharan Africa.

http://dx.doi.org/10.19103/AS.2016.0003.26

As there is little scope for easy expansion of agricultural land or irrigation – except for some areas in Latin America and sub-Saharan Africa – increased rice production mainly needs to be derived from existing agricultural land (Young, 1999; Ramankutty et al., 2002; You et al., 2011). This could be accomplished by reducing the 'yield gap', which is defined as the difference between potential yield and average on-farm yield obtained by farmers (Saito et al., 2013; van Ittersum et al., 2013). 'Potential yield' is defined as the maximum yield that can be obtained from a crop in a given environment as often determined by crop simulation models with plausible physiological and agronomic assumptions (Evans and Fischer, 1999). Under irrigated conditions, potential yield is determined by climate (solar radiation and temperature), varietal characteristics and crop establishment methods, including sowing date and density. Under rainfed conditions, potential yield is also affected by water availability, and refers to water-limited yield (Fischer, 2015).

There are other yield benchmarks such as 'economic yield', 'experimental yield' and 'best farmers' yield' which lie between potential yield (or water-limited potential yield) and average on-farm yield (Fig. 1). 'Economic yield' is defined as the yield attained by farmers with normal natural resources when economically optimal practices and levels of inputs have been adopted. Because of diminishing returns on investment and risk of financial loss when actual yield is closer to potential yield, economic yields in farmers' fields do not exceed a certain portion of the potential yield. Economic yields can lie 20–30% below the potential yield under favourable weather conditions (Lobell et al., 2009; Fischer, 2015). However, as economic yield depends largely on local rice price and production cost and changes over time, we will not deal with this benchmark in this paper. 'Experimental yield' can be defined as maximum yield obtained with best management practices in researchers' managed trials (Lobell et al., 2009; Nhamo et al., 2014). 'Best farmers' yield' can be defined as the average yield of the upper 10th percentile among a sizeable sample of farmers in an area of interest (Laborte et al., 2012; Tanaka et al., 2015); others, however, define it as the maximum yield produced (e.g. Becker et al., 2003).

Comparing the difference in yield between experimental yield or best farmers' yield and average on-farm yield might be alternative options for assessing yield gaps in areas where potential yields cannot be estimated because validated crop-simulation models and input values to run them, such as long-term weather data, are lacking. However, there have been a few studies comparing different approaches to estimating yield gaps. In cases where this has been done, some approaches were criticized without considering how the different approaches may complement each other.

There are two broad categories of studies on yield gap analysis, focusing on (1) quantification of yield gaps and (2) identification of causes for the yield gaps. The first type of study is generally conducted on a large scale to identify areas where large increases in rice yield can be achieved or to quantify progress in yield gap closure in the long term. On the other hand, where the yield gap is small, the target may be to improve potential yield (or water-limited potential yield) through genetic improvement and alternative cropping systems and crop management practices. Furthermore, spatially explicit estimates of yield gaps are used as inputs to economic models to assess food security and support development of regional and national policies, and to identify areas where research and development interventions should be given high priority (van Ittersum et al., 2013; van Oort et al., 2015b). The second type of study is often conducted with limited spatial coverage in order to identify opportunities for yield improvement in farmers' fields. Closing these yield gaps may require adoption of alternative crop management options with improved rice varieties and/or capital investment in, for example, irrigation and drainage facilities. The two types of study can be complementary: the first identifies large-scale regions with large yield gaps (i.e. with high potential for yield increase) and the second follows with detailed studies in target regions to investigate how the yield gap can be closed. While they can in theory be complementary to each other, we observe that in practice this rarely happens. Very often, large-scale studies lack detailed follow-up (Sumberg, 2012) and detailed local studies (e.g. Tanaka et al., 2013, 2015; Stuart et al., 2016) lack a solid scientific justification for the selection of a specific study site, with the risk that they focus on regions of limited relevance for food security at the national or continental level. As a first step towards bringing these different approaches to yield gap analysis together, we review the most important rice yield gap analyses that have been conducted in the past decade at different scales.

In this chapter, we provide case studies of yield gap analyses using different benchmarks (potential yield, experimental yield and best farmers' yield) for estimating yield gaps. We start with studies focusing on the quantification of yield gaps and then describe studies with a focus on identification of the causes of yield gaps. Our case studies are mainly from sub-Saharan Africa, where local rice supply has never caught up with demand and many governments are keen to close the gap by enhancing local rice production, following their experience with the food crisis of 2007–8 (Wopereis et al., 2013). In this region, publications on rice yield gaps show similar results to those from South Asia, and higher yield gaps than in East and Southeast Asia, where review papers on local rice yield gap studies were recently published (Stuart et al., 2016). Then, we reflect on the ways in which different approaches can be used and how they can complement each other. At the end of the chapter, we conclude with a discussion of the challenges to achieving better quantification of yield gaps and their causes, and the implications of yield gap studies for sustainable agricultural development to meet future rice demand.

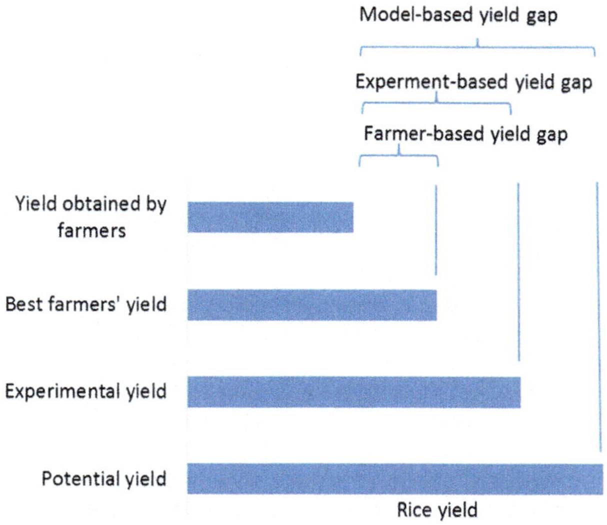


**Figure 1** Framework for estimation of yield gap based on three different potential yields (model-based 'potential yield', maximum experimental field 'experimental yield' and on-farm maximum 'best farmers' yield') (adapted from Lobell et al., 2009).

## 2 Studies on quantification of yield gaps

Following Fischer (2015), this review expresses the absolute difference between potential yield, experimental yield or best farmers' yield, and on-farm actual yield – the yield gap – as a percentage of average on-farm yields obtained by farmers, that is, (potential yield, experimental yield or best farmers' yield/on-farm actual yield – 1) x 100%.

This is because desirable increases in actual grain yields are inevitably expressed as a percentage of current actual yield, not current potential yield, experimental yield or best farmers' yield. On-farm yield has been reported from farmers' yield measurements, local experts, surveys, and/or local or national statistics. In recent years, on-farm yield can also be estimated from satellite images (Lobell, 2013). Depending on the type of study, different data sources have been used. Studies conducted at a large scale on the quantification of yield gaps (the first type of study discussed above) generally use sub-national or national statistics from FAOSTAT (van Ittersum et al., 2013), while studies on the causes of yield gaps (the second type of study discussed above) are often conducted at limited spatial scale using data from farmers' reports, crop cut through field survey, data estimated from satellite images or data-estimated from crop simulation models combined with satellite images (e.g. Tanaka et al., 2013, 2015; Lobell, 2013). As described by Grassini et al. (2015), data source selection for large-scale yield gap studies depends on data availability. Desirable data sources are in the following order: (1) most recent annual values reported at a fine spatial level; (2) annual yields reported at coarser spatial levels (e.g. sub-national level) or from census, surveys, trials and so on; and (3) yield reported by local experts or a national-level average. A major issue with sub-national or national statistics, especially for rice, is that the production system (e.g. irrigated lowland, rainfed lowland or rainfed upland rice) is often not classified (see Saito et al., 2013, for more information). It is well known that surface-water regimes and water sources (e.g. irrigation, rainfall and water table) determine rice production systems and potential or water-limited yield, as well as actual yield levels. As surface-water regimes, water sources and share of production systems at local or national levels are very variable (e.g. Diagne et al., 2013a), data from local or national statistics, which do not provide farm yield for each existing production system, are the least preferred option for yield gap studies on rice.

Opportunities for the application of remote sensing in yield gap analysis are still limited for rice (Lobell, 2013; Sadras et al., 2015). However, recent studies (Setiyono et al., 2015; Homma et al., 2016; Maki et al., 2016) successfully estimated on-farm yield by combining remote sensing with crop simulation models. In both studies, model uncertainties were minimized by recalibrating model parameters based on remotely sensed crop status on the ground (e.g. leaf area index). These methodologies need to be urgently studied in different rice-growing environments to examine whether they can provide a new approach for yield gap assessment. The main challenges which currently require innovative solutions are the availability and quality of remote sensing data and the potential misclassification of crops (Lobell, 2013). Furthermore, if rice is infested by weeds, remote sensing data cannot easily distinguish rice from weed coverage (Hirooka et al., 2015). The data also cannot detect damage from diseases and other pests in rice fields, and in these cases rice yield can be overestimated. As existing models do not consider deep flooding environments or flash flooding, rice yields under those conditions cannot be estimated.

Other issues for estimating on-farm yield for yield gap studies (e.g. number of years needed and sample size) have been discussed in several studies (Lobell et al., 2009; van

Ittersum et al., 2013; Fischer et al., 2014; Grassini et al., 2015) and will therefore not be described here.

## 2.1 Quantifying yield gaps using potential yield

We start with global studies on yield gap assessment (e.g. Mauser et al., 2015; Pradhan et al., 2015; West et al., 2014). These generally use empirical, statistical approaches or generic crop growth models and a grid-based approach using global datasets on weather, soils and general crop calendars. van Ittersum et al. (2013) critically reviewed these studies (Stehfest et al., 2007; Deryng et al., 2011; Mueller et al, 2012; Licker et al., 2010; Neumann et al., 2010; Monfreda et al., 2008), and indicated that yield gap estimates in existing global studies seem to be unreliable because of poorly validated methodologies and large uncertainties in the data, with the result that potential yield and water-limited yield differ between studies in the same areas. Furthermore, Van Wart et al. (2013a) argued that studies that rely on gridded weather databases to simulate agricultural productivity in current and future climates result in large uncertainties in estimates of potential yield or water-limited yield, because gridded weather data have a strong bias towards weather data measured through weather stations. Reasons for this bias were described in Van Wart et al. (2013a). However, if the bias can be corrected in gridded weather data and if local data are available, it is possible to estimate potential yield or water-limited yield at a finer scale. Together with existing rice cultivation maps and maps of soil and water resources, estimates of potential yield or water-limited yield in non-rice cultivation areas also help to identify areas where rice cultivation can be expanded, especially in sub-Saharan Africa.

Based on their literature review, van Ittersum et al. (2013) pointed out that a 'bottom-up' approach is needed to estimate locally relevant yield gaps, and recommended key methodologies that are transparent, reproducible and based on the best available science and data. These start with the application of locally validated crop simulation models to relatively homogenous climate zones (Van Wart et al., 2013b) for which measured weather data are available for estimating potential yield and water-limited yield. Within such zones, simulations are performed for the dominant soils using accurate agronomic input data and cropping systems. A location-specific yield gap is then estimated using current yield data. The use of robust crop simulation models permits an assessment to be made of yield stability and risk due to high or low temperatures and variability in rainfall. Location-specific estimates of yield gap are scaled up to regional and national levels by ensuring that all major climate zones, soil types and cropping systems are covered by the location-specific analyses, and by aggregating location-specific estimates to larger spatial scales by weighting for harvested crop area (van Bussel et al., 2015).

Using this 'bottom-up' approach, The Global Yield Gap and Water Productivity Atlas (GYGA) is producing an atlas covering maize (19 countries), wheat (17 countries), rice (12 countries), millet (11 countries), barley (10 countries), sorghum (10 countries), potato (2 countries), soyabean (1 country) and sugarcane (1 country) (http://www.yieldgap.org/yieldgaps, accessed at 30 March 2016). Where input data are of poor quality or not available, GYGA utilizes the best existing data, or proxy data, with the goal of improving them over time (see Table 3 in Grassini et al., 2015). Where measured weather data are scarce, a new propagation method based on limited years of observed daily maximum and minimum temperature and gridded NASA and TRMM weather data is used to generate long-term synthetic weather data for crop modelling (Van Wart et al., 2015).

In the following subsections, we describe what we learnt from rice modelling for the GYGA project and the results from GYGA, focusing on rice.

### 2.1.1 Rice modelling: lessons learnt

Scientists from the Africa Rice Center (AfricaRice) have been involved in GYGA (van Oort et al., 2015; van Bussel et al., 2015). Here, we describe the results from yield gap assessments following the GYGA approach in sub-Saharan African countries. We follow GYGA's approach to estimating yield gap based on potential yield in irrigated systems and water-limited yield in rainfed systems, using the best available data and models. We recognize the major weaknesses in the current version of yield gap assessment in GYGA, which include (1) the absence of reliable data sources on ground water level for estimating water-limited yield in rainfed rice production systems; (2) the large inconsistency in functions in cold-induced spikelet sterility among different crop simulation models for rice; and (3) use of fixed sowing date across years for estimating water-limited yield in rainfed rice production systems, which does not reflect rainfall pattern before sowing (van Oort et al., 2015; Bouman et al., 2003). These limitations all apply to the estimates of potential yield or water-limited yield.

van Oort et al. (2015b) reported the absence of a reliable data source on ground water level. Inconsistent results on the relationship between minimum water temperature during the most critical period and cold-induced spikelet sterility for a given variety across different locations were also observed by Dingkuhn et al. (2015). We suspect this is a result of cold hardening (i.e. if the crop gradually adapts to cold in the vegetative phase, then it is less sensitive in the reproductive phase). Further investigation of the physiological mechanisms underlying this adaptation strategy and improvements to the model are needed.

Our study, as well as many other yield gap studies, simulates water-limited yield with a fixed sowing date. In reality, rainfed farmers plant crops at the start of the wet season, or in the middle of the season, if there are two crops per season. Wolf et al. (2015) developed rainfall rules for more accurately estimating sowing date. With these rules, a crop is sown on a different date each year: this is the first possible date within a specified time window when there has been a total of 'y' mm of rain in the last 'x' days. While this makes logical sense from a biophysical perspective, these rules are yet to be validated with farmers' actual sowing dates and will not work for the second crop if two crops are grown in one wet season.

Several models, including ORYZA2000 (Bouman et al., 2003), have the additional problem that they do not include the effect of early drought on early crop establishment. For example, when drought occurs, the number of days from sowing to emergence should increase and emergence rate should reduce, but this is not reflected in the models. Percentage survival at the early vegetative stage during dry spells should also be reduced, but this is not included in the models. As a result, such models quite consistently showed higher water-limited yield in earlier sowing dates than delayed sowing dates. Recommendations for sowing dates based on the models are given much earlier than farmers' actual practice – but perhaps the models are wrong and the farmers are right. The models need to be improved in these aspects, especially if they are to be used for decision support on sowing dates, based on historical weather data or weather forecasts.

Thus in addition to providing quantitative estimates of yield gaps at a large scale (which we present in the next subsection), our experience in the GYGA project has contributed to our scientific understanding. Limitations of the models that are not apparent when applied and calibrated at local scale become visible when applied at large scales. These insights directly shape the crop modelling research agenda and eventually improved models may be used for decision support.

## 2.1.2 Results of Global Yield Gap and Water Productivity Atlas (GYGA)

Figures 2 and 3 show the absolute yield gap for rice in different climate zones and at national level, for eight sub-Saharan African countries (Burkina Faso, Ghana, Mali, Niger, Nigeria, Tanzania, Uganda and Zambia). Each dot in the map indicates a weather station. Yield gap was estimated in the major rice-growing climate zones for rainfed rice (including both lowland and upland rice) and irrigated lowland rice production systems (Fig. 4). This was aggregated to national level (Fig. 2; van Bussel et al., 2015). In rainfed rice, yield gap refers to the difference between water-limited yield and actual yield, whereas in irrigated rice, yield gap refers to the difference between potential yield and actual yield. At national level, actual yield and water-limited yield of rainfed rice ranged from 1.2 to 2.9 t/ha and from 4.1 to 7.9 t/ha, respectively. Similarly, actual yield and potential yield of irrigated rice ranged from 2.7 to 4.5 t/ha and from 7.6 to 10.8 t/ha, respectively. The relative yield gap of rainfed and irrigated rice ranged from 110 to 558% and from 122 to 218%, respectively, indicating that yield can potentially be doubled. In rainfed rice, the largest relative and absolute yield gaps were obtained in Zambia, where rainfed lowland rice is mainly grown and crop duration is very long, resulting in the highest water-limited yield among seven countries, followed by Ghana. In irrigated rice, Tanzania had the largest absolute yield gap, while Nigeria had the largest relative yield gap. On the basis of yield gap alone, these four countries – Ghana, Nigeria, Tanzania and Zambia – will be given high priority for interventions for enhancing rice yield.

These spatially explicit estimates of yield gap have been used to support the development of national policies. The use of crop growth models to quantify yield gap supports more realistic quantification of the extent to which self-sufficiency in rice can be achieved through yield gap closure. van Oort et al. (2015b) analysed the prospect for self-sufficiency in eight

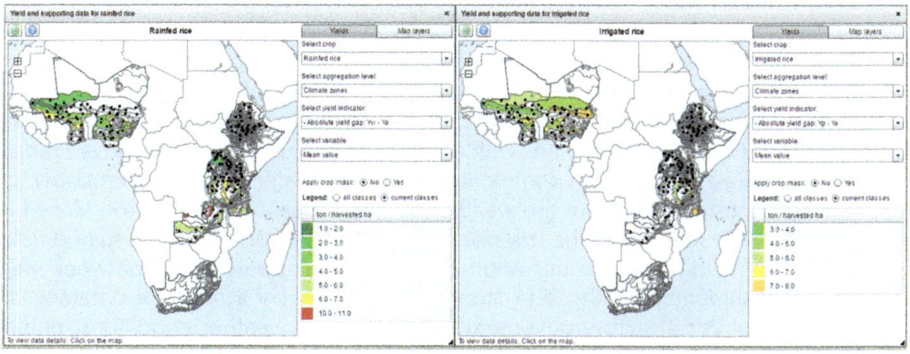


**Figure 2** Absolute yield gap for rice in different climate zones in eight sub-Saharan African countries (van Oort et al., 2015b; http://www.yieldgap.org/gygamaps/app/index.html, accessed at 30 March 2016).

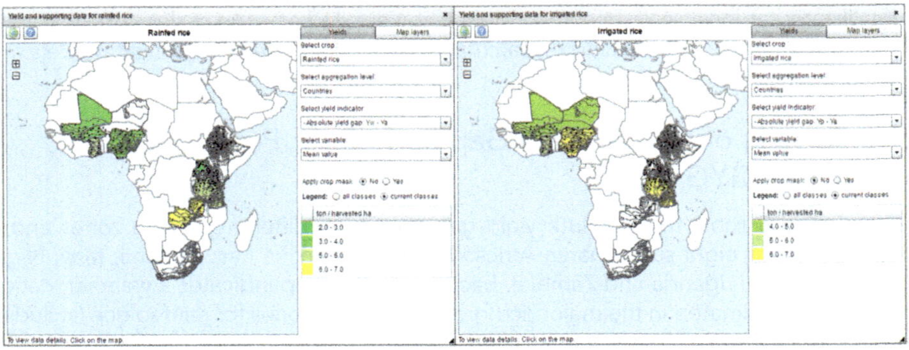


**Figure 3** Absolute yield gap for rice at national level in eight sub-Saharan African countries (van Oort et al., 2015b; http://www.yieldgap.org/gygamaps/app/index.html, accessed at 30 March 2016).

African countries by 2025 (Burkina Faso, Egypt, Ghana, Mali, Nigeria, Tanzania, Uganda and Zambia), using different scenarios of trends in consumption and yield, along with the quantified yield gap shown above. In a scenario using current consumption levels per capita and an enhanced yield growth rate, which is similar to that achieved in the Green Revolution period in Asia (Saito et al., 2015a), six of the eight countries (all except Tanzania and Uganda) cannot become fully self-sufficient in rice by 2025 by reducing the yield gap alone. This implies that a mixture of cropping intensification (moving from a single rice crop per year to two or three crops), area expansion and imports will be needed, as well as yield gap closure. Further research is needed to identify the most suitable new land for rice expansion and areas that should be protected.

## 2.2 Quantifying yield gaps using experimental yield

Fischer et al. (2014) estimated yield gap using experimental yield and actual yield in six major rice mega environments (Fig. 4, this chapter; and Table 4.5, page 177, Fischer et al., 2014). Yield gaps were estimated in Brazil, China, Egypt, India, Indonesia, Japan, the Philippines, Thailand and Vietnam. The relative yield gap ranged from 25 to 150%. Lower yield gaps were generally associated with irrigated rice environments compared with rainfed environments. The lowest relative yield gap was found in Egypt, which is well known for its high potential yield due to high solar radiation, cool temperature and low humidity, and high actual yield due to intensified cropping. About 25% of data points show a relative yield gap of more than 100%. Regions with a higher yield gap included Madhya Pradesh, India and North East India, which commonly experience water shortages. Probably as a result of limited information about the weather during a season, farmers experience high levels of uncertainty, leading to the low use of inputs (e.g. fertilizer), and resulting in low actual yield. The main issue with this approach is that the apparent gap between yields obtained in experiments and those in farmers' fields may be a result of differences in field-specific biophysical factors governing yields, such as weather conditions, nutrient deficiencies and diseases (Lobell et al., 2009; van Ittersum et al., 2013). Therefore, using experimental yield may not be adequate to derive robust estimates of yield gap that are representative of the dominant weather and soil conditions in a given region (van Ittersum et al., 2013). However, data on experimental yield along with data on crop management,

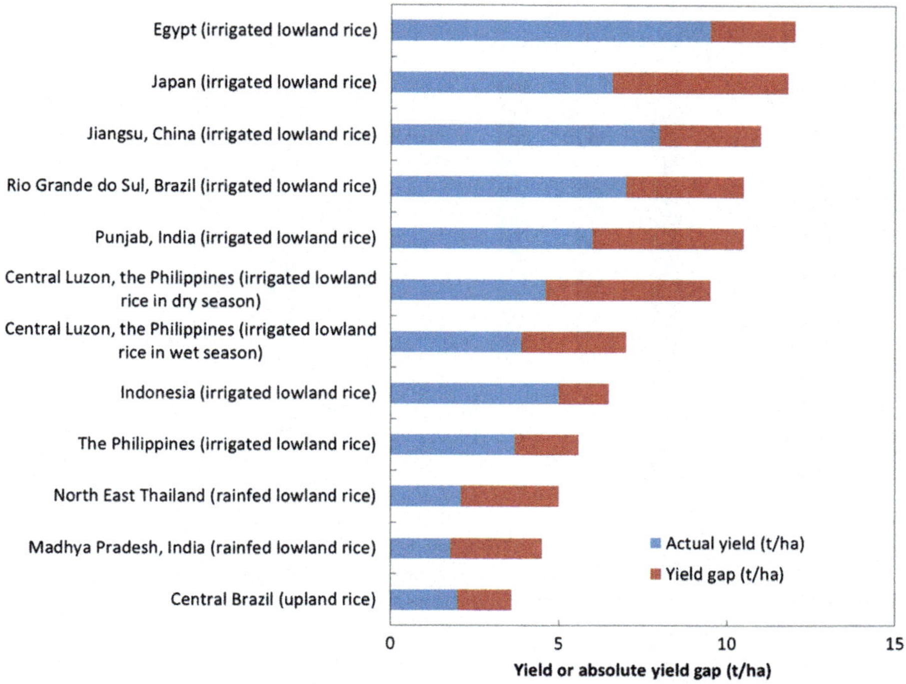


**Figure 4** Yield gap using experimental yield and actual yield (from Table 4.5, page 177, Fischer et al. (2014)).

weather and soils will certainly be useful for validating crop simulation models, which can be used for estimating yield gaps.

When yield gaps from Lobell et al. (2009) (potential yield based on crop simulation model) and Fischer et al. (2014) (experimental yield) are compared in the same countries, yield gaps identified by Fisher et al. (2014) are generally lower than those reported by Lobell et al. (2009). This is because Fischer et al. used experimental yield, in which it is hard to eliminate all yield-limiting and yield-reducing factors. However, this is not the case for Japan, where the yield gap was higher when estimated using experimental yield (80%) than using potential yield (40%). This might have been due to the fact that the experimental yield in Fischer et al. (2014) was achieved using recently released high-yielding varieties, while such high-yielding varieties were not considered in Lobell et al. (2009). It is noted that the variety used in determining experimental yield should be the same as, or have similar potential yield and water-limited yield to, the varieties used by the majority of farmers to avoid bias.

## 2.3 Calculating yield gap using best farmers' yield

Using historical data from surveys, Laborte et al. (2012) calculated yield gap between average yield and best farmers' yield (average yield of the upper 10th percentile) over five decades (mid-1960s, 1970s, 1980s, 1990s and 2000s) in Central Luzon, the Philippines

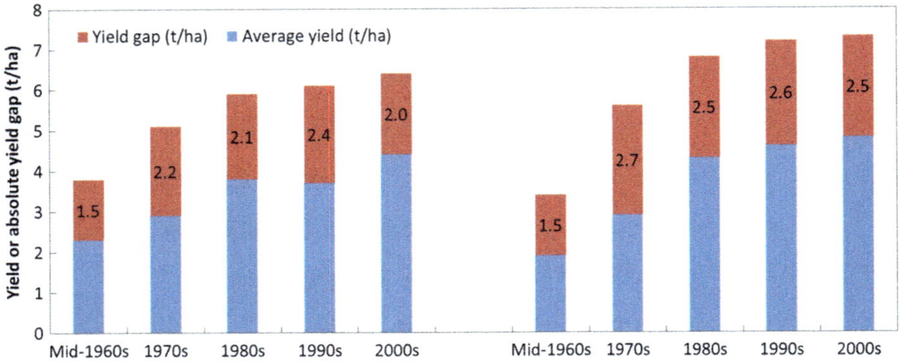


**Figure 5** Yield gap using best farmers' yield and average yield (average yield of the upper 10th percentile) in five decades (mid-1960s, 1970s, 1980s, 1990s and 2000s), in wet and dry seasons in Central Luzon, the Philippines (Laborte et al., 2012).

(Fig. 5). Since the 1970s, absolute yield gaps in both wet and dry seasons have not changed greatly as both average yield and best farmers' yield have simultaneously increased. However, relative yield gaps have been reduced as average yield has increased. The reason for the increase in both average and best farmers' yield may be that climatic conditions have improved for rice cultivation, resulting in higher best farmers' yield. However, experimental yield in the 1960s in IRRI in the Philippines had already reached about 6 t/ha and 9 t/ha in wet and dry seasons, respectively (De Datta et al., 1968), and such yield levels are still observed (Fischer et al., 2014). A more plausible explanation for the yield increases over time may be that limited access to inputs, markets, technologies and knowledge in the mid-1960s and 1970s prevented farmers from achieving higher yields. Consequently, best farmers' yields would be lower than those in the recent years, leading to relatively low yield gaps in the mid-1960s and 1970s. This indicates that best farmers' yield may have limitations for yield gap assessment, as it may not represent what is biophysically possible, but is also affected by socio-economic factors.

Laborte et al. (2012) also analysed rice yield gaps in farmers' fields in wet and dry seasons in four locations in Southeast Asian countries, including Central Luzon in the Philippines, West Java in Indonesia, Suphan Buri in Thailand and Can Tho in Vietnam. Figure 6 shows a comparison of the yield gaps using potential yield and farmers' best yield. The difference between yield gaps using potential yield and best farmers' yield varied across seasons and locations. There was a smaller difference in the wet season than in the dry season in West Java, whereas there was a larger difference in the wet season than in the dry season in Can Tho, Vietnam.

Similar results have been reported in West Africa (Fig. 6; Becker et al., 2003), showing a large variation in the difference between yield gaps using potential yield and best farmers' yield (in Becker et al., highest farmers' yield was used). These results support the limitation on yield gap assessment using best farmers' yield noted above. Other constraints that are not tackled by farmers, such as low pH, micronutrient deficiencies or biotic constraints, may also contribute to these results. If none of the farmers used lime, micronutrient fertilizers or pesticides, then yield would be restricted by these stresses even in the best farmers' fields. Nevertheless, the assessment using best farmers' yield aids our understanding of

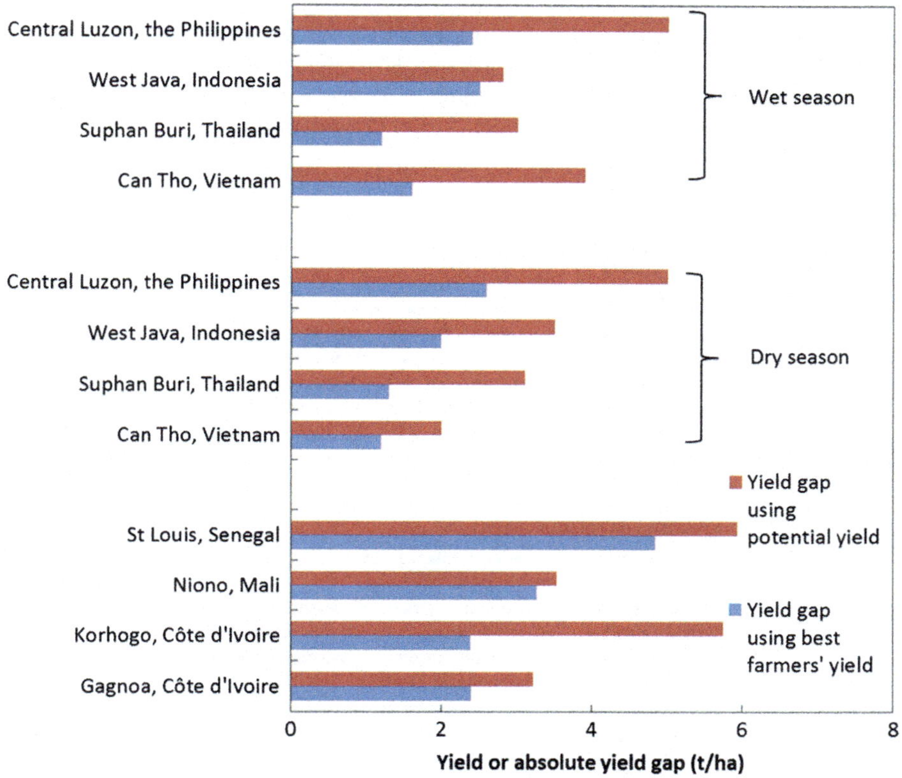


**Figure 6** Comparison of yield gaps using potential yield and farmers' best yield (from Laborte et al., 2012 and Becker et al., 2003).

the yields that could be achieved, based on location-specific environmental and socio-economic conditions with current management practices (Stuart et al., 2016).

In 2011, AfricaRice launched the Africa-wide Rice Agronomy Task Force, an agronomy research and development network in sub-Saharan Africa (Africa Rice Center, 2013). To date, 22 countries are member countries of this network, and these are Benin, Burkina Faso, Cameroon, Chad, Côte d'Ivoire, Democratic Republic of the Congo, Ethiopia, Ghana, Guinea, Guinea-Bissau, Madagascar, Mali, Maulitania, Niger, Nigeria, Rwanda, Senegal, Sierra Leone, Tanzania, Gambia, Togo and Uganda. The Task Force has launched a collective effort to analyse major determinants of rice productivity in different rice systems through yield gap assessment and diagnostic surveys. Yield gap was estimated using average yield and best farmers' yield (average yield of the upper 10th percentile) in 12 locations within 11 countries for irrigated lowland rice, in 18 locations within 13 countries for rainfed lowland rice and in 7 locations within 7 countries for rainfed upland rice production systems. In each production system in a given location, yields were determined in at least 20 farmers' fields over 1–3 seasons.

Mean yield was 3.9, 2.6 and 1.6 t/ha in irrigated lowland rice, rainfed lowland rice and rainfed upland rice production systems, respectively, with an average yield in the upper

10th percentile of 5.9, 4.8 and 3.5 t/ha (Fig. 7; authors, unpublished data). Thus, the average absolute yield gap ranged from 1.9 to 2.2 t/ha and relative yield gap ranged from 57 to 109%. In 50% of the irrigated lowland rice, 50% of rainfed lowland rice and 43% of rainfed upland rice locations, the absolute yield gap was larger than 2 t/ha. Similarly, in 50, 78 and 86% of the locations in the respective production systems, relative yield gap was larger than 50%.

A further limitation in this approach, especially for rainfed rice, is the large variation in water availability within small distances in rainfed rice fields (e.g. Homma et al., 2003;

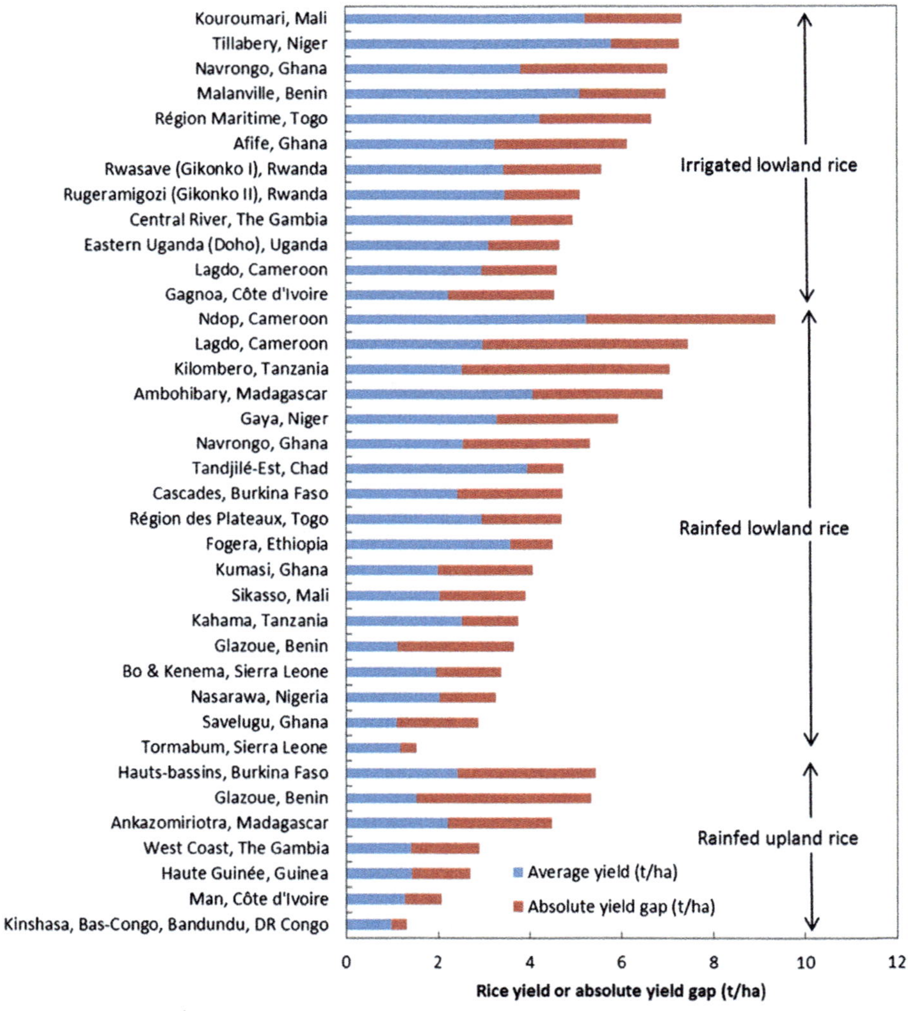


**Figure 7** Yield gap using best farmers' yield (average yield of the upper 10th percentile) and average yield in 12 locations within 11 countries for irrigated lowland rice, 18 locations within 13 countries for rainfed lowland rice and in 7 locations within 7 countries for rainfed upland rice production systems.

Boling et al., 2010; Worou et al., 2013). Generally, rainfed rice yield is higher under favourable water conditions. Consequently, best farmers' yields would come from fields with favourable water conditions. It would not be surprising if farmers applied more fertilizer in such good fields. As a result, the yield gap between best farmers' yield and average yield may be mainly attributable to the difference in a combination of water availability and fertilizer input. This is also the reason why large yield gaps are found in rainfed lowland and upland rice production systems in some locations in Fig. 7. Furthermore, variations in soil type may also create a bias in yield gap analysis using best farmers' yield. Best farmers might have better soils than other farmers, who may have problematic or toxic soils, which cannot be fully overcome. Further effort might be needed to group farmers' fields according to water availability and soil type in order to estimate yield gaps that might not be related to water availability and soil type. One approach is a boundary function approach, which is discussed later. Similarly, to avoid bias in the yield gap resulting from different rice varieties, farmers might be grouped according to the varieties used.

# 3 Studies to identify causes of yield gaps

Several groups of constraints which contribute to the yield gap are often reported, ranging from the biophysical to institutional (Dixon et al., 2001):

- Biophysical (weather, soil, water, pest pressure, weeds);
- Technical/management (labour availability, timing of operations, tillage, variety or seed selection, management of water, nutrients, weeds and pests, and harvesting);
- Socio-economic (social and economic status, farmers' cultural traditions, commitments and obligations, knowledge, family size and farm profitability);
- Institutional and policy (government policy, rice price, credit, input supply, land tenure, markets, research, development and extension);
- Technology learning and linkages (competence and resources of extension staff, knowledge and skills, linkages among public, private and NGO extension staff).

These constraints have been identified through various approaches like expert surveys, household surveys, on-farm field surveys, on-farm experiments, crop simulation models and remote sensing and a combination of approaches (e.g. Becker et al., 2003; Boling et al., 2010; Lobell, 2013; Tanaka et al., 2013, 2015; Fischer et al., 2014). In some cases, some constraints may not be considered as causes for yield gap. For example, if potential yield is used for calculating yield gap, variety and weather may not contribute to yield gaps, as they are considered in crop simulation models as well. In the next section, we describe methods and case studies for the quantification of the causes of yield gap.

## 3.1 Crop simulation model

Crop simulation models permit an assessment to be made of the relative contribution of key yield-limiting factors such as extreme temperature, water and nitrogen, as detailed crop responses to these factors are often included in crop simulation models such as ORYZA2000 (Bouman et al., 2003). We note that the current ORYZA2000 version does not deal with any other limiting factors, such as submergence, and soil-related constraints such as salinity and phosphorus, except nitrogen. The development of such modules will require substantial efforts.

The water limitation index is defined as the gap between the water-limited yield and potential yield, which assumes no limitations due to water supply. It is expressed as a percentage of potential yield:

Water limitation index = (1 – water-limited yield/potential yield) × 100%

(http://www.yieldgap.org/web/guest/glossary, accessed at 30 March 2016). A higher index (in per cent) indicates a higher risk of water limitation. Fig. 8 shows the water limitation index in seven sub-Saharan African countries (Burkina Faso, Ghana, Mali, Nigeria, Tanzania, Uganda and Zambia). Uganda has the highest water-limitation index among these countries, partly because rainfed upland rice cultivation dominates, where levels of drought stress are relatively severe, leading to relatively low water-limited yields. Innovations that can address drought risk, such as drought-tolerant varieties or supplemental irrigation facilities, would be likely to enhance actual yield and potential yield in Uganda.

The water limitation index and nitrogen (N) limitation index [(1 – N-limited yield/potential yield) × 100%] were used in a field study of rainfed lowland fields in Indonesia, where water conditions differ along toposequences (Boling et al., 2010). The relative limitations of water and N ranged from 0 to 28% and 35 to 63%, respectively, with large temporal and spatial variability. Based on this assessment, field-specific and time-specific management strategies to increase yield and yield stability were designed.

Spikelet sterility due to extreme temperatures has been reported to significantly reduce rice yield (van Oort et al., 2014, 2015a), and there is a potential risk of spikelet

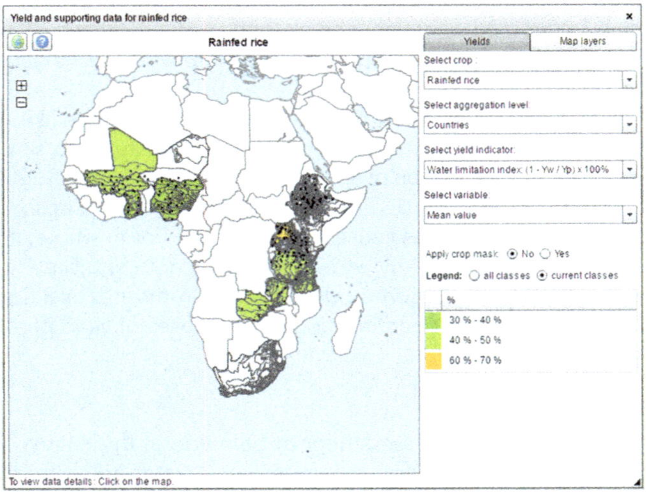


**Figure 8** Water limitation index in seven sub-Saharan African countries (http://www.yieldgap.org/web/guest/glossary, accessed at 30 March 2016).

sterility resulting from both high and low temperatures in semi-arid and arid zones in West Africa and in the highlands of East and Southern Africa (low temperature) (Saito et al., 2013). The effect of extreme temperature on spikelet sterility was quantified using the GYGA (Global Yield Gap Atlas) as a framework. Among selected West African countries in GYGA, Niger has a relatively higher risk of spikelet sterility due to high temperature than other countries (data not shown, authors, unpublished data). This is consistent to some extent with a previous study (Dingkuhn, 1995). For low temperature, all three East and Southern African countries (Tanzania, Uganda and Zambia) had a risk of cold-induced spikelet sterility (authors, unpublished data). By changing the parameters in crop models, we also identified the introduction of heat- or cold-tolerant varieties, along with alternative sowing times, as ways to reduce the risk of spikelet sterility for these countries.

## 3.2 Expert survey, household survey and on-farm field survey

Waddington et al. (2010) conducted a survey of 672 experts representing a diversity of backgrounds and experience to determine the most important production constraints for six major food crops (cassava, chickpea, cowpea, rice, sorghum and wheat). Yield gaps were found to be smallest for rice. The most severe and widespread specific constraints for rice included N fertilizer problems, soil fertility depletion, various leaf, stem and head pests and diseases, weed competition and inadequate water management. Socio-economic issues were especially prominent for rice in sub-Saharan Africa. Although this approach could identify key factors affecting yield gap, as the authors indicated, it was not able to fully describe or quantify many of the interactions, and some cause-and-effect relationships among important constraints. At a local level, more detailed follow-up surveys may be helpful.

At a local level, household surveys have been conducted using interviews with farmers (Laborte et al., 2012; Tanaka et al., 2015), and on-farm yield surveys, including field observations, measurements (e.g. soil sampling) and crop cut along with interviews (e.g. Becker et al., 2003; Africa Rice Center, 2013). With these approaches, farmers are directly asked about their perceptions of the biophysical constraints to rice production (e.g. Diagne et al., 2013b). The root causes of yield gaps may be identified, which may be beyond farmers' crop management ability, and may be related to socio-economic constraints, institutional and political arrangements, technology transfer (e.g. Tanaka et al., 2015) and constraints at the whole farming system level (e.g. Fujisaka, 1991).

There are several different methodologies for identifying the causes of yield gaps based on survey data. A simple descriptive approach compares factors between farmers with best farmers' yield and with average yield in comparable environments. With the minimum input and greatest simplicity, this allows for limited but useful benchmarks. With this method, it is often shown that yield gaps can be primarily attributed to differences in management. Laborte et al. (2012) revealed that best-yielding farmers were generally more educated than average farmers, and used production inputs more efficiently.

Multiple regression analysis including linear regression, and classification and regression tree (CART) analysis has been also used for yield gap analyses (e.g. Tanaka et al., 2013, 2015). However, most surveys on yield gap analysis deal with identifying the factors that affect variation in farmers' yields rather than variation in yield gaps, which were quantified for individual farmers' fields. Through regression analysis, Becker and Johnson (1999) found that age of seedlings at transplanting, timeliness of operations and application of P fertilizer could explain 60% of the observed variability in rice yield under irrigated systems in the forest zone of Côte d'Ivoire. Tanaka et al. (2013) revealed that 75% of the variation in yields under lowland irrigation in Benin could be explained by five agronomic factors (fallow residue management, ploughing method, water stress, rat damage and N application rate) and two edaphic factors (sloped surfaces and sand content in the soil). Similarly, in rainfed lowlands, Becker and Johnson (2001a) found that bunding, higher N application rate, higher splits and P application contributed to higher yield.

In the Senegal River Valley (SRV), a repeated farm survey was conducted over the period 2002–10 to monitor farmers' agricultural practices during the wet season in the Delta and Middle Valley (Tanaka et al., 2015). The primary factor was sowing time, and delayed sowing resulted in a yield reduction of around 1 t/ha in both zones. The major reasons behind delayed sowing were related to availability of credit, machinery and irrigation water. The two other factors identified were fertilizer management and bird control. The authors concluded that it is essential to create an enabling environment through institutional arrangements for improving access to resources and machinery, in order to enhance the adoption of integrated crop management options.

Machine learning methods have been proposed for ecological modelling (e.g. Recknagel, 2001; Prasad et al., 2006; Cutler et al., 2007). These methods are capable of detecting complex relationships among model variables without making a priori assumptions about the type of relationship, such as a linear dependence on predictors, and are able to process complex and noisy data. One of these methods is the random forest method (Vincenzi et al., 2011), which has not yet been used for yield gap analysis, but its testing would be warranted.

Crop yield is a function of capture and efficiency of resource use, and non-resource constraints governing crop development and growth. A boundary function approach has been used to evaluate the maximum yield responses to resources such as water, soil, N input, weed and soil-borne pests (e.g. Casanova et al., 1999; Shatar and McBratney, 2004; Affholder et al., 2013; Wairegi et al., 2010; Tittonell and Giller, 2013; van Ittersum et al., 2013; Sadras et al., 2015). Figure 9 shows the relationship between N fertilizer application rate and average yield based on data from Fig. 7. In Fig. 9, a boundary curve could be fitted for each production system separately, but as our main objective here is to describe this boundary approach, we draw the curve across three rice production systems.

Boundary curves were fitted using cubic smoothing splines to identify maximum yield at the given N application rates (Webb, 1972; Shatar and McBratney, 2004; Daouia et al., 2016). The yield gap due to resource limitation (here, N application rate) can be calculated based on the difference between the highest yield and yield at the given N application level, according to the boundary regression curve shown in Fig. 9. This yield gap can be a result of different farmers' objectives and access to inputs. The difference between actual yield and yield derived from the boundary function at the same level of N application rate can be considered as the yield gap due to poor N use efficiency. Data points close to the boundary curve have smaller yield gaps in Fig. 9, and poor N use efficiency might be due to poor rainfall, poor soil fertility, sub optimal timing of

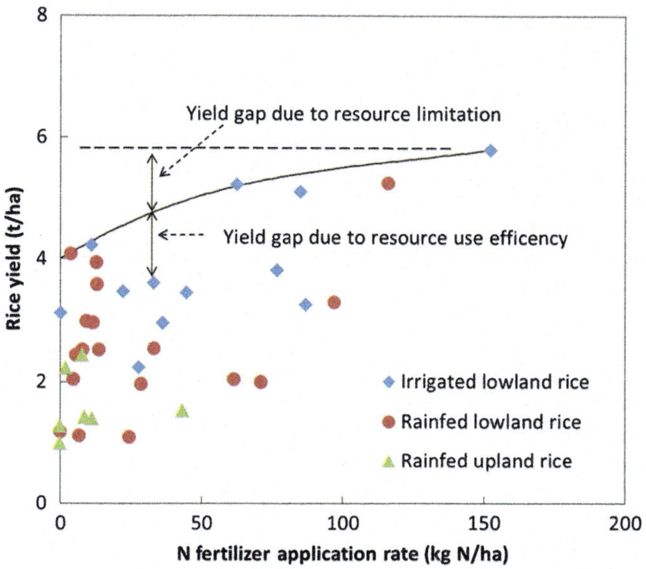


**Figure 9** Relationship between N fertilizer application rate and rice yield in sub-Saharan African countries. Data are derived from Fig. 7 and averaged over locations. Boundary curves were fitted using cubic smoothing splines to identify maximum yield at the given N application rates (Webb, 1972; Shatar and McBratney, 2004; Daouia et al., 2016).

application, labour shortage, and knowledge and information about how to deal with these limitations.

On average across locations, the yield gap due to N limitation was 0.9, 1.2 and 1.7 t/ha in irrigated lowland, rainfed lowland and rainfed upland rice production systems, respectively. Rainfed upland rice has the largest yield gap due to N limitation, followed by rainfed lowland rice, as N application rate is limited in rainfed systems. In fact, in both irrigated lowland and rainfed lowland rice, the yield gap is larger in East and Southern African countries than in West and Central African countries, as the N application rate is lower in East and Southern African countries. The yield gap due to poor N use efficiency is 1.1, 1.9 and 2.7 t/ha in irrigated lowland, rainfed lowland and rainfed upland rice production systems, respectively. Again, the largest yield gap is found in rainfed upland rice, where drought may have occurred and soil fertility might be poor. Although further analysis is needed to identify the causes of poor N use efficiency, especially for rainfed systems, this example clearly indicates that this method can offer a reasonable approach to estimating yield gap when crop simulation models and required data inputs are not available.

A major strength of this approach is that estimates of best farmers' yield are not 'static or non-spatially explicit' values like those derived from upper yield percentiles or maximum yields, and boundary functions can provide estimates of best farmers' yield across a wide range of resources (here, N application rate) to estimate yield gap (van Ittersum et al., 2013). The use of boundary functions is limited by the focus of the yield gap analysis on a single factor: additional analysis is needed to identify other factors that cause variation in yield gap at the same level of resource (here, N application rate) such as N application timing, and variations in rainfall, solar radiation and temperature.

## 3.3 On-farm experiment

On-farm experiments involving one or more factors as treatments can help quantify the determinants of yield gap (De Datta et al., 1978). Treatments can be based on experts' knowledge, survey results and/or literature review, and include management practices for nutrients, water, weeds and pests, as well as land preparation practices and crop establishment times and methods (e.g. Becker and Johnson, 2001a,b). For example, Becker and Johnson (2001a) studied the effects of improved water control and crop management on lowland rice productivity in West Africa. Retaining flood-water with field bunds increased rice yield by about 40% and improved weed control (with about 25% less weed biomass in bunded than in open plots). The application of mineral N fertilizer increased rice yields by almost 20% in bunded fields, but resulted in no increase in open fields. Thus, land levelling together with bunding facilitated improved water management, which decreased weed growth and increased nutrient use efficiency.

On-farm nutrient omission trials, which often include plots minus N, P and K, and which quantify indigenous nutrient supply, can also be considered as an approach for identifying the causes of yield gap (e.g. Dobermann et al., 2003a,b; Mussgnug et al., 2006). In fact, all experiments on the introduction of new technologies with potential to reduce yield gap could also be considered here, and the yield gain resulting from the introduction of new technologies can represent the yield gap between practices with new technologies and farmers' current practices (e.g. Saito et al., 2015b). However, it is noted that most on-farm experiments, especially those targeting smallholder farmers, have attempted to introduce adaptive technologies and innovations to farmers' fields rather than testing treatments to maximize yield in farmers' fields (e.g. Becker and Johnson, 2001a,b). Thus, in most cases, new technologies are unlikely to produce an 'experimental yield', which can be defined as the maximum yield obtained with best management practices in researchers' managed trials. Consequently, yield gaps across such studies should not be directly compared.

# 4 Future trends and conclusion

This chapter has provided various case studies of yield gap analysis with special focus on sub-Saharan Africa using different benchmarks consisting of maximum yield (potential yield, experimental yield and best farmers' yield) and average yield for estimating yield gaps at different spatial scales. It also described different methodologies for identifying determinants of yield gap. Table 1 summarizes the yield gap studies that have been discussed in this chapter. Different benchmarks for maximum yield have advantages and disadvantages. Rapid advances are being made in remote sensing technology for estimating various parameters such as soil quality, cropping season and water availability. Combined with the expected increase in sources of reliable weather data and improved bias correction for satellite-based weather data, this will enable crop model simulations of potential yield and water-limited yield that are applicable to a wide range of spatial scales to become major indicators of maximum yield in the near future. Experimental yield in representative sites will probably not be directly used for yield gap assessment, but the yield along with other information related to the experiments will serve to improve various functions in crop simulation models and introduce genotype-specific parameters of new varieties to the models. In particular, these experiments should include new high-yielding varieties and improved crop management practices that could overcome current potential yield barriers, especially where the yield gap is expected to be small. Best farmers' yield can be used for understanding what could be achieved based on location-specific environmental and socio-economic conditions with current management practices. Simple practical methods might be needed for removing major biophysical factors, such as soil and water-related constraints in rainfed systems. These may include a boundary function approach and stratified sampling techniques based on biophysical conditions. Furthermore, remote sensing techniques could also help estimating best farmers' yield at larger spatial scales.

Data from local or national statistics often do not provide farm yield for each production system, especially in sub-Saharan Africa. Thus, they are not useful for detailed yield gap assessment. It is clear that there is an urgent need to improve local or national statistics. Remote sensing and improved sampling methods, such as dot sampling involving crop cut, could improve information on actual yield obtained in farmers' fields (Srivastava, 2015).

There is a wide range of approaches for identifying determinants of yield gap. Depending on spatial coverage, different types of approaches can be considered. Machine learning methods, such as random forest, and the use of remote sensing combined with field surveys are warranted in future studies. A multidisciplinary approach is needed for identifying opportunities to reduce yield gap, as there are a wide range of biophysical, socio-economic and institutional constraints.

Yield gap analysis can also be used for monitoring progress in the dissemination of technologies and innovations in agricultural development projects, rather than monitoring yield only. Year-to-year climate variability could mask yield gains obtained by the introduction of technologies and innovations. If the weather conditions are unfavourable, potential yield or water-limited yield also will be lower, compared to seasons with more favourable weather conditions. A smaller yield gap may be detected and attributed to the introduction of new technologies even under bad weather conditions.

The quantification of yield gaps and the identification of their causes are not the end of the story, but just the first steps towards enhancing rice yield and production and consequently improving food security. It is important to examine the extent to which yield gaps can be reduced in an economically and environmentally sustainable manner, through technical and institutional innovations, as potential yield and economically optimal yield can differ across areas, especially for rainfed systems. Such an assessment is rarely done after a yield gap assessment, but if it is done, it would assist investment in agricultural development. Furthermore, yield gap studies to identify areas with the greatest potential for enhanced rice production need to be complemented by data and analyses of socio-economic factors such as distance to markets, road conditions and land-tenure issues, and the existing area for rice cultivation in Africa, where expansion for rice cultivation will be needed for achieving self-sufficiency (Erenstein et al., 2006; You et al., 2011; Saito et al., 2013; van Oort et al., 2015b).

**Table 1** Summary of yield gap studies described in this chapter.

| | Spatial coverage | Maximum yield: Potential yield/water-limited yield simulated by crop model | Maximum yield: Experimental yield | Maximum yield: Best farmers' yield | Data source on actual yield | Approach for identifying causes for yield gap |
|---|---|---|---|---|---|---|
| Quantification of yield gaps | Large | High applicability | Limited applicability | Limited applicability | National-level average from national statistics<br>Sub-national-level average<br>Estimated yield from satellite images combined with crop modelling<br>Local experts' estimation<br>Interview with farmers<br>Crop cut | Overlaying with other GIS information (e.g. soil map)<br>Sensitive analysis using crop simulation models<br>Remote sensing (e.g. crop growing season)<br>Expert survey<br>Household survey<br>On-farm field survey<br>Field experiments |
| Identification of causes for the yield gaps | Limited | High applicability | High applicability | High applicability | | |
| Advantage | | Can estimate maximum yield which is considered as biophysical limitation | Uncertainty in crop-simulation models and input data<br>Biophysical constraints that cannot easily be resolved through management are taken into account | Helps understanding of what could be achieved | | |

| | | | |
|---|---|---|---|
| Disadvantage | Validated crop-simulation models and input values to run them, such as long-term weather data, are often lacking<br><br>Uncertainty due to model quality and risk of unreliable results dependent on crop modeller's skills<br><br>No-high resolution groundwater data available for rainfed rice to estimate water-limited yield at large scale<br><br>Limited ability of crop models to estimate cold sterility risk | It is hard to eliminate all yield-limiting and yield-reducing factors in field experiments to achieve maximum yield<br><br>Apparent yield gap between yields obtained in experiments and in farmers' fields could be attributed to differences in field-specific biophysical factors governing yields such as weather condition, nutrient deficiencies, diseases, etc. | Biophysical and socio-economic constraints vary spatially, and even best farmers might produce much lower yield than what can be achieved biophysically<br><br>Required sample size is not known to estimate best farmers' yield<br><br>In rainfed rice, yield gap might be mainly related to biophysical limitation – water and soil-related constraints (alternative option would be to use a boundary function approach) |

# 5 Where to look for further information

## Introductions to the subject and seminal articles

Connor, D. J., Loomis, R. S. and Cassman, K. G. (2011). *Crop Ecology: Productivity And Management In Agricultural Systems*. 2nd ed. University Press, Cambridge, UK.

Fischer, R. A., Byerlee, D. and Edmeades, G. O. (2014). Crop yields and global food security: will yield increase continue to feed the world? ACIAR Monograph No. 158. Australian Centre for International Agricultural Research: Canberra. xxii + 634 pp.

Sadras, V. O., Cassman, K. G., Grassini, P., Hall, A. J., Bastiaanssen, W. G. M., Laborte, A. G., Milne, A. E., Sileshi, G. and Steduto, P. (2015). Yield gap analysis of field crops – Methods and case studies, FAO Water Reports No. 41, Rome, Italy.

Van Ittersum, M., Cassman, K. G., Grassini, P., Wolf, J. Tittonell, P. and Hochman, Z. (2013). Yield gap analysis with local to global relevance – A Review. *Field Crops Research*, 143, 4–17.

Wopereis, M. C. S., Johnson, D. E., Ahmadi, N., Tollens, E. and Jalloh, A. (2013). *Realizing Africa's Rice Promise*. CAB International, Wallingford, UK.

## Key journals

*European Journal of Agronomy*
*Field Crops Research*
*Global Food Security*

## Major international research projects

Africa-wide Rice Agronomy Task Force (http://www.africarice.org/)
Closing Rice Yield Gaps in Asia (CORIGAP) (http://corigap.irri.org/)
Global Rice Science Partnership (GRiSP) (http://www.grisp.net/main/summary)
The Global Yield Gap and Water Productivity Atlas (http://www.yieldgap.org/)

## Top five research centres

Africa Rice Center
CIRAD (French Agricultural Research Centre for International Development)
International Rice Research Institute
University of Nebraska–Lincoln
Wageningen University and Research Centre

# 6 References


Affholder, F., Poeydebat, C., Corbeels, M., Scopel, E. and Tittonell, P. (2013). The yield gap of major food crops in family agriculture in the tropics: assessment and analysis through field surveys and modeling. *Field Crops Research*, 143, 106–18.

Africa Rice Center (AfricaRice) (2013). Africa Rice Center (AfricaRice) Annual Report 2012: Africa-wide rice agronomy task force. Cotonou, Benin, 100 pp.

Alexandratos, N. and Bruinsma, J. (2012). World agriculture towards 2030/2050. The 2012 Revision. ESA Working paper No. 12-03, April 2012. Food and Agriculture Organization of the United Nations, Rome.

Becker, M. and Johnson, D. E. (1999). Rice yield and productivity gaps in irrigated systems of the forest zone of Côte d'Ivoire. *Field Crops Research*, 60, 201–8.

Becker, M. and Johnson, D. E. (2001a). Improved water control and crop management effects on lowland rice productivity in West Africa. *Nutrient Cycling in Agroecosystems*, 59, 119–27.

Becker, M. and Johnson, D. E. (2001b). Cropping intensity effects on upland rice yield and sustainability in West Africa. *Nutrient Cycling in Agroecosystems*, 59, 107–17.

Becker, M., Johnson, D. E., Wopereis, M. C. S. and Sow, A. (2003). Rice yield gaps in irrigated systems along an agro-ecological gradient in West Africa. *Journal of Plant Nutrition and Soil Science*, 166, 61–7.

Boling, A. A., Tuong, T. P., van Keulen, H., Bouman, B. A. M., Suganda, H. and Spiertz, J. H. J. (2010). Yield gap of rainfed rice in farmers' fields in Central Java, Indonesia. *Agricultural Systems*, 103, 307–15.

Bouman, B. A. M., Kropff, M. J., Tuong, T. P., Wopereis, M. C. S., ten Berge, H. F. M. and van Laar, HH. (2003). *ORYZA2000 Growth Model*. International Rice Research Institute, Los Baños, Philippines.

Casanova, D., Goudriaan, J., Bouma, J. and Epema, G. F. (1999). Yield gap analysis in relation to soil properties in direct-seeded flooded rice. *Geoderma*, 91, 191–216.

Cutler, D. R., Edwards Jr., T. C., Beard, K. H., Cutler, A., Hess, K. T., Gibson, J. and Lawler, J. J. (2007). Random forests for classification in ecology. *Ecology*, 88, 2783–92.

Daouia, A., Noh, H. and Park, B. U. (2016). Data envelope fitting with constrained polynomial splines. *Journal of the Royal Statistical Society: Series B (Statistical Methodology)*, 78, 3–30.

De Datta, S. K., Gomez, K. A., Herdt, R. W. and Barker, R. (1978). *A Handbook On The Methodology For An Integrated Experiment-Survey On Rice Yield Constraints*. The International Rice Research Institute, Los Baños, Laguna, Philippines.

De Datta, S. K., Tauro, A. C. and Balaoing, S. N. (1968). Effect of plant type and nitrogen level on the growth characteristics and grain yield of Indica rice in the tropics. *Agronomy Journal*, 60, 643–7.

Deryng, D., Sacks, W. J., Barford, C. C. and Ramankutty, N. (2011). Simulating the effects of climate and agricultural management practices on global crop yield. *Global Biogeochemical Cycles*, 25, 1–18.

Diagne, A., Alia, D., Amovin-Assagba, E., Wopereis, M. S. C. and Saito, K. (2013b). Farmer perceptions of the biophysical constraints to rice production in sub-Saharan Africa, and potential impact of research. In Wopereis, M. C. S., Johnson, D. E., Ahmadi, N., Tollens, E. and Jalloh, A. (Eds), *Realizing Africa's Rice Promise*. CAB Iinternational, Wallingford, UK, pp. 46–68.

Diagne, A., Amovin-Assagba, E., Futakuchi, K. and Wopereis, M. C. S. (2013a). Estimation of cultivated area, number of farming households and yield for major rice growing environments in Africa. In Wopereis, M. C. S., Johnson, D. E., Ahmadi, N., Tollens, E. and Jalloh, A. (Eds), *Realizing Africa's Rice Promise*. CAB International, Wallingford, UK, pp.35–45.

Dingkuhn, M. (1995). Characterizing irrigated rice environments using the rice phenology model RIDEV. In Miezan, K. M., Wopereis, M. C. S., Dingkuhn, M., Deckers, J. and Randolph, T. F. (Ed.), *Irrigated Rice in the Sahel: Prospects for Sustainable Development*. West Africa Rice Development Association, Bouaké, Côte d'Ivoire, pp. 343–60.

Dingkuhn, M., Radanielina, T., Raboin, L. M., Dusserre, J., Ramantsoanirina, A., Sow, A., Manneh, B., Balde, A. B., Soulié, J. C., Shrestha, S., Ahmadi, N. and Courtois, B. (2015). Field phenomics for response of a rice diversity panel to ten environments in Senegal and Madagascar. 2. Chilling-induced spikelet sterility. *Field Crops Research*, 183, 282–93.

Dixon, J., Gulliver, A. and Gibbon, D. (2001). *Farming Systems and Poverty: Improving Farmers'* Livelihoods in a Changing World. FAO & World Bank, Rome and Washington, DC, 412 pp.

Dobermann, A., Witt, C., Abdulrachman, S., Gines, H. C., Nagarajan, R. Son, T. T., Tan, P. S., Wang, G. H., Chien, N. V., Thoa, V. T. K., Phung, C. V., Stalin, P., Muthukrishnan, P., Ravi, V., Babu, M., Simbahan, G. C., Adviento, M. A. A. and Bartolome, V. (2003a). Estimating indigenous nutrient supplies for site-specific nutrient management in irrigated rice. *Agronomy Journal*, 95, 913–23.

Dobermann, A., Witt, C., Abdulrachman, S., Gines, H. C., Nagarajan, R., Son, T. T., Tan, P. S., Wang, G. H., Chien, N. V., Thoa, V. T. K., Phung, C. V., Stalin, P., Muthukrishnan, P., Ravi, V., Babu, M.,

Simbahan, G. C. and Adviento, M. A. A. (2003b). Soil fertility and indigenous nutrient supply in irrigated rice domains of Asia. *Agronomy Journal*, 95, 924–35.

Erenstein, O., Oswald, A. and Moussa, M. (2006). Determinants of lowland use close to urban markets along an agro-ecological gradient in West Africa. *Agriculture, Ecosystems and Environment*, 117, 205–17.

Evans, L. T. and Fischer, R. A. (1999). Yield potential: its definition, measurement, and significance. *Crop Science*, 39, 1544–51.

Fischer, R. A. (2015). Definitions and determination of crop yield, yield gaps, and of rates of change. *Field Crops Research*, 182, 9–18.

Fischer, R. A., Byerlee, D. and Edmeades, G. O. (2014). Crop yields and global food security: will yield increase continue to feed the world? ACIAR Monograph No. 158. Australian Centre for International Agricultural Research: Canberra. xxii + 634 pp.

Fujisaka, S. (1991). A set of farmer-based diagnostic methods for setting post 'green revolution' rice research priorities. *Agricultural Systems*, 36, 191–206.

Grassini, P., van Bussel, L. G. J., Van Wart, J., Wolf, J., Claessens, L., Yang, H., Boogaard, H., de Groot, H., van Ittersum, M. K. and Cassman, K. G. (2015). How good is good enough? Data requirements for reliable crop yield simulations and yield-gap analysis. *Field Crops Research*, 177, 49–63.

GRiSP (2016). RICE CRP proposal version March 3. http://www.grisp.net/main/summary accessed at 1 April 2016.

Hirooka, Y., Homma, K., Maki, M. and Sekiguchi, K. (2015). Applicability of synthetic aperture radar (SAR) to evaluate leaf area index (LAI) and its growth rate of rice in farmers' fields in Lao PDR. *Field Crops Research*, 176, 119–22.

Homma, K., Horie, T., Shiraiwa, T., Supapoj, N., Matsumoto, N. and Kabaki, N. (2003). Toposequential variation in soil fertility and rice productivity of rainfed lowland paddy fields in mini-watershed (Nong) in northeast Thailand. *Plant Production Science*, 6, 147–53.

Homma, K., Maki, M. and Hirooka, Y. (2016). Development of a rice simulation model for remote-sensing (SIMRIW-RS). *Journal of Agricultural Meteorology* (in press).

Laborte, A. G., de Bie, K. C., Smaling, E., Moya, P. F., Boling, A. A. and Van Ittersum, M. K. (2012). Rice yields and yield gaps in Southeast Asia: Past trends and future outlook. *European Journal of Agronomy*, 36, 9–20.

Licker, R., Johnston, M., Foley, J. A., Barford, C., Kucharik, C. J., Monfreda, C. and Ramankutty, N. (2010). Mind the gap: how do climate and agricultural management explain the 'yield gap' of croplands around the world? *Global Ecology and Biogeography*, 19, 769–82.

Lobell, D. B. (2013). The use of satellite data for crop yield gap analysis. *Field Crops Research*, 143, 56–64.

Lobell, D., Cassman, K. and Field, B. C. (2009). Crop yield gaps: their importance, magnitudes, and causes. *Annual Review of Environment and Resources*, 34, 179–204.

Maki, M., Homma, K., Hirooka, Y. and Oki, K. (2016). Development and practice of advanced basin model in Asia: Estimation of rice yield by assimilating remote sensing data into crop growth model (SIMRIW-RS). *Journal of Agricultural Meteorology* (in press).

Mauser, W., Klepper, G., Zabel, F., Delzeit, R., Hank, T., Putzenlechner, B. and Calzadilla, A. (2015). Global biomass production potentials exceed expected future demand without the need for cropland expansion. *Nature Communications*, 6, 8946.

Monfreda, C., Ramankutty, N. and Foley, J. A. (2008). Farming the planet: 2. Geographic distribution of crop areas, yields, physiological types, and net primary production in the year 2000. *Global Biogeochemical Cycles*, 22, 1–19.

Mueller, N. D., Gerber, J. S., Johnston, M., Ray, D. K., Ramankutty, N. and Foley, J. A. (2012). Closing yield gaps through nutrient and water management. *Nature*, 490, 254–7.

Mussgnug, F., Becker, M., Son, T. T., Buresh, R. J. and Vlek, P. L. G. (2006). Yield gaps and nutrient balances in intensive, rice-based cropping systems on degraded soils in the Red River Delta of Vietnam. *Field Crops Research*, 98, 127–40.

Neumann, K., Verburg, P. H., Stehfest, E. and Müller, C. (2010). The yield gap of global grain production: a spatial analysis. *Agriculture Systems*, 103, 316–26.

Nhamo, N., Rodenburg, J., Zenna, N., Makombe, G. and Luzi-Kihupi, A. (2014). Narrowing the rice yield gap in East and Southern Africa: using and adapting existing technologies. *Agricultural Systems*, 131, 45–55.

Pradhan, P., Fischer, G., van Velthuizen, H., Reusser, D. E. and Kropp, J. P. (2015). Closing yield gaps: how sustainable can we be? *PLoS ONE* 10(6): e0129487.

Prasad, A. M., Iverson, L. R. and Liaw. A. (2006). Newer classification and regression tree techniques: bagging and random forests for ecological prediction. *Ecosystems*, 9, 181–99.

Ramankutty, N., Foley, J. A., Norman, J. and McSweeney, K., (2002). The global distribution of cultivable lands: current patterns and sensitivity to possible climate change. *Glob. Ecol. Biogeogr.* 11(5), 377–92.

Recknagel, F. (2001). Applications of machine learning to ecological modeling. *Ecological Modeling*, 146, 303–10.

Sadras, V. O., Cassman, K. G., Grassini, P., Hall, A. J., Bastiaanssen, W. G. M., Laborte, A. G., Milne, A. E., Sileshi, G. and Steduto, P. (2015). Yield gap analysis of field crops – Methods and case studies, FAO Water Reports No. 41, Rome, Italy.

Saito, K., Diack, S., Dieng, I. and Ndiaye, M. K. (2015b). On-farm testing of a nutrient management decision-support tool for rice in the Senegal River valley. *Computers and Electronics in Agriculture*, 116, 36–44.

Saito, K., Dieng, I., Toure, A., Somado, E. A. and Wopereis, M. C. S. (2015a). Rice yield growth analysis for 24 African countries over 1960–2012. *Global Food Security*, 5, 62–9.

Saito, K., Nelson, A., Zwart, S. J., Niang, A., Sow, A., Yoshida, H. and Wopereis, M. C. S. (2013). Towards a better understanding of biophysical determinants of yield gaps and the potential for expansion of the rice area in Africa. In Wopereis, M. C. S., Johnson, D. E., Ahmadi, N., Tollens, E., Jalloh, A. (Eds), *Realizing Africa's Rice Promise*. CAB International, Wallingford, UK, pp. 188–203.

Setiyono, T., Nelson, A. and Holecz, F. (2015). Remote Sensing based Crop Yield Monitoring and Forecasting. In Crop monitoring for improved food security: Proceedings of the Expert Meeting Vientiane, Lao People's Democratic Republic, 17 February 2014. Editor: Mukesh K. Srivastava. The Food and Agriculture Organization of the United Nations and the Asian Development Bank, Bangkok.

Shatar, T. M. and McBratney, A. B. (2004). Boundary-line analysis of field-scale yield response to soil properties. *Journal of Agricultural Science*, 142, 553–60.

Srivastava, M. K. (2015). Crop Monitoring for Improved Food Security: Proceedings of the Expert Meeting Vientiane, Lao People's Democratic Republic, 17 February 2014. Food and Agriculture Organization of the United Nations and Asian Development Bank.

Stehfest, E., Heistermann, M., Priess, J. A., Ojima, D. S. and Alcamo, J. (2007). Simulation of global crop production with the ecosystem model DayCent. *Ecological Modeling*, 209, 203–19.

Stuart, A. M., Pame, A. R. P., Silva, J. V., Dikitanan, R. C., Rutsaert, P., Malabayabas, A. J. B., Lampayan, R. M., Radanielson, A. M. and Singleton, G. R. (2016). Yield gaps in rice-based farming systems: Insights from local studies and prospects for future analysis. *Field Crops Research*, in press.

Sumberg, J. (2012). Mind the (yield) gap(s). *Food Secur.* 4(4), 509–18. doi: 10.1007/s12571-012-0213-0.

Tanaka, A., Diagne, M. and Saito, K. (2015). Causes of yield stagnation in irrigated lowland rice production system in the Senegal River Valley. *Field Crops Research*, 176, 99–107.

Tanaka, A., Saito, K., Azoma, K. and Kobayashi, K. (2013). Factors affecting variation in farm yields of irrigated rice in southern-central Benin. *European Journal of Agronomy*, 44, 46–53.

Tittonell, P. and Giller, K. E. (2013). When yield gaps are poverty traps: The paradigm of ecological intensification in African smallholder agriculture. *Field Crops Research*, 143, 76–90.

van Bussel, L. G. J., Grassini, P., Van Wart, J., Wolf, J., Claessens, L., Yang, H., Boogaard, H., de Groot, H., Saito, K., Cassman, K. G., van Ittersum, M. K. (2015). From field to atlas: Upscaling of location-specific yield gap estimates. *Field Crops Research*, 177, 98–108.

Van Ittersum, M., Cassman, K. G., Grassini, P., Wolf, J., Tittonell, P. and Hochman, Z. (2013). Yield gap analysis with local to global relevance – A Review. *Field Crops Research*, 143, 4–17.

van Oort, P. A. J., de Vries, M. E., Yoshida, H., Saito, K. (2015a). Improved climate risk simulations for rice in arid environments. *PLoS ONE*, 10(3). e0118114.

van Oort, P. A. J., Saito, K., Tanaka, A., Amovin-Assagba, E., Van Bussel, L. G. J., van Wart, J., de Groote, H., van Ittersum, M. K., Cassman, K. G., Wopereis, M. C. S. (2015b). Assessment of rice self-sufficiency in eight African countries in 2025. *Global Food Security*, 5, 39–49.

van Oort, P. A. J., Saito, K., Zwart, S. J. and Shrestha, S. (2014). A simple model for simulating heat induced sterility in rice as a function of flowering time and transpirational cooling. *Field Crops Research*, 156, 303–12.

Van Wart, J., Grassini, P., Cassman, K. G. (2013a). Impact of derived global weather data on simulated crop yields. *Global Change Biology*, 19, 3822–34.

Van Wart, J., Grassini, P., Yang, H., Claessens, L., Jarvis, A., Cassman, K. G. (2015). Creating long-term weather data from thin air for crop simulation modeling. *Agricultural and Forest Meteorology*, 209–10, 49–58.

Van Wart, J., van Bussel, L. G. J., Wolf, J., Licker, R., Grassini, P., Nelson, A., Boogaard, H., Gerber, J., Mueller, N. D., Claessens, L., van Ittersum, M. K., Cassman, K. G. (2013b). Use of agro-climatic zones to upscale simulated crop yield potential. *Field Crops Research*, 143, 44–55.

Vincenzi, S., Zucchetta, M., Franzoi, P., Pellizzato, M., Pranovi, F., De Leo, G. A. and Torricelli, P. (2011). Application of a Random Forest algorithm to predict spatial distribution of the potential yield of Ruditapes philippinarum in the Venice lagoon, Italy. *Ecological Modeling*, 222, 1471–8.

Waddington, S. R., Li, X., Dixon, J., Hyman, G., de Vicente, M. C. (2010). Getting the focus right: production constraints for six major food crops in Asian and African farming systems. *Food Security*, 2, 27–48.

Wairegi, L. W. I., van Asten, P. J. A., Tenywa, M. M., and Bekunda, M. A. (2010). Abiotic constraints override biotic constraints in East African highland banana systems. *Field Crop Research*, 117, 146–53.

Webb, R. A. (1972). Use of the Boundary Line in the analysis of biological data. *Journal of Horticultural Science*, 47, 309–19.

West, P. C., Gerber, J. S., Engstrom, P. M., Mueller, N. D., Brauman, K. A., Carlson, K. M., Cassidy, E. S., Johnston, M., MacDonald, G. K., Ray, D. K. and Siebert, S. (2014). Leverage points for improving global food security and the environment. *Science*, 345(6194), 325–8.

Wolf, J., Ouattara, K. and Supit, I. (2015). Sowing rules for estimating rainfed yield potential of sorghum and maize in Burkina Faso. *Agricultural and Forest Meteorology*, 214–15, 208–18.

Wopereis, M. C. S., Johnson, D. E., Ahmadi, N., Tollens, E. and Jalloh, A. (Eds) (2013). *Realizing Africa's Rice Promise*. CAB International, Wallingford, UK.

Worou, N., Gaiser, T., Saito, K., Goldbach, H. and Ewert, F. (2013). Spatial and temporal variation in yield of rainfed lowland rice in inland valley as affected by fertilizer application and bunding in North-West Benin. *Agricultural Water Management*, 126, 119–24.

You, L., Ringler, C., Wood-Sichra, U., Robertson, R., Wood, S., Zhu, T., Nelson, G., Zhe, G. and Yan, S. (2011). What is the irrigation potential for Africa? A combined biophysical and socioeconomic approach. *Food Policy*, 36, 770–82.

Young, A. (1999). Is there really spare land? A critique of estimates of available cultivable land in developing countries. *Environment, Development and Sustainability*, 1(1), 3–18. doi: 10.1023/A:1010055012699.

Chapter 7

# Developments in the system of rice intensification (SRI)

*Norman Uphoff, Cornell University, USA*


1 Introduction: understanding the protean nature of SRI
2 The evolution of SRI: from practices to principles to fundamentals
3 The spread of SRI to other countries and crops and overall effects
4 Contributions of SRI to achieving climate-smart agriculture
5 Developments in SRI methodology
6 Assessing SRI yields
7 SRI in relation to other production systems
8 Summary and future trends
9 Where to look for further information
10 References


## 1 Introduction: understanding the protean nature of SRI

It is time to take stock of how the system of rice intensification (SRI), constructed in Madagascar some 30 years ago, has been evolving and spreading since the ideas and practices on which it is based were first validated outside that country. In the past 15 years, SRI 'proof of concept' has been validated in more than 50 countries. This chapter considers the dynamics and realities of the SRI phenomenon in its multiple dimensions over time.

SRI is better understood as a set of ideas than as a set technology. It can be characterized, for better and for worse, as 'protean' in that it is versatile and can do many different things. This creates tension with reductionist efforts to describe and assess SRI as 'only X' or as 'always Y' or as 'nothing more than Z'. Such thinking obscures the significance and potentials of SRI, which is not a thing, and certainly not a single thing. It has many aspects – biological, ecological, material, mental, economic, social, aspirational, and even philosophical.

This multiplexity has made it difficult for many people, and particularly some scientists, to grasp and accept SRI as a phenomenon different from what they are accustomed to

http://dx.doi.org/10.19103/AS.2016.0003.27

dealing with. Its multifaceted nature often gets interpreted incorrectly as ambiguity when the phenomenon is unavoidably complex while still very concrete. What connects the various elements that now come under the rubric of SRI is their common descent from the work and writings of Henri de Laulanié (1993).

SRI was not developed as a technology like the Green Revolution, but rather as a way of solving various existential problems of farmers in Madagascar, by tapping into productive potentials that exist in rice plants and in the soil systems that they grow in. SRI knowledge differs from much current agricultural technology in that it is not proprietary, instead being open-access, freely available to anyone and owned by no one. The theory and practice of SRI are not at all abstract, but they extend beyond the biophysical realm and are linked to human resource development for farmers and others. SRI has given impetus to what can be called a 'movement' (Bouman, 2013); it is also a paradigm shift that can be described in terms of an emerging 'post-modern agriculture' (Uphoff, 2007a).

Since it is better to be more concrete than abstract, this chapter considers specific developments in SRI theory and practice that have emerged since the productivity of SRI's methods was first demonstrated in China and Indonesia a decade and a half ago (Gani et al., 2002; Wang et al., 2002; Yuan, 2002; Zhu et al., 2002). While the core insights and concerns of SRI have remained the same, its scope is now broader and more diversified than when SRI was developed in Madagascar, its country of origin, in the 1980s.

# 2 The evolution of SRI: from practices to principles to fundamentals

The SRI was developed inductively by Fr. Henri Laulanié, knitting together various concrete practices, some of them quite counterintuitive, like getting higher crop yields by reducing plant populations in the field by 80–90% (Laulanié, 1993). Laulanié found that making certain changes in rice-growing practices, when used together, could induce more productive phenotypes of rice from whatever varieties the Malagasy farmers were already using, whether improved high-yielding varieties or unimproved local cultivars.

The effect of getting more profuse tillering and more prolific root systems could be explained at least in part by an understanding of the concept of phyllochron, developed by the Japanese scientist T. Katayama some 50 years earlier in Japan but little known outside of Japan. This periodicity in the emergence of tillers (and concomitantly of roots) in rice along with wheat, barley and other grass-family plants has been explained in English by Nemoto et al. (1995).

Laulanié found serendipitously that rice seedlings transplanted before the start of their fourth phyllochron of growth, usually about 15 days after sowing, retained more potential for tiller and root growth than did older transplants, as explained in Uphoff (2015: 154–62). This was, however, just one of a number of growth-promoting management practices that when combined can produce more productive and robust rice phenotypes. For example, wider spacing, which reduces the inhibition of root and tiller growth caused by dense planting, has more beneficial effects when young seedlings are used (Mishra and Salokhe, 2008).

## 2.1 Practices

SRI was initially presented and defined by certain specific changes in practice:

- Plant *single seedlings* in each hill, rather than three, four and five or more plants in a clump.
- Transplant them *quickly and carefully* to minimize stress to the roots.
- Locate the hills in a *square pattern*, usually 25 × 25 cm, rather than planting helter-skelter or in rows. This provides more room above- and below-ground for the profuse growth of rice plants' tillers and roots.
- *Refrain from continuous flooding* of SRI rice paddies, just keep the soil mostly wet, either with small daily applications of water or with alternative wetting and drying. In either case, the soil is allowed to dry out for short periods during the growing season.
- *Control weeds with a mechanical hand weeder.* This actively aerates the top layer of soil as it buries the weeds to decompose as green manure.
- *Enrich the soil with compost*, as much as available. This last practice was added to the preceding ones in the late 1980s when the Madagascar government ended its fertilizer subsidies, and poor, smallholding farmers could no longer afford this input.

With this empirically assembled methodology, there was no need for farmers to purchase either seeds or fertilizer. Also, it greatly reduced irrigated rice crops' water requirements.

SRI's distinct departure from Green Revolution technology, which required new varieties supplied with more fertilizer and more water, made many people (including the author at first) quite sceptical that such methods could succeed. However, they did succeed. Malagasy farmers who had been producing just 2 tons $ha^{-1}$ were able to achieve – with SRI crop management on the same poor soils and with the same varieties – average yields of 8 tons $ha^{-1}$, with some yields even higher (Uphoff, 1999). This was possible without having to rely on external inputs. Such results begged for an explanation.

Over the past 15 years, a growing scientific literature has investigated the mechanisms involved, probing the evident synergy among practices, such as better root systems supporting better canopies and vice versa. Evaluating one practice at a time does not adequately address the SRI phenomena as these are best assessed by multifactorial analyses, for example, Lin et al. (2009), Mishra and Salokhe (2008), Thakur et al. (2010, 2013, 2015), Uphoff and Randriamiharisoa (2002) and Zhao et al. (2009).

## 2.2 Principles

As a result of extensive research, SRI effects are now better understood and can be presented more meaningfully in terms of various principles that underlie the respective practices rather than focusing just on the practices (SRI-Rice, 2015). These principles can be summarized as follows:

- *Early crop establishment with healthy plants*, taking care to minimize root trauma during transplanting to avoid 'transplant shock'. As noted below, direct seeding has been developed as an alternative practice with other SRI techniques. But the principle of careful crop establishment and avoiding root trauma applies generally.

- *Minimize competition between plants* below- and above-ground by establishing single plants in a square pattern. This promotes profuse growth of root systems and canopies with more nutrient uptake and photosynthesis supporting grain development.
- *Enhance soil fertility with organic matter* to improve the structure and functioning of soil systems, with more abundance and diversity of beneficial soil organisms. Inorganic fertilizers can be used with SRI, especially if the availability of organic matter is a constraint, but building up the soil's fertility is a key component of SRI.
- *Manage water to avoid anaerobic soil conditions*, so that plant roots and aerobic soil organisms do not suffocate. Also undertake *active soil aeration* through mechanical weeding that controls weeds, which are more numerous with no continuous flooding.

The SRI practices that operationalize these principles are expected to be adjusted to best suit local conditions, varying factors like age of seedlings, distance between hills, timing of water issues, and frequency of weeding. To the extent that farmers understand these SRI principles, they will be better able to adjust and apply the respective practices.

## 2.3 Fundamentals

These principles pointed to simple but essential understandings of the foundations on which SRI productivity is based. It has been seen that the effectiveness of SRI methods, assembled in a purely inductive way, can be attributed to – and summarized in terms of – two factors that give theoretical and practical explanations for SRI results. SRI principles and practices are beneficial for rice plants' growth and health by supporting

- *larger, deeper and more robust root systems* (Barison and Uphoff, 2011), and
- *more beneficial soil biota* which becomes more abundant and more diverse due to root exudation and in aerobic soil with organic soil amendments (Anas et al., 2011).

It is understandable but unfortunate that because these effects are manifested underground, they are easily neglected. Root growth can be assessed by visual inspection, but the evaluation of microorganisms in the rhizosphere and of symbiotic endophytes that live mutualistically within plants is difficult.

The theory of SRI has followed rather than preceded its practice, much as the science of aeronautics came after the development of the aeroplane. SRI was assembled pragmatically, combining beneficial practices that were found to be productive. The fact that SRI practices resulted in much larger, more long-lived root systems was not difficult to document (Barison and Uphoff, 2011). Making the connections between SRI practices and changes in the soil biota has taken more time and effort (Randriamiharisoa et al., 2006; Zhao et al., 2010; Anas et al., 2011; Lin et al., 2011). The soil biota and its connections to rice productivity and improvement warrant considerably more attention than they have received so far.

## 2.4 Contributions of soil biology

Of the advances in knowledge about SRI over the past 15 years, the most important have come in the domain of soil microbiology and microbial ecology. This parallels what has happened in the realm of human biology during this time. Researchers have come to understand better, although still quite incompletely, the myriad contributions that the human microbiome – the plethora of microorganisms that inhabit our bodies – makes to our own growth and health.

In similar ways, we are learning that microbes living around and on rice plants – as well as inside their tissues and cells – affect plants' root growth below-ground, their canopies above, key parameters such as levels of chlorophyll and rates of photosynthesis, physiological water-use efficiency, resilience to drought, salinity and other stresses, and consequently yield (Yanni et al., 2001; Chi et al., 2005, 2010; Redman et al., 2011; Uphoff et al., 2013).

Whereas at first SRI could be characterized just as a collection of synergistic agronomic practices, it has become more evident over time that symbiotic microbes in the root zone and in the plants themselves are making some key contributions to the observed SRI effects.

A study by Chinese scientists has shown that microbes living as symbiotic endophytes in rice plant organs and cells affect the expression of plants' genes in their roots, sheaths and leaves, up-regulating the expression of some genes and down-regulating others (Chi et al., 2010).

Unfortunately, no research has been done on any epigenetic influences of microbes on rice plant growth and performance in response to specific SRI management practices. However, the evident and measurable improvements in rice plant phenotypes that can be evoked from a given genotype by making SRI modifications in cultural practices, such as that documented by Thakur et al. (2010), suggest such effects. Also, SRI plants' greater ability to withstand the stresses of drought, salinity and extreme temperatures parallels the resilience that is reported for rice plants that have been inoculated with symbiotic fungi (Redman et al., 2011). So this could be a productive area for future rice science research.

Both plant root systems and the soil biota have profound effects on crops' productivity, yet they were largely ignored in the science and practice of the Green Revolution. SRI experience has shown that its principles and associated practices converge on growing better plant root systems and on nurturing the soil biota. SRI methodology is thus more than just 'best management practices' as has sometimes been said. Since explanation is the essence of theory, in fact, SRI proponents have been developing cogent theoretical foundations for its effectiveness.

# 3 The spread of SRI to other countries and crops and overall effects

## 3.1 The dissemination of SRI ideas and practices

SRI principles, with appropriate adaptations of operational practices, have been applied productively in many and diverse regions of the world, from the high mountainous regions of Afghanistan (Thomas and Ramzi, 2011) to the arid conditions of Timbuktu on the edge of the Sahara desert in Mali (Styger et al., 2011) to the tropical climate of Panama (Turmel et al., 2011).

During the first 3 years after SRI began spreading beyond Madagascar, it was evaluated in 15 countries with results being reported at the first, and so far only, international conference on SRI held in China in 2002 (CIIFAD, 2002; Yuan, 2002). The number of countries where the productivity of SRI principles and practices has been observed is now, 12 years later, up to 58, most of them very different from the country where SRI methodology was first devised (http://sri.cals.cornell.edu/countries/).

An early assessment that SRI would be a 'niche innovation', suitable only for improving low-productivity rice cropping systems on marginal land with an abundance of cheap labour (Dobermann, 2003), has proven to be incorrect. This became evident from an evaluation by researchers at the Andhra Pradesh state agricultural university in India in the 2003 kharif season. When they conducted on-farm comparison trials in all 22 districts of the state, they found SRI practices giving an average yield advantage of 70% across widely varying soil types and topographies and with 19 different varieties evaluated (Satyanarayana et al., 2006). Surprisingly, they recorded the largest yield increases with SRI management in the interior region of the state (Rayalaseema) rather than in the lowland delta regions along the coast, which had always been thought of as having the best soil for growing rice. Better-drained upland soils, if their organic matter levels were enhanced, gave bigger yield responses to SRI management than did low-lying paddy soils, usually thought to be better for growing rice (ibid.).

The yield results with SRI management practices vary considerably, however. To some, this implies that none of the results should be taken seriously. However, the variation is attributable in large part to the role that the soil biota plays in influencing crop yield, as discussed above. This biota can differ greatly over time, and even within short periods, as well as across space in different locations. So this contributes to making yields quite variable.

As SRI knowledge and practices diffuse, the phenomenon itself has become more diversified and more difficult to characterize in simple terms. This is frustrating for those who favour reductionist views and methods for evaluation. But other scientists, and most practitioners, are more inclined to try to take advantage of the variability that SRI methods can create, rather than dismiss it.

## 3.2 Aggregated impacts on yield and income

The increasing pluralism of SRI makes assessing its impacts ever more difficult. As seen above, there are certain ideal-type practices that can characterize SRI and that have been validated in factorial trials (Uphoff and Randriamiharisoa, 2002). However, SRI is better understood as a matter of *degree* than in terms of any binary classification of SRI versus not-SRI. A framework to evaluate SRI as a matter of degree has been presented in Wu et al. (2015), offering a way to quantify and possibly classify SRI practice. But comparing SRI impacts with current practice becomes more difficult over time as recommended SRI practices get disseminated to and adapted and modified by other farmers so that the distinction between 'SRI' and conventional practice becomes blurred. It is better to analyse sets and combinations of particular practices under specified circumstances rather than compare nominal classifications anyway, since the purpose is not to prove or exalt SRI, but rather to help farmers improve the productivity of their available resources.

This said, it is reasonable to expect some quantified assessments of productivity gains, if any. In 2014, the SRI-Rice Center at Cornell University compiled an estimate of both the extent of uptake of SRI practices and the impact that this was having on yields. The countries polled were China, India, Indonesia, Vietnam and Cambodia as governments in these five countries have begun promoting SRI methods based on their own evaluations and experience, and because together they produce two-thirds of the world's rice.

Resources have never been available for any agency to collect and compile comparable data internationally on the numbers of farmers and area where SRI methods are being used. From some of these countries, government statistics could be reported; elsewhere only best estimates could be obtained from knowledgeable persons. Where no direct data were available, conservative estimates were requested to have numbers that represented at least the minimum likely levels of uptake and impact. The data gathered are shown in Table 1. The data reported on yield are more precise than those for area as they are based on measurements at field level. But these are still aggregate numbers, not distinguishing, for example, between irrigated and rain-fed rice production. The purpose of aggregation was to arrive at some tenable estimates giving orders of magnitude.

The totals regarding SRI management in these five countries by the end of 2013 came to at least about 9.5 million farmers using many or all SRI methods on over 3.4 million hectares. Actual totals would be higher than this because of the farmer-to-farmer spread of SRI knowledge.[1] Also, SRI-Rice does not have contact with all of the organizations and agencies that are now promoting SRI methods, SRI knowledge having been purposefully kept in the public domain.

Calculating the exact value of additional rice produced with SRI management methods would take a huge amount of time and resources. The following is an estimation based on the information in hand for the five countries. It does not consider the value of the irrigation water saved in producing paddy rice. A meta-analysis discussed in section 4.3 has calculated an average irrigation water reduction of 35% with SRI (Jagannath et al., 2013). In China, this water saving was calculated as 25% (Zheng et al., 2013). Considering the value of this water would greatly augment the total value associated with paddy production using SRI methods.

Also, the calculation below assumes no increase in farmers' costs of production. Actually, economic assessments have usually found some *reduction* in costs with SRI management, which would further raise the value of its economic returns. A study of SRI impacts in Tamil Nadu, a state in India, calculated an average reduction in production costs per hectare of 26% (Barah, 2009), while a larger study across 13 states of India reported the costs of production

[1] In China, for example, a former director of the Crop Research Institute of the Sichuan Academy of Agricultural Sciences, Dr. Zheng Jiaguo, estimates that about one-third of the rice farmers in China are now using some if not all of the SRI-recommended practices; but this is not referred to as 'SRI'. Dr. Zhu Defeng, a senior rice scientist at the China National Rice Research Institute in Hangzhou, has reported that 'SRI is becoming the main rice cultivation system in most of southern China' (IRIN, 2012). However, such assessments are not quantifiable. In Sichuan province, where the Provincial Department of Agriculture has kept records of SRI use and effects since it began promoting these new methods in 2004, the area under SRI management increased from 1133 ha in that first year to >300 000 ha by 2010. Comparing average SRI yields with non-SRI yields across the province, the department calculated that the incremental increase in paddy production achieved with SRI methods was 1.66 million tons, worth > $330 million at the current rate of exchange (Zheng et al., 2013). This was achieved with some localizing adaptations of SRI methods, which is appropriate and even expected for SRI.

**Table 1** Data from China, India, Vietnam, Indonesia and Cambodia on SRI area, farmer use and yields from 2013 surveys or reports

| Country | SRI area (ha) | SRI farmers (no.) | Comparison yield (t $ha^{-1}$) | SRI yield (t $ha^{-1}$) | % Increase | Sources and comments |
|---|---|---|---|---|---|---|
| **China** | 1 000 000 | 3 000 000 | 6.6 | 9.5 | 44% | Estimate from Institute of Crop Sciences, CAAS, cited in UN-IRIN report on SRI, 4/24/12 |
| *Sichuan* | *385 000* | *1 283 000* | *7.9* | *9.3* | *18%* | Sichuan Provincial Dept. of Agriculture, as reported by Sichuan Academy of Agricultural Sciences |
| **India** | 1 760 537+ | 3 514 165+ | **3.7** | **5.6** | **52** | These numbers represent only the states covered; total numbers for all-India will be at least 50% more |
| *Tamil Nadu* | *1 000 000* | *2 000 000* | *5.15* | *6.6* | *28* | Tamil Nadu Agricultural University report from Dept. of Agriculture and its own data |
| *Bihar* | *616 000* | *1 200 000* | *2.53* | *4.07* | *61* | Area and yield: Bihar Dept. of Agriculture; number of farmers is extrapolated from average size of holding |
| *Andhra Pradesh* | *121 815* | *220 000* | *5.87* | *6.44* | *10* | AP Dept. of Agriculture data and from JAI-SRI network survey; increase in net income/ha is 50% |
| *Orissa* | *20 880* | *82 617* | *3.88* | *5.23* | *36* | SRI Orissa Learning Alliance survey of member NGOs and government programmes and OCTMP |
| *Uttarakhand and Himachal Pradesh* | *1 177* | *9437* | *2.3* | *4.0* | *73* | Reports from People's Science Institute, Dehradun; the total numbers for these states are higher |
| *Gujarat* | *665* | *2011* | *2.46* | *3.93* | *60* | AKRSP-India report on second-year results working with tribal farmers |
| **Vietnam** | 366 951 | 1 813 201 | 5.4 | 5.8 | 7 | Data from the Ministry of Agriculture and Rural Development, Hanoi; profitability $ha^{-1}$ increased 70% |
| **Indonesia** | 300 000+ | 1 000 000+ | 5.00 | 6.75 | 35 | Estimate for all methods: organic, inorganic and mixed; 28 out of 33 provinces; comparative yields for 12 133 farmers (9429 ha) were 4.27 tons $ha^{-1}$ conventional and 6.75 tons $ha^{-1}$ with SRI (Sato and Uphoff, 2007) |
| **Cambodia** | 40 222+ | 200 000+ | 2.13 | 3.43 | 61 | Area and yields from CEDAC, Phnom Penh; most SRI is used for rain-fed rice production; total number of farmers using SRI methods is probably >250 000; |
| | 3 466 710+ | 9 527 366+ | 4.56 | 6.22 | 40 | Unweighted average |

were reduced by $29 per ton of paddy (Palanisami et al., 2013). Savings come mostly from reduced purchases of fertilizer and agrochemicals. If the costs of labour are higher at first (and in most of Asia they are lower), labour requirements diminish over time as skill and confidence are gained, with SRI usually becoming labour saving (section 4.4). Factoring in such cost reductions will further increase the net value from farmers' SRI paddy production.

| | |
|---|---|
| (A) Average increase in paddy yield based on using some or all of the recommended SRI practices | 1.66 tons $ha^{-1}$ |
| (B) Area of paddy production for which this increase is being achieved in China, India, Indonesia, Vietnam and Cambodia | 3 466 710 ha |
| (C) Increased production from higher SRI yield (A × B) | 5 754 738 tons |
| (D) Price received for increased paddy production based on current information on farmgate prices* | $150 $ton^{-1}$ |
| (E) Value of increased paddy production (C × D) | $862.5 million |

*This varies widely among countries, for example, $140 $ton^{-1}$ in Cambodia; $144 $ton^{-1}$ in Nepal; in India the government support price is $230 $ton^{-1}$, but the prevailing market price in Bihar state, for example, which produces the largest share of SRI rice in India, is $170 $ton^{-1}$. In Kenya, on the other hand, the current paddy price is $500 $ton^{-1}$. The farmgate price in the Philippines is currently $360. These figures suggest $150 $ton^{-1}$ as a reasonable average figure for estimating total amount.

This calculation of increase in the value of production is for the five countries in just one year, 2013. Overall, if adding in the value from other countries where SRI methods are being used, this figure would be well over $1 billion, with reductions rather than increases in the cost of producing this added value. Moreover, this figure does not consider the economic value of large reductions in water use. As the area under SRI management has been increasing by 50% or more per year, this value should continue to rise for years to come, with less incremental expenditure on its promotion than the resulting value of production.

## 3.3 Application to rain-fed rice production

The first major modification in the original version of SRI came from farmers and NGOs in the Philippines (BIND, 2003), then in Myanmar, Cambodia and India, who began experimenting with SRI practices under unirrigated, upland conditions, making appropriate adaptations. They found that they could get more productive and robust rice plants by transplanting single, young, widely spaced seedlings timed with the onset of the rains, while also enhancing soil organic matter and actively aerating the soil with mechanical weeding. Upland rice varieties responded favourably to such crop management much as did varieties that had been developed for irrigated conditions.

The best-documented results of upland SRI were from Myanmar, where SRI was demonstrated and disseminated through farmer field schools in Kachin and Shan states. Using SRI practices as recommended, farmers got yields on their farmer field school plots three times higher than they were achieving with their usual rain-fed practices. On their own fields, where the practices were used less fully or less rigorously, farmers were able to double their yields with no increase in production costs (Kabir and Uphoff, 2007).

In India, an evaluation of rain-fed SRI in West Bengal state done for the International Water Management Institute (IWMI) found that SRI yields were 32% higher on average, even with one of the two villages studied in-depth having suffered drought conditions. Net economic returns per hectare were calculated to be 67% higher, with an 8% reduction in the amount of labour required per hectare (Sinha and Talati, 2007). The NGO that introduced rain-fed SRI in West Bengal found that average upland paddy yields were raised to 7.5 tons $ha^{-1}$ when SRI alternative management methods were used, compared to 5 tons $ha^{-1}$ achieved with farmers' usual rain-fed rice growing methods (PRADAN, 2007).

A farmer adaptation contributing to these results in West Bengal was to plant several small SRI seedling nurseries, rather than just one, and sequentially, two weeks apart, so that farmers would have 'young seedlings' available for transplanting whenever the rains began. This meant that only one of the nurseries would be planted out and the others would be discarded. Such sacrifice of seeds and seedlings was at first hard for farmers to accept, but it became agreeable once they saw how much their transplanting young seedlings, less than 15 days old, could increase their yield. Since an SRI nursery requires only 10% as much seed as does conventional crop establishment, farmers' seed saving could still be about 70%, rather than the 90% that they could achieve if they planted and used just one SRI nursery, whose seedlings would be younger or older than optimal if the rains came early or late.

It was also difficult at first to persuade farmers not to hoard standing water in their rice paddies when the monsoon rains started. After several seasons, however, they understood that this prevailing practice suffocated their plants' roots, making them less able to tolerate water stress later in the season. Managing rainwater so that the soil is kept wet for as long as possible but not inundated, in order to promote growth of larger root systems, gave farmers hardier plants that could better withstand subsequent water shortages and stresses.

With appropriate modifications of practices, farmers have found that the SRI principles can significantly improve their rain-fed rice production systems. Unirrigated SRI has been the predominant form of SRI practice in Cambodia and in much of eastern India, where rice production necessarily depends upon rainfall rather than upon pump or gravity irrigation. This was an important extension of SRI ideas and practices since a disproportionate share of poverty and food insecurity in the world is found in upland, unirrigated regions. As seen next, combining SRI with water harvesting and crop diversification can significantly improve the productivity of land and water resources in rain-fed areas.

## 3.4 Applications to more complex farming systems

Although SRI was developed for more productive irrigated rice cropping as a monoculture, in recent years it has been integrated into more diversified farming systems that give farmers both more income and better distribution of income throughout the year. Higher rice yields achieved with SRI methods permit farmers to meet their basic food needs with less land area, permitting redeployment of some of their land resources to other, higher-value agricultural production like horticulture and aquaculture. Thus, an important development for SRI theory and practice has been its integration into households' total production systems, not using it just as an improved cropping system.

In Cambodia, small-holding farmers, with rain-fed farms averaging just two-thirds of a hectare, by tripling their paddy yields with SRI can afford to put 15% of their holding into ponds and canals for capturing rainwater and distributing it. Also, 25–30% of their area can be converted to horticultural and other production, including maintaining small livestock. This also gives supplementary water supply for their rice once the dry season begins.

With investments averaging just $300 spaced over several years, they can diversify their production and triple the net income from the same land and water resources. This gives households not only more income but also year-round sources of income, plus a more diversified diet for better nutrition (Lim, 2007).

Indian scientists have conducted trials with alternative management regimes to assess the productivity of 'integrated SRI' that combines aquaculture and horticultural production with rain-fed SRI. They have documented greatly increased productivity of land, labour and especially water from diversified rice-based production systems. Plots of 350 $m^2$ were evaluated comparing (a) whole plots under standard rain-fed paddy production, with (b) plots that used rain-fed SRI methods, (c) rain-fed SRI plots served by installation of a groundwater pump, and (d) rain-fed SRI plots with a water-harvesting pond constructed on 10% of the area, stocked with Asian carp and planting bananas and papayas on bunds around the pond that captured and held water, taking up 13% of the area. The catchment pond in (d) provided supplementary water for the rice crop similar to that from the pump with (c).

With such integrated management using SRI methods, paddy yield was more than doubled (data on changes in SRI plant phenotypes reinforced this finding). Because of the multiple income sources, net plot income could be raised from USD 2.25 to USD 142. (Conventional rain-fed paddy production in the area is little more than a break-even operation.) The productivity of total water per hectare (rainfall plus harvested water) was increased in this integrated system from less than 1 cent $m^{-3}$ to 30 cents $m^{-3}$ (Thakur et al., 2015). These numbers are perhaps hard to believe, but they come from replicated trials with randomized block design that were conducted over two years, one of them a drought year.

Integrating the ideas and methods of SRI with the production of other crops represents an important evolution beyond its initial 'cropping system' way of thinking to work within a 'farming systems' perspective.

## 3.5 Applications to other crops

After the ideas and methods of SRI, developed from improving irrigated rice production, were extended to upland rice (section 3.3), they have been further extrapolated, with appropriate modifications, to a variety of other crops, starting with finger millet, then wheat, then sugarcane and mustard, and then legumes and a number of vegetables (Behera et al., 2013). These methodologies have been grouped under the broad rubric of the System of Crop Intensification (SCI) (Abraham et al., 2014; SRI-Rice, 2014).

So far, much less research has been done and published on these extrapolations than on SRI. The most rigorous evaluation has been done on wheat by researchers who assessed the System of Wheat Intensification (SWI) in two years of trials at the Indian Agricultural Research Institute (IARI) in New Delhi, using standard randomized, replicated-plot design. By circumstance, in the second year there was unfavourable weather (high temperatures and excessive rainfall at critical times in the crop cycle) which generally lowered wheat yields across much of northern India, so resilience to climate effects (next section) was included in the evaluation.

In the first year, yields on SWI plots were 30% higher than from standard recommended practices (SRP) endorsed by IARI scientists. In the second year, SWI plants were seen to be more resilient to climate stresses as the yield advantage of SWI over SRP increased to 46%. Economic returns with SWI averaged 35% higher than SRP for the two years. Of relevance for sustainability concerns, the soil nutrient levels of N, P and K were found to be enhanced in the SWI plots over the course of the season, whereas they were diminished under SRP management, despite more application of inorganic nutrients (Dhar et al., 2015).

More research remains to be done on this extension of SRI thinking to other crops: modifications in crop establishment, spacing, water applications, nutrient management, soil aeration and other practices. How far this extrapolation can be productively extended is not known, but experiences in India, Pakistan, Nepal, Ethiopia, Mali and Cuba are indicating that there are benefits to be reaped from making changes in the management of plants, soil, water and nutrients, beneficially affecting root growth and the life in the soil. So SRI experience and research are taking SRI ideas and methods well beyond rice production.

# 4 Contributions of SRI to achieving climate-smart agriculture

## 4.1 Adaptations to the stresses of climate change

An unanticipated benefit from SRI crop management has been growing evidence that this can give farmers not just more productive rice plants, but also plants that are hardier and more resilient – better able to withstand the stresses of drought and storm damage, and also to some extent the effects of flooding and extreme temperatures (Uphoff, 2011). The sources of this ability are the larger, stronger root systems promoted by SRI practices, which keep roots from being degraded by the hypoxia of continuous flooding (Kar et al., 1974), as well as the resulting concentrations and diversity of soil organisms that enhance soil systems' capacity to absorb and retain water and to support more robust plant growth.

With climate change, crop losses to pests and diseases also tend to increase as plants under greater stress become more vulnerable to predation and infection. The first evidence on this came from thesis research at the Tamil Nadu Agricultural University, shown in Table 2. Further studies in India reinforced this observation that SRI rice plants are better able to withstand various pests and diseases, although not all of them, for example, Kumar et al. (2007), Gopal et al. (2010) and Karthikeyan et al. (2010). Evaluations of on-farm comparison trials across eight districts by the Vietnam Ministry of Agriculture and Rural Development's IPM programme showed reductions in the incidence of four major rice diseases and pests, 55% in the spring season and 70% in the summer season (Dung, 2007). Some of the protective influence of SRI management can be ascribed to wider spacing between plants, less humid conditions in the above-ground canopy, and reduced or no application of inorganic N fertilizer, for example.

How SRI management can enhance rice crops' ability to withstand both biotic and abiotic stresses is seen from a picture of two adjacent paddy fields given to the author by an SRI farmer in East Java, Indonesia in 2011 (Fig. 1). On the right, the farmer was growing a traditional aromatic variety of rice with organic SRI methods, while her neighbour's adjoining paddy field on the left was planted with a modern variety and using chemical fertilizers. The picture was taken after their village had been hit first by a brown plant hopper attack and then by a tropical storm. Hopper burn and crop lodging left her neighbour's field with almost no yield, while she harvested 800 kg from her 1000 $m^2$, a yield of 8 tons $ha^{-1}$. Some of the difference would be attributable to varietal differences in respective resistances to pest attack, but management methods surely also played a role. These cannot protect a rice crop against all climatic hazards, but they can give some benefit. For a previous report on SRI resistance to brown plant hopper, see Ragukkanu et al. (2007).

**Table 2** Pest abundance in nursery and in main field with different methods of cultivation, from evaluations done at Tamil Nadu Agricultural University, Coimbatore, India, 2002

| Insects and their damage per unit | SRI cultivation (mean ± SE) | Conventional cultivation (mean ± SE) | % decrease with SRI | t value |
|---|---|---|---|---|
| | | In nursery | | |
| CUT WORM – % damaged leaves per seedling | 0.0 ± 0.0 | 20.4 ± 4.8 | ∞ | 16.1** |
| THRIPS – no. per seedling | 0.5 ± 0.2 | 6.1 ± 0.5 | – 92% | 19.3** |
| GREEN LEAF HOPPER – no. per seedling | 0.1 ± 0.0 | 0.4 ± 0.1 | – 75% | 14.8** |
| BROWN PLANT HOPPER – no. per seedling | 0.0 ± 0.0 | 0.2 ± 0.0 | ∞ | 11.5** |
| WHORL MAGGOT – % damaged leaves per seedling | 0.8 ± 0.2 | 9.3 ± 2.6 | – 91% | 12.5** |
| | | In main field | | |
| CUT WORM – % damaged leaves per seedling | 17.9 ± 1.9 | 23.2 ± 2.0 | – 22% | 6.6** |
| THRIPS – no. per seedling | 6.6 ± 0.1 | 20.2 ± 2.0 | – 67% | 12.1** |
| GREEN LEAF HOPPER – no. per seedling | 0.6 ± 0.1 | 1.1 ± 0.2 | – 55% | 10.7** |
| BROWN PLANT HOPPER – no. per seedling | 1.1 ± 0.2 | 2.7 ± 0.2 | – 59% | 14.4** |
| WHORL MAGGOT – % truncated leaves per seedling | 5.6 ± 1.8 | 8.8 ± 1.4 | – 36% | 4.5** |
| GALL MIDGE – % silver shoot per hill | 5.0 ± 1.2 | 11.0 ± 1.5 | – 55% | 9.3** |
| STEM BORERS – deadhearts/white ears per hill | 11.7 ± 1.3 | 7.3 ± 1.0 | +60% | 10.1** |
| LEAF FOLDER – scraped leaves per hill | 20.3 ± 1.6 | 6.5 ± 1.0 | +212% | 15.4** |
| EARHEAD BUG – no. per hill | 0.9 ± 0.1 | 0.9 ± 0.1 | 0 | $0.4^{NS}$ |

**Significant difference ($P < 0.001$); NS: not significant; *Source*: Ezhil Rani (2004).

**Figure 1** Two rice paddy fields in Ngawi district, East Java, after exposure to the stresses of brown plant hopper infestation and a tropical storm in 2011.

## 4.2 Reducing greenhouse gas emissions

Irrigated rice production is one of the most significant contributors to the emission of greenhouse gases (GHG) that drive global warming and induce climate change. Flooded rice paddies are estimated to contribute as much as 20% of the methane ($CH_4$) coming from anthropogenic sources. Methane is about 25 times more powerful than carbon dioxide ($CO_2$) in terms of its global warming potential (GWP). Ceasing the continuous flooding of rice fields will, it is well understood, reduce methane emissions, so it is agreed that SRI water management can certainly help to counter climate change in this respect.

However, when the paddy soils growing rice are aerobic, it is expected that there will be an increase in the emission of nitrous oxide ($N_2O$), which has about 12 times more GWP than comparable amounts of methane. A small increase in $N_2O$ can thus offset the gains achieved from large reductions in $CH_4$. However, when experimental measurements have been made, such offsetting has not been found under SRI crop management, which favours organic soil nutrient supplementation over inorganic fertilization.

A study of GHG emissions from rice fields in Andhra Pradesh state done by Oxford and Indian researchers found that GWP from $CH_4$ and $N_2O$ emissions from SRI-managed rice paddies was >25% less per hectare (in $CO_2$ equivalence). Further, since SRI yields were 58% higher, the GWP of GHGs emitted from SRI practices compared to farmers' present management was >60% less per kilogram of rice produced. While the emissions of $N_2O$ $ha^{-1}$ were somewhat higher with SRI methods, they were lower per kilogram of rice produced (Gathorne-Hardy et al., 2013).

Even greater reductions in GWP from SRI paddies have been measured by researchers at Kangwon National University in Korea. GHG emissions in $CO_2$ equivalence from paddies with conventional management were 14.2 tons $ha^{-1}$ while those with SRI management were only 4.0 tons $ha^{-1}$ (Choi et al., 2014). In Vietnam, measurements of GHG emissions found a significant 20% decrease in the $CH_4$ emitted from SRI fields compared to control fields ($p < 0.01$). Somewhat surprisingly, there was an accompanying small *decrease* in

$N_2O$ emissions (1.4%). While this latter decrease was not statistically significant ($p > 0.1$), it had been expected that $N_2O$ emissions would increase when flooding of rice paddies was curtailed (Dill et al., 2013).

GHG measurements will always have considerable variability and volatility, being affected by temperature, soil moisture, pH, soil type and other factors. But most measurements of GHG emissions with SRI management have showed net reductions in GWP, something that was not considered or thought about 15 years ago.

## 4.3 Evidence of water saving

Contrary to general opinion, rice is not an aquatic plant. Rice plants can survive under flooded conditions, but their roots degenerate under hypoxic soil conditions (Kar et al., 1974). Managing irrigation so that rice plants have just enough water to meet their needs, but not an excess, has productivity benefits as well as benefits for the environment, by reducing extractions of water from surface or underground sources and well as by decreasing methane emissions.

A meta-analysis has been done of 29 published evaluations that compared SRI practices with standard crop and water management for irrigated rice, reporting results from 251 comparison trials conducted in eight countries (Jagannath et al., 2013). With SRI methods, applications of irrigation water per hectare were on average lower by 35% (7.2 million litres $ha^{-1}$ vs. 11.1 million litres $ha^{-1}$), and total water requirements, including rainfall, were 22% less per hectare (12.03 million litres $ha^{-1}$ vs. 15.33 million litres $ha^{-1}$).

With SRI practices giving higher yields with less water, water use efficiency (WUE) – the grams of grain produced per litre of water – was increased on average by 52% for total WUE and by 78% in terms of irrigation WUE (Jagannath et al., 2013). Since only 30% of the SRI trials in these published evaluations had used >80% of the recommended SRI practices (and the average SRI yield advantage in these trials was only 11%), the improvements in WUE with SRI could be even greater with better and fuller use of its recommendations.

## 4.4 Evidence of labour saving

When the labour intensity of SRI was first evaluated in Madagascar (Moser and Barrett, 2003), no consideration was given to the fact that traditional rice production in that country is relatively labour-*extensive*, since farmers invest little work in their operations such as transplanting and weeding and accordingly get low yields. When SRI has been introduced in most Asian countries where methods of rice production are labour-intensive because of the higher population density per unit of arable land there, farmers have usually found SRI methods to be labour-saving (Uphoff, 2007b; Mrunalini and Ganesh, 2008). Even in Madagascar, once farmers have gained experience, skills and confidence in the new methods, SRI has been found to become labour saving there as well (Barrett et al., 2004).

Labour-intensity is always a relative matter, to be assessed in comparison with other modes or kinds of production. The initial reports that SRI required more labour per hectare reflected in part the way that rice was produced in Madagascar. When the first reports came out, it was accepted that higher labour inputs were a necessary cost for obtaining the other benefits of SRI – higher yield, water saving, cost reduction, enhanced soil fertility, etc.

However, even when the labour inputs $ha^{-1}$ have been greater with SRI, the *productivity* of labour, that is, the kilograms of rice produced per day or per hour of labour input, has been increased because of the synergy among practices. No studies have shown labour productivity to be lowered by SRI use even when the inputs of labour $ha^{-1}$ were higher. So far, unless rice production is currently rather labour-extensive, as in Madagascar, SRI practices are likely to be labour-saving, even from their first use.

In 2004, the Tamil Nadu Agricultural University in India had 100 farmers in the Tamiraparani river basin grow the same varieties of rice on half-acre plots side by side, with SRI methods used on one and their usual methods on the other. The average yield increase with SRI methods was 27%, accomplished with an 8% reduction in farmers' labour inputs $ha^{-1}$ (Thiyagarajan, 2004). Similarly, an evaluation of rain-fed SRI methods in West Bengal state by a team from the IWMI found labour inputs $ha^{-1}$ reduced by 8% with SRI methods, while farmers' average paddy yield, even with drought conditions, went up by 32% and their net income $ha^{-1}$ was 67% higher (Sinha and Talati, 2007).

An evaluation of SRI use in a village in Sichuan province of China by a team from the China Agricultural University, where SRI use had gone from seven farmers in 2003 to 398 farmers in 2004, found that the factor most prompting farmers to take up the new methods was their saving of labour, to them of more importance than the 48% increase in yield or the 45% reduction in water requirements (Li et al., 2005).

Increased labour requirements for practising SRI have been seen as a major obstacle to farmer acceptance since it was initially characterized as being more labour-intensive. Certainly some additional labour time will be needed while the new methods are first being learnt. But once farmers move up the learning curve, the question whether more or less labour is required for SRI will depend mostly on how labour-intensive (or extensive) was their previous practice. Economic and environmental benefits may not be enough to get farmers to modify their practices. But if labour can be reduced as well as seeds, water and production costs, while getting a higher yield, this methodology should be attractive to most farmers, giving them incentive to take up practices that have the environmental benefits discussed in sections 4.1, 4.2 and 4.3 above.

# 5 Developments in SRI methodology

## 5.1 Direct-seeding for crop establishment

For farmers, reducing their labour requirements for rice production is always an important consideration. Since nursery management and transplanting can take as much as 40% of the labour time needed for transplanted lowland rice production, these operations offer the most opportunity to cut labour inputs if sufficiently productive rice crops can be established directly from seed, rather than by transplanting seedlings.

SRI farmers in a number of countries have experimented with alternative practices for direct-seeding while maintaining as much as possible the SRI practices of single seedlings per hill and wide spacing of hills, which reflect the second principle in section 2.2 above. Principles 3 and 4 in that section are also applied, while application of the first principle presented there takes a different form. Direct seeding can be done by hand, with dibble sticks, or with specifically designed implements, as shown in Fig. 2. This is an important departure from the original practices of SRI, which is likely to spread because of frequent shortages or higher cost of agricultural labour in many rural areas.

**Figure 2** Germinated seeds direct-seeded at 25 × 25 cm spacing for SRI at an agricultural demonstration centre in Andhra Pradesh state of India (Hussain Reddy et al., 2009).

## 5.2 Mechanization of operations

Because SRI practices were first assembled working with farmers having low incomes and very little land, the initial factor proportions for SRI involved relatively more labour than capital. But as the principles that account for the practices' productivity became better understood, it was apparent that SRI principles could be extended to larger scales of operation. Farmers everywhere like to increase the productivity of their labour, to reduce the time and drudgery required for production, and also to reduce their costs of operation.

The soil-aerating mechanical hand weeder recommended with SRI became popular because it could decrease both the time and drudgery involved in weed control (Mrunalini and Ganesh, 2008). Various implements for marking fields for the transplanting of seedlings at specified distances have also been designed to speed up this operation. Since SRI was first developed, many efforts have gone into mechanizing and motorizing its operations, such as land preparation and levelling, transplanting, and especially weeding.

In Costa Rica, one farmer adapted a Japanese mechanical rice transplanter to transplant single ten-day seedlings at 22 × 30 cm spacing because agricultural labour in his region was in short supply, making transplanting of seedlings prohibitively expensive. His yield of 8 tons $ha^{-1}$ the first season was an 86% improvement over his previous yields and with a reduction in labour (http://sri.cals.cornell.edu/countries/costarica/ElPedregalEng.html). In Pakistan, more ambitious mechanization was undertaken by a larger farmer on a laser-levelled test plot of 8 hectares. The machinery that he designed or adapted for this enabled him to reduce his labour requirements by 70%, and his water requirements by a similar amount, with a yield of 12 tons $ha^{-1}$ (Sharif, 2011).

Mechanization of SRI operations is just beginning to spread. A good example is mechanized direct seeding of SRI in Vietnam (https://www.youtube.com/watch?v=51uNFQL1zMw). This utilizes SRI principles along with the direct-seeding innovation discussed in section 5.1.

A justification given for this modification and use of SRI practices in Vietnam is their contribution to the reduction of GHG emissions (section 4.2 above).

## 5.3 Value-chain development

Initially, the focus of SRI was entirely on production, seeking to increase the supply of rice with less dependence on external inputs and lowering farmers' costs of production. Most households using the new methods were food-insecure and in food deficit, so there was little consideration of the marketing of SRI rice or of meeting urban consumers' demand.

It has been seen that SRI methods work well with many traditional (indigenous, heirloom) varieties of rice, which often command a higher market price because of consumer preferences. With SRI management, a growing number of farmers have become able to produce a marketable surplus of high-quality rice that matches consumers' demand for healthier food grain, produced in environmentally friendly ways and under socially desirable conditions, meeting organic and fair-trade criteria, and rewarding farmers with premium prices for their product.

Accordingly, researchers, NGOs and private-sector actors involved with SRI are now thinking and acting in terms of value-chain development, to better remunerate producers for their higher-quality grain, to give consumers a more desirable product that merits a higher price, and to benefit the environment by reducing agrochemical usage and also irrigation water consumption.

'Value-chain SRI' was started in Cambodia, Madagascar and Indonesia through a U.S. company that has imported certified organic and fair-traded heirloom SRI rice to North American markets (Vent et al., 2015; Randrianarivelo et al., 2013; Lotus Foods, 2015). The farmers' cooperative producing SRI rice in Indonesia also exports its rice to Europe, Australia and Southeast Asia through arrangements with a local company in Jakarta (https://en.wikipedia.org/wiki/Bloom_Agro). SRI value chains have been subsequently developed in Ghana, India, Kenya, Sri Lanka, Thailand, Vietnam and Zambia and are being developed currently also in Bhutan, Cuba and the Philippines. This development – beyond a production focus to include marketing and meeting consumer demand – represents a considerable reworking from the initial subsistence orientation of SRI.

# 6 Assessing SRI yields

SRI was developed to raise the average yields of poor, resource-limited farmers in Madagascar, with zero interest in setting yield records. It was learnt that with SRI management methods, farmers in the area around Ranomafana National Park were able to increase their average yields to 8 tons $ha^{-1}$ on the same poor soils and with the same varieties from which they had previously produced only 2 tons $ha^{-1}$ (Uphoff, 1999). Such an average yield increase under these circumstances was remarkable, but not unreasonable.

Some farmers who used the recommended methods fully and carefully, however, building up the fertility of their soil with SRI practices, were able to get some super-yields that were twice or more this average. These exceptional yields were reported in a factual, not boastful manner. But because they were beyond what was being achieved with conventional crop management methods, they attracted much attention and some emphatic rejection. They were said to be beyond the 'biological maximum' yield for rice which had been calculated to be 15.9 tons $ha^{-1}$ based on crop modelling (Yoshida, 1981; Evenson et al., 1996).

Such modelling depended, however, on physiological coefficients derived from rice plants that had been grown with standard practices: older seedlings, close spacing and flooded paddies (Dobermann, 2003). These practices constrain the growth of rice root systems, and indeed induce degeneration of roots due to hypoxia (Kar et al., 1974). They also limit beneficial soil biota, which are mostly aerobes rather than anaerobes. The numbers and services of mycorrhizal fungi, for example, which support the growth of most terrestrial plants, are suppressed in flooded, hypoxic soils.

There is thus little basis for assuming that the modelling calculations for maximum possible yield that are based on measurements from conventionally grown rice plants will necessarily apply to the more productive phenotypes achieved with the growing conditions of SRI (Sheehy et al., 2004). The kind of contrasting phenotype that can be grown with SRI management is seen in Fig. 3.

The highest yields that have been reported with SRI practices, over 20 tons $ha^{-1}$, are indeed outliers. They are exceptions that are of interest only because they indicate the output attainable when SRI practices are all used well, and when the growing conditions for rice plants are optimized. Super-yields are not considered very important by SRI proponents because it is average yields, rather than outliers, which raise people's incomes and reduce food insecurity. What should have received attention from the Madagascar

**Figure 3** These two rice plants of same age (52 days) and same variety (VN 2084) were planted in the same nursery on a farm in Cuba. The SRI plant on right – transplanted when 9 days old into an SRI growing environment with single plants widely spaced and with more soil organic matter and soil aeration – has 43 tillers compared to the five tillers on its 'sister' plant that had just been removed from the nursery for transplanting. The difference in size and colour of the root systems is as great as the difference in tillering. This differentiation over a growing season can be seen at: http://sri.cals.cornell.edu/countries/cuba/SICAenglish.wmv.

results were the increases in *average* yield attained. But unfortunately, contention over exceptional yields distracted attention from more typical SRI yields, which were remarkable and beneficial.

Controversy over maximum attainable rice yields flared up again when an Indian farmer was reported to have produced a yield of 22.4 tons $ha^{-1}$ in the 2011 kharif season in Bihar state, which was disputed by Prof. Yuan Long-ping, holder of the previous paddy yield record (Vidal, 2013a,b). How the new record was achieved and measured has been documented based on reports from technicians of the Bihar state Department of Agriculture and the Indian government's Directorate of Rice Development (Diwakar et al., 2012). That season, four other farmers in the same village also surpassed the previous record; however, this received no attention, which focused only on the highest yield. That these super-yields were all obtained with hybrid varieties and with integrated nutrient management (some inorganic fertilizer complemented the mostly organic soil amendments) seems not to have impressed the proponents of higher-tech agriculture, who could have claimed some share of credit for the accomplishment.

When these five farmers grew these same hybrid varieties with their usual practices that season on the same soils and with the same climate, they got yields around 7 tons $ha^{-1}$. This was about three times the Bihar state average, but only about one-third as much as they got when growing these hybrid varieties with SRI practices. This showed the importance of adopting better crop and water management compared to making increases just in genetic potential. In the 2013–14 season when a farmer in Tamil Nadu turned out an even higher yield of 23.8 tons $ha^{-1}$ with SRI methods (Times of India, 2014), there was hardly any reaction in the press or elsewhere.

Exactly how high such super-yields are, almost an order of magnitude greater than typical yields in their area, is of relevance and interest mostly to agronomists, plant breeders and journalists. Super-yields are not very salient for development agencies or for most farmers, who are more concerned with what is probable than with what is possible.

Even if the super-yields reported were only 18 tons, or even 15 tons, they represent opportunities for raising rice productivity that merit further investigation, especially because they are obtained with *reductions* in water and other inputs. It has been unfortunate for rice science and for rice cultivation that so much attention has been focused, mostly from sceptics and critics, on super-yields rather than on averages. Contention over exactly how high are the highest yields has impeded cooperation for assessing and benefiting from higher yields with SRI management, that is, for raising the factor productivity from the resources (land, labour, seeds, water, capital) utilized for growing rice.

# 7 SRI in relation to other production systems

## 7.1 Comparisons with best management practices

While much evidence has accumulated on the contribution of SRI to improving yields, some early evaluations suggested that SRI offers no significant advantages compared to what rice scientists consider as 'best management practices' (BMP). Such comparisons should be considered critically and carefully. From their comparison between the two sets of practices, Sheehy et al. (2004) concluded that SRI management can have 'no major role in improving rice production generally'. A subsequent study by McDonald et al. (2006)

came to a similar conclusion, suggesting an 11% yield differential in favour of BMP. As has been noted earlier, comparison between SRI and other systems is not a simple matter, and care should be taken in correctly following SRI protocols and ensuring that comparisons are realistic. This section points to some problematic issues with these evaluations, leaving it to readers to reach their own conclusions based on the written record and available evidence.

The conclusion of Sheehy et al. (2004) was based on results from just three small comparison plot trials in China which did not follow a recognizable SRI protocol. First, there was no soil-aerating weed control as is recommended for SRI, just the use of herbicides. This can affect adversely the beneficial soil biota, which are a crucial element in SRI performance. Further, although SRI methodology recommends organic soil fertilization, in the three trials there was excessive use of inorganic nitrogen fertilizer on the SRI plots (180–240 kg N $ha^{-1}$). It was not accounted for that one of the co-authors of the article had recommended that just 80 kg N $ha^{-1}$ was optimal under most Chinese conditions (Peng, 2004).

This excess N application caused the plants in one of the SRI plots to lodge, something seldom seen when recommended SRI practices are used, which lowered that plot's yield. Despite the deviations from recommended SRI methodology, the average yield from the SRI plots was higher than from the control (BMP) plots by 0.1 ton $ha^{-1}$. While this was not a statistically significant difference, given the small sample size, it was not evidence of SRI inferiority.

An important omission in the Sheehy article was its not considering the results of any of the dozens of studies that had already been done by Chinese rice scientists and which had shown favourable results from SRI management, such as Yuan (2002). Thus, the extensive Chinese literature evaluating SRI such as reviewed in Wu et al. (2015) was not considered, even though a majority of the co-authors of the article were Chinese and could read that language.

In a subsequent analysis of 40 data sets of comparisons between SRI and BMP yields from 10 countries, McDonald et al. (2006) correctly characterized the SRI protocol and went on to state that their analysis excluded 'experiments with [SRI] treatments that did not closely approximate the principles of SRI (e.g., included only one or two SRI elements)'. This means, however, that they included as SRI trials any trials where only three of the six SRI elements were used, thus considering 50% as 'approximating' 100%. One of the co-authors (Hobbs) subsequently acknowledged in a personal communication that probably half of the comparisons in their data base should not have been included in the statistical analysis because they did not meet the authors' stated criteria for inclusion. Further, the five data sets from Madagascar, which had showed a clear SRI yield advantage, were excluded from the analysis on the grounds that this country's soils were 'distinctive'. In fact, these five yield comparisons were based on more rigorous methodology, with random block design and replications, than were most of the other data sets included in the statistical analysis.

A more systematic meta-analysis of yield results comparing SRI with BMP has been done on evaluations made by Chinese scientists. This was based on a comprehensive literature search of three internet databases to identify all the studies done in China which had compared SRI practices with whatever the researchers considered as current BMP (Wu et al., 2015). Most of the 17 published studies did not evaluate the full SRI protocol, but whatever actual practices were used for comparison were discernible from these studies. These contained results from 26 sets of comparative field trials from eight of China's main rice-growing provinces.

Analysis of the whole pool of trial results showed an overall yield advantage of 10.9% for SRI methods over BMP, rather than the 11% disadvantage calculated by McDonald et al. More informative results came from a disaggregated analysis of the data set, considering the number of SRI practices that had been used in the respective trials, so that how completely the SRI recommendations had been followed could be evaluated. Comparison of trials in which there had been 'good' use of SRI methods showed a 20.3% yield advantage for SRI over BMP. 'Median' use of SRI methods gave an advantage of 12.0%; while with 'minimal' (partial) use of SRI practices, there was a 3.9% yield advantage for BMP over SRI (Wu et al., 2015).

For most of the past 15 years, there has been controversy over BMP versus SRI comparisons, much as over the information on super-yields discussed in the preceding section. The recent meta-analysis of Chinese results, based on better definitions and sampling of sites for comparison, may damp down the contention. Comparative scientific evaluations that focus on the different practices and their effects, both respectively and collectively, should continue, but such evaluations should not be preoccupied with names, focusing instead on the effects of actual practices. At this stage, studies addressing questions of *why* there are differences are likely to be more informative than *whether* there are differences. The latter question has become less salient as the evidence of differences in favour of SRI has continued to accumulate.

## 7.2 Convergence with conservation agriculture (CA)

Essentially, SRI is practised by making modifications in the current management of rice plants, soil, water and nutrients so as to have desirable effects on plant roots' growth and functioning and on the structure and functioning of soil systems, including their soil biota. SRI is related to a number of other agricultural production methods that are grounded in agro-ecological principles and dynamics, such as integrated pest management and agroforestry. SRI is particularly relevant to conservation agriculture, and in recent years there have been efforts to converge these production methodologies.

Conservation agriculture (CA) has itself been moving over the past 15 years from its initial emphasis on (a) zero or minimum tillage, to include also (b) maintenance of continuous ground cover using organic matter that is living, such as cover crops, or dead, such as mulch, plus (c) rotation of crops rather than continuous monocropping with the same cultivar. Biodiversity in cropping is associated with biodiversity below- and above-ground which makes cropping systems more robust, resilient and productive.

Since both CA and SRI promote practices that enhance soil fertility and the life in the soil, they have a common foundation. Much as transplanting seedlings can give way to direct-seeding when there are economic and other advantages, land preparation with SRI which has involved tillage can be succeeded by permanent raised beds whose fertility is enhanced with organic matter applications. Weed control on raised beds is managed by mulching or other practices rather than by soil-aerating mechanical weeding. When populations of micro-organisms, earthworms and other flora and fauna are built up in the beds, season by season, soil systems become better aggregated and better structured and the soil-aeration functions of mechanical weeding can be fulfilled by biological processes and activity.

The mechanized version of SRI developed in Pakistan discussed above (section 5.2) was purposefully designed to combine both SRI and CA systems of cultivation, also integrating the practices of organic agriculture (Sharif, 2011). After the mechanical construction of permanent raised beds, there is no further tillage of the field, although a mechanical weeder is run along the raised beds several times to control weeds and aerate the soil, breaking up the crust that forms readily on these soils. This enables both air and water to reach the root zone more easily. Irrigated rice crops are alternated with other crops to enhance the soil's fertility. Furrow irrigation provides supplementary water to the soil as needed, with siphons used to minimize energy costs.

In Sichuan province of China, given the soil, temperature and other conditions there, a system for growing rice on permanent raised beds using SRI methods has been evolved and evaluated where covering the beds with plastic film as an inorganic mulch both suppresses weeds and reduces losses of water through evaporation (Lu et al., 2013). Crop residues are also put on the beds for their protection and for nutrient enrichment, rotating rice with crops such as mustard (rapeseed) in the winter season. Average yield increases have been 2383 kg $ha^{-1}$, a 31% increase compared with farmers' usual practice. This integrated technology gives higher yields with less water, fertilizer, seeds and labour, with reduced soil and water pollution and thus better environmental quality.

Since SRI is regarded as a system of ideas and associated practices rather than as a set technology – always a work in progress – these kinds of adaptations and permutations are to be expected, and welcomed. As the convergence of SRI and CA practices is relatively recent, it is not possible to assess how far or how fast this development will proceed. But in considering the various evolutions of SRI over the past 15 years, this is one of the most important.

# 8 Summary and future trends

Summarizing the many changes discussed above and projecting likely or even possible future developments is difficult because of the nature of these innovations discussed in section 2, and because new and more things are being learnt every season. SRI and its derived SCI must still be – and probably always will be – described as a work in progress. This work needs to be thoroughly empirical, evidence based, with efforts being made at the same time to also contribute to a better theoretical understanding of the phenotypic effects being registered.

Crop modelling exercises, while worth doing, should not be considered as delineating innate limits. It is being seen that in the well-known nominal equation of $P = G + E + [G \times E]$ where P = phenotype, G = genetic potential and E = environment, E should be regarded more as a variable than as a given, to be modified by optimizing management practices. While G is the crucial starting point in the growth and development of rice and any other crops, more attention should be paid to E and to its interaction with G which affects plants' expression of their genetic potential. We are learning that E includes the plant–soil microbiome, which shapes phenotypes (P) to still inadequately understood extents.

One safe prediction is that ten years from now we should have a much clearer appreciation of the contributions that the microbial realm, together with the rest of their encompassing ecosystems, makes to rice and other production. Microbiology, ecology and epigenetics will play larger roles in rice science and practice than now.

This will be important for achieving greater sustainability of agricultural production systems because it will help us to understand how 'more can be produced from less,' that is, how to enhance production with reduced exogenous inputs (Uphoff, 2003; Uphoff et al., 2006; FAO, 2016). This is being achieved through SRI management by mobilizing and capitalizing on endogenous potentials and processes. But much more remains to be researched, analysed, critiqued and envisioned as the input-dependent agriculture which has served us reasonably well in the twentieth century makes way for what can be characterized as 'rebiologized agriculture', which will be essential for this twenty-first century.

# 9 Where to look for further information

There are a number of websites and pages that can readily provide further information:

- *SRI-Rice*: http://sri.cals.cornell.edu – the most complete source of information
- *World Bank*: http://info.worldbank.org/etools/docs/library/245848/
- *International Fund for Agricultural Development*: http://www.ifad.org/english/sri/
- *Oxfam America*: http://policy-practice.oxfamamerica.org/work/in-action/system-of-rice-intensification/ – and Africare/Oxfam/WWF (2010).
- *International Rice Research Institute*: http://irri.org/news/hot-topics/system-of-rice-intensification-sri

# 10 References


Abraham, B., Arayu, H., Berhe, T., Edwards, S., Gujja, B., Khadka, R. B., Koma, Y. S., Sen, D., Sharif, A., Styger, E., Uphoff, N. and Verma, A. (2014), The system of crop intensification: Reports from the field on improving agricultural production, food security, and resilience to climate change for multiple crops, *Agriculture & Food Security*, 3: 4. http://www.agricultureandfoodsecurity.com/content/3/1/4 (accessed 22 September 2015).

Africare/Oxfam/WWF (2010), *More Rice for People, More Water for the Planet: The System of Rice Intensification.* https://www.africare.org/wp-content/uploads/2014/04/more-rice-for-people-more-water-for-the-planet-sri.pdf (accessed 22 September 2015).

Anas, I., Rupela, O. P., Thiyagarajan, T. M. and Uphoff, N. (2011), A review of studies on SRI effects on beneficial organisms in rice soil rhizospheres, *Paddy and Water Environment*, 9: 53–64.

Barah, B. C. (2009), Economic and ecological benefits of system of rice intensification (SRI) in Tamil Nadu, *Agricultural Economics Research Review*, 22: 209–14.

Barison, J. and Uphoff, N. (2011), Rice yield and its relation to root growth and nutrient-use efficiency under SRI and conventional cultivation: An evaluation in Madagascar, *Paddy and Water Environment*, 9: 65–78.

Barrett, C. B., Moser, C. M., McHugh, O. V. and Barison, J. (2004), Better technology, better plots or better farmers? Identifying changes in productivity and risk among Malagasy rice farmers, *American Journal of Agricultural Economics*, 86: 869–88.

Behera, D., Chaudhury, A. K., Vutukutu, V. K., Gupta, A., Machiraju, S. and Shah, P. (2013), *Enhancing Agricultural Livelihoods through Community Institutions in Bihar.* South Asia Livehoods Learning Note, Series 3, No. 1, The World Bank, New Delhi. http://documents.worldbank.org/curated/

en/2013/01/17517688/enhancing-agricultural-livelihoods-through-community-institutions-bihar-india (accessed 22 September 2015).
BIND (2003), Growth and yield response of traditional upland rice on different distance of planting using *Azucena* variety. Broader Initiatives for Negros Development, Bacalod City, Philippines. http://sri.cals.cornell.edu/countries/philippines/binuprst.pdf (accessed 22 September 2015).
Bouman, B. (2013), SRI: Why the animosity? http://irri.org/blogs/bas-bouman-s-blog-global-rice-science-partnership/sri-why-the-animosity (accessed 22 September 2015).
Chi, F., Shen, S. H., Cheng, H. P., Jing, Y. X., Yanni, Y. G. and Dazzo, F. B. (2005), Ascending migration of endophytic rhizobia, from roots to leaves, inside rice plants and assessment of benefits to rice growth physiology, *Applied and Environmental Microbiology*, 71: 7271–8.
Chi, F., Yang, P. F., Han, F., Jing, Y. X. and Shen, S. H. (2010), Proteomic analysis of rice seedlings infected by *Sinorhizobium meliloti* 1021, *Proteomics*, 10: 1861–74.
Choi, J. D., Kim, G. Y., Park, W. J., Shin, M. H., Choi, Y. H., Lee, S., Kim, S. J. and Yun, D. K. (2014), Effect of SRI water management on water quality and greenhouse gas emissions in Korea, *Irrigation and Drainage*, 63: 263–70.
CIIFAD (2002), *Assessments of the System of Rice Intensification: Proceedings of an Intl. Conference held in Sanya, China, April 1–4, 2002* (eds. N. Uphoff, E. C. M. Fernandes, L. P. Yuan, J. M. Peng, S. Rafaralahy and J. Rabenandrasana), Cornell International Institute for Food, Agriculture and Development, Ithaca, NY. http://sri.cals.cornell.edu/proc1/index.html (accessed 22 September 2015).
Dhar, S., Barah, B. C., Vyas, A. K. and Uphoff, N. (2015), Evaluation of System of Wheat Intensification (SWI) practices as compared to other methods of improved wheat cultivation in the north-western plain zone of India, *Archives of Agronomy and Soil Science*, http://www.tandfonline.com/doi/full/10.1080/03650340.2015.1101518.
Dill, J., Deichert, G. and Le, T. N. T., eds (2013), *Promoting the System of Rice Intensification: Lessons Learned from Trà Vinh Province, Vietnam*, GIZ and IFAD, Hanoi. http://infoagro.net/archivos_Infoagro/Regatta/biblioteca/VN-GIZreportonLesson.pdf (accessed 22 September 2015).
Diwakar, M. C., Kumar, A., Verma, A. and Uphoff, N. (2012), Report on the world record SRI yields in kharif season 2011 in Nalanda district, Bihar state, India, *Agriculture Today* (New Delhi), 15: 54–6. http://www.independentsciencenews.org/wp-content/uploads/2012/11/India-Bihar-Paddy-Record-Yield-SRI.pdf (accessed 22 September 2015).
Dobermann, A. (2003), A critical assessment of the system of rice intensification (SRI), *Agricultural Systems*, 79: 261–81.
Dung, N. T. (2007), SRI application in rice production in northern ecological areas of Vietnam. Report of National IPM Program to Ministry of Agriculture and Rural Development, Hanoi. http://sri.cals.cornell.edu/countries/vietnam/vndungipmrpt06.pdf.
Evenson, R. E., Herdt, R. W. and Hossain, M. (1996), *Rice Research in Asia: Progress and Priorities*, CAB International and International Rice Research Institute, Wallingford, UK.
Ezhil Rani, K. (2004), Relative abundance of insects in SRI and conventional rice. M.Sc. thesis, Tamil Nadu Agricultural University, Coimbatore, TN, India. http://tnau.egranth.ac.in/cgi-bin/koha/opac-detail.pl?biblionumber=61732 (accessed 8 November 2015).
FAO (2016), *Save and Grow: Maize, Rice and Wheat – A Guide to Sustainable Crop Production*, UN Food and Agriculture Organization, Rome.
Gani, A., Kadir, T. S., Jatiharti, J., Wardhana, I. P. and Las, I. (2002), The system of rice intensification in Indonesia, in *Assessments of the System of Rice Intensification*, pp. 58–63, CIIFAD. http://sri.cals.cornell.edu/proc1/sri_14.pdf (accessed 22 September 2015).
Gathorne-Hardy, A., Narasimha Reddy, D., Venkatanarayana, M. and Harriss-White, B. (2013), A Life Cycle Assessment (LCA) of greenhouse gas emissions from SRI and flooded rice production in SE India, *Taiwan Water Conservancy*, 61: 110–25.
Gopal, R., Jalaluddin, M. and Jayaraj, T. (2010), Pest scenarios and IPM options in system of rice intensification, in *Rice for Future Generations: Proceedings of 3rd International Rice Congress, Hanoi*, 8–12 November. ID #565.

Hussain Reddy, P. B., Sreenivasalu, S. and Manohar, C. (2009), *Direct Seeding with Drum Seeder – Future Prospects*, RASS-Acharya Ranga Krishi Vigyan Kendra, Tirupati, AP, India. http://sri.ciifad.cornell.edu/countries/india/AP/InAPDrumSeederRpt_KVKBalaReddy09.pdf (accessed 22 September 2015).

IRIN (2012), Analysis: Why rice intensification matters in Asia. IRIN press release, 24 April. UN Office for Coordination of Humanitarian Affairs, Bangkok. http://www.irinnews.org/fr/report/95342/analysis-why-rice-intensification-matters-in-asia (accessed 22 September 2015).

Jagannath, P., Pullabhotla, H. and Uphoff, N. (2013), Meta-analysis evaluating water use, water saving, and water productivity in irrigated production of rice with SRI vs. standard management methods, *Taiwan Water Conservancy*, 61: 14–49.

Kabir, H. and Uphoff, N. (2007), Results of disseminating the system of rice intensification (SRI) with farmer field school methods in northern Myanmar, *Experimental Agriculture*, 43: 463–76.

Kar, S., Varade, S. B., Subramanyam, T. K. and Ghildyal, B. P. (1974), Nature and growth pattern of rice root system under submerged and saturated conditions, *Il Riso*, 23: 173–9.

Karthikeyan, K., Jacob, S. and Purushotaman, S. M. (2010), Incidence of insect pests and natural enemies under SRI method of cultivation, *Oryza*, 47: 154–7.

Kumar, R. M., Padmavathi, Ch., Surekha, K., Latha, P. C., Krishnamurthy, P. and Singh, S. P. (2007), Influence of organic nutrient sources on insect pests and economics of rice production in India, *Indian Journal of Plant Pathology*, 35: 264–7.

Laulanié, H. (1993), Le système de riziculture intensive malgache, *Tropicultura* (Brussels), 11: 110–14; English synopsis of article in *Tropicultura* (2011), 29: 183–7.

Li, X. Y., Xu, X. L. and Li, H. (2005), A socio-economic assessment of the system of rice intensification (SRI): A case study of Xinsheng Village, Jianyang County, Sichuan Province. Report for Center for Integrated Agricultural Development, China Agricultural University, Beijing; published in *Chinese Rural Economy* (2006), 13–22 (in Chinese). http://sri.cals.cornell.edu/countries/china/cnciadeng.pdf (accessed 22 September 2015).

Lim, S. (2007), *Experiences in Multi-Purpose Farm Development: Raising Household Incomes in Cambodia by Utilizing Productivity Gains from the System of Rice Intensification*, Cambodian Center for Study and Development of Agriculture (CEDAC), Phnom Penh. http://sri.cals.cornell.edu/countries/cambodia/cambSidMPREng.pdf (accessed 22 September 2015).

Lin, X. Q., Zhu, D. F., Chen, H. Z., Cheng, S. H. and Uphoff, N. (2009), Effect of plant density and nitrogen fertilizer rates on grain yield and nitrogen uptake of hybrid rice (*Oryza sativa* L.), *Journal of Agrobiotechnology and Sustainable Development*, 1: 44–53.

Lin, X. Q., Zhu, D. F. and Lin, X. J. (2011), Effects of water management and organic fertilization with SRI crop practices on hybrid rice performance and rhizosphere dynamics, *Paddy and Water Environment*, 9: 33–9.

Lotus Foods (2015), Website: http://www.lotusfoods.com/index.php/more-crop-drop/about-more-crop-drop/ (accessed 22 September 2015).

Lu, S. H., Dong, Y. J., Yuan, J., Lee, H. and Padilla, H. (2013), A high-yielding, water-saving innovation combining SRI with plastic cover on no-till raised beds in Sichuan, China, *Taiwan Water Conservancy*, 61: 94–109.

McDonald, A., Hobbs, P. R. and Riha, S. J. (2006), Does the system of rice intensification outperform conventional best management? A synopsis of the empirical record, *Field Crops Research*, 96: 31–6.

Mishra, A. and Salokhe, V. M. (2008), Seedling characteristics and the early growth of transplanted rice under different water regimes, *Experimental Agriculture*, 44: 1–19.

Moser, C. M. and Barrett, C. B. (2003), The disappointing adoption dynamics of a yield-increasing low external-input technology: The case of SRI in Madagascar, *Agricultural Systems*, 76: 1085–1100.

Mrunalini, A. and Ganesh, M. (2008), Work load on women using cono weeder in SRI method of paddy cultivation, *Oryza*, 5: 58–61.

Nemoto, K., Morita, S. and Baba, T. (1995), Shoot and root development in rice related to the phyllochron, *Crop Science*, 35: 24–9.

Palanisami, K., Karunakaran, K. R., Amarasinghe, U. and Ranganathan, C. R. (2013), Doing different things or doing it differently? Rice intensification practices in 13 states of India, *Economic & Political Weekly*, 48: 51–8.

Peng, S. (2004), Improving fertilizer-nitrogen use efficiency of irrigated rice: Progress of IRRI's RTOP Project in China. Paper presented at International Year of Rice Conference on Sustainable Rice Production, China National Rice Research Institute, Hangzhou, October.

PRADAN (2007), Report on SRI Experience in Purulia District, West Bengal, to IWMI for 2006 Kharif Season (2nd Season), Professional Assistance for Development, Purulia, West Bengal. http://sri.ciifad.cornell.edu/countries/india/WBengal/inpuruliakh0607.pdf (accessed 22 September 2015).

Ragukkanu, K., Ramadass, P. and Jecitha, J. (2007), Keeping the brown plant hopper at bay with SRI, *LEISA Magazine*, 23: 4, December. http://www.agriculturesnetwork.org/magazines/global/ecological-pest-management/keeping-the-brown-plant-hopper-at-bay-with-sri (accessed 8 November 2015).

Randriamiharisoa, R., Barison, J. and Uphoff, N. (2006), Soil biological contributions to the system of rice intensification, in *Biological Approaches to Sustainable Soil Systems* (eds. N. Uphoff, A. Ball, E. C. M. Fernandes, H. Herren, O. Husson, M. Laing, C. A. Palm, J. N. Pretty, P. Sanchez, N. Sanginga and J. E. Thies), pp. 409–24, CRC Press, Boca Raton, FL.

Randrianarivelo, J., Randrianarivelo, A., Abhukara, R., Fitzgerald, W., Pargee, C. and Vent, O. (2013), Organic production of pink rice in Madagascar, in *Organic Agriculture: African Experiences in Resiliency and Sustainability* (eds. R. Auerbach, G. Rundgren and N. E. Scialabba), pp. 112–35, UN Food and Agriculture Organization, Rome.

Redman, R. S., Kim, Y. O., Woodward, C. J. D. A., Greer, C., Espino, L., Doty, S. L. and Rodriguez, R. J. (2011), Increased fitness and adaptation of rice plants to cold, drought and salt stress via habitat adapted symbiosis: A strategy for mitigating impacts of climate change, *PLoS ONE*, 6(7): e14823. http://journals.plos.org/plosone/article?id=10.1371/journal.pone.0014823 (accessed 22 September 2015).

Satyanarayana, A. S., Thiyagarajan, T. M. and Uphoff, N. (2006), Opportunities for water saving with higher yield from the system of rice intensification, *Irrigation Science*, 25: 99–115.

Sharif, A. (2011), Technical adaptations for mechanized SRI production to achieve water saving and increased profitability in Punjab, Pakistan, *Paddy and Water Environment*, 9: 111–19.

Sheehy, J. E., Peng, S. B., Dobermann, A., Mitchell, P. L., Ferrer, A., Yang, J. C., Zou, Y. B., Zhong, X. H. and Huang, J. L. (2004), Fantastic yields in the system of rice intensification: Fact or fallacy?, *Field Crops Research*, 88: 1–8.

Sinha, S. K. and Talati, J. (2007), The impact of system of rice intensification (SRI) on paddy productivity: Results of a study in Purulia District, West Bengal, India, *Agricultural Water Management*, 87: 55–60.

SRI-Rice (2014), *The System of Crop Intensification: Agroecological Innovations for Improving Agricultural Production, Food Security, and Resilience to Climate Change*, SRI International Network and Resources Center (SRI-Rice), Cornell University, Ithaca, New York, and the Technical Centre for Agricultural and Rural Cooperation (CTA), Wageningen, Netherlands. http://publications.cta.int/en/publications/publication/1795/management-entrepreneurship/.

SRI-Rice (2015), Website page: http://sri.ciifad.cornell.edu/research/JournalArticles.html (accessed 22 September 2015).

Styger, E., Aboubacrine, G., Attaher, M. A. and Uphoff, N. (2011), The system of rice intensification as a sustainable agricultural innovation: Introducing, adapting and scaling up a system of rice intensification practices in the Timbuktu region of Mali, *International Journal of Agricultural Sustainability*, 9: 67–75.

Thakur, A. K., Mohanty, R. K., Singh, R. and Patil, D. U. (2015), Enhancing water and cropping productivity through integrated system of rice intensification (ISRI) with aquaculture and horticulture under rainfed conditions, *Agricultural Water Management*, 161: 65–76.

Thakur, A. K., Rath, S. and Mandal, K. G. (2013), Differential responses of system of rice intensification (SRI) and conventional flooded-rice management methods to nitrogen fertilizer, *Plant and Soil*, 370: 59–71.

Thakur, A. K., Uphoff, N. and Antony, E. (2010), An assessment of physiological effects of system of rice intensification (SRI) practices compared to recommended rice cultivation practices in India, *Experimental Agriculture*, 46: 77–98.

Thiyagarajan, T. M. (2004), On-farm evaluation of SRI in Tamiraparani command area. Presentation at World Rice Research Congress, Tsukuba, Japan, 1–4 November. http://sri.cals.cornell.edu/conferences/wrrc/wrrcppts/wrintnseveraju.ppt (accessed 22 September 2015).

Thomas, V. and Ramzi, A. M. (2011), SRI contributions to rice production dealing with water management constraints in northeastern Afghanistan, *Paddy and Water Environment*, 9, 101–9.

Times of India (2014), Farmer reaps high yield by combining organic, inorganic fertilisers, 2 May. http://timesofindia.indiatimes.com/city/madurai/Farmer-reaps-high-yield-by-combining-organic-inorganic-fertilisers/articleshow/34506408.cms (accessed 22 September 2015).

Turmel, M-S., Espinoza, J., Franco, L., Perez, C., Hernández, H., González, E., Fernández, G., Rojas, C., Sánchez, D., Fernández, E., Barrios, M., Whalen, J. K. and Turner, B. L. (2011), On-farm evaluation of a low-input rice production system in Panama, *Paddy and Water Environment*, 9, 155–61.

Uphoff, N. (1999), Agroecological implications of the system of rice intensification (SRI) in Madagascar, *Environment, Development and Sustainability*, 1: 297–313.

Uphoff, N. (2003), Higher yields with fewer external inputs? The system of rice intensification and potential contributions to agricultural sustainability, *International Journal of Agricultural Sustainability*, 1: 38–50.

Uphoff, N. (2007a), Agricultural futures: What lies beyond 'modern agriculture'? *Tropical Agricultural Association Newsletter*, 27(3): 13–19. 2nd Hugh Bunting Memorial Lecture, University of Reading, UK. http://www.taa.org.uk/assets/pubs/Sept2007NLWeb-1.pdf (accessed 22 September 2015).

Uphoff, N. (2007b), Reducing the vulnerability of households through agroecological practice: Considering the system of rice intensification, *Mondes en Développement*, 35: 85–100.

Uphoff, N. (2011), Agroecological approaches to 'climate-proofing' agriculture while raising productivity in the 21st century, in *Sustaining Soil Productivity in Response to Global Climate Change* (eds. T. Sauer, J. Norman and M. V. K. Sivakumar), pp. 87–102, Wiley-Blackwell, Hoboken, NJ.

Uphoff, N. (2015), *The System of Rice Intensification: Responses to Frequently Asked Questions*, Amazon Books. http://www.amazon.com/The-System-Rice-Intensification-Frequently/dp/1515022056.

Uphoff, N. and Randriamiharisoa, R. (2002), Reducing water use in irrigated rice production with the Madagascar system of rice intensification, in *Water-wise Rice Production: Proceedings of the International Workshop, 8–11 April 2002* (eds. B. A. M. Bouman, H. Hengsduik, B. Hardy, P. S. Bindraban, T. P. Tuong and J. K. Ladha), pp. 71–88, International Rice Research Institute, Los Baños, Philippines.

Uphoff, N., Ball, A., Fernandes, E., Herren, H., Husson, O., Laing, O., Palm, C. A., Pretty, J. N., Sanchez, P., Sanginga, N. and Thies, J., eds (2006), *Biological Approaches to Sustainable Soil Systems*, CRC Press, Boca Raton, FL.

Uphoff, N., Chi, F., Dazzo, F. B. and Rodriguez, R. J. (2013), Soil fertility as a contingent rather than inherent characteristic: Considering the contributions of crop-symbiotic soil microbiota, in *Principles of Sustainable Soil Management in Agroecosystems* (eds. R. Lal and B. Stewart), pp. 141–66, CRC Press, Boca Raton, FL.

Vent, O., Koma Y. S., Levine, C. and Uphoff, N. (2015), Market incentives for eco-friendly SRI rice production in Cambodia, in *Shades of Green: Multi-Stakeholder Initiatives to Reduce the Environmental Footprint of Commercial Agriculture* (eds. M. Swadeh and S. Jaffee), pp. 73–80. EcoAgriculture Partners, Washington, DC.

Vidal, J. (2013a), India's rice revolution, *The Guardian Sunday Observer*, 16 February, London. http://www.guardian.co.uk/global-development/2013/feb/16/india-rice-farmers-revolution (accessed 22 September 2015).

Vidal, J. (2013b), India's rice revolution: Chinese scientist questions massive harvests, *The Guardian Sunday Observer*, 23 February, London. http://www.theguardian.com/world/2013/feb/23/india-rice-revolution-questioned (accessed 22 September 2015).

Wang, S. H., Cao, W. X., Jiang, D., Dai, T. B. and Zhu, Y. (2002), Physiological characteristics and high-yield techniques with SRI rice, in *Assessments of the System of Rice Intensification*, pp. 116–24, CIIFAD. http://sri.cals.cornell.edu/proc1/sri_27.pdf (accessed 22 September 2015).

Wu, W., Ma, B. L. and Uphoff, N. (2015), A review of system of rice intensification in China, *Plant and Soil*, 393, 361–81.

Yanni, Y. G., Rizk, R. Y., Abd El-Fatah, F. K., Squartini, A., Corich, V., Giacomini, A., De Bruijn, F., Rademaker, J., Maya-Flores, J., Ostrom, P., Vega-Hernández, M, Hollingsworth, R., Martinez-Molina, E., Mateos, P., Velásquez, E., Wopereis, J., Triplett, E., Umali-Garcia, U., Rolfe, B., Ladha, J. K. and Dazzo, F. B. (2001), The beneficial plant growth-promoting association of *Rhizobacterium leguminosarum* bv. *trifolii* in rice roots, *Australian Journal of Plant Physiology*, 28: 845–70.

Yoshida, S. (1981), *Fundamentals of Rice Crop Science*, International Rice Research Institute, Los Baños, Philippines.

Yuan, L. P. (2002), A scientist's perspective on experience with SRI in China for raising the yields of super hybrid rice, in *Assessments of the System of Rice Intensification*, pp. 23–5, CIIFAD. http://sri.cals.cornell.edu/proc1/sri_06.pdf (accessed 22 September 2015).

Zhao, L. M., Wu, L. H., Li, Y. S., Animesh, S., Zhu, D. F. and Uphoff, N. (2010), Comparisons of yield, water use efficiency, and soil microbial biomass as affected by the system of rice intensification, *Communications in Soil Science and Plant Analysis*, 41: 1–12.

Zhao, L. M., Wu, L. H., Li, Y. S., Lu, X. H., Zhu, D. F. and Uphoff, N. (2009), Influence of the system of rice intensification on rice yield and on nitrogen and water use efficiency with different N application rates, *Experimental Agriculture*, 45: 275–86.

Zheng, J. G., Chi, Z. Z., Li, X. Y. and Jiang, X. L. (2013), Agricultural water savings possible through SRI for water management in Sichuan, China, *Taiwan Water Conservancy*, 61(4): 50–62.

Zhu, D. F., Cheng, S. H., Zhang, Y. P. and Lin, X. Q. (2002), Tillering patterns and the contribution of tillers to grain yield with hybrid rice and wide spacing, in *Assessments of the System of Rice Intensification*, pp. 125–31, CIIFAD. http://sri.cals.cornell.edu/proc1/sri_28.pdf (accessed 22 September 2015).

# Chapter 8

# Assessing the sustainability impacts of rice cultivation

*Wyn Ellis, Sustainable Rice Platform, Thailand*

1 Introduction
2 Trends in the production and consumption of rice
3 Sustainability concerns: resource use and climate change impacts
4 Assessing sustainability in rice systems: introduction to the Sustainable Rice Platform (SRP)
5 Application of the SRP standard for sustainable rice cultivation
6 Conclusions
7 Where to look for further information
8 Acknowledgements
9 References

## 1 Introduction

The food commodity spike of 2008 brought the range and severity of threats to our global food supply into sharp focus. The crisis highlighted the fragility of the balance between food supply, global security, energy and environment and drew attention to the increasingly global nature of supply chain linkages. It also led to heightened concerns over longer-term threats to the environment and overall sustainability.

This chapter explores the environmental and broader sustainability challenges posed by rice production, and is organized along five main themes. Beginning with the strategic role of rice to global food security, we proceed to discuss a range of key environmental and sustainability challenges posed by rice. Since climate change represents a fundamental dynamic in many of these challenges, we devote the following section to exploring both the contribution of rice to climate change, and some of the likely impacts on productivity, food security and the environment.

We then discuss the challenge of quantifying these impacts, which has been approached from diverse theoretical and practical perspectives, using different functional units, temporal scope and impact indicators. With a plethora of 'best-practice' rice production systems introduced and advocated in recent years as 'sustainable' or 'environmentally friendly', the wider impacts of introduction of such farm management practices are often still unclear, particularly in the absence of clearly defined system boundaries.

http://dx.doi.org/10.19103/AS.2016.0003.29

We then introduce one approach that combines best practice adoption, and measurement of sustainability impacts in rice. The Sustainable Rice Platform (SRP)'s Standard and Performance Indicators for Sustainable Rice Cultivation was launched in 2015 as a framework for assessing environmental and other dimensions of sustainability, and aimed primarily at smallholders in developing countries. The chapter concludes with a call for wider adoption of a 'Sustainable Food Systems' approach to gain a higher level of understanding of the wider ramifications of best practice interventions in rice.

# 2 Trends in the production and consumption of rice

## 2.1 Strategic importance

Rice is produced on 160 million hectares of land, with a large part cultivated by 144 million smallholders with, on average, less than one hectare of land, producing a minimal marketable surplus. Over 90% of the world's rice is produced and consumed within Asia (Dawe et al., 2010), with China and India alone accounting for approximately 50% of all rice grown and consumed. Figure 1 shows the top 10 producing countries. USDA estimates 2014/15 global rice production at 712.9 million tonnes of paddy, or 478.2 million tonnes on a milled basis (USDA, 2015).

Rice is the daily staple for half of the global population – more than 3.5 billion people, many of them classified as food-insecure, and is the staple food for the 70% of the world's poor living in Asia (Gulati and Narayanan, 2002). Figure 2 shows the geographical correspondence between rice production and the incidence of poverty. For many of the world's people living on below US$ 1.25 per day, rice is the only safety net that stands between them and penury or starvation. Rice provides up to 50% of the dietary caloric supply for millions living in poverty in Asia and is also becoming an important food staple in both Latin America and Africa.

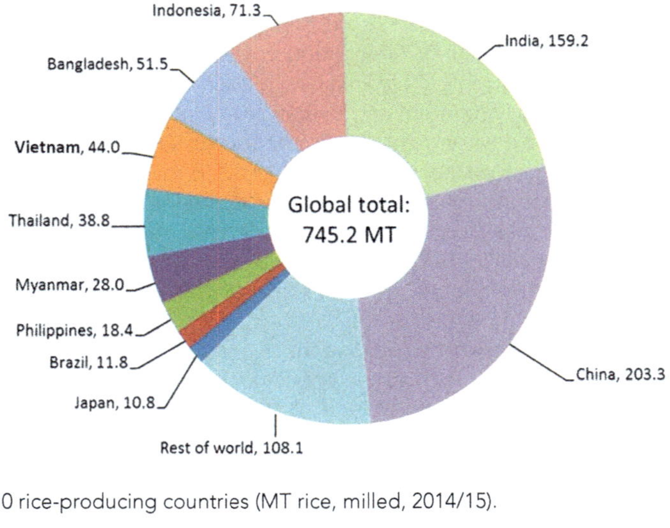


**Figure 1** Top 10 rice-producing countries (MT rice, milled, 2014/15).

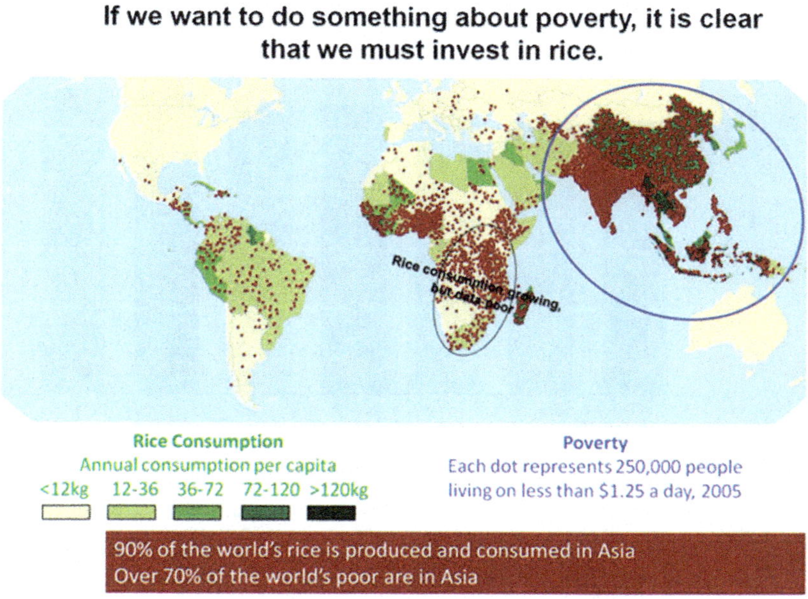


**Figure 2** Poverty is where rice is grown.
Source: www.irri.org/global-effort/poverty-is-where-rice-is-grown.

## 2.2 Production and consumption: projections to 2050

Production and consumption trends in rice markets have important implications for poverty, food security and economic development, especially in Asia. Because rice accounts for a large share of income for a large segment of the world's poor, price increases will have disproportionate impacts on families who are net buyers, reducing their net disposable income. Since 2009, global consumption has steadily increased from 437 MT in 2008/9 to 481.2 MT milled rice in 2013/14 (Fig. 3). According to the International Rice Research Institute (IRRI) (IRRC, 2015), around the world, about 2.7 billion rice farmers and consumers depend upon the sustainable productivity of irrigated rice ecosystems.

Global rice consumption is projected to continuously increase despite dietary shifts in Asia's megacities. Projections for future global rice consumption (demand) vary widely and depend on computation methodology and underlying assumptions about future rice yield growth; population growth; shifts in food preferences; income developments and elasticities of supply, demand and price of rice and substitutable food items such as wheat (among others). A review by Abdullah et al. in 2005 (cited in Timmer, 2010) revealed estimated global rice consumption in 2035 to vary between 380 and 540 million tonnes milled rice, while the IRRI estimate is somewhat higher, at 550 MT (Fig. 4). IRRI estimates that from a 2010 baseline, an additional 114 MT of milled rice will be needed to satisfy the projected demand by 2035 (Zeigler, 2012).

Longer-tem global consumption projections also vary significantly. According to Rejesus et al. (2012), global consumption will reach 650 million tonnes by 2050, further widening the shortfall in global supply.

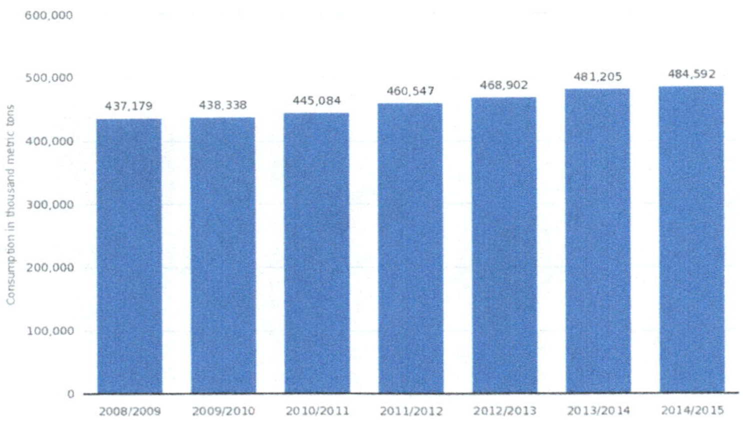


**Figure 3** Global rice consumption 2008/9 to 2014/15 (1000 MT).
Source: US Department of Agriculture; USDA Foreign Agricultural Service, cited by Statistica.

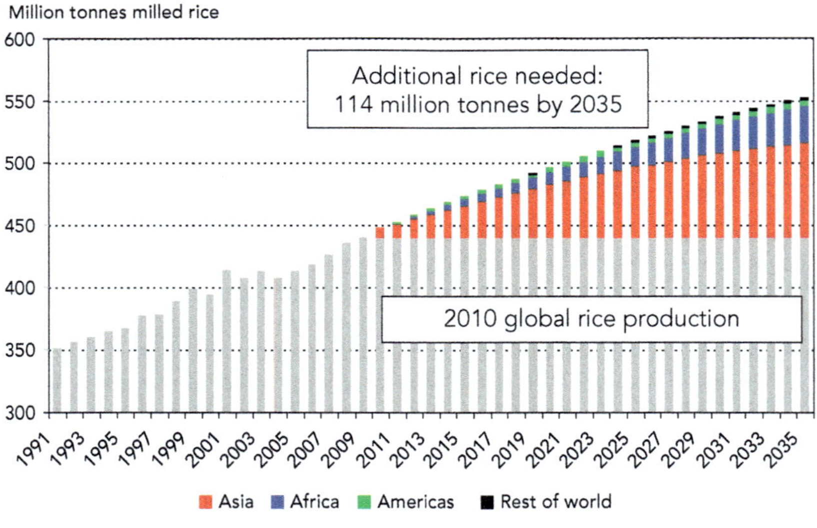


**Figure 4** Projected rice supply shortfall to 2035.
Source: IRRI (2015).

So, how will future demand be met? In the past, increases in global rice production have come from both increases in harvested area and in yield (Mohanty et al., 2010). From 1961 to 1977, the global harvested rice area increased by 1.38% per year, but supply growth has since slowed to just 0.33% per year (Dawe et al., 2010; Mohanty et al., 2010). Reasons for this slowdown include factors such as limited availability of new land, conversion of existing rice land, salinization and increased water scarcity. Pardey et al. (2011) estimate that with little or no new land available for area expansion, and concurrent decline in yield growth to less than 0.8% p.a., the projected global supply in 2050 is unlikely to meet demand growth. Climate change impacts are expected to further widen this gap (Li and Wassmann, 2011). Pisante et al. (2010) estimate that in developing countries, 80% of future crop production increases will need to be delivered through intensification via higher yields, multiple cropping and shorter fallow periods, rather than by area expansion.

Urbanization, industrialization and liberalization of markets add further pressure to grow more from a declining agricultural resource base; these trends carry significant environmental implications as rice producers strive to intensify production on existing land. With limited opportunity for area expansion and expected impacts of climate change on productivity and resilience, intensification of production on existing cropland thus seems to offer the only solution. Pisante et al. (2010) estimate that in developing countries, as much as 80% of future crop production increases will need to be delivered through intensification of cropping systems. How, therefore, might we assess the environmental and sustainability implications of such intensification?

# 3 Sustainability concerns: resource use and climate change impacts

Increasing rice production presents complex sustainability challenges. While flooded paddy rice production is considered an essentially sustainable system that can maintain long-term soil fertility and continuous high productivity, boosting productivity by intensification of cropping systems can create negative environmental externalities resulting from increased use of external inputs such as water, agrochemicals and energy, and lead to adverse impacts on health, biodiversity and water quality. According to Dobermann et al. (2004), sustainability concerns in rice production systems can be generally categorized as follows:

- Resource-use efficiency (land, water, agrochemicals, labour);
- climate change impacts;
- impacts on ecosystem services;
- soil impacts (e.g. salinization, arsenic, organic matter);
- disease impacts (e.g. water-borne pathogens).

These issues have been extensively documented in the literature (see overview by Bouman et al., 2007a), and we, therefore, draw attention here to two key concerns: resource-use efficiency (especially for water and fertilizers) and climate change impacts.

## 3.1 Resource-use efficiency

### 3.1.1 Water

Rice uses two to three times more water than other food crops such as wheat or maize (Barker et al., 1999). Globally, about 100 million ha of harvested rice is irrigated (Dawe et al., 2010), and irrigated rice accounts for 34–43% of the world's irrigation water (Bouman et al., 2006).

The imperative for improving water-use efficiency is sharpened by increasing incidence of local water scarcity, even in irrigated rice areas. Tuong and Bouman (2003) estimated that by 2025, 15–20 million ha of irrigated rice is expected to suffer some degree of water scarcity.

Conventional rice production systems require an average of 2500 L water per kg rice produced, although this varies from 800 to 5000 L/kg (including transpiration, evaporation, percolation and seepage). Developing methodologies that improve water-use efficiency or water productivity in rice production will allow for saving water and for its reallocation to other uses. A reduction of 10% in water used in irrigated rice would free up 150 000 million $m^3$, corresponding to about 25% of the total fresh water used globally for non-agriculture purposes (Klemm, 1999).

There are a number of options for farmers to save water in irrigated rice production: reduced tillage, land levelling, use of alternate wetting and drying (AWD), use of raised beds, mulching to provide soil cover and cultivating aerobic rice, among others (Bouman et al., 2007a,b; Thakur et al., 2011). However, as rice is highly sensitive to water stress, attempts to reduce water use may reduce yields as well as result in trade-offs with nitrogen-use efficiency weed management and rodent control.

### 3.1.2 Agrochemicals

The world's rice fields receive some 15% (or 24.3 million tonnes) of global fertilizer use; health and environmental impacts of excessive fertilizer use in intensive systems are compounded by low-use efficiencies; poor application timing can result in only 20–40% or less of applied nitrogen fertilizer captured by the crop (Islam et al., 2007).

Total fertilizer consumption has increased for most countries in the past 50 years, and this is also true for rice (Gregory et al., 2010). Based on the most recent crop-specific fertilizer consumption statistics, the world's rice fields receive some 15% (or 24.3 million tonnes) of global fertilizer ($N+P_2O_5+K_2O$) use – the same amount as the world's wheat and maize fields combined. Total N use stands at 15.7 million tonnes, $P_2O_5$ at 4.8 million tonnes and $K_2O$ at 3.8 million tonnes (Gregory et al., 2010). Country averages vary from more than 200 kg/ha in countries such as China and Vietnam, to less than 100 kg/ha in the Philippines and Thailand (IFA, 2009; cited in Gregory et al., 2010). The health and environmental impacts of excessive use of fertilizers in intensive systems are compounded by low fertilizer-use efficiencies; poor application timing can result in only 20–40% or less of applied nitrogen fertilizer captured by the crop (Islam et al., 2007).

Pesticide use, too, has grown. Using recent pesticide sales data, Norton et al. (2010) estimated that pesticide use in rice roughly doubled from 1980 to 1996, but has levelled off since then in real terms. In the period 1994–9, pesticide use on rice ranged from 0.4 kg active ingredients (a.i)/ha (Tamil Nadu, India) to 4.2 kg ai/ha (Zhejiang, China).

## 3.2 Overview of climate change impacts

Flooded rice cultivation (irrigated, rainfed and deep-water) represents an important source of atmospheric methane and, to a lesser extent, nitrous oxide. According to the Inter-governmental Panel on Climate Change (IPCC), the world's rice fields emit 31–112 Teragram (Tg) of methane per year, about 12–26% of the anthropogenic methane sources or 9–19% of global anthropogenic methane emissions (IPCC, 2007; cited in Wassmann et al., 2010).

However, rice production not only contributes to climate change, it is also highly vulnerable to its impacts, through changing hydrological cycles, rising temperatures and natural disasters. Over the past decade, the Asia-Pacific region has experienced increased frequency and severity of extreme weather events such as cyclones, floods and drought; water, soil and land resources also continue to decline. Forecasts project climate-induced declines in productivity over the coming 25 years. For example, in Indonesia, annual mean temperatures are projected to rise by 4.8°C by 2100, and the global mean sea level will increase by 70 cm during the same period (ADB, 2009). Changes in temperature regimes greatly influence not only the growth duration, but also the growth pattern and the productivity of rice crops. A decrease of 10% in rice yield has been found to be associated with every 1°C increase in temperature (Masutomi et al., 2009).

## 3.3 Case study: the impact of climate change on rice in Indonesia

In recent years, rice production systems in Indonesia as well as across Asia have become increasingly threatened by the effects of climate change (IPCC, 2007; Masutomi et al., 2009), particularly as a large proportion of the rice-growing areas is located in especially vulnerable regions. Climate change represents a major constraint to future rice production and farmer livelihoods, particularly through droughts, floods, irregular rainfall, rising sea levels and outbreaks of insect pests and disease. The impacts can be either direct (e.g. through changes in temperature and salinity) or indirect, through impacts on ecosystem functions that support rice production (e.g. water regulation, natural pest control, soil processes). Moreover, adaptive capacity is low among farmers who are heavily dependent on rice systems, with few or no alternative sources of income on which to depend in bad years.

The Indonesian government (Indonesian Govt, 2010) estimates that by 2050, mean temperatures may increase by up to 0.5–1.5°C, and the incidence of extreme weather events such as floods and droughts will increase in both frequency and severity. Climate change will affect all economic sectors, especially agriculture, health, energy and infrastructure. By 2050, without national and international efforts focusing on mitigation and adaptation, the combined impacts of climate change could bring serious impacts on economic growth.

Analysis of climate change impacts in rice production in Java suggests that between 2025 and 2050, rice production is likely to decrease by about 1.8 and 3.6 million tonnes from the current production level, respectively (Boer et al., 2009). By including the impact of rice field conversion to non-agriculture lands in Java, that is, taking 0.77% of land out of production per annum, the production loss in 2025 and 2050 will increase to 5.2 and 13.0 million tonnes, respectively. To compensate for this loss, new rice areas of around 1.5 and 3.5 million ha will be needed by 2025 and 2050, respectively, outside Java. Alternatively, the loss can be compensated by increasing rice productivity by about 20% and 50% from the current levels. However, to achieve this, rice productivity would need to increase by approximately 1 t/ha and 2.5 t/ha from its current level (~5 t/ha).

Forecasted shorter wet seasons are also expected to further increase the vulnerability of rice farmers to crop failure, especially in the dry season. If the current rice–rice rotations are maintained, improvement of water storage and irrigation facilities will be required to compensate for decreased June, July and August rainfall. Temperature rise and changes in rainfall patterns and season length may also trigger pest and disease epidemics. In Java, field observations in a number of districts, including Indramayu, Magelang, Semarang, Boyolali, Kulonprogo and Ciamis, support this prediction (Wiyono, 2007). At Indramayu, the drought occurrence associated with the 2002/3 El Niño caused widespread crop losses, and in 2003, the number of households that could not meet their basic food needs increased by 14% compared to normal years (Boer et al., 2006).

It is important to note that greenhouse gas emissions in irrigated rice can be strongly influenced by management practices, offering considerable potential for effective mitigation, for example through adapted water management (such as AWD, mid-season drainage and aerobic rice), residue management and appropriate selection of the type and timing of organic and inorganic fertilizer application (Wassmann et al., 2010). However, despite the major contribution of paddy rice to total agricultural emissions, rice is considered by many national governments as an 'emitter for survival'. It is, thus, important to demonstrate that interventions to mitigate climate change impacts do not compromise productivity or small farmer livelihoods.

Facing such formidable challenges, what can we do, as researchers, development practitioners, farmers and downstream supply chain actors, to assess and enhance sustainability at farm level and throughout the value chain, while protecting our environment and enhancing smallholder livelihoods?

# 4 Assessing sustainability in rice systems: introduction to the Sustainable Rice Platform (SRP)

## 4.1 Background: voluntary sustainability standards

Voluntary sustainability standards (VSS) have emerged as a transformative tool in our global agri-food system. By providing an objective basis for assessing sustainability, most VSS regimes harness market factors to create financial or other incentives for producers to adopt 'best-practice' regimes. VSS schemes today certify over 10% of global production of key global commodities (Potts et al., 2010). As a progressively more pervasive and potent influence on how global companies do business, VSS offer supply chain actors trustworthy mechanisms that add value, enhance accountability and contribute to social and environmental outcomes, as well as strengthen corporate branding and 'licence to operate'.

Despite the strong role of VSS in other agri-food commodities, and regardless of the central role of rice in national and global food security, market-based initiatives targeting rice value chains have until recently received surprisingly little attention. This may be explained by a combination of factors. First, transaction costs of establishing VSS-based certification schemes must be borne by value chain actors; farmers have, over time, been obliged to carry an increasing proportion of both costs and risks. In a low-value bulk commodity such as rice, such a compliance burden cannot easily be borne by small, resource-poor farmers. Moreover, because of the inherent challenge of farm-level traceability in rice, in the absence of a system for identity preservation, costs cannot readily be passed on to consumers in the form of retail premiums.

**Figure 5** Global rice trade flows.

This is especially problematic in markets where there is no consumer willingness to pay for intangible 'sustainability' attributes, and low levels of social and environmental activism (which have been a major driver of VSS in other commodities such as oil palm, coffee and seafoods). Only 5–7% of global rice production enters global trade, much of it via South–South bulk trade (e.g. from Southeast Asia to West Africa, Fig. 5). With less than 2% of global production destined for high-value markets, these markets do not, in general, serve as a demand driver for sustainable market transformation.

Alternative incentive mechanisms are, therefore, needed to stimulate adoption whilst avoiding the marginalization of small farmers through the creation of insurmountable compliance burdens. At the same time, the system also needs to create new value (e.g. financial or reputational) for buyers in destination markets.

Based on this rationale, the following needs were identified by an expert multi-stakeholder group convened by the United Nations Environment Programme (UNEP) and the International Rice Research Institute (IRRI), together with major private sector rice supply chain actors:

- Provide a globally recognized objective definition for sustainability in rice systems.
- Develop technical, policy and market-based tools to drive broad-scale impact, including tools to allow supply chain actors to make credible sustainability claims (value addition).
- Establish a broad-based platform for a multi-stakeholder approach to share knowledge and best practice, and make available SRP and other tools to drive sectoral transformation, particularly for rice smallholders.

## 4.2 The SRP

To address these needs, the SRP was launched in 2011 and co-convened by UNEP and IRRI to promote adoption of sustainable climate-smart best practices in rice, while protecting the environment by boosting the use efficiency of resources such as water and agrochemicals (UNEP, 2011). As a global multi-stakeholder alliance, SRP and its 70 member institutions are helping to drive a transformation towards a more viable global rice sector-improving smallholder livelihoods, reducing environmental/climate change impacts and enhancing food security. SRP is open to governments; international agencies and development partners, private sector actors, research institutes, non-profits and producer groups.

SRP promotes resource-use efficiency and sustainability, both on-farm and throughout value chains. Through both public policy development and voluntary market transformation initiatives, SRP strives to provide essential tools to catalyse action among private, non-profit and public actors in the global rice sector: sustainable production standards, indicators, capacity-building support and outreach mechanisms. SRP is closely allied to, and is a member of, the Global Rice Science Partnership (GRiSP), coordinated by IRRI with the goal of providing a single strategic plan and unique new partnership platform for impact-oriented rice research for development.[1]

The alliance aims to promote a more viable rice sector with high impact on poverty reduction and food security, and a reduced environmental footprint. Its three specific objectives are as follows:

- Develop a context-dependent modular standard for sustainable rice production and processing (including decision-making tools and quantitative-impact indicators).
- Leverage supply chain mechanisms and public policy development to develop and promote outreach models that foster large-scale adoption of sustainable best practices.
- Establish a global knowledge hub to promote sustainability in the rice sector, with broad participation from value chain actors, public and private sectors, as well as research and non-profit organizations.

## 4.3 Technologies and their adoption

Farm management practices in rice offer significant potential for enhancing resource-use efficiency while also mitigating GHG emissions, for example through adapted water management (such as alternate wetting and drying (AWD), mid-season drainage and aerobic rice), residue management and appropriate selection and timing of fertilizer applications (Yan et al., 2009; Wassmann et al., 2010). IRRI has undertaken extensive research into a range of such technologies across a range of rice systems. The following is a non-exhaustive list (IRRI, 2012):

- Safe alternate wetting and drying (AWD)
- Digital decision tools for extension workers and farmers
- Multi-stress tolerant varieties (salinity, drought, flooding) for climate-smart agriculture

1. http://www.grisp.net/main/summary

- Site-specific nutrient management tools to improve N and P efficiency
- Sustainable intensification of rice cropping systems
- Improved irrigation efficiency
- Biofortification
- Low-cost post-harvest technologies to reduce food losses.

Other emerging approaches such as the System for Rice Intensification (SRI) are also increasingly advocated. However, driving wide-scale adoption of such technologies among millions of resource-poor, risk-averse smallholders themselves present immense extension challenges. Direct economic benefits for farmers may not always be apparent or compelling, so behavioural change needs to be incentivized through market mechanisms or policy interventions. However, in general, both markets and policy frameworks have so far largely failed to internalize social and environmental costs (e.g. by valuation of ecosystem services delivered by rice landscapes) as a means of aligning interests in order to stimulate uptake of sustainable best practice at farm level.

To drive large-scale adoption of technological innovation, incentives and outreach strategies need to convince farmers of their direct economic benefits. SRP focuses on development and testing of sustainability guidelines, standards and outreach models, including decision-making tools and quantitative sustainability impact indicators. Looking beyond certification and verification for premium-based segregated supply chains, SRP is also working with its members to explore innovative alternatives, including new verification and business models for smallholder farmers, and the concept of payments for ecosystem services (PES) as a policy instrument.

## 4.4 SRP Performance Indicators

Given the feasibility of practical technological interventions to enhance on-farm sustainability, a set of quantitative indicators (SRP Performance Indicators for Sustainable Rice Cultivation, PIs) were developed by SRP stakeholders. The indicators provide a system for objective measurement of the benefits and impacts of adopting new technological packages, whether introduced through compliance with the SRP Standard or catalysed by other external interventions such as training. A number of criteria were used to select each indicator, which cover key sustainability topics. The following criteria were used:

- Perceived relevance to key sustainability issues in the rice sector
- Applicability across diverse rice-farming systems
- Ability of farmers to improve on the indicator
- Ease of measurement (cost, effort, complexity)
- Ability to quantify performance
- Ability to measure indicators against agreed targets and thresholds.

The resulting framework allows researchers to collect objective benchmark data, offering a flexible tool to enhance our understanding of the effectiveness of individual interventions and also to provide a basis to communicate on progress towards sustainability in any rice system, at any production scale and within any socio-economic or agro-ecological context. The SRP Performance Indicators (PIs) and measurement methods are summarized in Table 1.

As noted above, the indicators enable objective evaluation of the sustainability of any rice system. By establishing such a benchmark for a given system, the indicators allow practitioners and researchers to monitor the effect over time of adoption of recommended best practices promoted under the Standard, as well as the effects of other interventions such as farmer training or even policy. Following field validation, the PIs will offer a foundation for building an evidence base to measure the effectiveness of recommended practices, and also as a basis to develop sector-wide benchmarks.

**Table 1** SRP Performance Indicators

| Name of indicator | Measurement | Source |
|---|---|---|
| **SRP guiding principle: improved livelihoods** | | |
| 1. Profitability: net income from rice | USD/ha/crop cycle | Farm records |
| | USD/ha/year | Household survey |
| 2. Labour productivity | kg paddy rice/no. of days | Farm records |
| | USD net income from rice/no. of days | Household survey |
| 3. Productivity: grain yield | kg paddy/ha | Farm records<br>Household survey |
| **SRP guiding principle: consumer needs** | | |
| 4. Food safety | kg safe milled rice/kg milled rice × 100 | Laboratory test |
| **SRP guiding principle: resource-use efficiency** | | |
| 5. Water-use efficiency: total water productivity | kg paddy/L (rainfall + irrigation) | Farm records<br>Household survey |
| 6. Nutrient-use efficiency: N | kg paddy/kg elemental N | Farm records |
| | kg elemental N removal/kg elemental N input | Household survey |
| 7. Nutrient-use efficiency: P | kg paddy/kg elemental P | Farm records |
| | kg elemental P removal/kg elemental P input | Household survey |
| 8. Pesticide-use efficiency | Balanced scorecard | Farm records<br>Household survey |
| **SRP guiding principle: climate change mitigation** | | |
| 9. Greenhouse gas emissions | Mg/$CO_2$ eq/ha | Farm records<br>Household survey |
| **SRP guiding principle: labour conditions** | | |
| 10. Health and safety | Balanced scorecard | Household survey |
| 11. Child labour | Balanced scorecard | Household survey |
| **SRP guiding principle: social development** | | |
| 12. Women's empowerment | Balanced scorecard | Household survey |

Whilst the PIs provide an important basis for assessment, a concise normative framework that can be used in supply chain projects to serve as a practical basis for verifying any claim to sustainability performance is also needed. The **SRP Standard for Sustainable Rice Cultivation** was designed to provide such a framework. Together, these two tools can help monitor and drive improvement and permit compliant users to make a credible sustainability claim once targets have been defined, for example, as part of local interpretation guidelines. The Standard is discussed in more detail in the following section.

# 5 Application of the SRP standard for sustainable rice cultivation

## 5.1 The SRP standard as a framework for assessment

Having articulated the critical need to address sustainability challenges in the rice sector and established a science-based foundation for action, the **SRP Standard for Sustainable Rice Cultivation** was developed and launched in 2015 as the world's first VSS for rice. The standard and indicators documents are publicly available through the SRP website.[2]

Together with the PIs, the Standard allows objective assessment and benchmarking of the level of compliance of any rice production system, enabling field and policy interventions to be focused and tailored to local needs, and monitored for impact.

The SRP Standard supports two main objectives:

- *To promote improvement*: The SRP recognizes that improving sustainability performance is a journey that itself deserves recognition. However, improvement must be on-going in order to maintain a claim of improvement.
- *To define what is sustainable*: In addition to allowing users to claim that rice is 'sustainably cultivated', the Standard offers a clear definition of sustainability in line with international norms. It can be the basis for enhanced assurance in supply chains and most importantly, as a benchmark for policy making.

In terms of scope, the SRP Standard applies to rice production, including on-farm post-harvest processes that remain within the farmer's control. The SRP Standard can be applied by individual farmers, smallholder groups or large farms. If applied by a group of smallholders, the Standard suggests the establishment of a Group Management System (internal control system), whose requirements will be developed according to an assurance mechanism that is appropriate to the local/national production context; these are currently under development.

The Standard was designed and developed to conform to a number of guiding principles. With a primary focus on addressing the needs and resource-constrained smallholder, the Standard is equally relevant to a much wider range of rice production systems. Versatility in its applicability to the diversity of rice agro-ecosystems, and to any

2. http://www.sustainablerice.org.

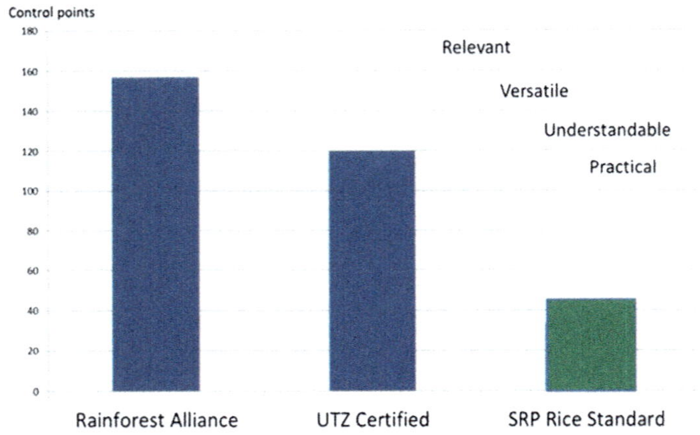


**Figure 6** A compact and practical standard.

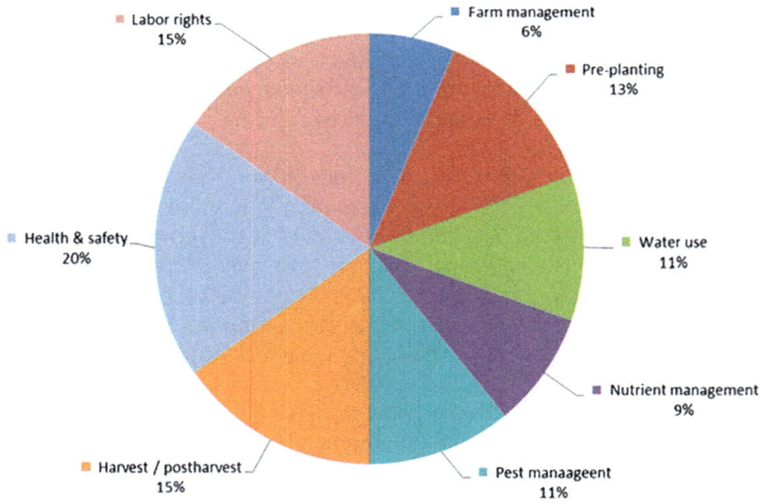


**Figure 7** SRP Standard weighting of requirements.

production scale, was also a key design consideration. Most importantly, and especially in reference to small farmers, the Standard needed to be easily implemented, in order to encourage participation, drive adoption of best practice and minimize the compliance burden for smallholders. The compactness of the SRP Standard is compared in Fig. 6 with prevailing standards for other commodities.

The Standard's 46 requirements (control points) are structured under eight themes, each aimed at achieving a specific desired on-farm impact. These requirements are weighted as shown in Fig. 7.

## 5.2 Scoring

By incorporating a scoring system, the Standard allows for stepwise improvement in order to encourage and reward progress towards full compliance. Most requirements within the Standard have several possible levels of performance to allow its use for both assessment and as a directional improvement tool to encourage and reward adoption by farmers. Following field validation, the Standard will serve as a basis for verification in supply chain governance and as an input for policymaking.

These different levels are developed in recognition that improving farmer performance takes time and can be a challenging process. The performance levels help to guide and reward the improvement process. The total score of a farmer on the Standard is presented on a 0–100 scale. This score is based on the total number of points a farmer has scored, divided by the maximum number of points that can be scored.

$$\textit{Score on Standard}\ (0\text{–}100) = \frac{\text{Total number of points corresponding to actual performance}}{\text{Maximum number of points possible}} \times 100$$

The auditor will exclude from the scoring any requirements that are non-applicable in any specific farm context.

## 5.3 Objectives

The SRP Standard supports two objectives:

- *To promote improvement*: The SRP recognizes that improving sustainability performance is a journey that itself deserves recognition. However, improvement must be on-going in order to maintain a claim of improvement.
- *To define what is sustainable*: The SRP Standard enables users to claim that rice is 'sustainably cultivated'. It recognizes that such a claim should correspond to a certain level of performance. To support this claim, the SRP has defined for each requirement an essential performance level that should be achieved before a claim can be made. This is indicated for each requirement by an asterisk (*) next to the level of performance.

The SRP also recognizes that some flexibility in performance should be allowed, taking into account the different contexts, farmer capacities and priorities. In line with this, the SRP Standard allows two types of claim to be made, as set out in Table 2:

If all requirements apply, all essential performance levels add up to a score of 67. The farmer is free to choose what requirements are used to bridge the gap to 90 or beyond. Figure 8 shows the scoring and claiming mechanism schematically.

**Table 2** Allowable claims and conditions under the SRP Standard

| Claim | Conditions |
|---|---|
| Working towards sustainable rice cultivation | • A farmer scores between 10 and 89, but does not meet the essential performance level of one or more requirements.<br>• A farmer has increased his/her score by 10 points compared with the previous year. |
| Sustainably cultivated rice | • A farmer scores at least 90 and meets all essential performance levels for all applicable requirements. |

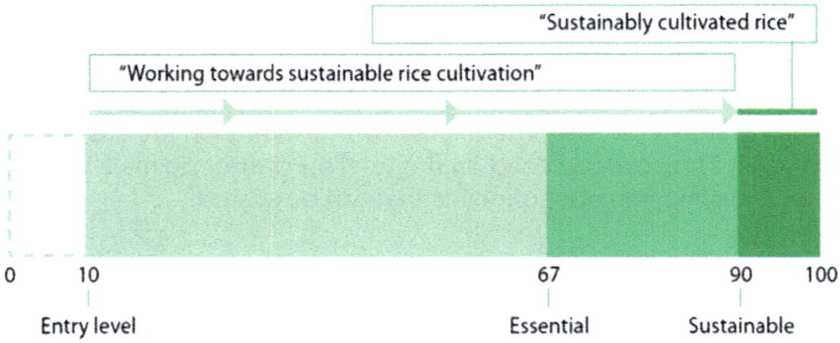


**Figure 8** SRP scoring and claims.

## 5.4 Target groups and associated benefits

The Standard and Indicators have been designed for use by policymakers, extension agencies, farmer organizations, supply-chain actors and development practitioners, as well as by researchers, to promote widespread farm-level adoption of climate-smart sustainable best practices. By promoting adoption among producers, supply-chain actors benefit from reduced supply risks and market confidence. The Standard also offers companies a route to meeting their obligations to shareholders under their corporate commitments to sustainable sourcing. For example, Mars Food, a leading SRP member and owner of the Uncle Ben's Rice brand, has announced its commitment to 100% sustainable sourcing of its rice by 2020, based on the requirements of the SRP Standard.

Ultimately, farmers will benefit directly through reduced input costs, improved stability of production, greater market access and enhanced livelihoods. By improving management of farm chemicals, compliance will also contribute to public health, diversity and integrity of agro-ecosystems and environmental health. Finally, improved water management in irrigated rice can make a significant contribution to mitigating climate change.

## 5.5 Next steps

The current Standard (Version 1.0) is intended as a practice-based instrument that is at the time of writing, undergoing an extensive multi-country field validation programme in farmers' fields, coordinated by IRRI. So far, implementation is planned in Brazil, Cambodia, China, India, Indonesia, Myanmar, Pakistan, Philippines, Thailand, Uganda and Vietnam. The findings of the validation programme will be used to revise the Standard and Indicators and define targets and voluntary/mandatory levels of compliance for each requirement. The body of empirical data generated by the fieldwork will also provide insights into assessing benefits to farmers, especially in terms of direct economic value derived from adoption of sustainable best practice. By taking guidance from the standard, farmers can optimize input efficiency while maintaining yield and net profits.

# 6 Conclusions

In an increasingly crowded, volatile and fragile world, rice represents a crucial driver of global food security. Our global food system faces mounting competition for finite resources as well as multiple risks threatening stability of supplies of strategic crops such as rice. While for other crops such as coffee, tea and oil palm, sustainability concerns have long emerged as key value chain drivers, sustainability of the global rice sector has only recently been addressed through a value chain approach, despite its environmental footprint and strategic importance for global food security. With proven technologies available today to enhance resource-use efficiency, close yield gaps, alleviate poverty and mitigate climate change impacts, the international development community needs to urgently prioritize the sector as part of its response to address targets defined in the UN Sustainable Development Goals, particularly SDG 2.4.[3]

Progress in reaching these targets will be anchored in the benchmarking and monitoring of sustainability impacts of new agronomic practices and policies. The SRP Standard and Indicators offer one of the first attempts to approach this task objectively and comprehensively, enabling farmers, managers, researchers and extension workers to focus on field interventions and training more effectively and to tailor them to the diverse on-farm needs. Success at farm level will, in turn, provide a foundation for change at landscape and policy levels. Only through a broad-based, scaled-up global response involving public and private partners can we hope to drive the required transformations in rice value chains and contribute to a truly sustainable food system for our global future.

# 7 Where to look for further information

Further information on rice and the Sustainable Rice Platform is available at www.sustainablerice.org and www.irri.org

# 8 Acknowledgements

The author extends his gratitude for the support and guidance of all colleagues at UNEP, IRRI, UTZ, GIZ and Aidenvironment. These organizations played pivotal roles in raising awareness of the critical need to drive sustainability in rice value chains as well as at policy level, through the establishment and development of the SRP. Kaveh Zahedi, Isabelle

3. Sustainable Development Goal 2.4: 'By 2030 ensure sustainable food production systems and implement resilient agricultural practices that increase productivity and production, that help maintain ecosystems, that strengthen capacity for adaptation to climate change, extreme weather, drought, flooding and other disasters, and that progressively improve land and soil quality.' See more at: https://sustainabledevelopment.un.org.

Louis, Sabin Basnyat, James Lomax, Matthew Morell, Bruce Tolentino, David Johnson, Grant Singleton, Reiner Wassmann, Sarah Beebout, Han de Groot, Dirk Straathof, Patricia Garcia Diaz, Joel Janiya and Jan Willem Molenaar continue to play crucial strategic roles in this initiative. Finally, I am indebted to SRP's members and dialogue partners in the private and public sectors, as well as from civil society, whose active support, contributions and guidance have made this work possible.

# 9 References

Asian Development Bank. 2009. The economics of climate change in Southeast Asia: a regional review. Manila.

Barker, R., Dawe, D., Tuong, T. P., Bhuiyan, S. I. and Guerra, L. C. 1999. In *Proceedings of the 19th Session of the International Rice Commission*. Food and Agriculture Organization, Rome, pp. 96–109.

Boer, R., Subbiah, A. R., Tamkani, K., Hardjanto, H. and Alimoeso, S. 2006. Institutionalizing Climate Information Applications: Indonesian Case. In R. Motha, M. V. K. Sivakumar and M. Bernardi (eds), *Strengthening Operational Agrometeorological Services at the National Level*. Proceedings of a Inter-Regional Workshop held 22–26 March 2004 in Manila, Philippines. AGM-9, WMO/TD-No. 1277. Available on line at http://www.wmo.ch/web/wcp/agm/Publications/html/AGM9.htm.

Boer, R. 2009. Impacts of climate change in agriculture. Country case study submitted to International Food Policy Research Institute, Washington, DC.

Bouman, B. A. M., Humphreys, E., Tuong, T. P. and Barker, R. 2006. Rice and water. *Adv. Agron*. 92: 187–237.

Bouman, B. A. M., Barker, R., Humphreys, E., Tuong, T. P., Atlin, G. N., Bennett, J., Dawe, D., Dittert, K., Dobermann, A., Facon, T., Fujimoto, N., Gupta, R. K., Haefele, S. M., Hosen, Y., Ismail, A. M., Johnson, D., Johnson, S., Khan, S., Lin Shan, Masih, I., Matsuno, Y., Pandey, S., Peng, S., Thiyagarajan, T. M. and Wassmann, R. 2007a. Rice: feeding the billions. In *Water for Food, Water for Life: A Comprehensive Assessment of Water Management in Agriculture*. Molden, D. (Ed), London, Earthscan and Colombo, International Water Management Institute, pp. 515–49.

Bouman, B. A. M., Lampayan, R. M. and Tuong, T. P. 2007b. Water management in irrigated rice: coping with water scarcity. Available at: http://www.knowledgebank.irri.org/ewatermgt/courses/course1/resources/books/water_management_1.pdf. Los Baños, Philippines: International Rice Research Institute (IRRI).

Dawe, D., Pandey, S. and Nelson, A. 2010. Emerging trends and spatial patterns of rice production. In S. Pandey, D. Byerlee, D. Dawe, A. Dobermann, S. Mohanty, S. Rozelle and B. Hardy (eds), *Rice in the Global Economy: Strategic Research and Policy Issues for Food Security*. Los Baños, Philippines: International Rice Research Institute (IRRI), pp. 15–37.

Dobermann, A., Witt, C. and Dawe, D., eds. 2004. *Increasing Productivity of Intensive Rice Systems through Site-Specific Nutrient Management*. Enfield, USA and Los Baños, Philippines: Science Publishers, Inc. and International Rice Research Institute (IRRI), p. 410.

Gregory, D. I., Haefele, S. M., Buresh, R. J. and Singh, U. 2010. Fertilizer use, markets, and management. In S. Pandey, D. Byerlee, D. Dawe, A. Dobermann, S. Mohanty, S. Rozelle and B. Hardy (eds), *Rice in the Global Economy: Strategic Research and Policy Issues for Food Security*. Los Baños, Philippines: International Rice Research Institute (IRRI), pp. 231–64.

Gulati, A. and Narayanan, S. 2002. *Rice Trade Liberalization and Poverty*. IFPRI Discussion Paper No. 51. Markets and Structural Studies Division. Washington, DC.: International Food Policy Research Institute.

IFA. 2009. Assessment of fertilizer use by crop at the global level: 2006/06–2007/08. International Fertilizer Association, p. 11.

Indonesian Government. 2010. Indonesia Second National Communication under the United Nations Framework Convention on Climate Change. Ministry of Environment, Republic of Indonesia, Jakarta.

IPCC. 2007. *Fourth Assessment Report of the Intergovernmental Panel on Climate Change*. In S. Solomon, D. Qin, M. Manning, Z. Chen, M. Marquis, K. B. Averyt, M. Tignor and H. L. Miller (eds), United Kingdom, New York and Cambridge: Cambridge University Press.

Irrigated Rice Research Consortium. 2015. Available at http://irri.org/networks/irrigated-rice-research-consortium, accessed 21 December 2015.

IRRI. 2012. *IRRI Annual Report 2012*. Available at http://irri.org/index.php?option=com_k2&view=itemlist&layout= category&task=category&id=652&Itemid=100095&lang=en.

Islam, Z., Bagchi, B. and Hossain, M. 2007. Adoption of leaf color chart for nitrogen use efficiency in rice: impact assessment of a farmer-participation experiment in West Bengal, India. *Field Crops Res.* 103: 70–5.

Klemm, 1999. Water saving in rice cultivation. In Assessment and orientation towards the 21st century, *Proceedings of 19th Session of the International Rice Commission*, Cairo, Egypt, 7–9 September 1998. FAO, pp. 110–17.

Li, T. and Wassmann, R. 2011. Modeling approaches for assessing adaptation strategies in rice germplasm development to cope with climate change. Available at: http://www.fao.org/fileadmin/templates/agphome/documents/IRRI_website/Irri_workshop/LP_16.pdf.

Masutomi, Y., Takahashi, K., Harasawa, H. and Matsuoka, Y. 2009. Impact assessment of climate change on rice production in Asia in comprehensive consideration of process/parameter uncertainty in general circulation models. *Agric. Ecosyst. Environ.* 131(3–4): 281–91.

Mohanty, S., Wailes, E. and Chaves, E. 2010. The global rice supply and demand outlook: the need for greater productivity growth to keep rice affordable. In S. Pandey, D. Byerlee, D. Dawe, A. Dobermann, S. Mohanty, S. Rozelle and B. Hardy (eds), *Rice in the Global Economy: Strategic Research and Policy Issues for Food Security*. Los Baños, Philippines: International Rice Research Institute (IRRI), pp. 175–87.

Norton, G. W., Heong, K. L., Johnson, D. and Savary, S. 2010. Rice pest management: issues and opportunities. In S. Pandey, D. Byerlee, D. Dawe, A. Dobermann, S. Mohanty, S. Rozelle and B. Hardy (eds), *Rice in the Global Economy: Strategic Research and Policy Issues for Food Security*. Los Baños, Philippines: International Rice Research Institute (IRRI), pp. 297–332.

Pardey, P. G. 2011. *African Agricultural Productivity Growth and R&D in a Global Setting*. Center on Food Security and the Environment. Stanford Symposium Series on Global Food Policy and Food Security in the 21st Century. Available at http://iis-db.stanford.edu/pubs/23811/Pardey_final.pdf.

Pisante, M., Corsi, S., Kassam, A. and Friedrich, T. 2010. The challenge of agricultural sustainability for Asia and Europe. Transition World Research Network. *Trans. Studies Rev.*17: 662–7. Central Eastern European University Network. Available at http://www.fao.org/ag/ca/CA-Publications/AgSustainability_Pisante_et_al.pdf.

Potts, J., van der Meer, J. and Daitchman, J., eds. 2010. *The State of Sustainability Initiatives Review 2010: Sustainability and Transparency*. A Joint Initiative of IISD, IIED, Aidenvironment, UNCTAD and ENTWINED. Available at: http://www.iisd.org/pdf/2010/ssi_sustainability_review_2010.pdf.

Rejesus, R. M., Mohanty, S. and Balagtas, J. V. 2012. Forecasting Global Rice Consumption. Purdue University Working Paper. Available at: http://www.agecon.purdue.edu/staff/balagtas/rice_timeseries_v6.pdf.

Thakur, A. K., Rath, S., Patil, D. U. and Kumar, A. 2011. Effects on rice plant morphology and physiology of water and associated management practices of the system of rice intensification and their implications for crop performance. *Paddy and Water Environ.* 9: 13–24.

Timmer, C. P. 2010. Behavioral dimensions of food security. *Proceedings of the National Academy of Sciences of the United States of America*, Early Edition, 20 September 2010.

Tuong, T. P. and Bouman, B. A. M. 2003. Rice production in water-scarce environments. In J. W. Kijne, R. Barker and D. Molden (eds), *Water Productivity in Agriculture: Limits and Opportunities for Improvement*. Wallingford, UK.: CABI Publishing, pp. 53–67.

UNEP. 2011. New Sustainability Targets for Rice – the World's Most Important Food Crop. UNEP launches global rice initiative. UNEP News Centre, Friday, 2 December 2011. Available at: http://www.unep.org/newscentre/default.aspx?DocumentID=2661&ArticleID=8967.

USDA. 2015. Rice: World Markets and Trade, December 2015. Office of Global Analysis, Foreign Agricultural Service, USDA. Available at: https://apps.fas.usda.gov/psdonline/circulars/grain-rice.pdf.

Wassmann, R., Nelson, G. C., Peng, S. B., Sumfleth, K., Jagadish, S. V. K., Hosen, Y. and Rosegrant, M. W. 2010. Rice and global climate change. In S. Pandey, D. Byerlee, D. Dawe, A. Dobermann, S. Mohanty, S. Rozelle and B. Hardy (eds), *Rice in the Global Economy: Strategic Research and Policy Issues for Food Security*. Los Baños, Philippines: International Rice Research Institute (IRRI), pp. 411–32.

Wiyono, S. 2007. Climate change and pests and diseases explosion. Paper presented in a Seminar on Biodiversity in the middle of global warming. KEHATI Foundation, Jakarta, 28 June 2007 (in Indonesian).

Yan, X., Akiyama, H., Yagi, K. and Akimoto, H. 2009. Global estimations of the inventory and mitigation potential of methane emissions from rice cultivation conducted using the 2006 Intergovernmental Panel on Climate Change Guidelines. *Global Biogeochemical Cycles*. 23(2): 3–25. doi: 10.1029/2008GB003299.

Zeigler, R. 2012. Cutting – Edge Rice Science for Food Security, Economic Growth and Environmental Protection. Available at: http://csis.org/files/attachments/121220_Zeigler_presentation.pdf.

Part 3

# Rice pests

Chapter 9

# Rice insect pests: biology and ecology

*E. A. Heinrichs, University of Nebraska-Lincoln, USA; F. Nwilene, The Africa Rice Center, Nigeria; M. Stout, Louisiana State University, USA; B. Hadi, International Rice Research Institute (IRRI), The Philippines; and T. Freitas, Universidade Federal Rio Grande do Sul, Brazil*

1 Introduction
2 Biology and ecology of rice-feeding insects: root and stem feeders
3 Biology and ecology of rice-feeding insects: stem borers
4 Biology and ecology of rice-feeding insects: rice gall midges
5 Biology and ecology of rice-feeding insects: leafhoppers and planthoppers
6 Biology and ecology of rice-feeding insects: foliage feeders
7 Biology and ecology of rice-feeding insects: panicle feeders
8 References

## 1 Introduction

Rice, the daily food of nearly half the world's population, is the foundation of national stability and economic growth in many developing countries. It is the source of one-quarter of global food energy and – for the world's poor – the largest food source. Rice production is the single largest use of land and is the biggest employer and income generator for rural people in the developing world. Rice production has been described as the single most important economic activity on earth. Because rice occupies approximately 9% of the planet's arable land, it is also a key area of concern – and of opportunity – in environmental protection.

As one of the world's major food crops, rice has a variety of uses and is adapted to a broad range of climatic, edaphic and cultural conditions. World rice production in 2013 was approximately 745 million t grown on more than 165 million ha (FAO 2016). Over 90% of this area lies in Asia, while the remainder is divided among Latin America, Africa, Australia, Europe and the USA. Annual production in Asia is 675 million t while it is only 36 million t in the Americas and 29 million t in Africa.

Rice cultivation has been the dominant land use in Asia for ages, but it is now playing an increasingly important role in Africa as well. In West and Central Africa – the most impoverished regions on earth according to the Food and Agriculture Organization

http://dx.doi.org/10.19103/AS.2016.0003.21

(FAO) – rice is grown under subsistence conditions by about 20 million smallholder farmers who are shackled to slash-and-burn farming and who lack rice varieties that are appropriate to local conditions. FAO statistics show the demand for rice in these regions is growing by 6% a year (the fastest-growing rice demand in the world), largely because of increasing urbanization. The increase in rice consumption is not limited only to Africa. World rice consumption is increasing.

To feed the growing world population, rice production must be increased. However, rice farmers face many abiotic and biotic constraints in their quest to increase rice production. In conjunction with the introduction of new high-yielding drought-and-flood-tolerant rice varieties, increasing yields will require a reduction in losses to insects and other stresses. As cropping intensity and cultural practices are changed to meet production needs, pest pressure is expected to intensify.

The rice plant is an ideal host for many insect species. All of the plant parts are vulnerable to insect-feeding from the time of sowing till harvest. There are over 800 insect species damaging rice in one way or another, although the majority of them do very little damage. In tropical Asia only about 20 species are of major importance and of regular occurrence (Grist and Lever 1969). In Africa, 15 species of insects are considered major pests of rice (Oteng and Sant'Anna 1999) and in the Americas about 20 species are considered major pests (Stout, pers. comm.).

To develop effective pest management strategies, it is essential to properly identify and to understand the biology and ecology of insect pests and the arthropods that help regulate their populations. This chapter effectively utilizes the unique knowledge and expertise of leading rice entomologists from Africa, Asia and the Americas to provide the first global coverage of rice insect pests. The discussion includes the geographical distribution; plant hosts other than rice; description and biology; and plant damage and ecology of the important rice insects in Africa, Asia and the Americas.

The insects are presented as based on feeding types:

1 Root and stem feeders,
2 Stem borers,
3 Rice gall midges,
4 Leafhoppers and planthoppers,
5 Foliage feeders, and
6 Panicle feeders.

In each case, the chapter provides an overview together with a case study of a particular insect, summarizing key information such as:

- Classification
- Distribution
- Host plants other than rice
- Description and biology
- Plant damage and ecology

An accompanying chapter discusses the current strategies to manage rice insect pests in an environmentally sustainable manner.

# 2 Biology and ecology of rice-feeding insects: root and stem feeders

There are many insects which feed on the roots and stem of rice plants. Some, such as white grubs and root aphids, attack exclusively the roots; others (e.g. mealybug, rice stem maggot) infest only the stems, while still others, such as mole crickets, damage both roots and stems. The infestation of the rice crop by different insect pests is related also to the growth stage of the plants. Seedlings are prone to attack by pests such as rice seedling flies, rice seed midges and mole crickets. Insect damage at the early stages of crop growth causes seedling death and leads to an uneven stand. Wilted plants may completely disappear from the field by being blown in the wind or by being consumed by saprophytic organisms. Heavily infested fields have many missing hills, leading to low plant density and low yield.

The subterranean environment in which root-feeding insects live limits mobility, especially in locating food. For this reason root feeders have adapted by (1) being long-lived either as individuals (beetles), as colonies of social insects (termites), or as dependent on social insects (mealybugs and aphids) and (2) having a wide host range (all species) (Litsinger et al. 1987).

Root-feeding insects include the mole crickets (family Gryllotalpidae), tobacco cricket (family Gryllidae), root aphids (family Aphididae), mealybugs (family Pseudococcidae), black bugs (Family Pentatomidae), stink bug (family Pentatomidae), seed midges (family Chironomidae), stem maggot (family Chloropidae), seedling flies (family Muscidae), white grubs (family Scarabaeidae), grape colaspis (family Chrysomelidae), termites (family Termitidae), root weevils (family Curculionidae), plant weevil (family Curculionidae), black beetles (family Scarabaeidae), and the rice water weevils (family Curculionidae).

## 2.1 Case study: rice water weevil, *Lissorhoptrus oryzophilus* Kuschel (Coleoptera: Curculionidae)

***Distribution:*** *Lissorhoptrus oryzophilus* Kuschel (Fig. 1) is native to North America east of the Rocky Mountains and north to Canada and is the most important invertebrate pest of rice in the United States. Populations of this species have invaded California (in the 1950s), Japan (1970s), Korea and China (1980s), and, most recently, Italy (2004), and this insect has become a serious pest of rice in all areas where it has been introduced (Aghaee and Godfrey 2014).

Various other species in this genus are reported to be pests of rice in Central and South America. *Lissorhoptrus brevirostris* (Fig. 2) and *L. isthmicus* are found in Cuba (Carbonell 1983) and Puerto Rico (Pantoja and Medina-Gaud 1988); *L. bosqi*, in Colombia (Pantoja et al. 1997); and *L. tibialis*, in Brazil (Ferreira and Martins 1984).

***Host plants other than rice*:** Many plants belonging to the families Alismataceae, Commelinaceae, Cyperaceae, Juncaceae and Poaceae.

***Description and biology*:** Invasive populations are parthenogenetic, whereas the species is sexual in its native range. Rice water weevils are semi-aquatic weevils that specialize on plants in the Poaceae and Cyperaceae (Tindall and Stout 2003, Lupi et al. 2012, O'Brien and Haseeb 2014). The biology of *L. oryzophilus* is representative of the genus.

**Figure 1** *Lissorhoptrus brevirostris* adult (Source: www.famu.edu).

**Figure 2** *Lissorhoptrus oryzophilus* adult (Source: LSU AgCenter).

Adults of *L. oryzophilus* (Fig. 2) are greyish brown and about 3 mm in length; markings are variable but most adults have a brownish V-shaped area on their backs. In the females, this area is more distinct and the abdomen is more swollen. Adult rice water weevils overwinter in bunch grasses, leaf litter and plant debris, often in riparian areas. Adult emergence from overwintering is influenced by temperature (Zou et al. 2004). Adults fly to rice fields and feed on leaves of young rice plants and other grassy weeds in proximity to rice fields.

Oviposition is triggered by standing water, and larval infestations usually commence when fields are flooded (Stout et al. 2002). The gravid female moves down the stem and lays eggs in or under the basal submerged portion of leaf sheath tissue and rarely in the roots. Maximum egg-laying occurs one or two weeks after flooding.

The egg is white, elongate and slightly curved. It is about 0.8 mm long. The egg stadium lasts for 4–9 days depending on temperature (Raksarart and Tugwell 1975).

Larvae are white, very small and legless. The head, which is brown, is small in relation to the rest of the body. The first instar grubs mine in the leaf sheath for about a day and then move down the plant to the soil, where they start feeding on roots. Larvae possess dorsal hooks that are modified abdominal spiracles. These structures facilitate larval movement in the soil and help in the acquisition of oxygen from the root aerenchyma (Everett 1966). Larvae develop through four larval instars (Cave and Smith 1983) and a pupal stage under flooded conditions in three to six weeks, depending on temperature (Agahee and Godfrey 2014). Larvae attain a maximum length of 8 mm in a period of approximately 21 days.

**Figure 3** *Lissorhoptrus oryzophilus* pupae on roots (Source: LSU AgCenter).

Pupation takes place in oval mud cells lined with a water-tight material which is attached to the roots (Fig. 3). The pupa is white and of the same size as the adult weevil. Adults that emerge from pupae may fly to overwintering sites or initiate a new generation. The number of generations per year varies from one in temperate areas to several in tropical and sub-tropical areas.

In Japan, only one or sometimes two generations are recorded in a year (Tsuzuki et al. 1982). But in southern Louisiana, under optimal conditions, up to 4 generations occur. Successive generations take place within the same rice field only when there is no crop in the seedling stage in the area.

By the end of August, the weevils move to overwintering sites such as woodland leaf litter, clumps of grass, or rice stubbles. Adult overwintering populations are generally higher in moist places than at dry sites. The weevils come out of overwintering areas and start feeding on grasses from late April to early May. They then fly into rice fields that are transplanted in May. The period of oviposition of the water weevil is mostly governed by the duration of daylight in the locality.

***Plant damage and ecology:*** The rice water weevil is the most important pest of rice in the USA. The adult feeds on the leaf epidermis of young rice plants. Feeding by adult rice water weevils on leaves is rarely damaging enough to warrant treatment. All stages of flooded rice are susceptible to oviposition, although preference is shown for rice in the tillering (vegetative) stage (Stout et al. 2013).

Economic losses stem from extensive injury to root systems that result in reductions in vegetative growth, tillering, panicle densities, grains per panicle, and grain weights of rice plants. Several analyses of weevil density–rice yield relationships suggest that each larva present on rice roots is associated with an approximate loss in yield of 1% (Agahee and Godfrey 2014). Yield losses up to 75% have been reported (Newsom and Swanson 1962; Grigarick 1963) due to the infestation of the water weevil.

A parthenogenetic strain of the pest is believed to have been introduced into Japan from California in 1976. *L. oryzophilus* is regarded as one of the most destructive pests of rice in Japan and probably one of the most difficult to control (Okada 1982).

The population densities in all stages of the pest are higher in those rice fields with standing water throughout the cropping season than in fields that are only intermittently irrigated (Yasuda et al. 1979). Shimohata and Kano (1982) observed that the weevils feed more and produce more eggs on younger rice seedlings.

The critical minimum temperature for adult female survival in Japan was found to be as low as 5 to −10°C (Tsuzuki et al. 1979). This demonstrates the ability of the pest to survive in low-temperature regions and to cause damage to rice crops there.

# 3 Biology and ecology of rice-feeding insects: stem borers

Rice stem borers are a key group of insect pests, mostly belonging to the two lepidopteran families of Pyralidae and Noctuidae. According to Pathak (1968), pyralid borers are the most common and destructive of all stem borers and usually exhibit a high degree of host plant specificity. The noctuid borers, on the other hand, are polyphagous and only occasionally cause economic losses. A list of common stem borers is provided in Table 1.

In Asia, *Scirpophaga incertulas* and *Chilo suppressalis* are the major stem borers and are widely distributed from India to Japan. Although there is a number of species that feed on rice in West Africa, five are of major importance: the dipterous stalk-eyed flies (*Diopsis longicornis* and *Diopsis apicalis*) (Togola et al. 2011) and the lepidopterous species including the white stem borer (*Maliarpha separatella*), the striped stem borer (*Chilo zacconius*) and the pink stem borer (*Sesamia calamistis*) (Akinsola 1975, 1979; ALam 1988; Alam et al. 1985).

Various species of Lepidopteran stem borers in the families Pyralidae, Noctuidae and Crambidae attack rice in North and South America. The most important stem-boring species are: *Diatraea saccharalis* (Fabricius), the sugarcane borer, which is the most widely distributed stem borer of rice in the New World and occurs from Argentina north to the southern United States; *Eoreuma loftini*, the Mexican rice borer, which is found in parts of northern Central America, Florida, and the western and southern United States; *Rupela albinella* (Cramer), the white rice stem borer, found from Mexico to Brazil; *Elasmopalpus lignosellus* (Zeller), the lesser cornstalk borer, a polyphagous insect, which is important as a rice pest in Mexico, Central and South America, especially in upland system regions. In Australia, rice is attacked by a species of *Phragmatiphila* (Hely 1958).

Damage caused by the various lepidopterous stem-boring species is similar. Feeding by early instar stem borers on leaves and within leaf sheaths produces characteristic orange-tan lesions but is not economically damaging. Feeding within the culm on the growing point and vascular tissue can sever the growing portion of the plant from the base of the plant. When feeding occurs during the vegetative stage of plant development, the tiller in which the larva is present often dies and fails to produce a panicle (deadheart) (Fig. 4). Rice is capable of partly or fully compensating for losses of tillers from stem borer attack of vegetative-stage rice by putting forth additional tillers (Lv et al. 2008).

When feeding occurs after panicle initiation, feeding by a larva within a stem results in drying of the panicle. Affected panicles may not emerge or, if they do, do not produce grains, remain straight, and appear whitish (whitehead) (Fig. 5). Some species, however, feed inside the stem without producing visible symptoms of deadhearts or whiteheads.

Infestations in rice fields often involve multiple species of borers, and yield losses under severe infestations can reach as high as 60% (Lv et al. 2008). For every percent of whitehead, 1–3% loss in yield may be expected (Pathak 1975). However, yield losses from stem borers have not been adequately characterized.

**Table 1** Global list of common rice stem borer species (Pathak 1975 modified)

| Scientific name | Common name | Distribution |
|---|---|---|
| *Ancylolomia chrysographella* Koll. | – | Southeast Asia |
| *Chilo agamemnon* Blez. | Small purple-lined borer | Africa |
| *Chilo auricilius* (Dudgeon) | Gold-fringed rice borer | Indian subcontinent, Southeast Asia |
| *Chilo loftini* (Dyar) | – | Mexico, USA |
| *Chilo partellus* (Swinhoe) | Pink borer<br>Sorghum stem borer | Oriental and Ethiopian regions |
| *Chilo phaeosoma* (Martin) | – | Ethiopian region |
| *Chilo plejadellus* (Zinken) | Rice stalk borer<br>American rice stem borer | Mexico, USA |
| *Chilo polychrysus* (Meyrick) | Dark-headed stem borer<br>Rice stalk borer | Indian subcontinent, Southeast Asia |
| *Chilo suppressalis* (Wlk.) | Striped stem borer<br>Rice striped borer<br>Pale headed striped borer<br>Asiatic rice borer | Asia, Italy, Spain |
| *Chilo zacconius* Bleszynski | African rice borer | Africa |
| *Diatraea saccharalis* (Fabricius) | Sugarcane borer<br>American sugarcane borer<br>Small moth borer | Southern USA, Central and South America |
| *Diopsis* spp. | Stalk-eyed borers | Africa |
| *Elasmopalpus lignosellus* (Zeller) | Lesser corn stalk borer | Neotropical, Trinidad, Brazil |
| *Eoreuma loftini (Dyar)* | Mexican rice borer | Mexico, USA |
| *Maliarpha separatella* (Ragonot) | White borer | Africa |
| *Niphadoses gilviberbis* (Zeller) | – | Burma |
| *Niphadoses palleucus* Common | – | Australia |
| *Phragmatiphila* sp. | – | Australia |
| *Proceras indicus* (Kapur) | Internodal borer | Indian subcontinent |
| *Rupela albinella* (Cramer) | South American white borer | South America |
| *Scirpophaga incertulas* (Wlk.) | Yellow stem borer | Asia |
| *Scirpophaga innotata* (Wlk.) | White rice borer<br>White stem borer | Asia, Australia |
| *Sesamia botanephaga* (Tams & Bowden) | – | Ethiopian region |
| *Sesamia calamistis* (Hampson) | African pink borer<br>Mauritius pink borer | Africa |
| *Sesamia inferens* (Walker) | Pink stem borer<br>Ragi stem borer<br>Purple stem borer | Asia |
| *Diataea lineolata* (Wlk.) | Neotropical corn borer<br>Maize borer | Venezuela |

**Figure 4** Deadheart damage (Source: F. Nwilene, Africa Rice Center).

**Figure 5** Whitehead damage (Source: F. Nwilene, Africa Rice Center).

## 3.1 Case study: rice striped borer, *Chilo suppressalis* (Walker) (Lepidoptera: Pyralidae)

***Distribution:*** **Europe:** France, Portugal, Spain; **Asia:** Bangladesh, Burma, Cambodia, China, Hong Kong, India, Indonesia, Iran, Japan, Korea, South Laos, Malaysia, Manchuria, Nepal, Pakistan, Philippines, Ryukyu Islands, Taiwan, Thailand and Vietnam; **USSR:** Bleszynski; **Australasia** and **Pacific Islands:** Australia, Hawaii, Papua New Guinea, Irian Jaya (CABI 2015).

***Host plants other than rice: Main hosts:*** *Sorghum bicolor* (L.) Moench. *Zea mays* L.; **Many wild hosts (Poaceae grasses)**, including *Echinochloa colona* (L.) Link, *Echinochloa* *crus-galli* (L.) Beauv., *Eleusine indica* (L.) Gaertn., *Panicum miliaceum* L., *Phragmites australis* (Cav.) Trin. ex Steud., *Sclerostachya fusca* (Roxb.) A. Camus, *Typha latifolia* L. and *Zizania aquatica* L. (CABI 2015).

***Description and biology***: The adult moths (Fig. 6) are 13–16 mm long and they have straw-coloured to light brown forewings. There are a number of silvery scales and usually five black dots at the tip of the forewing; the hind wing is yellowish white (Reissig et al. 1986). The females are lighter in colour than the males. Moths emerge from the pupa at about 7–8 pm and become active in the early morning. During the day they remain in dense foliage. Moths are capable of flying long distances; sometimes winds carry them to distant places (Dale 1994).

Male moths are strongly attracted to virgin females. Attraction is highest in the evening of their emergence, but declines as time passes. Mating usually takes place during the night of emergence, but egg-laying starts only on the next night. The mating frequency is highest among 2- or 3-day-old moths and it then decreases rapidly with age (Kanno and Sato 1978). Through laboratory studies, it has been shown that the threshold light intensity for mating initiation varies with temperature (Kanno 1980).

A female moth may lay 100–550 eggs in batches of 50–80, depositing one batch each night over a 3–5-day period. Eggs (Fig. 7) are naked and scale-like in appearance, pale to dark yellow, laid in masses formed of overlapping rows. Egg masses are found on the basal half of the leaves or occasionally on the leaf sheaths along the midrib of either the upper or the lower surface. The eggs hatch in 3–5 days. Maximum hatching takes place in the morning between 5 and 6 a.m. Generally, all eggs of a mass hatch together (Dale 1994).

The larvae live gregariously (Fig. 8) during the first three instars (Nozato 1981). If the young larvae are isolated from each other, they suffer high mortality. The newly emerged larvae crawl up the rice plant and then congregate beneath the leaf sheath. All larvae enter the stem through a common hole. Larval feeding generally occurs in the middle of the stem. The fully grown larva is approximately 26 mm long and 2.5 mm wide. It has a yellowish brown head and has three dorsal and two lateral brownish abdominal stripes.

**Figure 6** *Chilo suppressalis* adult (Source: Sylvia Villareal, IRRI).

**Figure 7** *Chilo suppressalis* eggmass (Source: Sylvia Villareal, IRRI).

**Figure 8** *Chilo suppressalis* larva (Source: Sylvia Villareal, IRRI).

The middle dorsal stripe is lighter in colour. Under ideal conditions there are 5–6 larval instars, while under adverse conditions, as many as 9 instars have been recorded. The larval stage normally lasts from 20 to 48 days (Dale 1994).

Pupation takes place within the stem. The last instar larva makes an exit hole in the internode for the moth to emerge. The pupa is reddish brown and is without a silken cocoon.

In temperate regions, the larvae enter into diapause in the winter and development resumes in the spring with warmer temperatures. Temperature, day length and the stage of growth of rice plants are principal factors inducing diapause. Exposure of larvae to short day lengths (8 to 14 hours) induces diapause, while longer days (14.5–16 hours) prevent it.

The rice striped borer is well adapted to a temperate climate. All stages of the pest can withstand low temperatures and development proceeds normally in cold habitats of temperate regions. The optimum temperature for larval development is between 22 and

33°C even though larvae are able to withstand temperatures as low as − 14°C for short periods of time. The rate of pupal development increases with temperature from 15 to 30°C. Beyond 35°C, the pupae suffer high mortality and the emerging moths are mostly deformed.

The pest can have one to five generations per year depending on the availability of host plants and the occurrence of favourable temperature conditions. The longer the period of host plant availability, the more the insect population is built up through increased generations in a year.

In Japan, distinct ecotypes of *C. suppressalis* have been reported: 'Shonai' in the north, 'Saigoku' in the southwest and 'Tosa' from Kochi prefecture. The Shonai ecotype is more tolerant to lower temperature than the Saigoku ecotype. The stem borer population between the areas distinctly occupied by these ecotypes is intermediate in nature (Dale 1994).

***Plant damage and ecology:*** The rice striped borer, *Chilo suppressalis*, prefers temperate regions and non-flooded conditions (Islam and Catling 2012) and has been considered as one of the most serious pests of rice in temperate and subtropical Asia (Dale 1994). However, its importance has gradually declined in many countries. The pest mainly attacks rice in the middle or late stage of plant growth. In addition to causing deadhearts and whiteheads, larval feeding can also cause reduced plant vigour, fewer tillers, unfilled grains and lodging of plants. High grain yield losses have been reported for this pest; up to 50% grain loss has been reported from the Philippines (Otanes and Sison 1947) and 20% in Taiwan (Ou 1959). In Spain, *Chilo suppressalis*, coupled with *Sesamia* sp., produces an estimated crop loss of about 2% (Grist 1965), but in Malaysia, no serious damage has been reported (Dale 1994). *Chilo suppressalis* occurs in Bangladesh but is rare (Islam and Catling 2012).

# 4 Biology and ecology of rice-feeding insects: rice gall midges

Globally, there are two major, closely related rice gall midge species attacking rice, the Asian species *Orseolia oryzae* (Wood-Mason) and the African species *O. oryzivora* Harris and Gagne. The occurrence of morphologically similar but distinct species of *Orseolia* on rice in Asia and Africa may be linked to the separate development of Asian- and African-cultivated rice. Both midge species are primarily pests of lowland irrigated rice. Also, the damage caused by the two species is similar. The main external symptom of attack is a 'silver shoot' or a 'gall' that resembles an onion leaf. The gall, which produces no grain, forms instead of the development of a panicle.

## 4.1 Case study: Asian rice gall midge, *Orseolia oryzae* (Wood-Mason) (Diptera: Cecidomyiidae)

***Distribution*: Asia:** Bangladesh, Burma, Cambodia, China, India, Indonesia, Laos, Malaysia, Myanmar, Nepal, Pakistan, Sri Lanka, Taiwan, Thailand, Vietnam (CABI 2016a).

***Host plants other than rice:*** *Cynodon dactylon* (L.) Pers., *Echinochloa colona* (L.) Link, *Heteropogon contortus* (L.) Beauv., *Ischaemum aristatum* L., *Leersia hexandra* Sw., *Oryza* *barthii* A. Chev., *Panicum miliaceum* L., *Paspalum scrobiculatum* L., *Sacciolepis interrupta* (Willd.) Stapf.

***Description and biology***: The adult gall midge (Fig. 9) is similar in appearance to a mosquito. Females have bright red abdomens. Mating takes place soon after emergence and egg-laying starts a few hours later. The female flies mate only once and the unmated ones lay sterile eggs. Adults are nocturnal and are highly attracted to light traps. Eggs are laid either singly or in groups on the underside and near the base of rice leaves or sometimes on leaf sheaths. A single female lays 100–200 eggs either singly or in groups. The males mostly die in 12–18 hours after emergence; females live up to three days (Dale 1994).

Eggs are long, tubular and shiny white, and sometimes with pink, red or yellow shades. They turn amber before hatching. Eggs require high humidity for development and hatching. The incubation period is around 3–4 days (Reissig et al. 1986).

The newly hatched maggots are grey-white, are fairly stout and have a pointed anterior end. They wiggle down the leaf blade in a film of dew and move between the leaf sheath and the stem until they reach the opening point of the apical or side bud at a node. The larva feeds inside the developing bud which is a zone of differentiation for new tillers. A hollow chamber called a 'gall' forms around the larva. As the larva feeds, the gall enlarges at the base and elongates, having the appearance of an onion leaf (Fig. 10).

**Figure 9** *Orseolia oryzae* adult (Source: www.flickr.com).

**Figure 10** Galls caused by *O. oryzivora* larvae (Source: O. E. Oyetunji).

There are three larval instars; total larval period is 15–20 days. Generally only one maggot is found in a tiller.

Pupation takes place inside the gall. The male and female pupae can be easily separated by their size and colour of the abdomen (Panda and Mohanty 1970). Male pupae are small and brown while the females are larger and pinkish. The colour turns darker before adult emergence. The pupa has several rows of abdominal spines that enable it to move up to the tip of the gall before adult emergence. The pupa makes a hole at the gall tip with its spines and the midge emerges through this hole, leaving the pupal skin behind. Adult emergence generally takes place at night or in the early morning. The pupal period varies from 2 to 8 days. The entire life cycle is completed in about 25–38 days (Reissig et al. 1986).

The gall midge remains inactive as a prepupa in wild rice or weeds during the dry season. At the onset of the monsoon seasons, it becomes active and completes one or two generations in grasses before it moves to the rice crop (Reddy 1967). In India, pest incidence is high during July and August if there is rainfall during the daytime. This period coincides with the maximum tillering phase in rice. The population then declines rapidly in December, primarily, because of the limited availability of suitable host plants. In some cases, the pest infests rice stubbles left after the first crop.

***Plant damage and ecology***: *Orseolia oryzae* is primarily a pest of lowland irrigated rice but has been reported in upland and deepwater rice (Venugopala 1975). The extent and severity of gall midge infestations has significantly increased during the last two decades. Favourable weather conditions, cultivation of high-tillering varieties, intensive management practices and low parasitization are conducive to the rapid multiplication of this pest.

The main external symptom of gall midge attack is a 'silver shoot' or 'gall' that resembles an onion leaf. A fully developed gall is a silvery white, hollow tube, about 1 cm wide and 10–30 cm long. Larval feeding suppresses leaf primordial differentiation at the growing tip. This, in turn, induces the development of radial ridges from the innermost leaf primordium followed by an elongation of the leaf sheath (Perera and Fernando 1970). Galls appear within a week after the larvae reach the growing point. Gall midge attack of rice seedlings leads to profuse tillering and stunting of the plants. A late infestation of the plants at the reproductive shoot apex causes malformed leaves and panicles. Israel et al. (1959) reported 0.5% loss in yield for every unit percent increase in incidence. But wide varietal differences exist in the effect of infestation on crop losses. In certain varieties, gall formation is not manifested and instead necrosis of the shoot apex results (Pieris 1977).

The gall midge attacks rice from the nursery to the end of tillering stage. Young maggots cannot survive in plants past the vegetative stage as there are no actively growing apical buds for them to infest.

The periodicity of infestation and extent of damage caused by the gall midge varies in different countries. In Thailand, the pest has long been reported to be associated with the rainfed rice of the North, Northeast and Eastern regions, but in recent years, outbreaks in the irrigated dry season rice of the Central Plains have also occurred (Katanyukul et al. 1980a).

The Asian rice gall midge, a major insect pest of rice in India (Bentur et al. 2003), was previously a pest only in the wet season crop but later it has been observed in the winter crop as well (Kalode and Kasiviswanathan 1976). In India, crop losses ranging from 10 to 100% have been reported (Siddiq 1991). Average annual yield loss due to the pest is estimated to be US$80 million (Ramaswamy and Jatileksono 1996).

In Sri Lanka, high infestations of gall midge were observed in the southwest monsoon season (April to July) in the wet zone while in the intermediate and dry zones, they were mostly seen during the northeast monsoon (October to November) (Kudagamage and Nugaliyadde 1981). The extent of damage caused by *O. oryzae* as reported by various workers is presented in Table 2.

Wongsiri et al. (1971) reported the ecological conditions that usually prevailed in Thailand at the times when gall midge populations were high. The sky was cloudy most of the days and the relative humidity was 75% on an average, the optimum mean temperature for adult emergence and oviposition ranged from 23 to 27°C. Hidaka et al. (1974) observed that seasonal fluctuations of gall midge populations were closely related to rainfall. Overcast skies and drizzling rains are favourable for rapid build-up of pest populations.

Israel et al. (1961) reported an increase in the pest population with high nitrogen levels. The incidence decreased with high levels of phosphate in combination with nitrogenous fertilizers. The infestation generally increased with higher levels of potash (Israel et al. 1963). Prakasa Rao (1972) observed an increased number of absolute tillers and infested tillers per unit area with the application of increased levels of nitrogen. Close spacing of transplanted rice resulted in a greater number of tillers and more leaves per unit area and an increase in relative humidity within the plant canopy, which was suggested as the reason for higher gall midge incidence than under wider spacing (Prakasa Rao 1975).

Variable reactions of differential rice varieties to *O. oryzae* have provided evidence of different biotypes in South and Southeast Asia (Heinrichs and Pathak 1981). The variability of the species is indicated by six biotypes in different regions of India (Bentur

**Table 2** Damage caused by *O. oryzae* in South and Southeast Asia rice

| Country | Zone | Season | Damage | Reference |
|---|---|---|---|---|
| India | Tamil Nadu | Late | 30% hills | Natarajan and Chandy (1978) |
| India | Uttar Pradesh | Late | 65% hills | Rizvi and Singh (1980) |
| India | Madhya Pradesh | Second crop | 45% hills | Kaushik et al. (1979) |
| India | Karnataka | – | 50% yield | Subramanian (1935) |
| India | Nizamabad | – | 30% yield | Khan and Murthy (1955) |
| India | – | – | 15% yield | Israel (1959) |
| India | – | – | 25% yield | Lever (1970) |
| Thailand | – | – | 50% yield | Grist (1965) |
| Thailand | Northern | Wet | 33% yield | Katanyukul et al. (1980b) |
| Thailand | Northeastern | Wet | 27% tillers | Katanyukul et al. (1980b) |
| Sri Lanka | Dry & Intermediate | Maha | 44% tillers | Wickremasinghe (1969) |
| Sri Lanka | Dry & Intermediate | Yala | 23% tillers | Wickremasinghe (1969) |
| Sri Lanka | Dry & Intermediate | Maha | 100% hills | Wickremasinghe (1969) |
| Sri Lanka | Dry & Intermediate | Yala | 18% hills | Wickremasinghe (1969) |
| Sri Lanka | Wet | Yala | 80% yield | Wickremasinghe (1969) |
| China | | October | 60% hills | IRRI (1978) |

et al. 2003). Behura et al. (1999) developed a polymerase chain reaction (PCR)-based assay that distinguished different biotypes of *O. oryzae*. This assay is faster, more reliable and unaffected by environmental factors as compared to the common use of plant host differentials and midge feeding behaviour for identifying biotypes.

# 5 Biology and ecology of rice-feeding insects: leafhoppers and planthoppers

The leafhopper (Cicadellidae) and planthopper (Delphacidae) genera occurring in Asia are also present in West Africa, but the species are different. The West African species are similar in appearance to Asian species but they are of only minor importance. In Asia, their importance has escalated with the intensification of rice production, especially the misuse of insecticides. Both African and Asian species not only cause direct plant damage, by removing the sap from leaves and stems, but several are also efficient vectors of rice viruses (Heinrichs and Barrion 2004, Dale 1994).

Leafhoppers attack all aerial parts of the plant but planthoppers occur primarily on the basal portion of the plant (Dale 1994). Oman (1949) classified the types of damage caused by leafhoppers and planthoppers into four categories: (1) removal of plant sap from the xylem or phloem, (2) damage to plant tissue and deformation of leaves and stems through oviposition, (3) transmission of pathogens, and (4) predisposition of the plants to pathogens that penetrate the oviposition and feeding punctures left by the hoppers in plant tissue. In Asia, leafhoppers are a threat to production mainly because of their efficiency in vectoring rice viruses, but generally do not cause severe direct damage by the removal of plant sap. The brown planthopper, *Nilaparvata lugens* (Stål), however, is a serious pest because it causes direct damage through the removal of large amounts of plant sap, and is a vector of several serious rice viruses (Heinrichs 1979, Cabauatan et al. 2009).

In Asia, the sudden increase in the importance of these pests in the 1970s was attributed to the changes in cultural practices accompanying the intensification of rice production during the Green Revolution. It appears that high levels of nitrogen fertilizer, monocultures, continuous cropping, and mainly the application of insecticides that cause hopper resurgence are some of the intensification practices contributing to leafhopper and planthopper outbreaks (Heinrichs and Mochida 1984, Cabauatan et al. 2009). In West Africa, the hoppers are still considered as only potential pests, which should be closely monitored as production practices are intensified.

Intensive surveys of the leafhoppers and planthoppers inhabiting West African rice environments are limited to those of Heinrichs and Barrion (2004) in Côte d'Ivoire. Based on the examination of museum specimens, Wilson and Claridge (1991) described the rice leafhoppers and planthoppers occurring on rice in the region south of the Sahara and have developed taxonomic keys for their identification.

Wilson and Claridge (1991) list three planthopper species of the Delphacidae family that occur on rice in West Africa, *Nilaparvata maeander* Fennah, *Tagosodes cubanus* (Crawford), and *Sogatella kolophon* (Kirkaldy). The literature and the WARDA Arthropod Reference Collection add *Delphacodes aglauros* and *Sogatella nigeriensis* to the list of delphacid species in West Africa. Closely related species attacking rice in Asia are the brown planthopper, *N. lugens* (Stål) and the whitebacked planthopper, *Sogatella furcifera* (Horvath). *Sogatella kolophon* is widely distributed in the world, occurring in West Africa,

Australia, the Orient, the Atlantic Islands, the Pacific, the Ethiopian region, the New World, and the eastern Palearctic. It is most commonly found in the tropics.

*Tagosodes cubanus* also occurs in South and Central America where it is a vector of the hoja blanca virus of rice, which can cause up to 50% yield loss (King and Saunders 1984).

The rice delphacid *Tagosodes oryzicolus* is a serious pest in tropical Central American, Caribbean, and South American rice-growing countries where it causes direct damage and transmits hoja blanca virus (Morales and Jennings 2010). This insect migrates and has occasionally been found (but not established) as far north as Texas and Louisiana, and as far south as Argentina.

*Nisia nervosa* (Motschulsky) is the only member of the Meenoplidae family attacking rice. The preferred host plant appears to be a sedge species, Cyperaceae (CABI 2016b), but rice is frequently used as a host (Wilson and Claridge 1991). Grist and Lever (1969) mention it as a minor pest of rice, and Huang and Qi (1981) have recorded it on rice and sugarcane in China.

*Locris* spp. (Cercopidae) are confined to Africa and can transmit RYMV (Koudamiloro et al. 2015). They are easily recognized by their large size and red, orange or brown colouration and patterning of the fore wings and head. Little is known about the biology of *Locris* spp. Akingbohungbe (1983) records *Locris rubens* Erichson, *L. maculata maculata* F., and *L. rubra* F. as minor pests of cereals in Nigeria. The most common *Locris* spp. occurring on rice are *L. erythromela* (Walker), *L. maculata maculata*, *L. rubra* and *L. rubens* (Akinsola, WARDA, 1992, pers. commun.). They prefer irrigated and lowland rice to upland rice. Severe plant damage caused by *Locris* spp. is not common but leaf bronzing and wilting can occur.

*Nephotettix* spp. are severe pests of rice in Asia, where *N. virescens* (Distant), *N. cincticeps* (Uhler) and *N. nigropictus* (Stål) are vectors of the virus diseases tungro, rice transitory yellowing, rice dwarf, rice gall dwarf and yellow dwarf. In West Africa, *N. afar* Ghauri, so far, is of only minor importance, and populations seldom reach levels where feeding injury causes economic damage to rice and it is not known to be a virus vector. *Nephotettix modulatus,* however, has recently been reported to be a vector of RYMV in Africa (Koudalimoro et al. 2015).

Two other cicadellid species, *Cofana spectra* (Distant) and *C. unimaculata* (Signoret) occur on rice in West Africa. Both are widely distributed in the Old World tropics from Africa to Australia. Recent studies by Koudalimoro et al. 2015 and Nwilene et al. 2009 have shown that *Cofana spectra* and *C. unimaculata* are RYMV vectors in Africa.

There are several cicadellid species of the genus *Recilia* that are found on rice throughout the world. Of these, the zigzagged leafhopper, *R. dorsalis* (Motschulsky), is an important rice pest in Asia where it transmits tungro, rice dwarf virus, rice gall dwarf viruses and orange leaf (Wilson and Claridge 1991). *Recilia mica* Kramer has only been recorded from West Africa (Kramer 1962), where it has been reported on rice (Zakra et al. 1986), but evidence of crop damage has not been reported.

## 5.1 Case study: brown planthopper, *Nilaparvata lugens* (Stål) (Homoptera: Delphacidae)

There are two major planthoppers in Asian rice production, the brown planthopper (BPH) *N. lugens* and the whitebacked planthopper (WBPH) *Sogatella furcifera*. The BPH is one of the most damaging insect pests in Asian rice because it not only reproduces and multiplies rapidly, causing plant damage known as 'hopperburn,' but also is an efficient vector of rice grassy stunt (RGSV) and ragged stunt (RRSV) viruses (Cabauatan et al. 2009). A minor pest in tropical Asia until the mid-1960s, it assumed major pest status in the 1970s when it caused severe losses in many Asian countries. Then, after keeping a low profile for a decade or so, the BPH again re-emerged as a major threat to rice in Asia, especially China, Thailand, Vietnam and Indonesia (Islam and Catling 2012)

***Distribution:* Asia** – Bangladesh, Bhutan, Brunei Darussalam, Cambodia, China, Fiji, India, Indonesia, Japan, Korea DPR, Korea ROK, Laos, Malaysia, Nepal, Pakistan, Philippines, Singapore, Sri Lanka, Taiwan, Thailand, Vietnam; **Oceania** – Australia, Fiji, Guam, Micronesia, New Caledonia, Northern Mariana Islands, Palau, Papua New Guinea, Solomon Islands (Dale 1994, CABI 2016c, Reissig et al. 1986).

***Host plants other than rice*:** *Arthraxon hispidus* (Thunb.), *Brachiaria mutica* (Forssk.) Stapf, *Digitaria ciliaris* (Retz.) Koeler, *Echinochloa crus-galli* (L.), *Isachne globosa* (Thunb.) Kuntze, *Leersia hexandra* Sw., *Leersia japonica* (Makino ex Honda) Honda, Poa *annua* L., *Saccharum officinarum* L., *Zizania latifolia* (Griseb.) Turcz. ex Stapf (Islam and Catling 2012).

***Description and biology:*** Adult emergence takes place at the basal part of the host plant. It begins at dawn and continues for 4–5 hours. The adult planthoppers are ochraceous brown dorsally and brown ventrally (Fig. 11). The tegmina are subhyaline with a dull yellowish tint. The female measures about 5 mm and the male 4.5 mm in length.

The adults are dimorphic, with fully winged 'macropterous' and truncate-winged 'brachypterous' forms. The macropters are potentially migrants and are responsible for colonizing new fields. At the time of colonization, the macropterous forms dominate in a rice field under both temperate (Kisimoto 1956, 1965; Kusakabe 1979) and tropical conditions (Dyck et al. 1979).

Adults usually mate on the day of emergence and the female starts laying eggs from the day following mating. Brachypterous females lay 300–350 eggs, while macropters lay fewer eggs. The eggs are thrust in a straight line generally along the mid-region of the leaf

**Figure 11** *Nilaparta lugens* adult (Source: Sylvia Villareal, IRRI).

**Figure 12** *Nilaparvata lugens* eggs (eggmass) (Source: Sylvia Villareal, IRRI).

sheath, though sometimes eggs are laid on the leaf midribs (Fig. 12). Eggs are covered with a dome-shaped egg plug secreted by the female. Only the tips of eggs protrude from the plant surface. The egg-laying sites appear as brownish streaks. The number of eggs laid at a site has been reported to vary in different countries – 2–3 in Japan, 4–10 in the Philippines and 2–12 in India.

The egg is banana shaped, about 1.0 × 0.2 mm, constricted towards the egg caps, whitish brown when freshly laid, darkening with age and developing two distinct red eyespots before hatching. Eggs hatch in about 6–9 days (Islam and Catling 2012).

The newly hatched nymph 0.9 × 0.37 is cottony white at first and turns purple brown within an hour. The fully developed nymph is 3.0 × 1.25 mm with a prominent median line from base of vertex to end of metathorax, where it is the widest. Feeding on the plant sap, it undergoes five instars to become an adult. The nymphal period varies widely depending on the food conditions, density during development and other environmental factors. In the tropics, it takes about 10–18 days from hatching until the first instar nymph reaches adulthood (Islam and Catling 2012).

*N. lugens* can have several generations during one cropping season on irrigated rice in the Asian tropics depending on the length of the crop. It has five generations on one rice crop in southern Japan (Mochida 1964), 5–6 in the central China (Lei and Wang 1958) and 4–6 in Indonesia (Mochida et al. 1977). There are usually three generations on the modern high-yielding intermediate duration varieties in the tropics (Heinrichs et al. 1986).

***Plant damage and ecology:*** The BPH is a serious insect pest of rice, especially in tropical Asia, where rice crops are continuously cultivated throughout the year (Dyck and Thomas 1979). The insect prefers rainfed and irrigated wetland fields to upland rice, and direct-sown fields to transplanted fields. It is known to primarily feed on rice and the weed *Leersia hexandra* (Heinrichs and Medrano 1984). It infests the rice crop at all stages of plant growth. Both BPH nymphs and adults damage rice plants through their extensive feeding and transmission of rice viruses (Cabauatan et al. 2009). *N. lugens* is a

**Figure 13** Hopperburn caused by *Nilaparvata lugens* feeding (Source: IRRI).

phloem-feeding insect. Removal of translocating nutrients from the plant sap reduces the net photosynthate available to the plant for sustenance and storage. The BPH secretes a solid feeding sheath into the phloem tissues, blocking the flow of plant sap while the feeding and ovipositional sites expose the plants to fungal and bacterial infection. The honeydew excreted by the nymphs and adults at the base of the plants is covered with sooty mould. At early infestation, round yellow patches appear in the field which soon turn brownish due to the drying of the plants. This condition is called 'hopperburn' (Fig. 13). The patches of infestation then may spread out and cover the entire field (Islam and Catling 2012). Hopperburn does not usually appear until the crop reaches the milk or dough stages.

The BPH transmits the viruses rice ragged stunt (RRSV) and rice grassy stunt (RGSV) (Hibino 1979, 1989, 1996; Cabauatan et al. 2009). Thus, increased levels of BPH occasionally accompany substantial losses of rice crops by the virus diseases. From 2005 to 2006, more than 485 000 ha of rice production area in southern Vietnam were severely affected by viral diseases seemingly spread by BPH, resulting in the loss of 828 000 t of rice valued at US$120 million (Du et al. 2007). The rice virus disease widely observed in southern Vietnam has been called 'yellowing syndrome' from the characteristic symptom of leaf yellowing. Rice yellowing syndrome was found associated with infection by RGSV or co-infection of RGSV with RRSV, both transmitted by the BPH (Du et al. 2007).

*N. lugens* is a long-range migrant. They make wind-assisted migratory flights each year to colonize the summer rice-growing areas of China. By capture of marked hoppers the insects are found to migrate 200–300 km in mainland China (Tu 1979). The hoppers are carried to Korea and Japan by prevailing winds from permanent breeding grounds in southern China. Influxes of the brown planthopper occur from late June to mid-July every year coinciding with the arrival of low-pressure fronts from the south. In the tropics, migrations also occur.

Saxena and Justo (1984) collected hopper adults in two successive inter-island voyages indicating that migrant planthoppers are reaching the Philippine archipelago from certain rice-growing areas lying to the southwest of the Philippines in the Indian Ocean. The inflowing warm and humid air currents facilitate this migration. Observations indicate that the brown planthoppers take off at dusk and that some are capable of continuously flying up to 26h if the temperature is more than 17°C (Rosenberg and Magor 1983). Others may fly for shorter durations, giving the transient population a chance of dispersal and colonization on rice plants on the way.

Several biotypes of *N. lugens* are known to occur throughout tropical Asia. Temporal change of the biotypes of various *N. lugens* populations indicated that there are three groups of populations in Asia: the East Asian, Southeast Asian and South Asian (Sogawa 1992). These populations have shown different properties of insecticide susceptibility (Nagata 2002), as well as differential feeding on resistant varieties (Sogawa 1992). The populations are distinguishable by their differential virulence, honeydew production and ability to survive on various rice varieties. The susceptibility of IR26 in India, the first brown planthopper-resistant variety released by IRRI, signalled the evolution of biotypes among hopper populations. Breeding for resistance to the BPH and the subsequent commercial release of numerous rice cultivars with genetic resistance to the brown planthopper has not provided a panacea for managing this pest because of the selection for biotypes that are capable of surviving on these cultivars (Pathak and Khush 1979, Claridge and den Hollander 1980).

The seasonal prevalence of the pest is mostly dependent on the availability of host plants. In India, in most regions, the peak population is observed during the late rainy season from October to November. Another peak appears during the dry season from April to May in the regions where double cropping is widely practised. In Japan and Korea, macropterous adults immigrate into rice fields from late June to early July every year. Afterwards, the pest spends several generations on rice and moves or dies at the end of the cropping season.

Temperature is a critical factor in the life activities of the BPH. The threshold temperatures of embryonic and post-embryonic development of the brown planthopper are 10.8 and 9.8°C respectively (Suenaga 1963). The hatchability and survival rate are the highest around 25°C. Eggs are very sensitive to desiccation and soon shrivel when the host plant starts wilting (Kisimoto 1977). The temperature conditions in the nymphal stage affect the longevity and oviposition of adult hoppers (Mochida 1964). The population growth of *N. lugens* is maximum at a temperature range of 28 to 30°C in the daytime and slightly lower at night. In warm and humid climates the planthoppers remain active throughout the year and their population fluctuates according to the availability of host plants, activity of natural enemies and other environmental factors prevailing in the locality.

Many factors have been attributed to the rapid development and widespread outbreaks of BPH populations in Asia. *N. lugens* is considered a secondary pest, which is often induced by the prophylactic use of insecticides that destroy basic ecosystem services that control them (Heong and Schoenly 1998, Gallagher et al. 1994). In fact, insecticides have been suggested as the major factor involved in population explosions (Heinrichs and Medrano 1985, Shepard et al. 1995). The promiscuous use of insecticides has been shown to promote resurgence of the BPH (Heinrichs and Mochida 1984) due to the poor control, destruction of the natural enemies that play a key role in suppressing BPH populations (Kenmore et al. 1984), and reproductive stimulation of the hopper at sublethal rates (Heinrichs and Mochida 1984).

In addition, increased fertilizer use in Green Revolution crops has been suggested to benefit BPH population growth. Laboratory studies of Sogawa (1971) have shown that the hoppers prefer and feed more on plants receiving high nitrogen rates. This leads to faster insect development and higher fecundity. These findings were confirmed in the work of Heinrichs and Medrano (1985), where results clearly showed that insect weight, feeding rate and population growth increased with the application of N fertilizer. Also, observations from the Philippines (IRRI 1979), Indonesia (Mochida et al. 1979) and Taiwan

(Cheng 1977) all concur that intensification of rice production through continuous irrigation favours outbreaks.

# 6 Biology and ecology of rice-feeding insects: foliage feeders

The 'foliage feeders' refers to a broad category which consists of a number of insect orders including the Hemiptera, Thysanoptera, Lepidoptera, Diptera, Coleoptera Orthoptera (grasshoppers) and mites that feed on rice foliage. The group has mouthparts which determine the type of feeding damage caused. The lepidopterous larvae have chewing mouthparts which cause defoliation. The Coleoptera larvae and adults and Orthoptera nymphs and adults also have chewing mouthparts and both are defoliators. The Thysanoptera have rasping-sucking mouthparts which are unique to this order. The Hemiptera and the mites have piercing-sucking mouthparts and remove plant sap. Table 3 lists the major insects, other than grasshoppers, that feed on rice foliage.

The whitefly, *Aleurocybotus indicus* (Hemiptera: Aleyrodidae) damages plants by sucking sap from the leaves. Honeydew, which is excreted on the leaves by the feeding of nymphs and adults, has a high sugar content and a black sooty mould fungus grows on it. Extensive feeding and high amounts of sooty mould may eventually lead to wilting and death of the plants (Alam 1989).

Spider mites, *Oligonychus pratensis, O. senegalensis,* O. *oryzae and Tetranychus neocaledonicus* (Acari: Tetranychidae) suck sap from the parenchymatous cells of the leaves and produce large masses of webbing. Leaves become discoloured with white patches and dry up, starting from the leaf tip. Plants become stunted with deformed panicles and empty spikelets (Heinrichs and Barrion 2004).

Rice thrips, *Stenchaetothrips biformis* (Thysanoptera: Thripidae) larvae and adults have rasping-sucking mouthparts that lacerate the green tissue of leaves. The maxillae and mouth cone, which form a tube, are used to suck leaf sap. Young plants, usually of 1 to 2 weeks after transplanting, are the most affected. Damage becomes evident as fine yellowish lines or silvery streaks on the leaves which later curl from the margin towards the midrib. In severe infestations, the plants become stunted and wither (Dale 1994).

The major rice leaffolder species are *Cnaphalocrocis medinalis*, *Marasmia patnalis*, *M. ruralis* and *Susumia exigua* (Lepidoptera: Pyralidae). Leaffolder larvae fold the rice leaves by spinning silk from one edge of the leaf to the other; as the silk shrinks, the leaf folds. Then the larvae feed on the green tissues of the leaf from within the fold. Feeding damage causes the leaf to dry which severely restricts photosynthetic activity of the plant (Heinrichs and Barrion 2004).

The green-horned caterpillar, *Melanitis leda ismene* (Lepidoptera: Satyridae), is among a number of defoliating insects that are minor feeders on rice in Africa and Asia. The pest occurs in all environments, but is most prevalent in rainfed rice. The larva feeds on the margins and tips of leaves and removes leaf tissue and veins. Damage is similar to that made by grasshoppers and armyworms (Heinrichs and Barrion 2004).

The larvae of the rice ear-cutting caterpillar, *Mythimna separata*; fall armyworm, *Spodoptera frugiperda*; rice swarming caterpillar, *Spodoptera mauritia*; *and the* common armyworm, *Mythimna unipuncta* (Lepidoptera: Noctuidae), are voracious feeders removing large areas of leaf blade or entire leaves and even completely stripping the plants. When

**Table 3** Rice foliage feeders and their geographical distribution

| Species | General distribution |
|---|---|
| Whitefly, *Aleurocybotus indicus* David and Subramaniam | Africa, Asia |
| Rice whitefly, *Aleurocybotus occiduus* Maria | Asia |
| Spider mite, *Oligonychus pratensis* Banks | Africa, S. America, Hawaii, S. Africa |
| Spider mite, *O. senegalensis* Gutierrez and Etienne | Africa |
| Spider mite, *O. oryzae* Hirst | Asia |
| Spider mite, *Tetranychus neocaledonicus* Andre | Africa |
| Rice thrips, *Stenchaetothrips biformis* (Bagnall) | Asia, Europe, S. America |
| Rice leaffolder *Cnaphalocrocis medinalis* (Guenée) | Asia |
| Rice leaffolder *Marasmia patnalis* (Bradley) | Asia |
| Rice leaffolder *Marasmia ruralis* (Walker) | Asia |
| Fijian rice leaffolder, *Susumia exigua* (Butler) | Asia, Australasia, Pacific Islands |
| Green-horned caterpillar, *Melanitis leda ismene* Cramer | Africa, Asia, Oceania |
| Rice skipper, *Parnara guttata* (Bremer et Grey) | Asia |
| Rice ear-cutting caterpillar, *Mythimna separata* (Walker) | N. & S. America, Australia, Pacific Islands |
| Fall armyworm *Spodoptera frugiperda* (J. E. Smith) | N. & S. America |
| Rice swarming caterpillar, *Spodoptera mauritia* (Boisd.) | Africa, Asia, Pacific Islands |
| Common cutworm, *Spodoptera litura* (Fabricius) | Asia, Australasia, Pacific Islands |
| Common armyworm, *Mythimna unipuncta* (Haworth) | Africa, Europe, N. & S. America |
| Rice green semilooper, *Naranga aenescens* (Moore) | Asia |
| Rice green semilooper *Naranga diffusa* Walker | Asia |
| Green hairy caterpillar, *Rivula atimeta* (Swinhoe) | Asia |
| Rice whorl maggot, *Hydrellia prosternalis* Deeming | Africa, Egypt |
| Rice leaf miner, *Hydrellia griseola* (Fallen) | Asia, Europe, N. & Central America |
| Rice whorl maggot, *Hydrellia philippina* Ferino | Asia |
| South American rice miner, *Hydrellia wirthi* Korytkowski | N., S. & Central America |
| Paddy stem maggot, *Hydrellia sasakii* Yausa et Isitani | Japan |
| Asian rice hispa, *Dicladispa armigera* (Oliver) | Asia, Oceania |
| African rice hispa, *Trichispa sericea* Guerin-Meneville | Africa |
| Rice blue beetle, *Leptispa pygmaea* Baly | Asia |
| Flea beetle, *Chaetocnema pulla* Chapuis | Africa |
| Flea beetle, *Chaetocnema pusilla* Laboissiere | Africa |
| Ladybird beetle, *Chnootriba similis* (Mulsant) | Africa |
| Rusty plum aphid, *Hysteroneura setariae* (Thomas) | Africa, Asia |
| Grain aphid, *Rhopalosiphum padi* (L.) | Worldwide |
| Short-horned grasshopper, *Hieroglyphus daganensis* Krauss | Africa |
| Large rice grasshopper, *Hieroglyphus banian* (F.) | Asia |
| Variegated grasshopper, *Zonocerus variegatus* (L.) | Africa |
| Small rice grasshopper, *Oxya chinensis* Thunberg | Africa, Asia |

large numbers are present, entire seedlings can be defoliated, resulting in severe stand loss (Dale 1994).

Larvae of the rice whorl maggots, *Hydrellia prosternalis* and *Hydrellia philippina* (Diptera: Ephydridae), feed on the mesophyll tissues within the leaf whorl. When the leaves emerge from the whorl, damage can be seen as pinholes in the leaves, which show white and yellowish lesions at the leaf edges. Severely damaged leaves break in the wind (Dale 1994).

Several species belonging to the Coleoptera – Chrysomelidae, subfamily Hispinae, commonly referred to as 'hispids' – feed on rice in Africa and Asia. The term 'hispa' means spiny and characterizes the adults of this group, which have numerous spines on the thorax and abdomen. Both grubs and adult beetles feed on rice plants. Grubs mine the leaves by feeding on the mesophyll between the veins. Adults feed on the leaf surfaces by removing chlorophyll between the veins of the lamina, giving the appearance of white parallel streaks and white irregular blotches on the leaves. Some African hispids are vectors of RYMV (Dale 1994, Heinrichs and Barrion 2004).

The term 'flea beetle' is applied to a group of small beetles (Coleoptera: Chrysomelidae) that have enlarged hind legs and that jump when disturbed. This group consists of several chrysomelid genera, including *Chaetocnema*, which make small holes in the leaf when feeding. Although *Chaetocnema* spp. are extremely abundant in upland rice, in West Africa, the feeding damage that they cause is minimal. However, they are vectors of RYMV (Banwo et al. 2001, Heinrichs and Barrion 2004).

Many aphid species (Hemiptera: Aphididae) infest the aerial parts of rice even though rice is not the primary host plant for most of them (Dale 1994). Most species prefer to feed on the panicles and stems rather than on the leaves. The rusty plum aphid *Hysteroneura setariae* is a pest on rice leaves in Africa and Asia. *Rhopalosiphum padi*, known as the bird cherry oat aphid and grain aphid, is a foliage feeder with worldwide distribution (Table 3). It is a vector of the virus disease 'giallume' (yellow disease or rice yellows) in Italy, where its primary host is the fruit tree *Prunus padus*. When *P. padus* is not available, the aphid overwinters on secondary hosts such as rice and various grasses (Dale 1994).

A number of species of grasshoppers are occasionally found in rice fields, but rarely cause significant damage other than along the field margins. They chew angular holes in leaves, causing an injury similar to that caused by armyworms. These insects are polyphagous; besides rice, they feed on maize, sugarcane, millets and various grasses.

Nearly 30 grasshopper species, belonging to the short-horned (Acrididae and Pyrgomorphidae) and the long-horned (Tettigoniidae) families, attack rice plants in West Africa. However, most species are not of economic importance on rice because they occur in low populations. Among the most important species are the short-horned hoppers – *Atractomorpha* spp. *Chrotogonus* spp., *Hieroglyphus africanus* Uvarov, *H. daganensis* Krauss, *Oxya hyla* Serville *Zonocerus variegatus* (L.) – and the long-horned hoppers, *Conocephalus* spp. In Asia, grasshoppers are of less importance than in Africa. Key species in Asia are *Hieroglyphus banian* (F.) and *Oxya chinensis* Thunberg.

## 6.1 Case study: African rice hispa, *Trichispa sericea* Guerin-Meneville (Coleoptera: Chrysomelidae)

Several species belonging to the subfamily Hispinae, commonly referred to as 'hispids' feed on rice in Africa. The term 'hispa' means spiny and characterizes the adults of this group, which have numerous spines on the thorax and abdomen. As reported by Breniere (1983), this group consists of *Trichispa sericea* Guerin-Meneville, *Dicladispa viridicyanea* (Kraatz) and *Dactylispa bayoni* Gestro. Additional Hispinae species in the WARDA Arthropod Reference Collection are *Agonita* sp., *Chysispa viridicyanea* (Weise), *Dactylispa spinigera* Gyllenhall, *Dicladispa paucispinis* (Weise) and *Dorcathispa bellicosa* (Guerin). There are likely additional Hispinae species in rice in West Africa whose taxonomy has not yet been determined. *Trichispa sericea is* the species described in detail below because it is a major pest of rice in West Africa, causing severe defoliation, and because it is a potential RYMV vector.

***Geographical distribution:*** Burkina, Faso, Cameroon, Cote d'Ivoire, Mali, Nigeria, Senegal, Sierra Leone, Togo. In addition to its occurrence in West Africa, *T. sericea* is found throughout Central, East and South Africa (Dale 1994).

***Host plants other than rice:*** *Chloris virgata* Sw., *Echinochloa holubii* (Stapf), *Eragrostis aethiopica* Chiov., *Eragrostis heteromera* Stapf, *Digitaria zeyheri* (Nees Hend.) and *Diplachne fusca* (L.) ex Stapf (Dale 1994).

***Description and biology:*** The *T. sericea* adult (Fig. 14) is a dark grey beetle covered with short spines and is about 3–4 mm long (Koudalimoro et al. 2015). Adult females live for 2 wk and lay about 100 eggs during this period. The eggs are white and boat-shaped and are about 1 mm long. The eggs are laid singly in slits made under the epidermis of the upper portion of the leaf. The wound left by the ovipositor is sometimes covered by a dark spot of excreta by the female. The larvae hatch in 3 or 4 d.

*Trichispa sericea* larvae, which are referred to as 'grubs', are slender, yellow, and about 6 mm long. The grubs mine from within the epidermal layer of the leaf. When infested leaves are held against the light, the dark spot of the larva or the pupa in the mine may be visible. The larval period lasts 10 d.

Pupation, which lasts about 6 d, takes place in the last mine bored by the larva and within those portions of the leaf lamina that are not submerged. After pupation, the emerging adults migrate to alternate host plants. In Madagascar, *T. sericea* completes a generation in about 1 mo (Ravelojaona 1970).

***Plant damage and ecology:*** Hispids are serious pests of rice in some countries. In addition to rice, a number of grassy weeds serve as hosts for *T. sericea* (Zongo 1993). In Swaziland, when rice leaves harden and become unattractive for adult feeding and oviposition, *T. sericea* adults migrate to other plants such as *Chloris virgata* Sw, Echinochloa *holubii* (Stapf), *Eragrostis aethiopica* Chiov., *Eragrostis heteromera* Stapf, *Digitaria zeyheri* (Nees Hend.), and *Diplachne fusca* (L.) Beauv. ex Stapf (COPR 1978). However, rice is the preferred host plant.

*Trichispa sericea* is generally most abundant during the rainy season. However, obtaining detailed data on the seasonal occurrence of *T. sericea* on the WARDA M'be Farm has been

**Figure 14** *Trichispa sericea* adult (Source: IRRI).

difficult because of the sporadic nature of its occurrence. However, in 1993, extremely high populations were observed during the rainy months of July and August.

*Trichispa sericea* attacks the rice crop in the early growth stages. In Cote d'Ivoire (Heinrichs and Barrion 2004) adults are observed in the rice field shortly after transplanting, when they attack small seedlings. Larval feeding occurs through the tillering phase. Both the adults and the larvae feed on the leaf tissues of young rice plants (COPR 1976, Dale 1994). The first attack in a field is highly localized, but the infested area spreads rapidly. Adults feed on the green portion of the leaves, leaving only the epidermal membranes. Adult feeding damage is evident by the characteristic narrow white streaks, or feeding scars, that run along the long axis of the leaf and the irregular brown patches.

The pale brown patches, which appear as blisters, are the larval mines within the leaf. The larvae mine into the leaf between the epidermal membranes and the mining results in brown, blister-like areas. Feeding results in a loss of chlorophyll and the plants wither and die. The most serious damage occurs in nurseries that may be completely destroyed. Severe infestations sporadically occur on transplanted rice and can kill the plants. When the plants survive, they usually recuperate and produce some grain. However, damaged plants often mature late.

*Trichispa sericea* is not a highly mobile insect and can spend several hours feeding on the same plant. It is able to retain and transmit RYMV for one to two days via the semipersistent transmission mode. *Trichispa sericea* was confirmed as an RYMV vector in Niger, Mali, Cameroon and Cote d'Ivoire (Koudalimoro et al. 2015).*Trichispa sericea,* along with *Dactylispa bayoni* and *Dicladispa viridicyanea,* has been reported to transmit rice RYMV in Kenya in East Africa (Bakker 1971). Reckhaus and Andriamasintseheno (1997) list *Dicladispa gestroi* (Chapman) and *T. sericea* as RYMV vectors in Madagascar, where the disease has significantly impacted rice yields.

RYMV was first reported on rice in West Africa in 1976 (Raymundo and Buddenhagen 1976). Symptoms of the virus disease are pale yellow mottled leaves, stunted growth, reduced tillering, asynchronous flowering, poor panicle exsertion, spikelet discolouration, and sterility. RYMV was reported in lowland irrigated rice and in mangrove and inland swamps in Guinea during 1982–6 (Fomba 1990), and in upland rice in Cote d'Ivoire in 1985 and Sierra Leone during 1987 (Awoderu et al. 1987). From a distance, infected fields appear yellow. Symptoms on the leaves are linear, chlorotic mottles that coalesce into broken or continuous pale green to yellowish streaks up to 10 cm long. Later, whole plants become light green and then turn pale yellow.

Observations on the WARDA M'be Farm and in farmers' lowland irrigated fields at Sakassou, near Bouake, Cote d'Ivoire, indicate that high populations of *T. sericea* are usually associated with high incidence of RYMV (Heinrichs and Barrion 2004). In plots planted at WARDA in July and August of 1993, *T. sericea* populations and RYMV incidence were high and grain yields were low. However, high incidence of RYMV is not always associated with high T. *sericea* populations. This is an indication that insects other than *T. sericea* may be RYMV vectors or that there are other means of mechanical transmission.

Populations of *T. sericea* adults are affected by the spacing of transplanted seedlings. Adults in sweep net collections in lowland rice on the WARDA M'be Farm were higher in close spacings of 10 x 10 cm (100 hills $m^2$) than in wider spacings of 20 x 20 (25 hills $m^2$) or 40 x 40 cm (6 hills $m^2$) (Heinrichs et al. 1993).

Hispids are prevalent in wetland environments, especially irrigated lowland fields (Reissig et al. 1986). Studies conducted on the continuum toposequence on the WARDA M'be Farm also indicated that damage caused by *T. sericea* is most prevalent in the lowlands.

Percent leaf area damaged by *T. sericea* was about 15% in the lowlands and 1% in the uplands of the toposequence (Heinrichs and Barrion 2004).

The effect of soil moisture on the extent of *T. sericea* damage was studied on the WARDA M'be Farm (Heinrichs 1991). An upland rice variety (IDSA 6) was planted at various intervals along the continuum toposequence from the valley bottom to the hydromorphic zone. *Trichispa sericea* damage was significantly (positively) correlated with soil moisture percentage. High leaf damage (100%) occurred only at high soil moisture levels. The results indicated that an upland variety growing in standing water constitutes an attractive host for ovipositing *T. sericea* adults.

# 7 Biology and ecology of rice-feeding insects: panicle feeders

Insects that attack rice panicles can be separated into those that feed on the floral parts (mostly the pollen) and the stink bugs that suck the milk-like sap from the developing grains. Insects that feed on the floral parts, such as earwigs, blister beetles and panicle thrips prevent the spikelet from filling and thus it remains empty and aborts.

Numerous hemipteran insect species belonging to the Alydidae, Coreidae, Pentatomidae and Pyrrhocoridae families suck the milk-like sap from the developing rice grains. Depending on the time of attack, in relation to stage of grain development, direct damage by milk (liquid endosperm) removal may completely or partially destroy the grain. In addition to the direct damage caused by the feeding of the bugs, secondary pathogens may enter the feeding wounds left by the bugs and cause a staining of the grain that cannot be removed in milling. This is referred to as 'pecky rice', which is prone to breakage in milling and lowering of the grain quality. The 'dirty-panicle' symptom often observed in rice fields is believed to be caused by a combination of bug feeding and fungal infection (Agyen-Sampong and Fannah 1980). If a sample of rice contains some pecky grains, the whole lot is considered contaminated, which lowers the market price in some countries.

About 60 stink bug species have been reported to infest rice fields and at least half of them are considered pests causing yield loss or pecky rice. The major bug species causing pecky rice are listed in Table 4. The rice bugs most commonly found in rice in Asia are *Leptocorisa acuta* (Thunb.), *L. chinensis* (Dallas), *L. varicornis* F. and *L. oratorius* (F.)

## 7.1 Case study: rice bug, *Leptocorisa acuta* (Hemiptera: Alydidae)

***Distribution:*** Asia, Australia, Oceania and several Central American countries (Hayashi 1997, Dale 1994, Reissig et al. 1986, Schaefer and Panizzi 2000).

***Host plants other than rice*:** More than 65 plant species (Srivastava and Saxena 1967)

***Description and biology:*** The description and biology of the various *Leptocorisa* spp. are similar. The following describes the life stages and biology of *L. acuta* (Serrano et al. 2014).

*Leptocorisa acuta* adults (Fig. 15) are long (14–17 mm) and slender (3–4 mm wide) and light yellowish green to yellowish brown in colour. The head is broad, often similar in length and width to the pronotum (upper surface of the first plate on the thorax) and the scutellum (triangular shaped plate on the thorax, posterior to the pronotum). These bugs

**Table 4** Bugs that cause pecky rice (Kisimoto 1983, Dale 1994, Heinrichs and Barrion 2004)

| Family | Species | Distribution |
|---|---|---|
| Alydidae | *Leptocorisa acuta* (Thunb.) | Asia, Australia, Oceania, Central America |
| | *L. chinensis* (Dallas) | Japan |
| | *L. varicornis* F. | India |
| | *L. oratorius* (F.) | Asia, Australia, Oceania |
| | *Mirperus jaculus* Thunberg | West Africa |
| | *Riptortus dentipes* (F.) | West Africa |
| | *Stenocoris apicalis* (Westwood) | West Africa |
| | *S. claviformis* (Ahmad) | West Africa |
| | *S. elegans* (Blote) | West Africa |
| Coreidae | *Cletus punctiger* Dallas | Japan |
| Lygaeidae | *Nysius plebejus* Distant | Japan |
| | *Togo hemipterus* Scott | Japan |
| Miridae | *Stenodema sibiricum* Bergroth | Japan |
| | *Trigonotylus coelestialium* Kirkaldy | Japan |
| Pentatomidae | *Aspavia armigera* F. | West Africa |
| | *A. acuminata* Montandon | West Africa |
| | *A. brunnea* (Signoret) | West Africa |
| | *A. hastator* (F.) | West Africa |
| | *Dolycoris baccanum* L. | Japan |
| | *Eysarcoris lewisi* Distant | Japan |
| | *E. parvus* Uhler | Japan |
| | *E. ventralis* Westwood | Japan |
| | *Lagynotomus elongatus* Dallas | Japan |
| | *Nezara antennata* Scott | Japan |
| | *N. viridula* (L.) | Worldwide |
| | *Oebalus pugnax* (F.) | USA, Caribbean |
| | *O. ornata* (Sailer) | Caribbean |
| | *O. poecilus* Dallas | South America |
| | *O. ypsilongreseus* (De Geer) | Caribbean and northern South America |
| | *O. insularis* (Stal) | Central America, Mexico, Colombia |
| | *Scotinophara lurida* Burmeister | S.E. Asia, China, Indonesia, India |
| Ropalidae | *Aeschynanthus maculatus* Fieber | Japan |

have globular, protruding eyes in addition to very small ocelli (simple eyes). The fourth antennal segment of the *Leptocorisa* spp. is longer than the third segment. *Leptocorisa* spp. have long legs and antennae and brown-green bodies.

Eggs (Fig. 16) are oval with flattened tops. Females lay eggs in batches of 10–20, symmetrically arranged in 2–3 rows, on the upper surface of the leaf blade. When they are freshly deposited, eggs are a cream-yellow colour, turning to reddish-brown after approximately one week.

**Figure 15** *Leptocorisa acuta* adult (Source: http://blog.yam.com).

**Figure 16** *Leptocorisa oratorius* eggs (Source: Sylvia Villareal, IRRI).

**Figure 17** *Leptocorisa oratorius* nymphs (Source: Sylvia Villareal, IRRI).

The eggs hatch a week following oviposition. They congregate around the empty egg cases for 3–4 hours before they disperse for feeding. Nymphs feed voraciously and gregariously until the fourth instar. There are five wingless nymphal instars, with a total nymphal period of 25–30 days. Nymphs are mostly pale yellow-green and have long antennae (Fig. 17). Each nymphal instar looks remarkably similar to the one before it, except each successive nymph is larger than the last and wing pad enlargement occurs. The final moulting takes place at night. Nymphs stop feeding about five hours prior to selecting a place to moult on the leaves or flowers of the host plant. The nymph then remains motionless in an extended position for about two hours before ecdysis. The actual process of moulting takes only two minutes.

Rice bugs are diurnal but are most active during the early morning and the evening. During the hotter parts of the day, the bugs are seen taking shelter under the leaves of rice plants or weeds near rice fields. Rice bugs fly short distances when disturbed. The females are stronger fliers than the males. The males are capable of mating shortly after emergence but the females start only one or two weeks after becoming adults. Mating generally takes place in the morning hours and rarely at night.

They complete 1–2 generations on grassy weeds before migrating to rice fields. A small number of adults usually appear in the rice fields as soon as the plants start flowering. When about 80% of the grains have ripened, the adults migrate to fields having rice plants of less maturity.

In India, teeming populations of *L. acuta* have been found to hibernate on trees during the relatively colder and drier months of December to March (Banerjee and Chatterjee 1965).

***Plant damage and ecology:*** Rice bugs are found in all rice environments, but are more prevalent in rainfed wetland or upland rice. Factors that favour high bug populations are adjacent woodlands, extensive weedy areas near rice fields and staggered rice planting (Reissig et al. 1986).

Both nymphs and adults are destructive to the rice crop, even though the damage by nymphs is more severe (Pathak 1975). Nymphs prefer feeding on grains at the milky stage, when the starches within the grains are not yet fully formed. They infest rice crops in large numbers at flowering. *Leptocorisa* adults and nymphs have piercing-sucking mouthparts and feed by sucking fluid from florets, soft (milky) rice grains and plant sap. They do not bore a hole through the rice hull but feed by the insertion of the proboscis at points where the glumes (lemma and palea) meet. During the process, the bug secretes a proteinaceous stylet sheath to form a feeding canal for its sucking mouthparts (Gyawali 1981). The point of insertion becomes visible when the white exudate turns into a brown spot.

Removal of stored assimilates from the developing grains may result in either unfilled or partially filled grains with the symptoms of damage. When the bugs feed on soft or hard dough endosperm, they inject enzymes to predigest the carbohydrates. In the process they contaminate the grain with microorganisms that cause grain discolouration or pecky rice, which is more liable to break during milling (Reissig et al. 1986). Feeding enzymes and microorganisms enhance feeding damage (Srivastava and Saxena 1967). Damaged florets may drop.

The quality of grains is also reduced by the insect attack. Pecky rice in Japan and the United States is downgraded and priced lower than undamaged grains. The foul odour of the infested grains also lowers market value. Damaged grains, even after cooking, retain the buggy smell. Sometimes even the straw has an off-flavour, which is unattractive to cattle.

The intensity and type of damage caused by rice bugs depend on the stage of the rice crop, population density of the pest and ecological conditions. In India, on an average, 5–10% of the grains are damaged due to infestations of the rice bug. In years of severe attacks, damage reaches 40–60% (Kalode 1982). In the case of early paddy, maturing in September, the loss is usually greater than in the late maturing varieties. *Leptocorisa* spp. cause significant damage in Thailand every year. In the Malay Peninsula, the yield loss in certain areas exceeds 25% (Dale 1994).

In addition to the direct feeding damage, *Leptocorisa acuta* is known to transmit *Sarocladium oryzae* and *Sarocladium attenuatum* (fungi), the cause of sheath rot disease. Sheath rot disease damages the panicle (branched arrangement of flowers) of the rice plant, which causes the plant to produce under-developed or damaged rice grains. In severe cases, the infected plant may not produce rice grains (Serrano et al. 2014).

Rice bugs are often seen on flowering wild grasses near canals. In Papua New Guinea, the number of *Leptocorisa* was observed to increase enormously when weeds in the adjacent paddies were left unmanaged (Sands 1977).

The bugs are attracted to the rice crop when the plants start flowering. Kainoh et al. (1980) have shown that adult bugs of *L. chinensis* can detect the odour of rice plants when they are close to the rice field. But their aggregation on the panicles seems to be due to the arresting effect of the flowers.

The bug reacts favourably to high humidity and frequent drizzles prevailing from April to June, the active season of the bugs in India (Singh and Chandra 1967). Srivastava and Saxena (1967) reported that it was the light intermittent rain which favoured the build-up of bug populations, while heavy rains had an adverse effect on the pest. In the Philippines, rice bug activity was at its peak during the rainy period in November-December, and dry season crops which attained the milky stage in March-May were less severely attacked (Uichanco 1921).

# 8 References


Aghaee, M. and Godfrey, L.D. 2015. A century of rice water weevil (Coleoptera: Curculionoidae): A history of research and management with an emphasis on the United States. *J. Int. Pest Man.* D1–D14.

Agyen-Sampong, M. and Fannah, S.J. 1980. Dirty panicles and rice yield reduction caused by bugs *Stenocoris* and *Aspavia* in West Africa. *Int. Rice Res. Newsl.* 5(1):11–12.

Akingbohungbe, A.E. 1983. Nomenclatural problems, biology, host plant and possible vector status of Auchenorryncha associated with crop plants in Nigeria. In Knight, W.J., Pant, N.C., Robertson, T.S. and Wilson, M.R. (eds), *Proceedings of the 1st International Workshop on Leafhoppers and Planthoppers of Economic Importance*, pp. 365–70. London: Commonwealth Institute of Entomology.

Akinsola, E.A. 1975. Present status of different rice stem borers in plants of Nigeria. *Rice Entomol. Newsl.* 3.

Akinsola, E.A. 1979. The biology and ecology of rice stem borers in Nigeria. PhD thesis, University of Ibadan, Nigeria.

Alam, M.S., John, V.T. and Zan, K. 1985. Insect pests and diseases of rice in Africa. In: Rice improvement in Eastern, Central and Southern Africa. Manila (Philippines): International Rice Research Institute. pp. 67–82.

Alam, M.S. 1988. Seasonal abundance of rice stem borer species in upland and irrigated rice in Nigeria. Insect Sci. Appl. 9:191–5.

Alam, M.S. 1989. Whitefly (Hemiptera: Aleyrodidae)—a potential pest of rice in West Africa. *Int. Rice Res. Newsl.* 14(3):38–9.

Awoderu, V.A., Alam, M.S., Thottapilly, G. and Alluri, K. 1987. Outbreaks and new records. Ivory Coast. Rice yellow mottle virus in upland rice. *FAO Plant. Prot. Bull.* 35(1):32–3.

Bakker, W. 1971. Three new beetle vectors of rice yellow mottle virus in Kenya. *Neth. J. Plant Pathol.* 77:201–6.

Banerjee, S.N. and Chatterjee, P.B. 1965. On the alternate host plants and hibernation of rice bug, *Leptocorisa acuta* Thunberg in North Bengal. *Sci. Cult.* 31:259–60.

Banwo, O.O., Makundi, R.H., Abdallah, R.S. and Mbapila, J.C. 2001. Newly recorded species of *Chaetocnema*, vector of rice yellow mottle virus in Tanzania. *New Zealand J. Crop Hortic.* 29:61–5.
Behura, S.K., Sahu, S.C., Rajamani, S., Devi, A., Mago, R., Nair, S. and Mohan, M. 1999. Differentiation of Asian rice gall midge, *Orseolia oryzae* (Wood-Mason), biotypes by sequence characterized amplified regions (SCARs). *Insect Mol Biol.* 8(3):391–7.
Bentur, J.S., Pasalu, I.C., Sharma, N.P., Prasada, R.U. and Mishra, B. 2003. Gall midge resistance in rice: Current status in India and future strategies. DRR Research Paper Series 01/2003. Directorate of Rice Research, Rajendranagar, Hyderabad, India.
Brenière, J. 1983. The principal insect pests of rice in West Africa and their control. Monrovia (Liberia): West Africa Development Association.
Cabauatan, P.Q., Cabunagan, R.C. and Il-Ryong, C. 2009. Rice viruses transmitted by the brown planthopper *Nilaparvata lugens* Stål. In Heong, K. L. and Hardy, B. (eds), *Planthoppers: New Threats to the Sustainability of Intensive Rice Production Systems in Asia*, pp. 357–68. Los Baños (Philippines): International Rice Research Institute.
CABI. 2015. *Chilo suppressalis*. Distribution Maps of Plant Pests 1977 No. December pp. Map 254 (Revised). http://www.cabdirect.org/abstracts/20056600254.html;jsessionid=763380931F6628D579D6AA5E0E33DE64.
CABI. 2016a. *Orseolia oryzae* (rice stem gall midge). Invasive Species Compendium. http://www.cabi.org/isc/datasheet/37921.
CABI. 2016b. *Nisia nervosa* (gray planthopper) Invasive Species Compendium. http://www.cabi.org/isc/datasheet/3634.
CABI. 2016c. *Nilaparvata lugens* (brown planthopper). Invasive Species Compendium. http://www.cabi.org/isc/datasheet/36301.
Carbonell, R. M. 1983. Pathogens and nematodes control of rice water weevil in Cuba. *Int. Rice Res. Newsl.* 8(6):16–17.
Cave, G.L. and Smith, C.M. 1983. Number of instars of rice water weevil, *Lissorhoptrus oryzophilus* (Coleoptera: Curculionidae). *Ann. Entomol. Soc. Am.* 76:293–4.
Cheng, C.H. 1977. The occurrence of control of the rice brown planthopper *(Nilaparvata lugens)* in Taiwan. Pacific Science Association Symposium, Denpasar, Indonesia.
Claridge, M.F. and Den Hollander, J. 1980. The 'biotype' of the rice brown planthopper, *Nilaparvata lugens. Entomol. Exp. Appl.* 27:23–30.
COPR (Centre for Overseas Pest Research). 1976. Pest control in rice. PANS Manual No. 3. 2nd edition. London: COPR.
COPR (Centre for Overseas Pest Research). 1978. Pest control in tropical root crops. PANS Manual No. 4. London: COPR.
Dale, D. 1994. Insect pests of the rice plant-their biology and ecology. In Heinrichs, E. A. (ed.), *Biology and Management of Rice Insects,* Wiley Eastern Limited, New Delhi and the International Rice Research Institute, pp. 363–485. Los Baños: Philippines.
Du, P. V., Cabunagan, R.C., Cabauatan, P.Q., Choi, H.S., Choi, I.R., Chien, H.V. and Huan, N.H. 2007. Yellowing syndrome of rice: etiology, current status, and future challenges. *Omonrice* 15:94–101.
Dyck, V.A., Misra, B.C., Alam, A., Chen, C.N., Hsieh, C.Y. and Rejesus, R.S. 1979. Ecology of the brown planthopper in the tropics. In *Brown Planthopper: Threat to Rice Production in Asia.* International Rice Research Institute, pp. 61–98. Los Baños: Philippines.
Dyck, V.A. and Thomas B. 1979. The brown planthopper problem. In *Brown planthopper: Threat to Rice Production in Asia.* International Rice Research Institute. pp. 3–17. Los Baños: Philippines.
Everett, T.R. 1966. Arthropod pests of rice fields in the United States. 11th session of the Working Party on Rice Production and Protection, Lake Charles, La. Intl. Rice Comm., FAO.
FAO. 2016. FAOSTAT. http://faostat.fao.org/site/567/DesktopDefault.aspx?PageID=567#ancor.
Ferreira, E. and Martins, J.F.S. 1984. Insetos prejudiciais ao arroz no Brasil e seu controle. *Embrapa Arroz e Feijão*, p. 67.
Fomba, S.N. 1990. Rice yellow mottle virus (RYMV) on swamp rice in Guinea. *Int. Rice Res. Newsl.* 15(6):21.

Gallagher, K.D., Kenmore, P.E. and Sogawa, K. 1994. Judicial use of insecticides deter planthopper outbreaks and extend life of resistant varieties in Southwest Asian rice. In Denno, R. F. and Perfect, T. (eds), *Planthoppers-Their Ecology And Management*, pp. 599–614. London: Chapman & Hall.

Grigarick, A.A. 1963. Rice plant injury by invertebrate pests. *Calif. Agric.* 17:6–7.

Grist, D.H. and Lever, R.J.A.W. 1969. *Pests of Rice.*. Harlow, London: Longmans: Green and Co Ltd.

Grist, D.H. 1965. *Rice.* 4th edition, Longmans: London.

Gyawali, B.K. 1981. Feeding behavior and damage assessment of rice bug, *Leptocorisa oratorius* (Fabricius) on rice. M.S. thesis, University of the Philippines. Los Banos: Philippines.

Hayashi, H. 1997. Historical changes and control of rice stink bug complex causing the pecky rice. *Plant Prot.* 51:455–61 (in Japanese).

Heinrichs, E.A., Aquino, G.B., Valencia, S.L., de Sagun, S. and Arceo, M.B. 1986. Management of the brown planthopper, *Nilaparvata lugens* (Homoptera: Delphacidae), with early maturing rice cultivars. *Environ. Entomol.* 15:93–5.

Heinrichs, E.A. and Barrion, A.T. 2004. *Rice-Feeding Insects and Selected Natural Enemies in West Africa: Biology, Ecology, Identifcation.* Los Baños (Philippines): International Rice Research Institute and Abidjan (Côte d'Ivoire) and WARDA-The Africa Rice Center, p. 242.

Heinrichs, E.A., Johnson, D.E. and Sy, A.A. 1993. Seedling age and plant spacing effects on insect pests in lowland irrigated rice. In WARDA annual report 1993. Bouaké (Côte d'Ivoire): West Africa Rice Development Association, pp. 43–4.

Heinrichs, E.A. and Medrano, F.G. 1984. *Leersia hexandra,* a weed host of the rice brown planthopper, *Nilaparvata lugens* (Stal). *Crop Prot.* 3:77–85.

Heinrichs, E.A. and Medrano, F.G. 1985. Influence of fertilizer on the population development of brown planthopper (BPH). *Intl. Rice Res. Newsl.* 10(6):20–1.

Heinrichs, E.A. and Mochida, O. 1984. From secondary to major pest status: The case of insecticide-induced rice brown planthopper, *Nilaparvata lugens* resurgence. *Prot. Ecol.* 7:201–18.

Heinrichs, E.A. and Pathak, P.K. 1981. Resistance to the gall midge, *Orseolia oryzae* (Wood-Mason) in rice. *Insect Sci. Appl.* 1:121–32.

Heinrichs, E.A. 1979. Control of leafhopper and planthopper vectors of rice viruses. In Maramorosch, K. and Harris, K.F. (eds), *Leafhopper Vectors of Plant Disease Agents*, pp. 529–60. New York: Academic Press.

Heinrichs, E.A. 1991. Effect of soil moisture on *Trichispa sericea* damage. In WARDA annual report 1991. Bouaké (Côte d'Ivoire): West Africa Rice Development Association, pp. 57–8.

Hely, P.C. 1958. Insect pests of the rice crop. *Agric. Gaz., N. S. W.* 69:29–32.

Heong, K.L. and Schoenly, K. 1998. Impact of insecticides on herbivore-natural enemy communities in tropical ecosystems. In Haskell, P.T. and McEwen, P. (eds), *Ecotoxicology: Pesticides and Beneficial Organisms*, pp. 381–403. London: Chapman and Hall.

Hibino, H. 1979. Rice ragged stunt, a new virus disease occurring in tropical Asia. *Rev. Plant Prot Res.* 12:98–110.

Hibino, H. 1989. Insect-borne viruses in rice. In Harris, K.F. (ed.), *Advances In Disease Vector Research*, pp. 209–41. New York: Springer-Verlag.

Hibino, H. 1996. Biology and epidemiology of rice viruses. *Annu. Rev. Phytopathol.* 34:249–74.

Hidaka, T., Vungsilabutr, P. and Kadkao, S. 1974. Study on ecology and control of the rice gall midge in Thailand. Tech. Bull. No. 6. Tropical Agriculture Research Centre, Tokyo, Japan.

Huang, P.K. and Qi, S.C. 1981. A list of insect pests of rice of Fujian province, China. *Wuyi Sci. J.* 1:79–92.

IRRI (International Rice Research Institute). 1978. *Rice Research and Production in China: An IRRI Team's View.* Los Baños, Philippines: International Rice Research Institute.

IRRI (International Rice Research Institute). 1979. *Brown Planthopper: Threat to Rice Production in Asia.* Manila (Philippines): IRRI.

Islam, Z. and Catling, D. 2012. *Rice Pests of Bangladesh: Their Ecology and Management.* Dhaka, Bangladesh: The University Press Ltd, p. 422.

Israel, P., Vedamoorthy, G. and Rao, Y.S. 1959. Assessment of field losses caused by pests in rice. *Proc.* IRC working party, Ceylon (mimeo.). FAO of U. N. Intl. Rice Comm. Working Party on Rice Production and Protection, 14–19 December 1959.

Israel, P., Vedamoorthy, G. and Rao, Y.S. 1961. Technical report of the Central Rice Research Institute. Entomology Division, pp. 48–51.

Israel, P., Vedamoorthy, G. and Rao, Y.S. 1963. Technical report of the Central Rice Research Institute. Entomology Division, pp. 58–71.

Israel, P. 1959. Latest and effective methods of controlling insect pests on rice. *Proc. Indian Acad. Sci. Sec. B.* 49:363–8.

Kainoh, Y., Shimizu, K., Maru, S. and Tamaki, Y. 1980. Host finding behavior of the rice bug, *Leptocorisa chinensis* Dallas (Hemiptera: Coreidae), with special reference to diet patterns of aggregation and feeding on rice plant. *Appl. Entomol. Zool.* 15:225–33.

Kalode, M.B. 1982. Rice gundhi bug. In *Recent advances in the Knowledge of Insect Pests of Paddy: Distribution, Life History and Control*, pp. 22–7. New Delhi, India: Indian Agricultural Research Institute.

Kalode, M.B. and Kasiviswanathan, P.R. 1976. Change in relative status of insect pests in rice. *Indian J. Plant Prot.* 4:79–91.

Kanno, H. and Sato, A. 1978. Mating behavior of the rice stem borer moth, *Chilo suppressalis* Walker (Lepidoptera: Pyralidae). I. Effects of moth age on mating activity. *Appl. Entomol. Zool.* 13:215–21.

Kanno, H. 1980. Mating behavior of the rice stem borer moth, *Chilo suppresalis* Walker (Lepidoptera: Pyralidae). IV. Threshold light intensity for mating initiation under various temperatures. *Appl. Entomol. Zool.* 15:372–7.

Katanyukul, W., Kadkao, S., Boonkerd, S. and Chandaraprapa, N. 1980a. Rice gall midge outbreaks in Thailand. *Intl. Rice Res. Newsl.* 5(2):13–14.

Katanyukul, W., Kadkao, S. and Boonkerd, S. 1980b. Rice yield losses due to gall midge infestation in Northern Thailand. *Intl. Rice Res. Newsl.* 5(5):10–11.

Kaushik, U.K., Shukla, D., Deshmukh, P.D., Pophali, D.J., Srivastava, S.K. and Gangrade, G.A. 1979. Resistance of paddy varieties to gall midge. *Intl. Rice Res. Newsl.* 4(4):8.

Kenmore, P.E., Carino, F.G., Perez, C.A., Dyck, V.A. and Gutierrez, A.P. 1984. Population regulation of the rice brown planthopper (*Nilaparvata lugens* Stal) within rice fields in the Philippines. *J. Plant Protect. Tropics* 1:19–37.

Khan, M.Q. and Murthy, D.V. 1955. Some notes on the rice gallfly, *Pachydiplosis oryzae* (W.M.). *J. Bombay Nat. Hist. Soc.* 53:97–102.

King, A.B.S. and Saunders, J.L. 1984. *The Invertebrate Pests of Annual Food Crops in Central America. A Guide to their Recognition and Control.* London: Overseas Development Administration.

Kisimoto, R. 1956. Effect of diapause in the fourth larval instar on the determination of wing-form of adult small brown planthopper, *Delphacodes striatella* Fallen. Studies on the polymorphism in the planthoppers (Homoptera, Araeopidae). *Oyo-Kontyu* 12:202–10.

Kisimoto, R. 1965. Studies on polymorphism and its role in the population growth of the brown planthopper, *Nilaparvata lugens* Stal. *Bull. Shikoku Agric. Exp. Stn.* 13:1–106.

Kisimoto, R. 1977. Bionomics forecasting of outbreaks and injury caused by the rice brown planthopper. In *The Rice Brown Planthopper*, pp. 27–41. Taipei, Taiwan: Food and Fertilizer Technology Centre for the Asian and Pacific Region.

Kisimoto, R. 1983. Damage caused by rice stink bugs and their control. *Jpn. Pestic. Inf.* 43:9–13.

Koudamiloro, A., Nwilene, F.E., Togola, A. and Akogbeto, M. 2015. Insect vectors of rice yellow mottle virus. *J. Insects*, Article ID 721751, p. 12. http://dx.doi.org/10.1155/2015/721751.

Kramer, J.P. 1962. New Liberian leafhopper of the genus *Recilia* (Homoptera: Cicadellidae: Deltocephalinae). *Proc. Biol. Soc. Lond.* 75:259–68.

Kudagamage, C. and Nugaliyadde, L. 1981. Gall midge infestation in a monthly-planted trial. *Intl. Rice Res. Newsl.* 6(2):7.

Kusakabe, S. 1979. Dispersal of the brown planthopper, *Nilaparvata lugens* Stal. (Hemiptera: Delphacidae) in relation to its population growth. *Appl. Entomol. Zool.* 14:224–5.

Lei, H.C. and Wang, C.H. 1958. Studies on *Nilaparvata lugens* Stal. in Hunan (in Chinese). *Acta Oec. Entomol. Sin.* 1:283–313.

Lever, R.J.A.W. 1970. Major insects and control. *World Farming* 12:16–25.

Litsinger, J.A., Barrion, A.T. and Soekarna, D. 1987. Upland rice insect pests: their ecology, importance and control. *IRRI Res. Pap. Ser.* 123:41.

Lupi, D., Guidici, M.L., Cenghialta, C., Jucker, C. and Colombo, M. 2012. Japonica cultivars' susceptibility to the rice water weevil *Lissorhoptrus oryzophilus* (Coleoptera: Curculionidae: Brachyceridae). *J. Appl. Entomol.* 137(5):355–64.

Lv, J., Wilson, L.T. and Longnecker, M.T. 2008. Tolerance and compensatory response of rice to sugarcane borer (Lepidoptera: Cambidae) injury. *Environ. Entomol.* 37:796–807.

Mochida, O., Suryana, T. and Sutarli, W. 1979. The effect of the application of nitrogen fertilizer to the rice plant and the irrigation of water on the population growth of the planthopper *Nilaparvata lugens* (Stal) (Hom. Delphacidae). In *Proceedings of the Kongress Entomology*, Jakarta, Indonesia, 9–11 January.

Mochida, O., Tatang, S. and Ayuk, W. 1977. Recent outbreaks of the brown planthopper in southeast Asia (with special reference to Indonesia). In *The Rice Brown Planthopper*, pp. 170–91. Taipei, Taiwan: Food and Fertilizer Technology Centre for the Asian and Pacific Region.

Mochida, O. 1964. On the oviposition in the brown planthopper, *Nilaparvata lugens* (Stal) (Hom., Auchenorrhyncha). *Jpn. J. Appl. Entomol. Zool.* 8:141–8.

Morales, F.J. and Jennings, P.R. 2010. Rice hoja blanca: a complex plant-virus-vector pathosystem. CAB Reviews: Perspectives in Agriculture, Veterinary Science, Nutrition, and Natural Resources 5, No. 043.

Nagata, T. 2002. Monitoring on insecticide resistance of the brown planthopper and the whitebacked planthopper in Asia. *J. Asia-Pacific Entomol.* 5:103–11.

Natarajan, K. and Chandy, K.C. 1978. Screening rice varieties for resistance to gall midge. *Intl. Rice Res. Newsl.* 3(6):10.

Newsom, L.D. and Swanson, M.C. 1962. Treat seed to stop the rice water weevil damage. *La Agri.* 5(2):4–5.

Nozato, K. 1981. Oviposition patterns and some characteristics of newly hatched larvae of three closely related pyralid moths, *Chilo suppressalis, C. lutullus,* and *Ancylolomia japonica* (Lepidoptera: Pyralidae). Kontyu, (Tokyo) 49:702–11.

Nwilene, F.E., Traore, A.K., Asidi, A.N., Sere, Y., Onasanya, A. and Abo, M.E. 2009. New records of insect vectors of rice yellow mottle virus (RYMV) in Côte d'Ivoire, West Africa. *J. Entomol.* 6(4):189–97.

O'Brien, C.W. and Haseeb, M. 2014. Revision of the 'rice water weevil' genus *Lissorhoptrus* LeConte (Coleoptera: Curculionidae) in North America North of Mexico. *The Coleopterists Bulletin* 68(2):163–86.

Okada, M. 1982. Recent expansion of the distribution of the rice water weevil, *Lissorhoptrus oryzophilus* Kuschel [in Japanese]. *Syokukurbutu-Bocki* 36:561–5.

Oman, P.W. 1949. The Nearctic leafhoppers (Hemiptera: Cicadellidae) – A generic classification and check list. *Wash.Entomol. Soc. Mem.* 3:1–253.

Otanes, F.Q. and Sison, P.L. 1947. Pests of rice. *Philipp. J. Agric.* 13:36–88.

Oteng, J.W. and Sant'Anna, R. 1999. Rice production in Africa: current situation and issues. In *Int. Rice Comm. Newsl.* 48. FAO. URL: http://www.fao.org/DOCREP/003/X2243T/x2243t05.htm.

Ou, S.H. 1959. Diseases and insect pests of rice in Taiwan and their control. In Rice Improvement in Taiwan. Plant, Ind. Ser. No. 12, pp. 50–7.

Panda, N. and Mohanty, M. 1970. Biological studies on the paddy gall midge *Pachydiplosis oryzae* (Wood-Mason) Mani in rice and its alternate hosts. *J. Orissa Univ. Agric. Tech.* (India) 2:8–18.

Pantoja, A., Fischer, A., Correa-Victoria, F. and Ramirez, A. 1997. MIP en arroz: Manejo integrado de plagas: artrópodos, enfermidades y malezas. CIAT, p. 147.

Pantoja, A. and Medina-Gaud, S. 1988. *Lissorhoptrus isthmicus* Kuschel (Coleoptera: Curculionidae): a new rice pest for Puerto Rico. *Journal of Agriculture of the University of Puerto Rico* 72(2):333.

Pathak, M.D. and Khush, G.S. 1979. Studies of varietal resistance in rice to the brown planthopper at the International Rice Research Institute. In *Brown Planthopper: Threat to Rice Production in Asia*, pp. 285–301. Los Baños, Philippines: International Rice Research Institute.

Pathak, M.D. 1968. Ecology of common insect pests of rice. *Ann. Rev. Entomol.* 13:257–94.

Pathak, M.D. 1975. *Insect Pests of Rice.* Los Baños, Philippines: The international Rice Research Institute.

Perera, N. and Fernando, H.E. 1970. Infestation of young rice plants by the rice gall midge, *Pachydiplosis oryzae* (Wood- Mason) (Dipt. Cecidomyiidae), with special reference to shoot morphogenesis. *Bull. Entomol. Res.* 59:605–13.

Pieris, I.D.R. 1977. Report of the Entomology Division, Peradeniya, Sri Lanka: Central Agricultural Research Institute, Department of Agriculture.

Prakasa Rao, P.S. 1972. Ecology and control of *Tryporyza incertulas* Walkerand *Pachydiplosis oryzae* Wood-Mason in rice. Ph.D. thesis, Utkal University, Bhubaneswar, India.

Prakasa Rao, P.S. 1975. Ecological studies on rice gall midge. *Intl. Rice Comm. Newsl.* 24:71–3.

Raksarart, P. and Tugwell, P. 1975. Effect of temperature on development of rice water weevil eggs. *Environ. Entomol.* 4:543–4.

Ramaswamy, C.J. and Jatileksono, T. 1996. Inter-country comparison of insect and disease losses, pp. 305–16. In Evenson, R.E., Herdt, R.W. and Hussain, M. (eds), *Rice Research in Asia: Progress and Priorities.* CABI Publisher, Wallingford, UK. p. 432.

Ravelojaona, G. 1970. Observations sur la dynamique des populations de *Trichispa sericea* Guerin, Coléoptère Hispinae nuisible au riz à Madagascar. *ComptesRendus Seances Soc. de Biol. de Madagascar* 164:474–6.

Raymundo, S.A. and Buddenhagen, I.W. 1976. A rice virus disease in West Africa. *Int. Rice Comm. Newsl.* 25(12):58.

Reckhaus, P.M. and Andriamasintseheno, H.F. 1997. Rice yellow mottle virus in Madagascar and its epidemiology in the northwest of the Island. *Z. Pflanzenkr. Pflanzenschutz.* 104:289–95.

Reddy, B.D. 1967. The rice gall midge, *Pachydiplosis oryzae* (Wood-Mason). In *The Major Insect Pests of the Rice Plant*, pp. 457–91. Baltimore, USA: Johns Hopkins Press.

Reissig, W.H., Heinrichs, E.A., Litsinger, J.A., Moody, K., Fiedler, L., Mew, T.W. and Barrion, A.T. 1986. *Illustrated Guide to Integrated Pest Managment of Rice in Tropical Asia.* Los Baños, Laguna, Philippines: International Rice Research Institute, p. 411.

Rizvi, S.M.A. and Singh, H.M. 1980. Gall midge occurrence in Uttar Pradesh, India. *Intl. Rice Res. Newsl.* 5(2):14.

Rosenberg, L.J. and Magor, J.I. 1983. Flight duration of the brown planthopper *Nilaparvata lugens* (Homoptera, Delphacidae). *Ecol. Entomol.* 8:341–50.

Sands, D.P.A. 1977. The Biology and Ecology of *Leptocorisa* (Hemiptera: Alydidae) in Papua New Guinea. Res. Bull. No. 18, Department of Primary Industry, Port Moresby, Papua New Guinea, p. 104.

Saxena, R.C. and Justo, H.D. 1984. Trapping airborne insects aboard interisland ships in the Philippine archipelago, with emphasis on the brown planthopper (BPH). *Intl. Rice Res. Newsl.* 9(5):16.

Schaefer, C.W. and Panizzi, A.R. 2000. *Heteroptera of Economic Importance.* CRC Press, Boca Raton, p. 856.

Serrano, A.C., Russell, F., Mizell, R.F. and Byron, M.A. 2014. Rice Bug (suggested common name) *Leptocorisa acuta* (Thunberg) (Insecta) EENY614. Entomology and Nematology Department, UF/IFAS Extension, Gainesville, FL 32611. http://edis.ifas.ufl.edu/pdffiles/IN/IN106700.pdf.

Shepard, B.M., Barrion, A.T. and Litsinger, J.A. 1995. *Rice-Feeding Insects of Tropical Asia.* Manila (Philippines): International Rice Research Institute.

Shimohata, T. and Kanno, M. 1982. The effect of the growth stages of rice seedlings on the oviposition and feeding of adult females of the rice water weevil (in Japanese). *Proc. Kansai Plant Prot. Soc.* 24:43.

Siddiq, E.A. 1991. Genes and rice improvement. *Oryza* 28:1–7.

Singh, M.P. and Chandra, S. 1967. Population dynamics of rice stink bug (*Leptocorisa varicornis* Fabr.) in relation to weather factors. *Indian J. Agric. Sci.* 37:112–19.

Sogawa, K. 1971. Effects of feeding of the brown planthopper on the components in the leaf blade of rice plant (in Japanese). *Jpn. J. Appl. Entomol. Zool.* 15:175–9.

Sogawa, K. 1992. A change in biotype property of brown planthopper populations immigrating into Japan and their probable source areas. *Proc. Assoc. Plant Prot. Kyushu* 38:63–8.

Srivastava, A. S. and Saxena, H. P. 1967. Rice bug *Leptocorisa varicornis* Fabricius and allied species. In *The Major Insect Pests of the Rice Plant*. Baltimore, USA: Johns Hopkins Press, pp. 525–48.

Stout, M.J., Hamm, J.C., Abbe, I. and Bergeron, C. 2013. The influence of plant age on susceptibility to the rice water weevil, *Lissorhoptrus oryzophilus*. *J. Appl. Entomol.* 137:241–8.

Stout, M.J., Riggio, M.R., Zou, L. and Roberts, R. 2002. Flooding influences ovipositional and feeding behavior of the rice water weevil (Coleoptera: Curculionidae). *J. Econ. Entomol.* 95:715–21.

Subramanian, C.K. 1935. Report of the section of Entomology for the year 1933–4. Department of Agriculture, Mysore, India, pp. 23–7.

Suenaga, H. 1963. Analytical studies on the ecology of two species of planthoppers, the whitebacked planthopper (*Sogata furcifera* Horvath) and the brown planthopper (*Nilaparvata lugens* Stal), with special reference to their outbreaks (in Japanese). *Bull. Kyushu Agric. Exp. Stn.* 3:1–52.

Tindall, K.V. and Stout, M.J. 2003. Use of common weeds of rice as hosts for the rice water weevil (Coleoptera: Curculionidae). *Environ. Entomol.* 32:1227–33.

Togola, A., Nwilene, F.E., Koné, B. and Chougourou, D. 2011. Sédentarisation des populations des mouches Diopsides dans les agro-systèmes rizicoles au Bénin. *Tropicultura* 29(2):101–6. ISSN: 0771–3312.

Tsuzuki, H., Asayama, T. and Takimoto, M. 1982. Studies on the rice water weevil. XII. The effect of daylength on the oviposition on newly emerged adult females (in Japanese). *Proc. Kansai Plant Prof. Soc.* 24:62.

Tsuzuki, H., Asayama, O. K. and Yamada, T. 1979. The expansion of the occurring areas of the rice water weevil in Aiti Prefecture (in Japanese). *Proc. Kansai Plant Prot. Soc.* 21:49.

Tu, Cheng-wen. 1979. The brown planthopper and its control in China. International Rice Research Workshop (IRTP and China collaboration), Guangzhou, PROC, 22–26 October 1979.

Uichanco, L. B. 1921. The rice bug, *Leptocorisa acuta* Thunberg in the Philippines. *Philipp. Agric. Rev.* 14:87–125.

Venugopala, R. A. 1975. Incidence of the gall midge in deepwater rice. *Rice Entomol.* 3:36.

Wickremasinghe, N. 1969. Observations on the rice gall midge and on its effect on rice plant. *Intl. Rice Comm Newsl.* 18(4):27–32.

Wilson, M.R. and Claridge, M.F. 1991. *Handbook for the Identification of Leafhoppers and Planthoppers of Rice*. Wallingford (UK): Centre for Agriculture and Biosciences International.

Wongsiri, T., Vungsilabutr, P. and Hidaka, T. 1971. Study on ecology of the rice gall midge in Thailand. *Trop. Agric.* 19–24 July 1971, pp. 267–90.

Yasuda, H., Shimohata, T., Umlbuchi, T., Asano, T., Hirose, O. and Yatsuhiro, O. 1979. The occurrence of the rice water weevil in late transplanted rice crop in Gihu Prefecture (in Japanese). *Proc. Kansai Plant Prot. Soc.* 21:52.

Zakra, A.N., Pomier, M. and de Taffin, G. 1986. Initial results of an intercropping experiment of coconut with food crops in the Middle Côte d'Ivoire. *Oléagineux* 41:381–9.

Zongo, J.O. 1993. Importance des hispides dans les plaines rizicoles de Karfiguela et Tingrela, Banfora, Burkina Faso. Sahel PV Info. no. 50:18–20.

Zou, L., Stout, M.J. and Ring, D.R. 2004a. Degree-day models for emergence and development of the rice water weevil (Coleoptera: Curculionidae) in Southwestern Louisiana. *Environ. Entomol.* 33(6):1541–8.

## Chapter 10

# Management of rice insect pests

*E. A. Heinrichs, University of Nebraska-Lincoln, USA; F. Nwilene, The Africa Rice Center, Nigeria; M. Stout, Louisiana State University, USA; B. Hadi, International Rice Research Institute (IRRI), The Philippines; and T. Freitas, Universidade Federal Rio Grande do Sul, Brazil*

1 Introduction
2 Overview of concepts and options for integrated pest management (IPM) of rice
3 Cultural practices in rice IPM
4 Promoting natural enemies of rice pests: conservation biological control
5 Augmentative biological control
6 Selective use of insecticides
7 Dissemination mechanisms for rice IPM
8 References

## 1 Introduction

Rice, the daily food of nearly half the world's population, is the foundation of national stability and economic growth in many developing countries. It is the source of one quarter of global food energy and – for the world's poor – the largest food source. It is also the single largest use of land for producing food and the biggest employer and income generator for rural people in the developing world. Rice production has been described as the single most important economic activity on Earth. Because rice occupies approximately 9% of the planet's arable land, it is also a key area of concern – and of opportunity – in environmental protection.

To feed the growing world population, rice production must be increased. However, rice farmers face many abiotic and biotic constraints in their quest to increase rice production. In conjunction with the introduction of new high-yielding drought-and–flood-tolerant rice varieties, increasing yields will require a reduction in losses to insects and other stresses. As cropping intensity and cultural practices are changed to meet production needs, pest pressure is expected to intensify.

The rice plant is an ideal host for many insect species. All of the plant parts are vulnerable to insect-feeding, from the time of sowing till harvest. More than 800 insect species damage rice, although the majority of them do very little damage. In tropical Asia, only about 20 species are of major importance and of regular occurrence (Grist and Lever 1969). In Africa, 15 species of insects are considered major pests of rice (Oteng and

http://dx.doi.org/10.19103/AS.2016.0003.37

Sant'Anna 1999), and in the Americas, about 20 species are considered major pests (Stout, pers. comm.).

To develop effective pest management strategies, it is essential to properly identify and understand the biology and ecology of insect pests and the arthropods that help regulate their populations. Chapter 21a effectively utilizes the unique knowledge and expertise of leading rice entomologists from Africa, Asia and the Americas to provide the first global coverage of rice insect pests. This chapter discusses the current strategies to manage rice insect pests in an environmentally sustainable manner.

# 2 Overview of concepts and options for integrated pest management (IPM) of rice

Pre-harvest yield losses due to rice insect pests can be significant. This is especially so under pest outbreak scenarios. Irrigated rice has been shown to be an agroecosystem with rich biodiversity and a redundant food chain (Schoenly et al. 1996, Wilby et al. 2006). Insecticide abuse negatively affects the natural enemy communities, shortens the food chain length and increases the propensity for pest outbreaks in irrigated rice ecosystems (Heong and Schoenly 1998, Settle et al. 1996, Way and Heong 1994). An alternative framework for insect pest management is needed to address the pest vulnerability of rice production, without introducing a higher risk of pest outbreaks.

Wyss et al. (2005) and Zehnder et al. (2006) proposed a conceptual framework for arthropod pest management in organic crop production (Fig. 1). In this framework, arthropod pest management strategies are classified into four 'phases'. The framework prioritizes pest management options that will prevent damaging levels of pests (phase 1 and 2) and will minimize the need for curative actions (phase 3 and 4). We will adopt and modify this conceptual framework to structure the discussion on various rice insect pest management options in rice ecosystems.

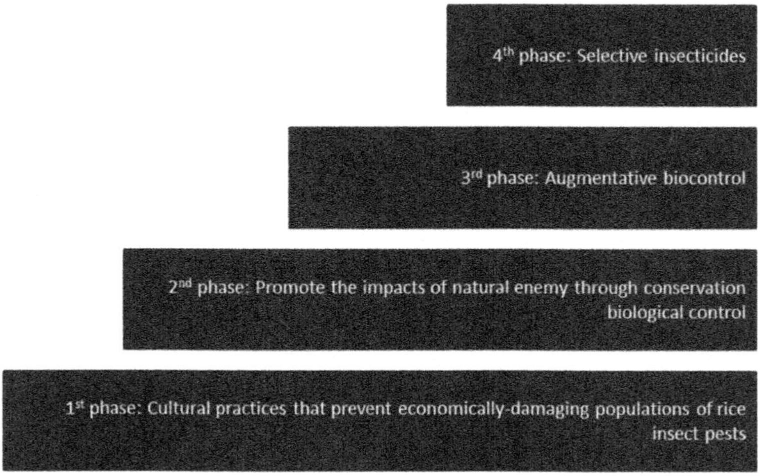


**Figure 1** Conceptual framework for rice IPM (adapted from Zehnder et al. 2006).

# 3 Cultural practices in rice IPM

Generally, two types of practices are found under this phase: (1) cultural practices that directly target insect pest populations (primary practices) and (2) crop husbandry practices with secondary effects on pest populations (secondary practices). Practices that directly target insect pest populations probably form the earliest methods of pest management in rice and include mechanical and physical removal of insect pest populations. These practices are typically labour intensive. In terms of the optimum spatial scale, these cultural practices can be further classified into two: (1) those that operate in single-field scale and (2) those that operate in community scale (i.e. covers multiple fields within a locale). Litsinger (1994) catalogued a number of cultural practices that prevent the development of economically damaging populations of rice pests, and we classify these practices according to their principal intent (e.g. primarily and secondary practices) and optimum spatial scale in Table 1.

**Table 1** Examples of cultural practices that prevent the development of economically damaging populations of rice insect pests (from Litsinger 1994)

| | Single-field scale | Community scale |
|---|---|---|
| Primary practices | • Deployment of resistant rice varieties against brown planthopper in Asia (Cohen et al. 1997) and rice water weevil in North America (e.g. Stout et al. 2001) | • Insect light traps using blue light were used to control stemborer moths, Chilo suppressalis, in paddy fields across Japan during WW II and the post-war years (Shimoda and Honda 2013) |
| Secondary practices | • Fertilizer management affects hopper and stemborer populations in Asia (e.g. Lu et al. 2004, Hou and Han 2010)<br>• Delayed planting date to avoid population peak of small rice stink bug in South America (Albuquerque 1993).<br>• Water and planting date management affects rice water weevil population in North America (e.g. Morgan et al. 1989, Stout et al. 2002, Thompson et al. 1994) | • Synchronous planting affects the incidence of tungro, a viral disease vectored by green leafhopper (Cabunagan et al. 2001) |

Of the cultural practices listed above, deployment of host plant resistance (a primary cultural practice) and fertilizer management (a secondary cultural practice) form basic building blocks for developing a rice ecosystem with reduced vulnerability against insect pests.

## 3.1 Host plant resistance: insect adaptation and deployment strategies

Heinrichs et al. (1985) documented standard methods for screening for resistance to multiple insect pests of rice. Much work has been done to identify resistance genes, especially against hemipteran pests, and breed them into elite varieties (Fujita et al. 2013, Jena and Kim 2010). Sources of resistance have been identified for most of the rice insect pests (Heinrichs et al. 1985).

Durability is the primary challenge in the utilization host plant resistance as a tool to manage rice insect pests. Rice pest populations showed rapid adaptation against the deployed resistant varieties. This is especially well documented in rice – brown planthopper systems (e.g. Alam and Cohen 1998).

Mundt (2014) listed several general deployment strategies to improve the durability of resistance genes against pests and diseases: gene rotation, field mixture of resistance genes and gene pyramiding. Gene rotation involves a regional deployment of a resistance gene, monitoring of pest virulence/adaptation to the deployed gene within the region and rotation to another resistance gene upon detection of pest adaptation. There is evidence that this deployment scheme is feasible in prolonging durability against rice blast disease (Crill et al. 1981). The challenge lies in the implementation of regional monitoring schemes to detect the target pest's adaptation to the deployed gene. Such regional monitoring schemes are likely to be expensive. Furthermore, in the event of detected adaptation, it may prove difficult to choose alternative genes to be deployed since cross adaptation may occur across several resistance genes.

While gene rotation subjects the same cropping space to two or more resistance genes over time, field deployment of a mixture of varieties essentially subjects various parts of the cropping space to two or more varieties with different levels of resistance in a given period of time. Large-scale trials in China showed that a mixture of resistant and susceptible varieties suppressed rice blast severity and reduced the need for fungicide application (Zhu et al. 2000). However, the report did not monitor the effect of the strategy in prolonging the durability of resistance against rice blast. Concern has been raised that this particular deployment strategy may select for pathogen race or pest populations that are adapted to multiple resistance genes (Mundt 2014).

A number of studies reported that pyramiding two or more resistance loci resulted in higher resistance compared to monogenic lines, as measured by the target pest's feeding, settling, survivorship, population growth and damage on host plants (e.g. Fujita et al. 2010, Hu et al. 2012, Qiu et al. 2012). However, Sharma et al. (2004) showed that a pyramided line (incorporating Bph1 and bph2) had the same resistance level to introgression lines carrying Bph1 alone, as measured by the effects of brown planthopper infestation on plant heights. None of these studies measured the effects of pyramiding resistance genes to the durability of the resistance.

Gene function and resistance mechanisms associated with identified resistance genes against rice pests are not fully understood (Fujita et al. 2013). Moreover, the mechanisms and the genetic/physiological basis for pest adaptation to resistant rice varieties are not always clear. Without this information, it is difficult to predict how the different deployment strategies will delay pest adaptation towards resistant rice varieties. Thus, a considerable amount of research is needed to evaluate the effects of deployment strategies described above to the durability of resistance genes against insect pests.

## 3.2 Fertilizer management

In general, fertilizer management affects rice insect pest populations by modifying rice plants' suitability for and attractiveness towards herbivorous insects (Ge et al. 2013, De Kraker et al. 2000, Chantaprapha et al. 1980, Swaminathan et al. 1985, Lu et al. 2004, Jian and Cheng 2003, Hou and Han 2010, Ranganathan et al. 2006, Hosseini et al. 2012, Djamin and Pathak 1967).

Increased nitrogen fertilizer rate was associated with a longer larval development duration and higher oviposition rate of the leaffolder (*C. medinalis*) (Ge et al. 2013). Leaffolder female moths preferred rice plants with higher fertilizer rates to lay their eggs compared to plants with lower fertilizer rates (De Kraker et al. 2000). A small-scale field experiment confirmed that the number of leaffolder eggs found on plots with high fertilizer rate (~150 kg/Ha) was higher than those found on plots with lower fertilizer rates (~0 and 75 kg/Ha). Furthermore, the ratio between the natural enemy and leaffolder larval abundance was correlated negatively with nitrogen fertilizer rate and, consequently, the survivorship of leaffolder larvae was the highest in plots receiving high amounts of nitrogen fertilizer (De Kraker et al.. 2000). Other field studies confirmed that higher severity of leaffolder injury was typically observed on plots with high nitrogen fertilizer rates ( $\geq$ 150 kg/Ha) (Chantaraprapha et al. 1980, Swaminathan et al. 1985).

Lu et al. (2004) reported a positive correlation between rice leaf nitrogen content and brown planthopper nymphal survivorship, longevity of adult females, female fecundity and egg hatching rate. Additionally, higher leaf nitrogen content was associated with faster nymphal development of brown planthopper. Taken together, these findings indicate that excessive nitrogen fertilization rate may lead to faster population development and higher maximum population levels of brown planthopper. Nitrogen fertilizer regimes may affect brown planthopper population's ability to grow on resistant or tolerant varieties. Heinrichs and Medrano (1985) recorded a positive correlation between nitrogen fertilizer rate and the number of brown planthopper populations developing on Triveni, a tolerant rice variety with some antibiosis activity, and IR60, a resistant variety with high antibiosis.

Nitrogen fertilizer regime affected rice plant's host suitability for stemborer and the plant's ability to compensate for tiller loss. In a screenhouse experiment, Jiang and Cheng (2003) reported that striped stemborer's larval weight gain and development were positively correlated with nitrogen fertilizer rates within a certain range (200–600 mg N/pot) and started to decrease with additional nitrogen fertilizer. Moreover, nitrogen, as a plant nutritional factor, contributes to the ability of rice to produce new tillers. Therefore the plants compensated for striped stemborer injury between the low and medium range of nitrogen fertilizer rate (200–400 mg N/pot) but not under a high or excessive range of nitrogen rate (600–800 mg N/pot) (Jiang and Cheng 2003).

Silicon amendment was associated with a longer time needed for striped stemborer's penetration into rice stalk, lower larval weight gain and lesser extent of damage (Hou and Han 2010). These effects were even more pronounced in the tested susceptible variety than the resistant variety. In greenhouse experiments, silicon amendment was negatively correlated with whitehead incidence caused by yellow and striped stemborers

(Ranganathan et al. 2006, Hosseini et al. 2012). Djamin and Pathak (1967) posit that varietal difference in accumulating silica in the leaves may affect the variety's suitability for stemborer larvae and partially contribute to their resistance against stemborer. Indeed, they observed a negative correlation between leaf silica content and dead heart incidence caused by striped stemborer. Through an observation of striped stemborer larval mandibles, Djamin and Pathak (1967) further inferred that high leaf silica content decreased stem-borer-related damage by mechanically grinding the incisor region of larval mandibles.

The form of fertilizer may have an impact on rice herbivorous insects. White-backed planthoppers feeding on rice with poultry-manure fertilizer had lower nymphal survivorship, reproductive rate and egg hatching rate compared to those feeding on plants with inorganic fertilizer (Kajimura 1995a,b). The same effect of lower nymphal survivorship on organically fertilized rice plants was observed with brown planthopper (Kajimura 1995a). Organically fertilized plants had lower nitrogen content compared to those with inorganic fertilizer. Specifically, the leaves of organically fertilized plants contained a significantly lower amount of the amino acid asparagine compared to those with inorganic fertilizer. Asparagine acts as a feeding stimulant for brown planthopper (Sogawa 1982) and its absence or low titre has been associated with planthopper resistance in various indica and japonica rice varieties (Sogawa and Pathak 1970, Shigematsu et al. 1982).

Apart from its impact on host plant suitability and attractiveness towards herbivorous insects, fertilizer may have effects on the natural enemy communities, exerting regulation function on pest populations. Jiang and Cheng (2004) reported higher abundance of collembolans in organically manured rice plots compared to those with inorganic fertilizer. While no statistical difference was observed in arthropod predator abundance between the two rice plot types, significantly higher egg predation rate of whitebacked planthopper was observed in organically manured rice plots.

# 4 Promoting natural enemies of rice pests: conservation biological control

## 4.1 Natural pest regulation in irrigated rice ecosystems

Among cultivated plants, irrigated rice constitutes a complex ecosystem with a relatively high species diversity and redundant food web (Schoenly et al. 1996, Wilby et al. 2006). Various life stages of herbivorous insects in rice ecosystems are preyed upon and parasitized by predators and parasitoids from a number of classes and orders, including araneae (Preap et al. 2001, Sigsgaard 2000), orthopterans (Rubia et al. 1990), coleopterans (de Kraker et al. 2000), aquatic and terrestrial heteropterans (Nakasuji and Dyck 1964, Reyes and Gabriel 1975), hymenopterans, strepsipterans and dipterans (Gurr et al. 2011).

Settle et al. (1996) showed that across irrigated rice production fields in Java, generalist predators (e.g. spiders, mirid bug) were always abundant early in the cropping season. An increase in the organic matter before the beginning of the season resulted in the increased decomposer abundance and led to even higher levels of predator populations. These correlations indicated that predators in rice ecosystems utilize decomposer communities as alternative prey before the arrival of phytophagous insects (Settle et al. 1996). The widely practised staggered planting in intensive rice production areas may further ensure the continuous availability of prey and hosts in rice fields and contribute to the abundance of the natural enemy populations in this ecosystem (Way and Heong 1994).

Conservation biological control in rice ecosystems capitalizes on and enhances the natural pest regulation provided by these prevalent the natural enemy communities in rice fields. Annual monoculture cropping systems are often associated with high disturbance regimes on the natural habitat for the natural enemies. High usage of pesticides and potential lack of adult food source and shelter are among the disturbances affecting the natural enemy populations in these cropping systems (Landis et al. 2000). Conservation biological control aims to remedy this situation by limiting insecticide use, promoting selective insecticides and altering crop habitats to allow better support for the natural enemy populations.

## 4.2 The effects of insecticide application on the natural enemy and pest populations

In a descriptive work on insect population dynamics as affected by early insecticide application (20–50 days after transplanting), Schoenly et al. (1996) found a stark difference in pest and natural enemy populations between sprayed and unsprayed fields. By mid-season, natural enemy abundance was higher in the unsprayed field compared to that in the sprayed field. On the other hand, herbivorous insect populations were higher by mid-season on the field receiving three early insecticide applications. Indeed when food webs were constructed for the two field types, the authors found that the mean length of the food chain was shorter in fields receiving early insecticide application. This food chain shortening continued for about 40 days after the first spray. The abundance of the natural enemy population eventually rose in the field receiving early insecticide application, yet the damage due to increased herbivore abundance during the mid-season might have already been done.

At least two mechanisms contribute to this dramatic increase in pest population due to early insecticide applications. Schoenly et al. (1996) linked a reduction in the natural enemy population immediately after insecticide application to a quick rise of herbivore abundance. The rise of herbivore population has indeed been attributed to the temporal release of herbivores from the natural enemies following early insecticide applications (Heinrichs and Mochida 1984, Fagan et al. 1998). Chelliah et al. (1980) and Chelliah and Heinrichs (1980) reported an increase in the reproductive rate of brown planthopper in response to a sub-lethal dose of insecticide application. This phenomenon in which a sub-lethal dose of select insecticide increases ecological fitness parameters of certain insects, termed hormesis, has received further attention recently with works published on different rice pest species using various active ingredients (Azzam et al. 2009, Bao et al. 2008, Hu et al. 2010, Ling et al. 2009, Ling et al. 2011, Suri and Singh 2011). Stimulation of ecological fitness parameters by a sub-lethal dose of insecticide forms the second mechanism explaining the rise in pest population as affected by insecticide application (Heinrichs et al. 1982).

What factors drive the practice of early insecticide application? In the Philippines, a survey of farmers' insecticide decision protocols revealed that a significant portion of early insecticide application coincided with fertilizer application, in part to provide an easy-to-remember guide to time a prophylactic application (Bandong et al. 2002). Heong et al. (1994) reported that early insecticide application is often driven by farmers' perception of defoliators' ability to affect yield. Indeed, defoliator pests which produce conspicuous injuries such as leaffolders often infest rice fields early in the season. Yet, empirical and simulation studies showed that early stage defoliation by leaf folder only resulted in the rare occurrence of minor yield loss (Graf et al. 1992, Heong 1994).

Consequently, much of the early season application of insecticides was based on the exaggerated estimation of plant damage due to leaffolder and was likely to be uneconomical (Heong et al. 1994). These observations led to the formulation of a simple rule of thumb: 'Insecticide spraying for leaf folder control in the first 40 days after sowing is not needed' (Heong et al. 1998). This simple message, supported by public media and an entertainment education campaign, has been shown to successfully change farmers' practice in the Mekong Delta, Vietnam (Huan et al. 1999, Heong et al. 2008).

A number of caveats need to be taken to complement this simple message. While irrigated rice may be an inherently resilient ecosystem against pests, it is possible that insect pest populations occasionally reach damaging levels early in the season, perhaps due to a particularly high immigration rate or a high degree of planting asynchrony. Unfortunately, economic injury levels for early rice stages that take into account the natural enemies' capacity to provide biological control are not available and this lack of knowledge hampers our ability to make judicious pest management decisions. More research is needed to identify these economic injury levels.

The effects of different active ingredients on rice natural enemies vary widely (Fabellar and Heinrichs 1986, Preetha et al. 2009, Preetha et al. 2010, Takashi and Kiritani 1973, Tanaka et al. 2000, Wang et al. 2008). 'Reduced risk' insecticides, due to their selectivity against the natural enemies and other non-target organisms, have recently been developed. These newer insecticides are not commonly used in rice production, partly due to their relatively high costs (Norton et al. 2010). Careful research needs to be conducted to test the effects of these reduced risk insecticides on field populations of natural enemies when applied early in the season. Information from these studies should then be used to evaluate whether these 'reduced risk' insecticides may become options for rare occasions of early insecticide application as dictated by pest population levels.

## 4.3 Habitat management for natural enemy conservation

Various habitat management schemes within agricultural fields, for example by intercropping and the introduction of wildflower strips or plants repellent to herbivores, were shown to significantly suppress herbivore populations, enhance the natural enemies and reduce crop damage (Letorneau et al. 2011). While the limitation on early insecticide application enhances rice natural enemies by reducing the known effect of a disturbance regime (insecticide), habitat management schemes typically provide resources (e.g. alternative or complementary food sources, shelter) to the natural enemy species.

In the Asian rice ecosystem, much effort has been invested in designing a habitat management scheme in which flowering plants are introduced on rice bunds to conserve the natural enemies (Gurr et al. 2011, 2012, 2016, Zhu et al. 2013, 2014, 2015). The introduction of flowering plants seemed to have a primary benefit of food provisioning (e.g. pollen, floral and extrafloral nectar) to members of the natural enemy community (Lu et al. 2014). Indeed many predators and parasitoids are known to utilize non-prey food sources, usually to satisfy their need for metabolic energy (Lundgren 2009). Laboratory studies showed that, compared to a water control, the presence of sesame flower (*Sesamum indicum*) as a food source increased the adult longevity of planthopper parasitoids (*Anagrus optabilis* and *Anagrus nilaparvatae*) (Zhu et al. 2015) and stemborer parasitoids (e.g. *Apanteles ruficrus, Cotesia chilonis* and *Trichogramma chilonis*) (Zhu et al. 2015). A similar laboratory study showed that the adult longevity of a predatory mirid bug, *Cyrtorhinus lividipennis*, was significantly increased in the presence of sesame as well as other flowers (e.g. *Tagetes erecta, Trida procumbens*, and *Emilia sonchifolia*) compared to water or nil controls (Zhu et al. 2014). Furthermore, these controlled experiments showed that the presence of sesame flowers increased the parasitism and predation rates of the natural enemies compared to water or nil controls (Zhu et al. 2013, 2014).

These benefits, observed in controlled experiments, may not translate into the enhancement of the natural enemies at the field level. There are a number of further considerations to be taken (Zhu et al. 2015, Lu et al. 2014): (1) the plant selected for a habitat management scheme will have to provide a resource *not already available in the field*, (2) the scheme will have to allow for a back-and-forth movement of the natural enemies between the flowering plants and the major crop, (3) the availability of food resource provided by the scheme will have to start as early as possible and to last as long as possible across the season to maximize the benefits for the natural enemies, (4) selection of plants to be included in the habitat management scheme must also consider potential adverse effects of the plant's introduction; for example, some plants may act as an alternative food source for pests as well as the natural enemies.

Yet, in the case of sesame flowers the natural enemy benefits reported in laboratory studies were reproduced at the field scale. In a factorial study with combinations of sesame borders and pesticide applications, fields with sesame borders and no pesticide applications had the highest predator and parasitoid abundances compared to all other combinations (e.g. no sesame border and no pesticide application, no sesame border with pesticide application, with sesame border and with no pesticide application) (Gurr et al. 2016). Indeed, the same report contains a multi-country comparison where rice fields with nectar-bearing flowers on their borders that had significantly lower pest populations, received less insecticide applications (as guided by farmers' decision), and produced higher yields compared to control fields (Gurr et al. 2016).

# 5 Augmentative biological control

As mentioned above, while the irrigated rice ecosystem constitutes a redundant food web it is plausible for the insect pest populations to occasionally reach economically damaging levels. Then, curative actions are needed and, whenever available, augmentative biological control should form the first curative option. In contrast to conservation biological control, the aim of which is to improve the impact of the indigenous natural enemy community, augmentative biological control typically aims for a limited-time increase in specific natural enemy species to temporally suppress pest populations and activities and avoid economic damage in yield. Augmentative biological control is usually achieved by field releases of predators, parasitoids or pathogens with expected effects to last either season-long (inoculative biological control) or within a short period of time in the season (innundative biological control).

Rombach et al. (1987) provided a bibliography on pathogens of rice insect pests from 1960 to 1985 and listed members of all fungal major groups, bacteria, viruses and nematodes as potential biological control agents of rice insect pests. Since then, a number of studies have continued to show the efficacy of primarily two entomopathogenic fungal genera, *Metarhizium* and *Beauveria*, on rice pests (Aguda et al. 1987, Jin et al. 2008, 2010, Rombach et al. 1987, Li et al. 2012a,b). Additionally, in a laboratory experiment using an artificial diet, toxins produced by *Bacillus thuringiensis* (Bt) have been shown to inhibit feeding by larvae of the Asiatic stemborer (Rombach et al. 1989).

One of the criticisms against the use of entomopathogens as augmentative biocontrol agents in rice production is the relatively long period it takes for control and the limited efficacy at field scale (Way 2003). Rombach et al. (1994), for example, recorded very low level of mortality after field application of five entomopathogenic species (including *Beauveria bassiana* and *Metarhizium anisopliae*) seven days after treatment (0–4% mortality rate). Yet the mortality rate reached 70–100% at 21 days after treatment. In an evaluation of *Metarhizium faloviridae* against a rice grasshopper, *Hieroglyphus daganensis*, Thomas et al. (1998) found that survivorship of a field population was ~40% at 10 days after application but the mortality rate dropped significantly afterward.

Further development of biological control agents will have to address these criticisms by identifying highly virulent fungal strains, optimizing formulation and, perhaps, considering a mixture of augmentative biocontrol with other control options. There is an obvious variability in biocontrol efficacy among strains of the same fungal species (e.g. Toledo et al. 2008, Jin et al. 2008). An evaluation of 35 global strains of a single species, *Metarhizium anisopliae*, for example, yielded nymphal mortality rates ranging from 6.5 to 64.2% (Jin et al. 2008). The density of conidia needed to kill 50% of a given target pest population (LC50) also varies with the formulation. When prepared as an emulsifiable formulation, the LC50 of a virulent strain of *Beauveria bassiana* was reduced by 38% compared to the unformulated conidial solution (Feng and Pu 2005). Finally, combining *M. anisopliae* application with low-rate buprofezin, an insect growth regulator, increased the brown planthopper control rate from ~60 to ~80% (Jin et al. 2010).

There are very few reports on the effects of augmentative biological control agents on rice natural enemies. Rombach et al. (1994) noted that while some entomopathogenic fungi infect beneficial insects, natural infection was always observed at low levels. Schoenly et al. (2003) reported no effect of Bt-spray on rice natural enemy's diversity and abundance at the field scale. Scant published data signals a need to conduct thorough studies on the effects of augmentative biocontrol on the populations of indigenous beneficial insects.

# 6 Selective use of insecticides

Insecticides should be regarded as the last resort in the arsenal of IPM options. There are a number of considerations to guide insecticide application in rice ecosystems: application timing, selectivity of active ingredients and resistance among pest populations associated with insecticide overuse. For the insecticide application to be profitable, application decision must be guided by an estimate of economic loss that may be incurred if no curative action is taken. This is usually done by introducing an economic threshold value for pest populations to guide decisions. Furthermore, unfortunate timing of insecticide application may prove harmful to the indigenous natural enemy community and its attendant pest regulation function (Schoenly et al. 1996). Some active ingredients were shown to be more toxic to the natural enemy than to the pest species (e.g. Fabellar and Heinrichs 1986). Finally, overuse of insecticides may lead to the development of insecticide resistance among insect pest populations (e.g. Su et al. 2013, Matsumura and Sanada-Morimura 2010).

## 6.1 Economic action thresholds

Economic injury level, defined as 'the lowest population density that will cause economic damage', is a keystone of IPM theory (Stern et al. 1959). The economic injury level relates the field population of insect pests to pest injuries and pest injuries to economic damage (e.g. yield loss). An economic action threshold to guide the timing of insecticide application is formulated based on the economic injury level.

A number of economic injury levels and economic action thresholds against individual rice pests have been developed and tested in Asia (e.g. Bautista et al. 1984, Litsinger et al. 2005, Litsinger et al. 2006a,b,c) and North America (e.g. Morgan et al. 1989). In large-scale testing against rice bugs and plant- and leafhoppers in the Philippines, Litsinger et al. (2005) reported that while implementation of action thresholds typically resulted in significantly higher rice yield compared to untreated controls (by about 0.3–0.5 ton/ha), they are not always profitable to farmers. A number of factors may contribute to this result:

- While the development of an economic injury level typically takes into account the commodity market value and the cost of management action, in practice the economic injury level is not adjusted to account for temporal fluctuation and spatial variation in rice price. This means that, at times, the action threshold will overestimate or underestimate the actual economic injury levels for a given season/location.
- Most of the economic injury levels on rice pests do not take into the natural pest regulation potential of the indigenous natural enemies. The only exception where the natural enemy population was somehow accounted for (by reducing the number of pest individuals by a certain factor for each natural enemy individual found) was the action threshold tested for brown planthopper (Litsinger et al. 2005). In a large-scale test, the population levels of brown planthopper never crossed this threshold. By not accounting for the natural enemy population in formulating action thresholds, we risk affecting the natural enemy community and pest regulation function it provides, which may lead to pest resurgence and total net loss for the farmers (as discussed under 'The effects of insecticide application on the natural enemy and pest populations').
- Most of the economic injury levels on rice were developed on a limited number of susceptible varieties (e.g. Bautista et al. 1984). The relationship between insect pest population, insect injuries and yield loss differs markedly between resistant and susceptible rice varieties. By basing the economic injury levels on susceptible varieties, we risk overestimating the impact of insect pest populations for rice landscapes that incorporate resistant and tolerant varieties.

Due to the limitations above, the practicality of using action thresholds in rice ecosystems, especially among smallholder farmers, has been questioned (Matteson 2000). Since the available economic threshold values do not typically take into account the local and temporal specifics (in terms of price variability, the natural enemy populations and varieties planted, for example), efforts to use economic thresholds may backfire and trigger unnecessary application of insecticides.

## 6.2 Selectivity against the natural enemies

Studies have shown that insecticidal active ingredients vary in their selectivity against the natural enemies in different Asian countries. In the Philippines, Fabellar and Heinrichs (1986) showed that deltamethrin, chlorpyrifos and endosulfan are more toxic to *Pardosa pseudoannulata*, a common predatory spider, compared to the brown planthopper. Interestingly, one of these active ingredients, deltamethrin, is much more toxic to green leafhopper compared to *P. pseudoannulata*. Indeed, a number of reports stated that while deltamethrin was associated with brown planthopper resurgence (e.g. Heinrichs et al. 1982, Reissig et al. 1982a,b), no such resurgence was observed for green leafhopper (Fabellar and Heinrichs 1986). Triazophos and carbaryl were shown to be more toxic to *Cyrtorhinus lividipennis*, a predatory mirid bug, compared to the brown planthopper and green leafhopper (Fabellar and Heinrichs 1986).

In Japan, BPMC, a common carbamate insecticide, was found to be more toxic to *P. pseudoannulata* and *Conocephalus maculatus*, a predatory grasshopper, compared to the green rice leafhopper *Nephotettix cincticeps* (Takahashi and Kiritani 1973). Additionally, in Japan, deltamethrin was found to be more toxic to a number of predatory spiders, mirid bugs and *Haplogonatopus apicalis*, a drynid wasp, compared to the brown planthopper (Tanaka et al. 2000). Using a risk quotient approach (calculated as the ratio between recommended field dose of a given active ingredient and the LC50 of the same active ingredient to a given natural enemy), Zhao et al. (2012) reported that all tested organosphosphates (chlorpyrifos, fenotrothion, phoxim, profenofos and triazophos) and carbamates (carbosulfan, carbaryl, isoprocarb, metolcarb and promecarb) posed high risks against *Trichogramma japonicum*, a parasitoid of rice lepidopteran pests, collected in China. This means that the recommended field rates of these active ingredients are much higher than their comparative LC50s against *T. japonicum*, in some cases more than 1000 times higher. In the same study, insect growth regulators (chlorfluazuron, fufenozide, hexaflumuron, tebufenozide) were reported to pose the lowest risk against *T. japonicum*.

Insecticides may also affect the biological control function of the natural enemies. In a laboratory study, Fabellar and Heinrichs (1986) reported a significantly lower predation rate of *L. pseudonannulata* on plants treated with azinphos ethyl, an organophosphate insecticide, compared to untreated spiders.

In summary, active ingredients of different members of the organophosphates, pyrethroids, organochlorines and carbamates seemed to pose a high risk against various members of the natural enemy communities in the rice ecosystem. Thus, screening among the different active ingredients for their risks against rice natural enemies should be taken seriously, especially at the insecticide registration phase. Heinrichs et al. (1982) added that such screening needs to be done for potential tendency to induce resurgence, likely caused by hormesis (discussed above), on target insect pests.

## 6.3 Insecticide resistance

Insecticide resistance to various active ingredients has been reported among different rice insect pest species across Asia. In 2006, resistance to imidacloprid was detected on brown planthopper populations collected in East Asia (Taiwan, China, and Japan) and Vietnam but not on those collected in the Philippines. Additionally, resistance to fipronil was detected among whitebacked planthopper populations from East Asia, Vietnam and the Philippines (Matsumura et al. 2008). Resistance to imidacloprid among brown planthopper populations collected in China was also reported by Wen et al. (2009) and Zhang et al. (2014).

Moderate resistance to buprofezin, an insect growth regulator, was widespread among white-backed planthopper populations in China collected in 2010–11 (Su et al. 2013). The same study reported lower frequency (30–60% of the samples) of low-to-moderate resistance to imidacloprid, thiamethoxam and chlorpyrifos among white-backed planthopper populations in China. An eleven-year survey showed a dramatic increase (up to 28 fold) in buprofezin resistance among Chinese brown planthopper populations between 1996 and 2006 (Wang et al. 2008).

Development of insecticide resistance is often associated with the overuse of a certain active ingredient in a given region over an extended period of time (Matsumura et al. 2008, Wang et al. 2008). Indeed, laboratory selection by constantly subjecting generations of the insect population to a given active ingredient commonly yield an insect strain with over 100-fold resistance compared to susceptible strains (e.g. Ding et al. 2013, Liu and Han 2006, Wang et al. 2008).

On the other hand, development of resistance in a given population may carry an ecological fitness cost. For example, a brown planthopper population selected for imidacloprid resistance had significantly lower life table parameters (e.g. fecundity, nymphal survivorship, and egg hatchability) compared to an imidacloprid-susceptible population (Liu and Han 2006). If indeed the insecticide-susceptible strain has an ecological fitness advantage, an environment with reduced or no pesticide exposure may lead to the selection of susceptible individuals. Indeed, a brown planthopper population collected in China was found to be resistant to imidacloprid, chlorpyrifos, fipronil and fenobucarb when compared to a susceptible strain. This field-collected population was then reared for 15 generations without exposure to any insecticide and their susceptibilities to the four active ingredients above were measured routinely. A consistent decrease of LD50 was detected across all tested active ingredients, indicating a reversal of resistance development occurring in the absence of insecticide exposure (Yang et al. 2014). This observation indicates that development of resistance to a given active ingredient may be slowed down or even reversed by drastic exposure reduction of the active ingredient towards the insect population.

# 7 Dissemination mechanisms for rice IPM

Over the last two decades, two major dissemination mechanisms for rice IPM in Asia have taken shape: Farmer Field School (FFS) and Mass Media Campaign (MMC). FFS is a participatory training approach that utilizes experiential and reflective learning principles. The approach improves farmers' capacities through a series of regular meetings with highly trained facilitators (typically extension workers or NGO staffers) in which the farmers conduct their own field experiments, learn to diagnose problems and formulate solutions (Van den Berg and Jiggins 2007). The use of FFS as a means for rice IPM dissemination was first developed in the Philippines as a collaboration between International Rice Research Institute (IRRI) and the Philippines' Bureau of Plant Industry (Pontius et al. 2002). The FAO later launched a large-scale FFS campaign in Indonesia in the late 1980s. Based on the success in Indonesia, FFS was adopted as a dissemination mechanism for IPM and other best agronomic practices across the globe (Van den Berg and Jiggins 2007). The impact and cost effectiveness of FFS as a means for IPM dissemination has been contested. Participation in FFS on rice IPM has been shown to reduce insecticide use and increase yield over time (e.g. Pincus 1999). FFS participants used much less pesticide and produced significantly more yield compared to control farmers (e.g. Tripp et al. 2005). Thus, participation in FFS had a significant impact on farmers' IPM practice and their respective yield. The major criticisms against FFS are its financial sustainability and effectiveness to spread information among non-participants. There are evidences that very limited, if any, diffusion of IPM knowledge and practices occurred between FFS alumnae and neighbouring farmers (e.g. Rola et al. 2002, Tripp et al. 2005). In light of this limited diffusion effects, out scaling of FFS to cover a significant farmer population may be crucial to achieving a large-scale impact of the IPM packages. However, Quizon et al. (2001) observed high upfront and overhead costs of FFS upscaling to cover a meaningful farming population and concluded that FFS was financially unsustainable.

As FFSs were launched in Vietnam in 1992, the project focused on farmers living around towns or main roads. Thus there was a concern that the method may not reach farmers who live in areas outside town proper or with limited road access. MMC was introduced in Vietnam specifically to reach these 'underserved' farmers (Rejesus et al. 2009). There has been an evolution of the messages disseminated through MMC in Vietnam. As described under the subsection 'The effects of insecticide application on the natural enemy and pest populations' above, a simple rule of thumb, 'insecticide spraying for leaf folder control in the first 40 days after sowing is not needed (No early spray – NES)', was formulated and the multimedia materials to communicate this message were designed through a participatory workshop that involved researchers, extension and agricultural communication specialists. The resulting mass media campaign involved the distribution of leaflets, posters, roadside billboards and radio drama broadcasts through local stations and played at coffee shops (Heong et al. 1998). After 31 months of the MMC-NES in Long Anh province, Vietnam, the proportion of farmers who believed that early season spraying was required dropped from 77 to 23%. Reported spray frequency was also significantly reduced from 3.35 sprays per season to 1.56 sprays per season (Heong et al. 1998).

Between 1992 and 1997 both FFS and MMC-NES coincided in Vietnam providing an opportunity to compare the impacts of both interventions. In terms of reach, FFS trained 108 000 farmers in the study sites across Mekong Delta while MMC reached 2.3 million farmer households (Huan et al. 1999). Both FFS and MMC-NES had effects on farmer beliefs. The percentage of farmers who believed that killing the natural enemies can cause more pests are the highest among those exposed to both FFS and MMC-NES compared to those exposed to only MMC-NES.

Furthermore, the percentage of farmers who believed that applying insecticides will increase yield is lower among those exposed to both interventions compared to those exposed to only MMC. An evaluation study of both the FFS and MMC-NES message was conducted in 2005 in three southern Vietnam provinces. The study showed that farmers exposed to either FFS and MMC-NES are more technically efficient in using pesticide compared to control farmers (i.e. those not exposed to either intervention). However, there is no statistical difference in the amount of insecticides used by farmers receiving MMC-NES compared to control farmers. This may mean that MMC-NES farmers were spraying the same amount of insecticides with better timing compared to control farmers, thus producing better technical efficiency. FFS farmers used statistically less insecticides compared to control farmers; however, there was no difference in spray frequency between FFS and control farmers. This may mean that FFS farmers spray as frequently as control farmers with lower rates, perhaps as the results of following label dosage and better recognition of pest and beneficial insects (Rejesus et al. 2009). The authors posit that the difference between their evaluation results of MMC-NES impact and previous studies (e.g. Heong 1998, Huan et al. 2008) may have stemmed from the chosen evaluation methodologies. Rejesus et al. (2009) pointed out that previous evaluation studies for MMC impacts did not take into account the possible selection bias and endogeneity problem which might affect the studies' inferences.

In 2002, a large-scale participatory experiment was conducted with 951 farmer collaborators in multiple Vietnamese provinces in the Mekong Delta to evaluate the effects of reducing pesticide, fertilizers and seed rates on rice farming productivity and profitability. This exercise showed that reducing input from the then current practice had no effect on productivity and improved profitability (Huan et al. 2008). This large-scale participatory evaluation formed the basis for a message expansion disseminated through MMC in Vietnam. The new campaign, locally termed 'Ba Giam Ba Tang' or 'Three reductions, Three Gains' promoted a reduction of pesticide, synthetic fertilizer and seed rates (i.e. three reductions) and promised savings in production costs, improved farmers' health and environmental quality (i.e. three gains). The campaign distributed leaflets and posters. Additionally, a radio drama and a 30s TV commercial were broadcast over local radio and TV stations. An evaluation of this MMC was conducted in Can Tho and Tien Gang provinces. Two months after the launch of the campaign, 81% and 56% of surveyed farmers had heard of the campaign in Can Tho and Tien Gang, respectively (Huan et al. 2008). The evaluation programme monitored the changes in farmers' belief attitudes and input use (self-report) as affected by the campaign. Reductions of insecticide spray (~13–33%), seed rate (~10%) and nitrogen fertilizer rate (~7%) were reported. Additionally, there was an overall change in belief attitudes favouring reduction of the targeted inputs (Huan et al. 2008). Huelgas et al. (2008) found an increase of $92–118/ha in net income from An Giang province in Vietnam. However, economic assessment data from Can Tho province showed lower net income among 'Ba Giam Ba Tang' adopters compared to the non-adopters.

In conclusion, FFS is an effective educational tool to improve the farmers' capacity to manage their agroecosystem. This improvement translates into a reduction in pesticide use and higher yield. However, the high cost and tendency for low diffusion effects to non-participants are the weaknesses of FFS. MMC is an effective method to communicate simple messages to a large farming audience. When designed correctly, the simple messages work to modify farmers' attitude and behaviour. Both FFS and MMC increased the technical efficiency of pesticide use by farmers; however, the household-level economic impact of MMC may vary with geographic location. FFS and MMC complement each other by creating a nucleus of farmers with an in-depth understanding of agro-ecosystem management and the ability to conduct their own field experiments while mass communicating simpler IPM messages towards a larger farming audience.

## 8 References

Aguda, R. M., Rombach, M. C., Im, D. J. and Shepard, B. M. 1987. Suppression of populations of the brown planthopper, *Nilaparvata lugens* (Stål) (Hom.: Delphacidae) in field cages by entomogenous fungi (Deuteromycotina) on rice in Korea. *J. Appl. Ent.* 104:167–72.

Alam, S. N. and Cohen, M. B. 1998. Durability of brown planthopper, *Nilaparvata lugens*, resistance in rice variety IR64 in greenhouse selection studies. *Entomol. Exp. et Appl.* 89: 71–8.

Albuquerque, G. S. 1993. Planting time as a tactic to manage the small rice stink bug, *Oebalus poecilus* (Hemiptera: Pentatomidae), in Rio Grande do Sul, Brazil. *Crop Prot.* 12: 627–30.

Azzam, S., Wang, F., Wu, J. C., Shen, J., Wang, L. P., Yang, G. Q. and Guo, Y. R. 2009. Comparisons of stimulatory effects of a series of concentrations of four insecticides on reproduction in the rice brown planthopper, *Nilaparvata lugens* Stål (Homoptera: Delphacidae). *Int. J. Pest Manag.* 55:347–58.

Bandong, J. P., Canapi, B. L., Cruz, C. G. and Litsinger, J. A. 2002. Insecticide decision protocols: A case study of untrained Filipino farmers. *Crop Prot.* 21:803–16.

Bao, H., Liu, S., Gu, J., Wang, X., Liang, X. and Liu, Z. 2008. Sublethal effects of four insecticides on the reproduction and wing formation of brown planthopper, *Nilaparvata lugens*. *Pest Manag. Sci.* 65: 170–4.

Bautista, R. C., Heinrichs, E. A. and Rejesus, R. S. 1984. Economic injury levels for the rice leaffolder *Cnaphalocrocis medinalis* (Lepidoptera: Pyralidae): insect infestation and artificial leaf removal. *Environ. Entomol.* 13:439–43.

Cabunagan, R. C., Castilla, N., Coloquio, E. L., Tiongco, E. R., Truong, X. H., Fernandez, J., Du, M. J., Zaragosa, B., Hozak, R. R., Savary, S. and Azzam, O. 2001. Synchrony of planting and proportions of susceptible varieties affect rice tungro disease epidemics in the Philippines. *Crop Prot.* 20:499–510.

Chantaraprapha, N., Ratanyukui, W., Semapetch, U. and Seetanun, W. 1980. Influence of ammonium phosphate levels on rice leaf roller incidence. *Int. Rice Res. Newsl.* 5:21.

Chelliah, S. and Heinrichs, E. A. 1980. Factors affecting insecticide induced resurgence of the brown planthopper. *Environ. Entomol.* 9:773–7.

Chelliah, C., Fabellar, L. T. and Heinrichs, E. A. 1980. Effect of sub-lethal doses of three insecticides on the reproductive rate of the brown planthopper, *Nilaparvata lugens*, on rice. *Environ. Entomol.* 9: 778–80.

Cohen, M. B., Alam, S. N., Medina, E. B. and Bernal, C. C. 1997. Brown planthopper, *Nilaparvata lugens*, resistance in rice cultivar IR64: Mechanisms and role in successful *N. lugens* management in Central Luzon, Philippines. *Entomol. Exp. Applic.* 85: 221–9.

Crill, P., Ham, Y. S. and Beachell, H. M. 1981. The rice blast disease in Korea and its control with race prediction and gene rotation. *Korean J. Breed.* 13:106–14.

De, Kraker, J., Rabbinge, R., Van, Huis, A., Van, Lenteren, J. C. and Heong, K. L. 2000. Impact of nitrogenous fertilization on the population dynamics and natural control of rice leaffolder (Lep.: Pyralidae). *Int. J. Pest Man.* 46: 225–35.

Ding, Z., Wen, Y., Yang, B., Zhang, Y., Liu, S., Liu, Z. and Han, Z. 2013. Biochemical mechanisms of imidacloprid resistance in *Nilaparvata lugens*: Over-expresion of cytochrome P450 CY6AY1. *Insect Biochem. Mol. Biol.* 43:1021–7.

Djamin, A. and Pathak, M. D. 1967. Role of silica in resistance to Asiatic rice borer, *Chilo suppressalis* (Walker), in rice varieties. *J. Econ. Entomol.* 60: 347–51.

Fabellar, L. T. and Heinrichs, E. A. 1986. Relative toxicity of insecticides to rice planthoppers and leafhoppers and their predators. *Crop Prot.* 5:254–8.

Fagan, W. F., Hakin, A. L., Ariawan, H. and Yuliyantiningsih. 1998. Interactions between biological control efforts and insecticide applications in tropical rice agroecosystems: The potential role of intraguild predation. *Biol. Control* 13:121–6.

Feng, M. G. and Pu, X. Y. 2005. Time-concentration-mortality modeling of the synergistic interaction of *Beauveria bassiana* and imidacloprid against *Nilaparvata lugens*. *Pest Manag. Sci.* 61:363–70.

Fujita, D., Kohli, A. and Horgan, F. G. 2013. Rice resistance to planthoppers and leafhoppers. *Crit. Rev. Plant Sci.* 32:162–91.

Fujita, D., Yoshimura, A. and Yasui, H. 2010. Development of near-isogenic lines and pyramided lines carrying resistance genes to green rice leafhopper (*Nephotettix cincticeps* Uhler) with the Taichung 65 genetic background in rice (*Oryza sativa* L.). *Breeding Sci.* 60:18–27.

Ge, L.-Q., Wan, D.-J., Xu, J., Jiang, L.-B. and Wu, J.-C. 2013. Effects of nitrogen fertilizer and magnesium manipulation on the *Cnaphalocrocis medinalis* (Lepidoptera: Pyralidae). *J. Econ. Entomol.* 106:196–205.

Graf, B., Lamb, R., Heong, K. L. and Fabellar, L. T. 1992. A simulation model for the population dynamics of rice leaffolders (Lepidoptera: Pyralidae) and their interactions with rice. *J. Appl. Ecol.* 29: 558–70.

Grist, D. H. and Lever, R. J. A. W. 1969. *Pests of Rice*. London and Harlow: Longmans, Green and Co. Ltd.

Gurr, G. M., Liu, J., Read, D. M. Y., Catindig, J. L. A., Cheng, J. A., Lan, L. P. and Heong, K. L. 2011. Parasitoids of Asian rice planthopper (Hemiptera: Delphacidae) pests and prospects for enhancing biological control by ecological engineering. *Ann. Appl. Biol.* 158:149–76.

Gurr, G. M., Lu, Z., Zheng, X., Xu, H., Zhu, P., Chen, G., Yao, X., Cheng, J., Zhu, Z., Catindig, J. L., Villareal, S., Chien, H. V., Cuong, L. Q., Channoo, C., Chengwattana, N., Lan, L. P., Hai, L. H., Chaiwong, J., Nicol, H. I., Perovic, D. J., Wratten, S. D. and Heong, K. L. 2016. Multi-country evidence that crop diversification promotes ecological intensification of agriculture. *Nature Plants*. DOI: 10.1038/NPLANTS.2016.14.

Gurr, G. M., Read, D. M. Y., Catindig, J. L., Cheng, J., Liu, J., Lan, L. P. and Heong, K. L. 2012. Parasitoids of the rice leaffolder *Cnaphalocrocis medinalis* and prospects for enhancing biological control with nectar plants. *Agric. Forest Entomol.* 14:1–12.

Heinrichs, E. A. and Medrano, F. G. 1985. Influence of fertilizer on the population development of brown planthopper (BPH). *Intl. Rice Res. Newsl.* 10(6):20–1.

Heinrichs, E. A., Medrano, F. G. and Rapusas, H. R. 1985. *Genetic Evaluation for Insect Resistance in Rice*. Los Banos, Philippnes: IRRI.

Heinrichs, E. A., Reissig, W. H., Valencia, S. and Chelliah, S. 1982. Rates and effect of resurgence-inducing insecticides on populations of *Nilaparvata lugens* (Homoptera: Delphacidae) and its predators. *Environ. Entomol.* 11:1269–73.

Heinrichs, E. A. and Mochida, O. 1984. From secondary to major pest status: the case of insecticide-induced rice brown planthopper, *Nilaparvata lugens*, resurgence. *Prot. Ecol.* 7:201–18.

Heong, K. L., Escalada, M. M., Huan, N. H., Ky, Ba, V. H., Quynh, P. V., Thiet, L. V. and Chien, H. V. 2008. Entertainment-education and rice pest management: a radio soap opera in Vietnam. *Crop Prot.* 27:1392–7.

Heong, K. L., Escalada, M. M., Huan, N. H. and Mai, V. 1998. Use of communication media in changing rice farmers' pest management in the Mekong Delta, Vietnam. *Crop Prot.* 17:413–25.

Heong, K. L., Escalada, M. M. and Mai, V. 1994. An analysis of insecticide use in rice: case studies in the Philippines and Vietnam. *Int. J. Pest Manag.* 40:173–8.

Heong, K. L. and Schoenly, K. 1998. Impact of insecticides on herbivore-natural enemy communities in tropical ecosystems. In Haskell, P. T. and McEwen, P. (eds), *Ecotoxicology: Pesticides And Beneficial Organisms*, pp. 381–403. London: Chapman and Hall.

Hosseini, S. Z., Jelodar, N. B. and Bagheri, N. 2012. Study of silicon effects on plant growth and resistance to stem borer in rice. *Comm. Soil Sci. Plant Analys.* 43:2744–51.

Hou, M. and Han, Y. 2010. Silicon-mediated rice plant resistance to the Asiatic rice borer (Lepidoptera: Crambidae): Effects of silicon amendment and rice varietal resistance. *J. Econ. Entomol.* 103: 1412–19.

Hu, J., Li, X., Wu, C., Yang, C., Hua, H., Gao, G., Xiao, J. and He, Y. 2012. Pyramiding and evaluation of the brown planthopper resistance genes Bph14 and Bph15 in hybrid rice. *Mol. Breeding* 1:61–9.

Hu, J.-H., Wu, J.-C., Yin, J.-L. and Gu, H.-N. 2010. Physiology of insecticide-induced stimulation of reproduction in the rice brown planthopper (*Nilaparvata lugens* (Stal)): Dynamics of protein in fat body and ovary. *Int. J. Pest Manag.* 56:23–30.

Huan, N. H., Chien, H. V., Quynh, P. V., Tan, P. S., Du, P. V., Escalada, M. M. and Heong, K. L. 2008. Motivating rice farmers in the Meking Delta to modify pest management and related practices through mass media. *Int. J. Pest Manag.* 54:339–46.

Huan, N. H., Mai, V., Escalada, M. M. and Heong, K. L. 1999. Changes in rice farmers' pest management in the Mekong Delta, Vietnam. *Crop Prot.* 18: 557–63.

Huelgas, Z. M., Templeton, D. and Castanar, P. 2008. Three Reductions, Three Gains (3R3G) technology in South Vietnam: Searching for evidence of economic impact. Contributed paper at the 52nd Annual Conference of the Australian Agricultural Resource Economics Society held at Rydges Lakeside, Canberra ACT, Australia, 5–8 February 2008.

Jena, K. K., Kim, S.-M. 2010. Current status of brown planthopper resistance and genetics. *Rice* 3:161–71.

Jiang, M. X. and Cheng, J. A. 2003. Interactions between the striped stem borer *Chilo suppressalis* (Walk.) (Lep., Pyralidae) larvae and rice plants in response to nitrogen fertilization. *J. Pest Sci.* 76:124–8.
Jin, S. F., Feng, M. G. and Chen, J. Q. 2008. Selection of global *Metarhizium* isolates for the control of the rice pest Nilaparvata lugens (Homoptera: Delphacidae). *Pest Manag. Sci.* 64:1008–14.
Jin, S. F., Feng, M. G., Ying, S. H., Mu, W. J. and Chen, J. Q. 2010. Evaluation of alternative rice planthopper control by the combined action of oil-formulated *Metarhizium anisopliae* and low-rate buprofezin. *Pest Manag. Sci.* 67:36–43.
Kajimura, T., Fujisaki, K. and Nakasuji, F. 1995a. Effect of organic rice farming on leafhoppers and planthoppers: 2. Amino acid content in the rice phloem sap and survival rate of planthoppers. *Appl. Entomol. Zool.* 30:17–22.
Kajimura, T., Widiarta, I. N., Nagai, K., Fujisaki, K. and Nakasuji, F. 1995b. Effect of organic rice farming on planthoppers: 4. Reproduction of the white backed planthopper, *Sogatella furcifera* Horváth (Homoptera: Delphacidae). *Res. Popul. Ecol.* 37:219–24.
Landis, D. A., Wratten, S. D. and Gurr, G. M. 2000. Habitat management to conserve natural enemies of arthropod pests in agriculture. *Annu. Rev. Entomol.* 45: 175–201.
Letourneau, D. K., Armbrecht, I., Rivera, B. S., Lerma, J. M., Carmona, E. J., Daza, M. C. Escobar, S., Galindo, V., Gutierrez, C., Lopez, S. D., Mejia, J. L., Rangel, A. M. A., Rangel, J. H., Rivera, L., Saavedra, C. A., Torres, A. M. and Trujilo, A. R. 2011. Does plant diversity benefit agroecosystems? A synthetic review. *Ecol. Appl.* 21:9–21.
Li, M. Y., Lin, H. F., Li, S. G., Xu, A. M. and Feng, M. F. 2012a. Efficiency of entomopathogenic fungi in the control of eggs of the brown planthopper *Nilaparvata lugens* Stål (Homoptera: Delphacidae). *Afric. J. Microbiol. Res.* 6:7162–7.
Li, M. Y., Lin, H. F., Li, S. G., Chen, P. R., Jin, L. and Yang, J. 2012b. Virulence of entomopathogenic fungi to adults and eggs of *Nilaparvata lugens* Stål (Homoptera: Delphacidae). *Afric. J. Agric. Res.* 7:2183–90.
Ling, S., Zhang, J., Hu, L. and Zhang, R. 2009. Effect of fipronil on the reproduction, feeding and relative fitness of brown planthopper, *Nilaparvata lugens. Appl. Entomol. Zool.* 44: 543–8.
Ling, S., Zhang, H. and Zhang, R. 2011. Effect of fenvalerate on the reproduction and fitness costs of the brown planthopper, *Nilaparvata lugens* and its resistance mechanism. *Pestic. Biochem. Physiol.* 101: 148–53.
Litsinger, J. A. 1994. Cultural, mechanical, and physical control of rice insects. In Heinrichs, E. A. (ed.), *Biology and Management of Rice Insects*, pp. 549–84. New Delhi, India: Wiley Eastern.
Litsinger, J. A., Bandong, J. P., Canapi, B. L., Dela, Cruz, C. G., Pantua, P. C., Alviola, A. L. and Batay-An, E. H. 2005. Evaluation of action thresholds for chronic rice insect pests in the Philippines: I. Less frequently occurring pests and overall assessment. *Int. J. Pest Manag.* 51: 45–61.
Litsinger, J. A., Bandong, J. P., Canapi, B. L., Dela, Cruz, C. G., Pantua, P. C., Alviola, A. L. and Batay-An, III, E. H. 2006a. Evaluation of action thresholds for chronic rice insect pests in the Philippines: II. Whorl maggot and defoliators. *Int. J. Pest Manag.* 52: 167–80.
Litsinger, J. A., Bandong, J. P., Canapi, B. L., Dela, Cruz, C. G., Pantua, P. C., Alviola, A. L. and Batay-An, III, E. H. 2006b. Evaluation of action thresholds for chronic rice insect pests in the Philippines: III. Leaffolders. *Int. J. Pest Manag.* 52:181–94.
Litsinger, J. A., Bandong, J. P., Canapi, B. L., Dela, Cruz, C. G., Pantua, P. C., Alviola, A. L. and Batay-An, III, E. H. 2006c. Evaluation of action thresholds for chronic rice insect pests in the Philippines: IV. Stem borers. *Int. J. Pest Manag.* 52:195–207.
Lu, Z. X., Heong, K. L., Yu, X. P. and Hu, C. 2004. Effects of plant nitrogen on ecological fitness of the brown planthopper, *Nilaparvata lugens* in rice. *J. Asia-Pacific Entomol.* 7:97–2004.
Lu, Z. X., Zhu, P. Y., Gurr, G. M., Zheng, X. S., Read, D. M. Y., Heong, K. L., Yang, Y. J. and Xu, H. X. 2014. Mechanisms for flowering plants to benefit arthropod natural enemies of insect pests: Prospects for enhanced use in agriculture. *Insect Sci.* 21: 1–12.
Lundgren, J. G. 2009. *Relationships of Natural Enemies and Non-Prey Foods.* The Netherlands: Springer International.

Matsumura, M. and Sanada-Morimura, S. 2010. Recent status of insecticide resistance in Asian rice planthoppers. *Jpn. Agric. Res. Q.* 44: 225–30.

Matsumura, M., Takeuchi, H., Satoh, M., Sanada-Morimura, S., Otuka, A., Watanabe, T. and Thanh, D. V. 2008. Species-specific insecticide resistance to imidacloprid and fipronil in the rice planthoppers *Nilaparvata lugens* and *Sogatella furcifera* in East and South-east Asia. *Pest Manag. Sci.* 64:1115–21.

Matteson, P. C. 2000. Insect pest management in tropical irrigated rice. *Ann. Rev. Entomol.* 45: 549–74.

Morgan, D. R., Tugwell, N. P. and Bernhardt, J. L. 1989. Early rice field drainage for control of rice water weevil (Coleoptera: Curculionidae) and evaluation of an action threshold based upon leaf-feeding scars of adults. *J. Econ. Entomol.* 82:1757–9.

Mundt, C. C. 2014. Durable resistance: A key to sustainable management of pathogens and pests. *Infect. Genet. Evol.* 27:446–55.

Nakasuji, F. and Dyck, V. A. 1984. Evaluation of the role of *Microvelia douglasi atrolineata* (Bergroth) (Heteroptera: Veliidae) as predator of the brown planthopper *Nilaparvata lugens* (Stål) (Homoptera: Delphacidae). *Res. Popul. Ecol.* 26: 134–49.

Norton, G. W., Heong, K. L., Johnson, D. and Savary, S. 2010. Rice pest management: issues and opportunities. In Pandey, S., Byerlee, D., Dawe, D., Dobermann, A., Mohanty, S., Rozelle, S. and Hardy, B. (eds), *Rice In The Global Economy: Strategic Research And Policy Issues For Food Security*, pp. 297–332. Los Banos, Philippines: IRRI.

Oteng, J. W. and Sant'Anna, R. 1999. Rice production in Africa: current situation and issues. In: International Rice Commission Newsletter Vol. 48 FAO, Roma, 1999.

Pincus, J. 1999. The impact of IPM farmer field schools on farmers' cultivation practices in their own fields. Report prepared for the FAO Programme for Community IPM in Asia.

Pontius, J. C., Dilts, R. and Bartlett, A. (eds). 2002. Ten years of IPM training in Asia: From farmer field school to community IPM. Bangkok, Thailand: FAO.

Preap, V., Zalucki, M. P., Jahn, G. C. and Nesbitt, H. J. 2001. Effectiveness of brown planthopper predators: Population suppression by two species of spider, *Pardosa pseudoannulata* (Araneae, Lycosidae) and *Araneus inustus* (Araneae, Araneidae). *J. Asia-Pacific Entomol.* 4:187–93.

Preetha, G., Stanley, J., Suresh, S., Kuttalam, S. and Samiyappan, R. 2009. Toxicity of selected insecticides to *Trichogramma chilonis*: Assessing their safety in the rice ecosystem. *Phytoparasitica* 37:209–15.

Preetha, G., Stanley, J., Suresh, S. and Samiyappan, R. 2010. Risk assessment of insecticides used in rice on mirid bug, *Cyrthorhinus lividipennis* Reuter, the important predator of brown planthopper, *Nilaparvata lugens* (Stål). *Chemosphere* 80: 498–503.

Qiu, Y., Guo, J., Jing, S., Zhu, L. and He, G. 2012. Development and characterization of japonica rice lines carrying the brown planthopper-resistance gene BPH12 and BPH6. *Theor. Appl. Genet.* 124:485–94.

Quizon, J., Feder, G. and Murgai, R. 2001. Fiscal sustainability of agricultural extension: the case of the farmer field school approach. *J. Int. Agric. Exten. Edu.* 8: 13–24.

Ranganathan, S., Suvarchala, V., Rajesh, Y. B. R. D., Srinivasa, Prasad, M., Padmakumari, A. P. and Violeti, S. R. 2006. Effects of silicon sources on its deposition, chlorophyll content, and disease and pest resistance in rice. *Biol. Plant.* 50: 713–16.

Reissig, W. H., Heinrichs, E. A. and Valencia, S. L., 1982a. Insecticide-induced resurgence of the brown planthopiper, *Nilaparvata lugens*, on rice varieties with different levels of resistance. *Environ. Entomol.* 11(1):165–8

Reissig, W. H., Heinrichs, E. A. and Valencia, S. L., 1982b. Effects of insecticides on *Nilaparvata lugens* and its predators: spiders, *Microvelia atrolineata*, and *Cyrtorhinus lividipennis*. *Environ. Entomol.* 11(1):193–9.

Rejesus, R. M., Palis, F. G., Lapitan, A. V., Chi, T. T. N. and Hossain, M. 2009. The impact of integrated pest management information dissemination methods on insecticide use and efficiency: Evidence from rice producers in South Vietnam. *Rev. Agric. Econ.* 31:814–33.

Reyes, T. M. and Gabriel, B. P. 1975. The life history and consumption habits of *Cyrtorhinus lividipennis* Reuter (Hemiptera: Miridae). *Philipp. Entomol.* 3:79–88.

Rola, A. C., Jamias, S. B. and Quizon, J. B. 2002. Do farmer field school graduates retain and share what they learn? An investigation in Iloilo, Philippines. *J. Int. Agric. Exten. Edu.* 9:65–76.

Rombach, M. C., Aguda, R. M., Picard, L. and Roberts, D. W. 1989. Arrested feeding by the Asiatic rice borer (Lepidoptera: Pyralidae) by *Bacillus thuringiensis*. *J. Econ. Entomol.* 82: 416–19.

Rombach, M. C., Roberts, D. W. and Aguda, R. M. 1994. Pathogens of rice insects. In Heinrichs, E. A. (ed.), *Biology and Management of Rice Insects*, pp. 613–56. New Delhi, India: Wiley Eastern.

Rombach, M. C., Rombach, G. M. and Roberts, D. W. 1987. Pathogens of insect pests of rice: A bibliography. *Insect Sci. Appl.* 8:197–210.

Rubia, E. G., Pena, N. B., Almazan, L. P. and Shepard, B. M. 1990. Efficacy of selected predators against some insect pests of rice. *J. Plant Protection Tropics* 7: 69–76.

Schoenly, K. G., Cohen, J. E., Heong, K. L., Arida, G. S., Barrion, A. T. and Litsinger, J. A. 1996. Quantifying the impact of insecticides on food web structure of rice-arthropod populations in a Philippine farmer's irrigated field: A case study. In Polis, G. A. and Winemiller, K. O. (eds), *Food Webs: Integration of Patterns And Dynamics*. New York, USA: Chapman and Hall.

Schoenly, K. G., Cohen, M. B., Barrion, A. T., Zhang, W., Gaolach, B. and Viajante, V. D. 2003. Effects of *Bacillus thuringiensis* on non-target herbivore and natural enemy assemblages in tropical irrigated rice. *Environ. Biosafety Res.* 3:181–206.

Settle, W. H., Ariawan, H., Astuti, E. T., Cahyana, W., Hakim, A. L., Hindayana, D. and Lestari, A. S. 1996. Managing tropical rice pests through conservation of generalist natural enemies and alternative prey. *Ecology* 77:1975–88.

Sharma, P. N., Torii, A., Takumi, S., Mori, N. and Nakamura, C. 2004. Marker assisted pyramiding of brown planthopper (*Nilaparvata lugens*) resistance genes Bph1 and bph2 on rice chromosome 12. *Hereditas* 140: 61–9.

Shigematsu, Y., Murofushi, N., Ito, K., Kaneda, C., Kawabe, S. and Takahashi, N. 1982. Sterols and asparagines in the rice plant, endogenousfactors related to resistance against the brown planthopper (*Nilaparvata lugens*). *Agric. Biol. Chem.* 46: 2877–9

Shimoda, M. and Honda, K. 2013. Insect reactions to light and its applications to pest management. *Appl Entomol Zool.* 48:413–21.

Sogawa, K. 1982. The rice brown planthopper: Feeding physiology and host plant interactions. *Annu. Rev. Entomol.* 27: 49–73.

Sogawa, K. and Pathak, M. D. 1970. Mechanisms of brown planthopper (Hemiptera: Delphacidae) resistance in mudgo variety of rice. *Appl. Entomol. Zool.* 5: 145–58.

Stern, V. M., Smith, R. F., Van, Den Bosch, R. and Hagen, K. S. 1959. The integrated control concept. *Hilgardia* 29:81–101.

Stout, M. J., Rice, W. C., Linscombe, S. D. and Bollich, P. K. 2001. Identification of rice cultivars resistant to *Lissorhoptrus oryzophilus* (Coleoptera: Curculionidae), and their use in an integrated management program. *J. Econ. Entomol.* 94:963–70.

Stout, M. J., Riggio, M. R., Zou, L. and Roberts, R. 2002. Flooding influences ovipositional and feeding behavior of the rice water weevil (Coleoptera: Curculionidae). *J. Econ. Entomol.* 95: 715–21.

Su, J., Wang, Z., Zhang, K., Tian, X., Tin, Y., Zhao, X., Shen, A. and Gao, C. F. 2013. Status of insecticide resistance of the whitebacked planthopper, *Sogatella furcifera* (Hemiptera: Delphacidae). *Fla. Entomol.* 96:948–56.

Suri, K. S. and Singh, G. 2011. Insecticide-induced resurgence of the whitebacked planthopper *Sogatella furcifera* (Horvath) (Hemiptera: Delphacidae) on rice varieties with different levels of resistance. *Crop Prot.* 30: 118–24.

Swaminathan, K., Saroja, R. and Raju, N. 1985 Influence of source and level of nitrogen application on pest incidence. *Int. Rice Res. News.* 10: 24.

Takahashi, Y. and Kiritani, K. 1973. The selective toxicity of insecticides against insect pests of rice and their natural enemies. *Appl. Ent. Zool.* 8: 220–6.

Tanaka, K., Endo, S. and Kazano, H. 2000. Toxicity of insecticides to predators of rice planthoppers: Spiders, the mirid bug and the drynid wasp. *Appl. Entomol. Zool.* 35: 177–87.

Thomas, M. B., Blandford, S., Gbongboui, C. and Lomer, C. J. 1998. Experimental studies to evaluate spray applications of a mycoinsecticide against the rice grasshopper, *Hieroglyphus daganensis*, in northern Benin. *Entomol. Exp. Applic.* 87:93–102.

Thompson, R. A., Quisenberry, S. S., N'Guessan, F. K., Heagler, A. M. and Giesler, G. 1994. Planting date as a potential cultural method for managing the rice water weevil (Coleoptera: Curculionidae) in water-seeded rice in southwest Louisiana. *J. Econ. Entomol.* 87:1318–24.

Toledo, A. V., Giambelluca, L., Marino de Remes, Lenicov, A. M., Lopez and Lastra, C. C. 2008. Pathogenic fungi of planthoppers associated with rice crops in Argentina. *Int. J. Pest Manag.* 54:363–8.

Tripp, R., Wijeratne, M. and Piyadasa, V. H. 2005. What should we expect from farmer field schools? A Sri Lanka case study. *World Dev.* 33:1705–20.

Van den Berg, H. and Jiggins, J. 2007. Investing in farmers: The impacts of farmer field schools in relation to integrated pest management. *World Dev.* 35:663–86.

Wang, H. Y., Yang, Y., Su, J. Y., Shen, J. L., Gao, F. C. and Zhu, Y. X. 2008. Assessment of the impact of insecticides on *Anagrus nilaparvatae* (Pang et Wang) (Hymenoptera: Mymaridae), an egg parasitoid of the rice planthopper, *Nilaparvata lugens* (Hemiptera: Delphacidae). *Crop Prot.* 27:514–22.

Way, M. J. and Heong, K. L. 1994. The role of biodiversity in the dynamics and management of insect pests of tropical irrigated rice: A review. *Bull. Entomol. Res.* 84:567–87.

Way, M. O. 2003. Rice arthropod pests and their management in the United States. In Smith, C. W. and Dilday, R. H. (eds), *Rice: Origin, History, Technology, And Production*, pp. 437–56. USA: John Wiley & Sons.

Wen, Y., Liu, Z., Bao, H. and Han, Z. 2009. Imidacloprid resistance and its mechanisms in field populations of brown planthopper, *Nilaparvata lugens* Stal in China. *Pest. Biochem. Physiol.* 94:36–42.

Wilby, A., Lan, L. P., Heong, K. L., Huyen, N. P. D., Quang, N. H., Minh, N. V. and Thomas, M. B. 2006. Arthropod diversity and community structure in relation to land use in the Mekong delta, Vietnam. *Ecosystems* 9:538–49.

Wyss, E., Luka, H., Pfiffner, L., Schlatter, C., Uehlinger, G. and Daniel, C. 2005. Approaches to pest management in organic agriculture: a case study in European apple orchards. *Organic Research* (May) pp. 33–6.

Yang, Y., Dong, B., Xu, H., Zheng, X., Heong, K. L. and Lu, Z. 2014. Susceptibility to insecticides and ecological fitness in resistant rice varieties of field *Nilaparvata lugens* population free from insecticides in laboratory. *Rice Sci.* 21:181–6.

Zehnder, G., Gurr, G. M., Kuehne, S., Wade, M. R., Wratten, S. D. and Wyss, E. 2006. Arthropod pest management in organic crops. *Annu. Rev. Entomol.* 52:57–80.

Zhang, X., Liu, X., Zhu, F., Li, J., You, H. and Lu, P. 2014. Field evolution of insecticide resistance in the brown planthopper (*Nilaparvata lugens* Stål) in China. *Crop Prot.* 58:61–6.

Zhao, X., Wu, X., Wang, Y., Cang, T. and Chen, L. 2012. Assessment of toxicity risk of insecticides used in rice ecosystem on *Trichogramma japonicum*, an egg parasitoid of rice lepidopterans. *J. Econ. Entomol.* 105: 92–101.

Zhu, Y., Chen, H., Fan, J., Wang, Y., Li, Y., Chen, J., Fan, J.-X., Yang, S., Hu, L., Leung, H., Mew, T. W., Teng, P. S., Wang, Z. and Mundt, C. C. 2000. Genetic diversity and disease control in rice. *Nature* 406:718–22.

Zhu, P. Y., Gurr, G. M., Lu, Z. X., Heong, K. L., Chen, G. H., Zheng, X. S., Xu, H. X. and Yang, Y. J. 2013. Laboratory screening supports the selection of sesame (*Sesamum indicum*) to enhance *Anagrus* spp. parasitoids (Hymenoptera: Mymaridae) of rice planthoppers. *Biol Control* 64:83–9.

Zhu, P., Wang, G., Zheng, X., Tian, J., Lu, Z., Heong, K. L., Xu, H., Chen, G., Yang, Y. and Gurr, G. M. 2015. Selective enhancement of parasitoids of rice Lepidoptera pests by sesame (*Sesamum indicum*) flowers. *Biol Control* 60: 157–67.

Chapter 11

# Plant protection products in rice cultivation: critical issues in risk assessment and management to promote sustainable use

*Maura Calliera and Ettore Capri, Università Cattolica Sacro Cuore di Piacenza, Italy*

1 Introduction
2 Scientific modelling for the risk assessment of PPPs
3 The role of ethical, social and behavioural issues in PPP risk assessment
4 The European SUD and IPM
5 Farmer involvement in IPM
6 Conclusion
7 Where to look for further information
8 References

## 1 Introduction

'Sustainability' can be defined in numerous different ways. In this chapter, we will use the definition of the Brundtland Commission (WCED, 1987)[1]: '... *Sustainable development is development that meets the needs of the present without compromising the ability of future generations to meet their own needs*'. In July 2015 the world's population was estimated at 7.3 billion people, and based on the United Nations' median projection, the global population is expected to reach 8.5 billion by 2030, 9.7 billion in 2050 and 11.2 billion in 2100, according to a new UN DESA report, 'World Population Prospects: The 2015 Revision'. The pressing need is to move towards more sustainable patterns of food production to supply this growing population (SOER EEA, 2010).[2] Food security continues to be one of the main indicators of overall prosperity. However, achieving sustainable development requires that we focus not just on ensuring that the world produces and consumes enough '*calories in bulk*' (Jessica Fanzo, 2015),[3] but also on ensuring that our efforts to produce enough food for the present generation do not compromise the natural capital required to support ongoing economic prosperity and human well-being.

The Green Revolution of the 1960s contributed significantly to alleviating hunger and poverty. However, the intensification that came with the Green Revolution also had negative social, environmental and economic consequences. Intensification aimed for yield increases through investment in technologies, fertilisers and pesticides. The resulting practices have produced increased environmental pressure; a loss of biodiversity; and the

http://dx.doi.org/10.19103/AS.2016.0003.22

loss of traditional farming practices, cultures and local knowledge. They have also had negative effects on human health and well-being.

Rice is a primary food source for over 3 billion people. Current rice production is heavily reliant on agricultural science and technology and features more intensive production methods than traditional rice cultivation (Qiqi Chen, 2015).[4] Plant protection products (PPPs) play an important role in modern agriculture and are extensively used in rice cropping systems in combination with other modern agricultural technologies and practices. In Europe, rice field weeds are considered the most significant harmful organisms affecting rice production[5] (Ferrero and Tinarelli, 2008), and weed control through the use of pesticides is considered important to avoid yield losses, although the implementation of improved weed control strategies is within the context of integrated pest management (IPM).

A CropLife Foundation study emphasises both the importance of rice as a crop and the role of pesticides in increasing production:

> [R]ice is an essential crop for feeding the populations of South and South East Asia. Rice provides more than 50% of the calories consumed in Bangladesh, Cambodia, Myanmar, Laos, and Vietnam and 20–44% in Thailand, Philippines, Malaysia, India, Nepal and Sri Lanka. South East Asian rice production increased about 40% from 2000 to 2013 while rice production in South Asia increased about 30%. Increased use of pesticides has been one of the significant contributing factors to these increases.[6] (Gianessi, 2014)

Over two hundred pesticide active ingredients are currently registered for use on rice in China, most of which are herbicides, insecticides and fungicides (Yang Cheng, 2015).[7]

The use of PPPs is therefore a key issue in evaluating the sustainability of rice production. It is essential to evaluate the risks these products pose to human health and the environment, and to determine what measures are required to keep the risks within acceptable limits.

In the next two sections of this chapter, we review how the risks posed by PPPs are currently evaluated, and consider where current risk assessment practices require improvement. Section 2 offers an overview of the evidence-based approach to risk assessment used to inform government regulation, and focuses on the importance of multi-tier assessments and the availability of a range of simulation models. Section 3 explains why ethical, social and behavioural issues must also be considered in order for risks to be effectively assessed and managed. In Sections 4 and 5, we move on to a discussion of managing the risks associated with pesticides and encouraging more sustainable rice cultivation. Section 4 considers the example of the European Sustainable Use Directive (SUD), and explains how IPM is an essential part of SUD's strategy. Section 5 then argues that for IPM to be implemented successfully, it is essential for farmers and other stakeholders to be involved in developing the IPM strategy for a given situation.

## 2 Scientific modelling for the risk assessment of PPPs

In rice cultivation (as in agriculture more generally), PPPs are subject to political regulatory interventions. These interventions aim to control the risks associated with these products as well as to ensure sustainability in their use.

Governments set high standards for the registration of new pesticides and require re-evaluation of pesticides that are already on the market to ensure that they meet current health, environment and safety standards. Cooperation in this area, for example through joint review processes and the sharing of existing evaluations, has increased among OECD countries.[8] In the past five years, progress has also been made in Asia among non-OECD member countries, as shown in the FAO publication 'Progress in pesticide risk assessment and phasing out of highly hazardous pesticides in Asia' (FAO, 2015).[9]

The objectives of the risk evaluation process for PPPs are twofold:

1 to develop an evidence-based approach which, at the end of the process, provides solid facts.
2 to provide answers based on appreciation of the ethical, social and behavioural aspects of a situation (e.g. including issues of responsibility, trust and reliance) as well as appreciation of features such as the economic significance of the issue.

These two objectives are not always easy to reconcile, and this is perhaps particularly true in the case of rice cultivation. Rice production situations are very diverse and involve complex agro-biosystems with both aquatic and terrestrial components. This complexity must be taken into account in an environmental risk evaluation, which aims to make recommendations for the use of PPPs so as to protect human health, avoid pollution of surface (SW) and groundwater (GW), and reduce exposure to non-target organisms.

In this section, we offer an overview of risk assessment procedures for evaluation of PPPs, with a focus on the use simulation models to determine predicted environment concentrations (PECs). We emphasise the importance of taking into account the diversity of rice production systems, and consider the limitations of current practices.

## 2.1 Overview of risk assessment procedures for evaluation of PPPs

According to the European Food Safety Authority (EFSA):

> Scientific assessments are evidence-based and demand rigorous methodologies to collect, evaluate and integrate scientific evidence, together with transparent and open communication of the processes and results of the assessment. A structured and clearly documented approach is essential if the outcome of the scientific assessment is to be communicated unambiguously to decision makers, the wider scientific community and stakeholders.[10]

In almost all countries, changes in the way pesticides are managed and regulated are largely determined by advances in scientific evidence and assessment of the risks pesticides pose.

Registration of pesticides is the process by which authorities (e.g. national governments or regional authorities) approve the sale and use of a PPP after evaluation of comprehensive scientific data, demonstrating that the product is effective for its intended purposes and does not pose an unacceptable risk to human or animal health or the environment. In common with other risk assessment procedures, PPPs to be used in rice production will go through a stepwise, tiered risk assessment approach. The lowest tier involves evaluating the product based on simple assumptions and conservative scenarios, and subsequent tiers work up to a more realistic and site-specific analysis. The need for further research or assessment at each step depends on the results from the previous steps: if there is an indication of a risk at a lower step, then additional, higher-tier tests should be performed in more realistic exposure conditions, or measures should be put in place to limit the proposed uses of the product in order to prevent the identified risks or address other uncertainties.

## 2.2 The need for multiple simulation models

Pesticide risk assessments in rice conducted for regulatory purposes largely depend on the use of simulation models to determine PECs of PPPs. In order to establish a realistic assessment and management procedure for environmentally friendly rice production, it is important to develop and validate mathematical models adapted to the diversity of rice production situations worldwide.

The European group 'Mediterranean-Rice' (MED Rice) was charged with developing a common system for the risk assessment of PPPs in rice. However, the group's review of the crop, water and pesticide management practices applied in European rice cultivation revealed significant differences among the various rice-producing countries (Med-Rice, 2003). As a result, it was concluded that the development of various well-defined national rice scenarios would in fact be more appropriate than the development of a single pan-European scenario, or even a few such general scenarios.

Even within a single country, several region-specific scenarios may be needed, as Yang Cheng argues for the case of China:

> As a large agricultural country, China exhibits significant differences in climate, soil, terrain conditions and water management practices among the various rice-cultivation regions. The existence of such differences suggests that it would be more appropriate to develop various region-specific scenarios instead of one or a few pan-China scenarios. (Yang Cheng, 2015).

Numerous models of varying complexity are available. It is not possible to provide a detailed review of all the models here. However, in the rest of Section 2 we offer a brief review of some tier 1 and higher-tier assessment models, as well as commenting on how models can be refined for higher-tier assessments and specific situations, and noting current limitations in the use of models. For more detailed consideration of different models, readers may consult the inventory provided in 'The Guidance Document for Environmental Risk Assessments of Active Substances used on Rice in the EU' (MED-Rice, 2003)[11] and Cheng et al. (2015), which covers Pesticide Concentration in Paddy Field (PCPF-1), PADDY (pesticide paddy field model), RICE Tier I, RICEWQ (RICE Water Quality model), and the model NIFE-AAP.

## 2.3 Tier-1 assessments

The PCPF-1 model was developed to simulate the transport of pesticides in paddy water and in the top 1 cm of paddy soil. The pesticide paddy field model (PADDY) was also developed to estimate pesticide concentrations in paddy water and surface soil. Both models have been validated with data obtained from laboratory experiments and field monitoring studies.[12]

The model NIFE-AAP (Tsuchida and Katou, 1992)[13] gives a description of the processes governing the distribution of pesticides after application. It is able to forecast the behaviour of the pesticides for several days after application, and is particularly useful for predicting inhalatory intake by humans in the surroundings of the treated area. The concentrations of pesticide in the SW, GW and soil near the application site can also be calculated using this model.

The MED-Rice group have developed a rather simple tier-1 spreadsheet, parameterised according to a conservative pan-Europe scenario (MED-Rice, 2003) which could be used for calculating PECs in GW and SW. Similarly, the Tier I Rice Model v1.0 was developed by the US Environmental Protection Agency[14] to estimate pesticide concentrations in rice paddies following their application.

## 2.4 Higher-tier risk assessments

According to the tiered approach for risk assessment of pesticide use in rice paddies, when a potential risk has been identified, a more complex model should be used to refine the results and effective risk mitigation measures should be suggested.

The MED Rice group propose using the model RICEWQ as a first option for developing a higher-tier exposure model for evaluating pesticide transport in rice production. RICEWQ[15] is a pesticide transport model for evaluating flowing water scenarios.[16] This model has also been developed in the United States for the purpose of pesticide registration.

For higher-tier exposure assessment, RICEWQ has been coupled with other models such as VADOFT (Vadose Zone Transport model, Capri and Miao, 2002),[17] EXAMS (Exposure Analysis Modeling System, developed by U.S. Environmental Protection Agency) RIVWQ[18] and TOXSWA (TOXic substances in Surface WAters used in the EU working group on Surface Water Modelling) to predict PECs in GW and the surrounding SW at a watershed level.

All these models are widely used and applied in various countries around the world. However, when models are applied to rice crop management scenarios that differ from the scenarios for which they were developed, the models need to be adapted appropriately. Refining a model to be used in a higher-tier risk assessment involves dropping the conservative assumptions that have been used in the previous risk assessment steps, and instead aiming to represent the actual agricultural situation in question (Karpouzas, 2008).[19] A common method for doing this is to include landscape information, using data from remote sensing and geographical information systems (GIS). This approach is particularly helpful for providing realistic, higher-tier modelling for rice-cultivated basins.

Excellent progress has been made in this direction. The EU project 'An Earth obseRvation Model based ricE information Service' (ERMES)[20] is an example of a project which supports the rice sector by integrating the information from satellite observations, in situ measurements and crop modelling. The European Space Agency Sentinel-1A, a radar imaging satellite, is an important tool for rice mapping and monitoring. It can provide detailed information on Asia's rice crop, such as planted area, seasonality, cropping intensity and areas damaged due to floods or drought, as well as information on rice crop growth.[21] These approaches have the potential to provide reliable real-time information on crop distribution, status and seasonal dynamics to support the implementation of agro-environmental policies and to promote sustainable management practices.

## 2.5 Limits on PPP risk assessments

All risk assessment methodologies, including methods to retrieve and analyse pertinent information, must be employed in a way that ensures the assessment is fit for purpose and appropriately tailored to answer the question posed. This means that different degrees of complexity and completeness, both of data examined and of methodologies applied, will be appropriate for different situations, depending on knowledge available on the topic, the time frame and the resources available for responding to a particular question. In other words, limiting factors mean that the most detailed, comprehensive approach is not always the right one, because, for example, it may deliver its verdicts too late to be of use to the relevant decision makers and risk managers (EFSA, 2015).[22]

There are also various factors which limit the effectiveness of real-life risk assessment processes, as examined by the HEROIC (Health and Environmental Risks: Organisation, Integration and Cross-fertilisation of Scientific Knowledge) EU FP7 project, with a stakeholder's consultation process developed within the project. The scenarios actually used in risk assessments are not always relevant to the contexts in which they are applied, and fail to represent important features of the real agricultural situation. The currently available models and computational tools still have limited flexibility in representing different, real-world scenarios. User knowledge of models and computational tools is also limited, resulting in problems both in the choice of methodology for a particular situation and the execution of the risk assessment. Finally, data required as the input for the risk assessment is not always available.

In conclusion, environmental policies based on the risk assessment practices discussed in this section have delivered many improvements, but even when we focus just on the first objective of risk assessment presented above, substantial challenges remain. In the next section, we focus on the second objective of risk assessment stated above: to provide answers based on appreciation of the ethical, social and behavioural aspects of a situation as well as appreciation of features such as the economic significance of the issue.

# 3 The role of ethical, social and behavioural issues in PPP risk assessment

Various factors make the case of risk assessment and regulation of pesticides an interesting test case when considering the role of ethical, social and behavioural issues in risk assessment. The regulation of PPPs affects, and is of interest to, a diverse range of parties, including regulatory risk assessors, risk managers and risk communicators; applicants for product authorisation; and the wider scientific community and the general public. In comparison to the situation with many other chemical classes, pesticide risk analysis has a long history. It is therefore comparatively well regulated, especially with respect to market authorisation procedures. It has been shown that information about how these substances are employed and how socio-behavioural factors can influence human or environmental exposure is useful in the task of avoiding risks to human health and the environment (Remoundou et al., 2014).[23] Finally, recent years have witnessed increased awareness from policy-makers on the adverse effects of occupational exposure to pesticides among agricultural workers.

Several of these features of the situation suggest the relevance of ethical and socio-behavioural issues. However, European risk assessment of the pesticide authorisation process does not include socioeconomic evaluation, except in very few cases e.g. the authorisation of so-called minor uses of pesticides, that concern high value speciality crops (as aromatic plants or fruits). Likewise, socio-behavioural aspects are not addressed.

It is commonly argued that engagement in unsafe pesticide use and disposal practices are the result of a lack of knowledge, and of misperceptions of the risks associated with pesticides among operators and workers. However, explaining the risks posed by pesticide use without exploring the underlying reasons for particular attitudes about pesticide risks could limit the efficacy of the risk analysis and management process. In this section, we underline the importance of analysing, from a social science perspective, what a relevant community knows about pesticides and how they perceive the risks associated with pesticides.

## 3.1 Labour force demographics

Employment growth in agriculture has slowed, and agricultural employment is falling in developed economies; EU agricultural labour is down by nearly 25% since 2000, according to data from Eurostat (2010), while in East Asia, Latin America and the Caribbean regions, nearly 30% of the world's labour force is employed in agriculture (World Bank, 2014).

In many countries, child labour is mainly an agricultural issue. Worldwide, 60% of all child labourers in the age group 5–17 years work in agriculture, including farming, fishing, aquaculture, forestry and livestock [Data extracted from International Labour Organization (ILO), http://www.ilo.org/ipec/areas/Agriculture/lang--en/index.htm]. Moreover, in many countries basic labour rights have not been extended to agriculture. Family farming is an anomalous category where child labour assumes an ambiguous character – part apprenticeship, part forced labour.

Children are more sensitive to the toxic effects of pesticide than adults for various biological and behavioural reasons, and it is very important to take this into account in those contexts where children are exposed to pesticides as workers. Where children are involved in agricultural labour, risk management strategies need to place a particular focus on improving their knowledge of and minimising their exposure to pesticides. FAO and ILO have developed a facilitation tool, 'Protect children from pesticides!', which can be used to raise awareness and build knowledge at a local level of the issue of child labour and exposure to pesticides.[24]

## 3.2 Risk perception and knowledge of proper use procedures

For all categories of workers employed in agriculture (including vulnerable people such as children and pregnant women), pesticides use can lead to substantial environmental and health risks, especially if the pesticides are not used in accordance with safety advice (Pimentel, 1992).[25] Previous works (Damalas and Eleftherohorinos, 2011[26]; Calliera, 2013[27]) have explored existing evidence that environmental and occupational exposure could occur due to accidents while mixing, loading or applying pesticides or through contact with treated crops. Exposure risk increases when operators and workers ignore safety instructions on how to properly use pesticides, as well as instruction on the use of personal protective equipment (PPE) and adoption of sanitation practices (Damalas and Eleftherohorinos, 2011; Remoundou et al., 2014).

Pesticide risk assessment, as currently performed, is a deterministic and also quantitative process. Mitigation measures are considered to reduce risk, and the authorisation is bound to so-called 'good agricultural practices'. Based on these assumptions, the results of the risk assessment are then extrapolated to the whole population, but may not be representative of global risk or societal needs. Several published studies show that in developing countries, high illiteracy rates contribute to farmers' (and farmworkers') difficulties in understanding and following pesticide use instructions and safety advice (Remoundou et al., 2014).

It should also be recognised that risks have social and psychological dimensions (Kasperson, 1988[28]; Slovic, 1987[29]) and are shaped by values, beliefs, attitudes, political systems and cultural factors. In order to address this, we must have an understanding of how different 'actors' involved in rice cultivation (operators, workers, children and women) perceive the risks associated with pesticide exposure. On the basis of such understanding, more effective risk communication strategies could be developed.

Traditionally, policies have focused on changing farmworkers' behaviour using financial incentives (e.g. reduced taxes, subsidies) and regulations (e.g. prohibiting certain actions, setting standards), rather than aiming to change perceptions and attitudes. However, these tactics have not always reduced the risks sufficiently, and further efforts are needed. Several studies show that the association between risk perceptions and behaviours is not straightforward, because other factors (work, economic and peer group pressures, and cultural factors) may also influence the extent to which farmers adopt self-protective behaviours (Remoundou et al., 2014). In recent years, a significant body of literature has emerged which examines how perceptions of risk and attitudes held by different pesticide users are linked to exposure. The level of formal knowledge is not the only factor that determines how actors will follow technical recommendations. For example, some research has suggested that training on pesticide risks will fail to have the desired effect if pesticide users are just recipients of formal knowledge and are not able to link this knowledge to their own skills or experiences (Rios Gonzales, 2013).[30]

A significant finding of a recent review on this topic is that higher risk perceptions and better levels of knowledge about the risks involved do not always translate into better use of, and adherence to, protective advice, guidelines and protective equipment. Central to the development of effective training and risk communication materials is an understanding of how different pesticide users perceive the risks associated with pesticide exposure, and how these perceptions are shaped by socio-economic, cultural, environmental and/or institutional factors (Remoundou et al., 2014).

# 4 The European SUD and IPM

In 2009 the Council of the European Union adopted Directive 128, which established a framework for community action to achieve the sustainable use of pesticides (SUD). As a consequence, member states (MSs) were required to develop specific measures to minimise environmental and occupational exposure to pesticides, and also to produce communication programs aimed at raising public awareness about the risks of pesticide exposure. This directive represents a recognition that improvements are needed in managing the risks associated with PPPs.

Both developing and developed countries, together with international organisations, have gradually moved their attention from the efficacy of a single tool, such as a particular pesticide, to the sustainability of a collection of agricultural practices. This is one of the key principles at the heart of the European Directive on the Sustainable Use of Pesticides, and also of IPM, which is incorporated into and recommended by the directive.

The objective of the EU directive is to achieve sustainable use of pesticides by implementing a plan for risk reduction, which includes promoting IPM and introducing effective non-chemical alternatives to traditional pesticides. MSs are required to transpose the directive into national legislation and to implement it through a national action plan (NAP).

MSs are required to set up their quantitative objectives, targets, measures and timetables to reduce the time trend risks and impacts of pesticide use on human health and the environment, and to encourage the development and introduction of IPM strategies and of alternative approaches or techniques in order to reduce risks and cover various areas of concern (such as worker protection, protection of the environment and reducing pesticide residues in food products). The plan requires reviewing, approval and periodical monitoring by the Commission to verify compliance and to identify best practices.

One of the official definitions of IPM provided by the International Organisation for Biological and Integrated Control of Noxious Animals and Plants (IOBC-OILB) states that '*Integrated control represents procedures which utilize all economically, ecologically, and toxicologically acceptable methods for keeping the pests under the threshold of harmfulness with preferential and meaningful utilization of natural restricting factors*'. IPM must therefore be considered as a broad ecological approach which draws on a variety of pest control techniques and targets an ecosystem as a whole, rather than as a specific measure. The directive includes a list of principles to be followed for an IPM approach, covering prevention of pests, monitoring of the pest situation and intervention using a broad range of measures, including anti-resistance strategies and biological control. The Directive shows how such principles can then be used to guide choices of PPPs and their conditions of application. Article 14 of the Directive requires MSs to take all necessary measures to promote IPM implementation essential for the success of reducing pesticide risks.

A clear statement of how the Directive understands IPM and aims to apply it for the sustainable use of pesticides is given in Article 3:

> *'[I]ntegrated pest management' means careful consideration of all available plant protection methods and subsequent integration of appropriate measures that discourage the development of populations of harmful organisms and keep the use of plant protection products and other forms of intervention to levels that are economically and ecologically justified and reduce or minimise risks to human health and the environment. 'Integrated pest management' emphasises the growth of a healthy crop with the least possible disruption to agro-ecosystems and encourages natural pest control mechanisms. (Directive 2009/128/EC — EU action to achieve the sustainable use of pesticides, http://eur-lex.europa.eu)*

From this, it is clear that the implementation of the principles of IPM can take many different forms. IPM solutions must be tailored to the specificities of a certain cropping system, climatic conditions and pest pressure, availability of efficient solutions or timing during a cropping cycle.

IPM should be distinguished from specific production systems, like organic farming, that by definition ban the use of some chemicals in the protection of crops. Completely avoiding the use of pesticides may be considered the easiest way to avoid the risks associated with them. However, such an approach cannot always deliver economic and social security. This is why some level of pesticide application may be included in the implementation of IPM.

An example of a pest control solution which is tailored for a specific situation can be found in the rice-producing territories of Piedmont and Lombardy in Italy. The Laboratory of Mycology and Plant Pathology, University of Pavia, has developed a surveillance system that allows active protection of the rice paddies in these areas from the pathogenic fungus *Pyricularia oryzae* only during the periods when there is an effective risk of infection. The risk is calculated based on the presence/abundance of the pathogen in the air, the phenological stage of the plant, the environment and cultural practices predisposing infection.

Another important aspect of SUD is regulating the information manufacturers provide about the proper application of PPPs. The product labels must specify the quantity necessary for the ideal vegetative stage of the plant/ha. Careful observation of these guidelines by those using PPPs is one of the most important preventative measures to minimise risks. Instructions on the label need to be interpreted to ensure a correct and precise application of a PPP for the specific vegetative stage of the crop, climate conditions and application period time frames. The labelling of products should also help users to make comparative assessments of products in order to choose the product with the least impact on the environment and human health in a given situation.

# 5 Farmer involvement in IPM

IPM has been accepted as a cardinal principle of plant protection under the National Agricultural Policy of the Government of India. The IPM package for rice[31] aims to combine knowledge of pests and crop interaction with knowledge of locally available pest management tools to identify the best tools for a given situation. One key component of the approach involves monitoring the number of pests present, and also monitoring the presence of beneficial insects. Prevention (both cultural and genetic) and timely intervention (biological and/or chemical) are the other two key components of the approach. The package gives recommendations for which chemical control measures to adopt for particular pests in its Generic IPM module based on vegetative stage (Prakash et al., 2014, 17–18). It recommends that chemical interventions should only be used as a last resort, once the pest population has surpassed the Economic Threshold Level defined for each pest in the package (Prakash et al., 2014, 9).

Beginning in the mid-1980s, Indonesia also introduced IPM as a new crop management approach, developing the Farmer Field School (FFS) model of empowering farmers to make informed crop management decisions. This was then piloted in China during the early 1990s with FAO assistance, with an initial focus on IPM for rice and cotton.[32]

The examples above help to show that IPM research programs can only be successful with involvement of farmers and other stakeholders. However, defining the ideal form for this engagement is not a simple task. Research, training and extension interactions between farmers and scientists are complex, and depend on the available financial and human resources in a given situation. Both farmers and scientists have a role to play in the knowledge generation process, as each group brings an important perspective.

IPM is a leading form of sustainable intensification, which offers to complement and restrain the use of synthetic pesticides. It has particular importance for tropical smallholders (Jules Pretty, 2015).[33] Due to the diversity of rice production situations worldwide, there is no single recipe for IPM. The location, season, weather conditions and many other factors affect its implementation and require flexible management decisions to adapt to the local agronomic, pest and on-farm circumstances.

A single intervention is rarely an effective way to control a pest, disease or weed. IPM recognises that a package of technologies and practices will be more successful. Moreover, the package for a given situation needs to be developed in partnership with local farmers, so that it is tailored to the important features of the local situation. By developing such packages in partnership with local farmers in a process of participatory research, the farmers' knowledge will of course also increase (Pretty, 2015).

In the paper 'Integrated Pest Management in Rice', the authors comment that although 'IPM is not for farmers but is by farmers' may be a common saying in IPM circles, nevertheless there are difficulties in getting IPM into the hands of farmers.[34] The usefulness of establishing general guidelines for the production of rules specific to a given situation has been demonstrated several times. The past 25 years have seen a substantial increase in understanding of how to increase farmers' knowledge, and the most significant innovation has been the ongoing development and deployment of FFS from the late 1980s. Agricultural experiments such as the International Rice Research Institute (IRRI)'s Long-Term Continuous Cropping Experiment (LTCCE) (which recently reached its 150th rice cropping season) help to make complex sustainability issues more tangible.

However, despite advances in the application of IPM, significant challenges remain. Ramsar Resolution XI.15 notes the following serious concerns:

> [I]n some rice-growing countries the regulation of rice pesticide use remains underdeveloped, the impacts of unsustainable pesticide usage are not adequately addressed, and the response strategies required are complex and wide ranging ... [T]he risks of the overuse or inappropriate use of such pesticides to human health, rice pest control by natural predators, and overall wetland biodiversity, including that depended upon by local communities for their livelihoods, such as from fisheries, are not always well recognized or dealt with by stakeholders.[35] (Ramsar COP11 Resolution)

Despite various efforts to promote the sustainable use of pesticides, these concerns are still valid. As noted in Weaver and Rotmans (2006),[36] what sustainability means in any given context is a question that needs to be interpreted by a range of stakeholders. Likewise, efforts to achieve sustainability must be implemented by the group of stakeholders in a given situation. When stakeholders in and end users of a process are not sufficiently involved in its development, there is an increased chance that the process will fail to achieve the desired outcomes.

# 6 Conclusion

The conventional assumption is that pesticide use correlates with increased crop yields. However, yield maximisation does not necessarily lead to maximum socioeconomic welfare (Ryoichi Doi and Supachai Pitiwut, 2014).[37]

Achieving a profitable crop may not allow for the elimination of pesticide treatments, and in order to reach a sustainable use of PPPs and to implement IPM effectively in rice cultivation, the following points must be taken into account:

- It is often difficult for farmers to adopt the ideas and technologies proposed by scientists, and it is also difficult for scientists to fully appreciate the individual characteristics of particular rice production areas, and the significance of these differences between sites (Ryoichi Doi and Supachai Pitiwut, 2014).
- There are limitations to the way pesticide risk assessments for the case of rice crops are produced. One of the conclusions of the HEROIC project is particularly important for rice cultivation: 'despite human resource constraints, the demand for RA remains high, regulatory timeframes remain ambitious and assessments are getting increasingly complex. The growing complexity and the multidisciplinarity of RA sciences, are creating major capability issues which are amplified by the lack of adequate training programs' (Pery et al., 2014).[38]
- One of the critical success factors in carrying out an adequate risk assessment is the availability and accessibility of the necessary data. It is globally recognised that regulatory science is moving forward very slowly compared to other scientific disciplines (Wilks et al., 2015).[39]
- When performing a risk assessment and developing strategies to limit the exposure risks caused by pesticide use, it is important to take into account socio-behavioural aspects of the situation where possible. The evaluation of occupational risk should focus more on real-life conditions, people's own experiences and local forms of knowledge transmission (Rios Gonzales, 2013). Such factors, once identified, could be useful in determining whether adjustments and eventual revisions of local risk assessment policy and processes are needed (Calliera et al., 2015).[40]
- There is the evidence that IPM is most successful when, rather than farmers being told how to implement it, they are enabled to learn about the rice agro-ecosystem and make their own decisions, especially about pesticide use, variety deployment, crop establishment and crop management.
- The further development of IPM requires the development of criteria for assessing the social benefits associated with sustainable agriculture, the development of tools to aid stakeholder and user involvement in a variety of contexts, and strong governmental support in the form of both policy support and subsidies.

# 7 Where to look for further information

For data on population and development:
http://www.un.org/en/development/desa/news/population/2015-report.html

For more information on ethics and food:
http://www.bioethicsinstitute.org/globalfoodethics/the-7-by-5-agenda-for-ethics-and-global-food-security#lit

For European procedure on pesticide risk assessment and evaluation:
http://www.efsa.europa.eu

The European legislation is available on:
http://eur-lex.europa.eu/homepage.html

For be better informed on labour and children employed in agriculture:
http://www.ilo.org/ipec/areas/Agriculture/lang--en/index.htm
http://www.fao.org/childlabouragriculture/en/

For IPM and sustainable use pesticide in rice culture we suggest:
http://www.iobc-wprs.org/ip_ipm/IOBC_IP_principles.html
http://irri.org
http://www.fao.org/nr/land/sustainable-land-management/farmer-field-school/en/

# 8 References


1 *Official Records of the General Assembly, forty-second Session, supplement No. 25* (A/42/25) From A/42/427. Our Common Future: Report of the World Commission on Environment and Development (1987).

2 SOER, EEA: The European Environment–State and Outlook 2010: Synthesis. European Environment Agency, Copenhagen (2010). doi: 10.2800/45773.

3 Jessica Fanzo, Ethical issues for human nutrition in the context of global food security and sustainable developement, *Global Food Security* 7 (2015), 15–23. doi:10.1016/j.gfs.2015.11.001.

4 Qiqi Chen, Junbiao Zhang and Lu Zhang, *Sustainability* 7 (2015), 563–83; Risk Assessment, Partition and Economic Loss Estimation of Rice Production in China. doi:10.3390/su7010563.

5 Ferrero and Tinarelli, Rice Cultivation in the E.U. Ecological Conditions and Agronomical Practices Pesticide in Risk Assessement in Rice paddy. E. Capri and D. Karpouzas (eds) (2008). ISBN 978-0-444-53087-5.

6 L. P. Gianessi, Importance of Pesticides for Growing Rice in South and South East Asia. International Pesticide.Benefit Case Study 108, October 2014, Crop life foundation, https://croplife.org/wp-content/uploads/pdf_files/Case-Study-108-Rice-in-Asia2.pdf.

7 Yang Cheng et al., A China rice paddy scenario and modeling system for pesticide aquatic ecological risk assessment. *Journal of Chemical Engineering and Chemistry Research* 2, no. 5 (2015), 589–98.

8 OECD Programme on Pesticides and Sustainable Pest Management, http://www.oecd.org/chemicalsafety/pesticides-biocides/OECD-Pest-Vision-Final.pdf.

9 FAO Regional office for Asia and the pacific, Bangkok, 2015 Progress in pesticide risk assessment and phasing out of highly hazardous pesticides in Asia. ISBN 978-92-5-108709-1.

10 A. Hardy, J. L. C. M. Dorne, E. Aiassa, J. Alexander, B. Bottex, Q. Chaudhry, A. Germini, B. Nørrung, J. Schlatter, D. Verloo and T. Robinson, Increasing robustness, transparency and openness of scientific assessments. *EFSA Journal* 13, no. 3 (2015), e13031.

11 MED-Rice, Guidance Document for Environmental Risk Assessments of Active Sub- stances used on Rice in the EU for Annex I Inclusion. Document prepared by Working Group on MED-Rice, EU Document Reference SANCO/1090/2000 – rev.1, Brussels, June 2003, 108p.

12 Yan Cheng, Junying Zhou, Zhengjun Shan, Deyang Kong, Min Guo, Wenlin Chen and Martin Williams, A China rice paddy scenario and modeling system for pesticide aquatic ecological risk assessment. *Journal of Chemical Engineering and Chemistry Research* 2, no. 5 (2015), 589–98.

13 Hirosji Tsuchida and Tatsuo Katou, Mathematical model of pollution caused by aerial application of pesticides to rice fields. *Chemosphere* 24, no. 7 (1992), 867–79. See the extended summary in *Pestic Science* 49 (1997), 210–12.

14 http://www2.epa.gov/pesticide-science-and-assessing-pesticide-risks/tier-i-rice-model-version-10-guidance-estimating.

15 W. M. Williams, A. M. Ritter, J. M. Cheplick and C. E. Zdinak, RICEWQ: Pesticide runoff model for rice crops – user's manual and program documents version 1.6.1. Waterborne Environment Inc., S.E. Leesburg, VA, USA (1999).

16 http://www.waterborne-env.com/model/ricewq-19/.

17 E. Capri and Z. Miao, Modeling pesticide fate in rice paddy. *Agronomie* 22 (2002), 363–71.

18 W. M. Williams, C. E. Zdinak, A. M. Ritter, J. M. Cheplick and P. Singh, RIVWQ: Chemical transport model for riverine environments – user's manual and program documentation version 2.02. Water-borne Environment Inc., S.E. Leesburg, VA, USA (2004).
19 D. G. Karpouzas and Z. Miao, Higher Tier Exposure Assessment in Rice paddy Areas: an European Perspective, in Risk Assessement in Rice paddy. E. Capri and D. Karpouzas (eds) (2008). ISBN 978-0-444-53087-5.
20 www.ermes-fp7space.eu.
21 http://www.esa.int/Our_Activities/Observing_the_Earth/Copernicus/Sentinel-1.
22 European Food Safety Authority (EFSA), Editorial: increasing robustness, transparency and openness of scientific assessments. *EFSA Journal* 13, no. 3 (2015), e13031, 3p. doi:10.2903/j.efsa.2015.e13031. ISSN: 1831-4732.
23 K. Remoundou, M. Brennan, A. Hart and L. Frewer, Pesticide risk perceptions, knowledge, and attitudes of operators, workers, and residents: a review of the literature. *Human and Ecological Risk Assessment: An International Journal* 20, no. 4 (2014), 1113–38.
24 FAO 2015, www.fao.org/childlabouragriculture/en/.
25 D. Pimentel, H. Acquay, M. Biltonen, P. Rice, M. Silva, J. Nelson, V. Lipner, S. Giordano, A. Horowitz and M. D'amore, Environmental and economic costs of pesticide use. *BioScience* 42 (1992), 750–60.
26 C. Damalas and I. Eleftherohorinos, Pesticide exposure, safety issues, and risk assessment indicators. *International Journal of Environmental Research and Public Health* 8, no. 5 (2011), 1402–19. doi:10.3390/ijerph8051402
27 C. Calliera, F. Berta, T. Galassi, F. Mazzini, R. Rossi, R. Bassi, P. Meriggi, A. Bernard, A. Marchis, A. Di Guardo et al., Enhance knowledge on sustainable use of plant protection products within the framework of the Sustainable Use Directive. *Pest Management Science* 69, no. 8 (2013), 883–8.
28 R. Kasperson, O. Renn, P. Slovic, H. Brown, J. Emel, R. Goble, J. Kasperson and S. Ratick, The social amplification of risk: a conceptual framework. *Risk Analysis* 8, no. 2 (1988), 177–87.
29 P. Slovic, Perception of risk. *Science* 236 (1987), 280–5.
30 A. Ríos-González, K. Jansen, H. J. Sánchez-Pérez, Pesticide risk perceptions and the differences between farmers and extensionists: towards a knowledge-in-context model. *Environmental Research* 124 (2013), 43–53.
31 A. Prakash, J. S. Bentur, M. Srinivas Prasad, R. K. Tanwar, O. P. Sharma, S. Bhagat, M. Sehgal, S. P. Singh, M. Singh, C. Chattopadhyay, S. N. Sushil, A. K. Sinha, R. Asre, K. S. Kapoor, K. Satyagopal and P. Jeyakumar (2014), Integrated Pest Management for Rice, http://www.ncipm.org.in/NCIPMPDFs/ipmpackages/Rice%20Bulletin.pdf.
32 http://www.vegetableipmasia.org/pages/6-china-national-ipm-programme.
33 Jules Pretty and Zareen Pervez Bharucha, Integrated pest management for sustainable intensification of agriculture in Asia and Africa. *Insects* 6 (2015), 152–82. doi:10.3390/insects6010152.
34 K. D. Gallagher, P. A. C. Ooi, T. W. Mew, E. Borromeo and P. E. Kenmore, Integrated pest management in rice. In International Rice Commission Newsletter (FAO), v. 51 (2002), 9–24, ftp://ftp.fao.org/docrep/FAO/004/y6159t/y6159t02.pdf.
35 11th Meeting of the Conference of the Parties to the Convention on Wetlands (Ramsar, Iran, 1971) 'Wetlands: home and destination' Bucharest, Romania, 6–13 July 2012. Resolution XI.15 Agriculture-wetland interactions: rice paddy and pest control.
36 P. M. Weaver and J. Rotmans, Integrated sustainability assessment: what is it, why do it and how? *International Journal of Innovation and Sustainable Development* 1 (2006), 284–303.
37 R. Doi and S. Pitiwut, From maximization to optimization: a paradigm shift in rice production in Thailand to improve overall quality of life of Stakeholders. *The Scientific World Journal* 2014 (2014), 604291. doi:10.1155/2014/604291.
38 A. R. Péry, G. Schüürmann, P. Ciffroy, M. Faust, T. Backhaus, L. Aicher, E. Mombelli, C. Tebby, M. Y. Cronin, S. Tissot, S. Andres, J. M. Brignon, L. Frewer, S. Georgiou, K. Mattas, J. C. Vergnaud,

W. Peijnenburg, E. Capri, A. Marchis and M. F. Wilks, Perspectives for integrating human and environmental risk assessment and synergies with socio-economic analysis. *Science Total Environment* 456–7 (2013), 307–16.
39 M. F. Wilks et al., White paper on the promotion of an integrated risk assessment concept in European regulatory frameworks for chemicals. *Science of the Total Environment* 521–2 (2015), 211–18.
40 Maura Calliera, Alex Marchis, Gabriele Sacchettini, Ettore Capri, Stakeholder consultations and opportunities for integrating socio-behavioural factors into the pesticide risk analysis process. *Environmental Science Pollution Research* 23 (2016), 2937–47.

Chapter 12

# Integrated pest management for sustainable rice cultivation: a holistic approach

*F. G. Horgan, University of Technology Sydney, Australia*

1 Introduction
2 Emerging rice production landscapes and 'crop health syndromes'
3 Increasing rice production
4 Increasing water productivity
5 Reducing labour costs
6 Increased availability and marketing of agrochemicals
7 Pest management for rice: scientific approaches
8 Knowledge of the rice production system
9 Biocides, traps and barriers
10 Transgenic and herbicide-resistant rice varieties
11 Genetic approaches to resistance and tolerance
12 Enhancing natural regulation
13 Integrating management practices
14 Education and extension
15 Future trends and conclusion
16 Where to look for further information
17 References

## 1 Introduction

Integrated pest management (IPM) has been the dominant paradigm for crop protection since the 1960s. In theory, farmers practicing IPM will make decisions about pest management interventions that are economically justified, knowledge-based and that minimize risks to the environment and human health (Parsa et al., 2014). A key feature of IPM is an active monitoring of the crop for pest populations or damage that is weighed

http://dx.doi.org/10.19103/AS.2016.0003.23

against some pre-established damage threshold above which farm profits would be negatively affected. Farmers are expected to balance the costs of intervention against the potential economic losses if no action were taken (Matteson, 2000). Although elegant in theory, it has been difficult to put IPM into practice on rice farms (Morse and Buhler 1997; Palis, 1998; Bandong et al., 2002) as farmers are pulled between opposing pest management paradigms and vested interests (Matteson, 2000; Matsuno et al., 2006; Spangenberg et al., 2015).

This chapter aims to critically examine IPM in rice. It assesses directions and advances in pest management science, and identifies opportunities and obstacles for IPM to form a basis for truly sustainable pest management. The chapter focuses on rice production from land preparation to rice grain storage, mainly in Asia.

# 2 Emerging rice production landscapes and 'crop health syndromes'

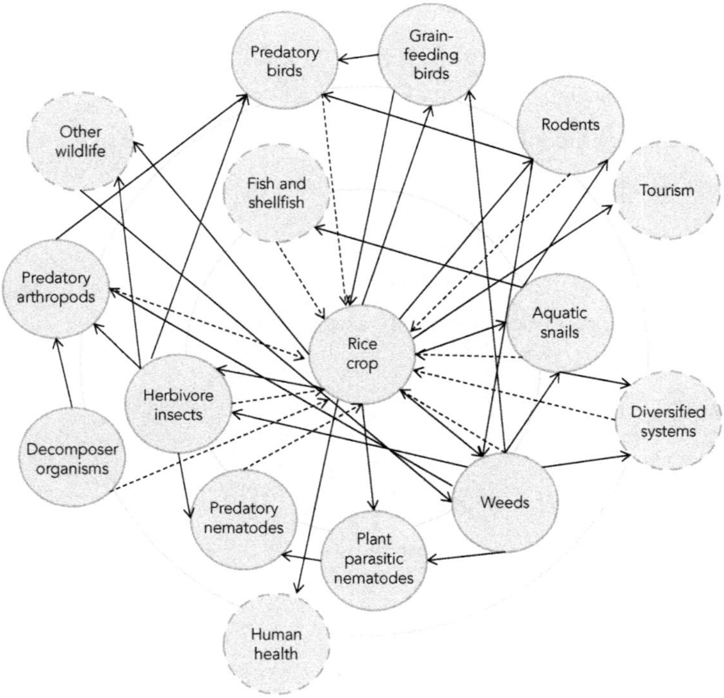


**Figure 1** Relations between some components of the rice ecosystem. Components located closer to the rice crop (at the centre of the diagram) include permanent residents in the crop. Components located away from the centre are often occasional visitors to the crop that occur in the rice production landscape. Circles with dashed outlines represent other sectors that interact with rice production. Solid arrows indicate resource flows or benefits (in the direction of the arrow) to or from components. Dashed arrows are indirect benefits to the rice crop.

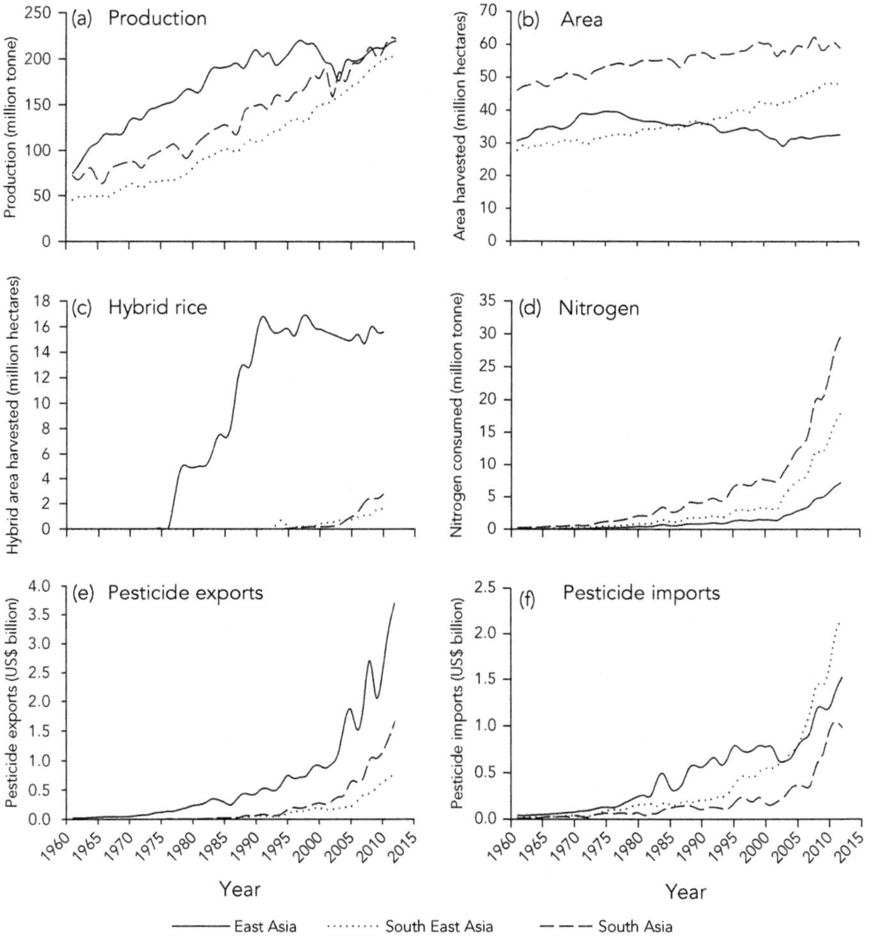


**Figure 2** Trends in rice production and agricultural inputs in East (China, Japan and South Korea), South East (Cambodia, Indonesia, Malaysia, Myanmar, Philippines, Thailand and Vietnam) and South (Bangladesh, India, Nepal, Pakistan and Sri Lanka) Asia between 1961 and 2012. Graphs indicate trends in (a) rice production, (b) the area of rice harvested, (c) the area of hybrid rice planted (East = China; South East = Indonesia, Philippines and Vietnam; South = Bangladesh and India), (d) total nitrogen use, (e) total pesticide exports and (f) total pesticide imports for each region. Graphs a, b, d, e and f were drawn using open access data from FAO (2016). Graph c was redrawn from Horgan and Crisol (2013).

As farming practices evolve over time, associated pest problems will also change with them. It is useful therefore to approach pest management issues not as dealing with a series of individual problems, but in more holistic terms as approaching generalized 'crop health syndromes' (Savary et al., 2005; Reddy et al., 2011). This approach not only draws attention to production systems and associated management practices as drivers of pest problems, but also better reflects the several interrelations between different species components of rice ecosystems – some of which can become economically challenging (Moody and Drost, 1983; Pathak and Khan, 1994) (Fig. 1).

The idea of crop health syndromes also acknowledges trends in ecological succession acting on rice fields. For example, several herbivores, including insects, snails and rodents are associated with paddy field weeds (Drost and Moody, 1982; Tindall and Stout, 2003; Horgan et al., 2014a), sometimes facilitating the weeds by reducing shade from the rice canopy. Weeds may in turn lead to increased damage to rice from birds, as has been noted from recent studies in Africa (Rodenburg et al., 2014). The concept of crop health syndromes also recognizes that under different management systems, certain 'pests' may become beneficial 'allies' because of their predatory or herbivore roles, for example, apple snails (*Pomacea* spp.: Ampullariidae) may increase rice production costs by damaging young seedlings in flooded rice systems, but can play a major role in reducing weed biomass during later crop stages (Okuma et al., 1994).

Four of the major drivers of recent changes in rice production practices have been demands and policies for greater food security, demands for water conservation, increasing labour costs and an increasing availability of agrochemicals (Savary et al., 2005; Godfray et al., 2010; FAO, 2016; Fig. 2). Synergies in reducing greenhouse gas emissions or reducing energy consumption have helped favour certain changes of practice (Cai et al., 1997). Some examples of possible trade-offs and synergies for pest management are explained in the following sections.

# 3 Increasing rice production

Increasing rice production has involved (i) increasing land areas for agriculture and (ii) intensifying production per unit area by adopting high-yielding varieties (HYVs), applying higher amounts of fertilizer and reducing yield gaps.

## 3.1 Increasing areas of production

Many countries are faced with agricultural land shortages as farmland is progressively converted to make space for housing and industry. As land is reduced close to urban areas, natural areas and marginal lands face increasing pressures from expanding agricultural fronts (Balasubramanian et al., 2007; De Koninck et al., 2011; Cockfield et al., 2012; Grecchi et al., 2014; Laurance et al., 2014). New areas for rice production face distinct challenges compared to traditional rice-growing regions including damage from wildlife and weeds; for example, elephants (*Elephas maximus* L.), peafowl (*Pavo cristatus* L.) and wild boars (*Sus scrofa* L.) are often among the principal rice 'pests' indicated by farmers in the vicinity of wildlife sanctuaries in India and Sri Lanka. Such regions can become the focus of conflict between farmers and wildlife (Bandara and Tisdell, 2002; Santiapillai et al., 2010; Rana and Jain, 2013).

As rice production intensifies in expanding agricultural areas, and particularly where pesticide use is high, conflict between rice production and other economic sectors can arise. Such has been the case with aquaculture and rice farming in the Mekong Delta (Vietnam) (Pucher et al., 2013). Furthermore, despite clear advantages for adapting rice to climate change, the use of modern varieties with high tolerance to abiotic stresses such as submergence (Jagadish et al., 2012) also draws responsibilities for habitat management as marginal areas that were once a refuge for wildlife become more feasible for rice production. There is an urgent need for scientists and managers to develop sound strategies for coexistence of rice production, resource extraction activities (aquaculture and fishing) and wildlife conservation, particularly as biodiversity conservation is a major index of sustainable development.

## 3.2 Increasing fertilizer use

Increasing fertilizer use is the most likely pathway farmers will take to increase rice production (Stewart et al., 2005). Most plants and their herbivores are limited by nitrogen availability (Mattson, 1980); therefore, the addition of nitrogen to rice ecosystems affects a range of organisms including the rice plant, its weed competitors and their herbivores (Blackshaw et al., 2003; Hu et al., 2016).

Under high nitrogen conditions, rice plants grow faster, partition more biomass to shoots and grain, and gain tolerance to pests and diseases. However, they also become notably more attractive to pests and more susceptible to a variety of diseases (Horgan and Crisol, 2013; Horgan et al., 2016a; Hu et al., 2016). Field and laboratory studies have indicated higher levels of damage from insects, snails and weeds to rice produced under high nitrogen conditions (Eggum and Juliano, 1973; Ampong-Nyarko and Datta, 1993; Hu et al., 2016). The structure of weed communities (relative abundance and species richness) may also significantly change under high nitrogen conditions (Blackshaw et al., 2003). Furthermore, high levels of nitrogen have been shown to decrease the effectiveness of some anti-herbivore resistance genes in rice (Salim and Saxena, 1991; Vu et al., 2014; Horgan et al., 2016b), although this likely depends on the specific resistance mechanisms involved (Horgan et al., 2016c).

Rice plants under high nitrogen are predicted to better tolerate damage (can compensate for biomass loss) depending on the nature of the attacking herbivore or parasite (Horgan et al., 2016a; Prot et al., 1994). For example, yield losses to rice from certain insect herbivores can be avoided by treating affected plants with high nitrogen (Jiang and Cheng, 2003; Inthavong et al., 2004).

Under high nitrogen, weeds will also gain tolerance to their own herbivores and diseases (i.e. Suding et al., 2004). Furthermore, high nitrogen can lead to a greater carryover of weeds to successive cereal crops depending on the length and favourability of fallow periods for their germination and survival (i.e. Sexsmith and Pittman, 1963). The type of nitrogen fertilizer used may also reduce damage from certain pests. For example, ammonia-based nitrogen reduces root-knot nematode (*Meloidogyne graminicola* [Golden & Birchfield]) infestation of rice (Patil et al., 2013) and causes significant mortality of apple snails (Stuart et al., 2014). Compared to chemical fertilizers, organic fertilizers are also associated with lower incidences of certain pests, particularly phloem-feeding insects (Hu et al., 2016).

## 3.3 High-yielding varieties

High-yielding rice varieties (HYVs) play an important role in attaining food security. However, HYVs may be more susceptible to pests and diseases than traditional varieties because of ecological trade-offs between yield (plant reproduction) and defences (resistance) (Brown, 2002); this has led to a large research investment in improving resistance in HYVs (see Section 11). A major advantage of HYVs is the availability of certified seed, which reduces the incidence of weedy (red) rice, weeds and seed-borne diseases (Chin, 2001). However, HYVs are often indirectly affected by pests through associated management practices (Cheng, 2009; Horgan and Crisol, 2013).

Farmers who adopt HYVs, particularly hybrid rice varieties, will often use higher amounts of fertilizer to meet the yield potential of the varieties (Cheng, 2009; Horgan and Crisol, 2013). Hybrid rice has been associated with higher incidences of several pests and diseases, including stem borers and planthoppers (Horgan and Crisol, 2013). Several hybrid rice varieties, particularly varieties with a wild abortive cytoplasmic male sterile (WA-CMS) parent, have noted susceptibility to the white-backed planthopper (*Sogatella furcifera* [Horvath]: Delphacidae: Homoptera) and green leafhoppers (*Nephotettix* spp.) (Cicadellidae: Homoptera) (Sogawa et al., 2003; Horgan et al., 2016a). This includes an estimated 40% of modern hybrid varieties and most Chinese-bred hybrids (Horgan and Crisol, 2013).

However, hybrid rice may be more tolerant of herbivore damage, which represents a major advantage for natural pest regulation in well-managed rice production systems, but also a potential drawback because it creates a reservoir for diseases, including insect-vectored rice viruses (Horgan, 2012; Horgan et al., 2016a). Knowledge of the weaknesses of hybrid rice has led researchers to seek new sources of male sterility without associated planthopper susceptibility (Horgan and Crisol, 2013); however, faster growth rates and the nutrient demands of hybrids may continue to make these more attractive to herbivores (Price, 1991; Horgan et al 2016a).

Marketing of hybrids and other HYVs as part of a package with other inputs, including pesticides, could threaten their sustainability (Cheng, 2009; Horgan and Crisol, 2013). It is predicted that the high initial costs of purchasing hybrid and certified seeds and a general belief that hybrid rice is susceptible to insects and diseases will cause farmers to apply higher amounts of pesticides in an attempt to protect their investments (Pingali and Gerpacio, 1998). These pesticides are often supplied by the same agro-suppliers as the hybrid seed (orchestrated value chains: Hinterhuber, 2002). As will be seen in Section 6 of this chapter, high pesticide use can also result in increased pest and weed problems.

## 3.4 Cropping intensity

At landscape scales, issues around cropping systems, the synchrony of planting and the nature of the rice landscape in terms of crop diversity are important drivers of certain plant health syndromes. Increasing cropping intensity will increase the potential for insect herbivores to complete further generations and results in practical problems such as carry-over of pests and weeds from one crop through stubble or fallow periods to the next crop (Holt et al., 1996; Azzam and Chancellor, 2002).

During dry fallows, where rice plants die, specialist herbivores such as the brown planthopper (*Nilaparvata lugens* [Stål]: Delphacidae: Homoptera) will disperse and populations decline or disappear (Holt et al., 1996). Dry fallows are also important opportunities for wild birds to consume dropped grain and prevent volunteer rice or weedy rice in succeeding crops (Stafford et al., 2010). In some regions, for example California (the United States), rice fields are flooded during non-productive winter fallows to conserve wild ducks; this also reduces weed abundance and biomass including the abundance of weedy rice and can contribute to nutrient cycling in the rice fields (Elphick and Oring, 1998; Van Groenigen et al., 2003).

In tropical regions with abundant water, a ratoon crop may develop during the fallow period. Ratooning rice can support pest populations between crops, particularly populations of plant-sucking homopterans that are attracted to new, fast-growing tillers (Fig. 3) – these may include important virus vectors such as leafhoppers. However,

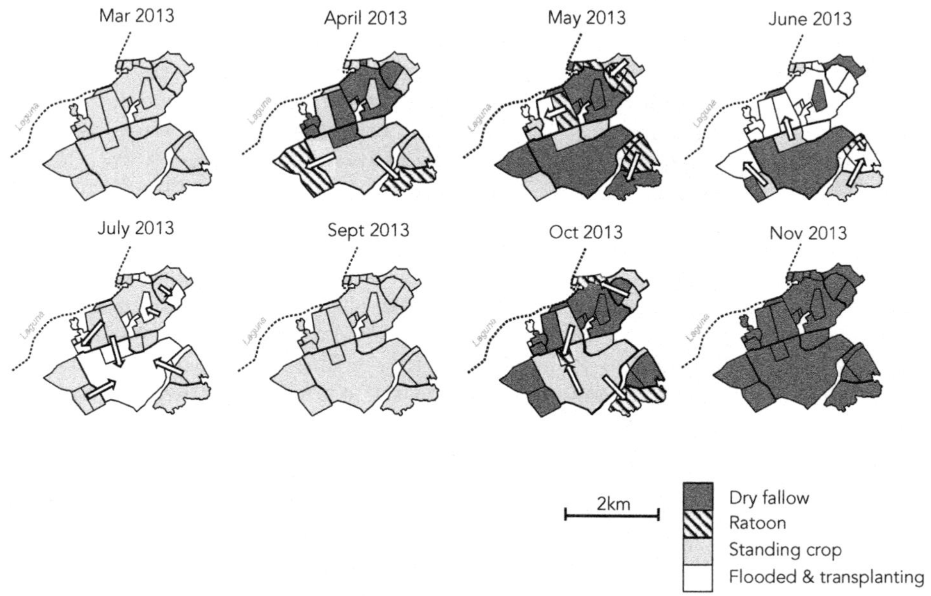


**Figure 3** Cropping pattern at Pila, Laguna (Philippines), during 2013. Shaded polygons are rice fields adjacent to Laguna Lake (shoreline indicated by the dotted line). Asynchrony in the rice landscape creates refuges for herbivorous insects and their natural enemies. Arrows indicate predicted movements of insects from older crop stages to younger plants (transplanted seedlings and ratooning/active tillering plants). Poor synchrony is expected to maintain monophagous herbivores and their specialized natural enemies through the dry season crop (January to May) to the wet season crop (June to October). Dry conditions in November prevent ratooning and break the cycle of re-infestation.

ratooning rice also maintains predator and natural enemy populations that can reduce densities of pests in succeeding crops (Dela Cruz and Litsinger, 1988).

Synchronous planting is regarded as an important factor in reducing pest-related yield losses. Synchrony will have its greatest impact in areas with less intensive cropping and after extended fallow periods. Synchrony reduces the incidence of viral diseases such as tungro, and the incidence of stem borer and rodent damage in rice crops (Holt et al., 1998; Jacob et al., 2010). However, attaining wide-area synchrony of planting is often limited by water availability and the management of water supplies by local authorities. Rice fields that are planted late in an otherwise synchronous rice crop landscape are particularly vulnerable to leafhoppers and the viral diseases they transmit (Assam and Chancellor, 2002).

# 4 Increasing water productivity

Water productivity is defined as crop yield per unit of water consumed. Increasing water productivity in rainfed and irrigated systems involves (i) increasing the marketable yield of the crop for each unit of water transpired; (ii) reducing outflows; and (iii) increasing the effective use of rainfall, stored water and water of marginal quality. This is achieved by a combination of reduced land preparation time, adoption of intermittent or no flooding, and increased use or efficiency of fertilizers (Hafeez et al. 2007).

The duration and depth of flooding during rice production are key determinants of different crop health syndromes. For example, deep-water rice is rarely affected by pests other than stem borers (Catling, 1992). However, deep-water rice is severely hampered in regions where apple snails have been introduced (Horgan et al., 2014b). Because deep-water rice is produced in environmentally vulnerable areas, it constitutes an important habitat for wildlife including fish, amphibians, reptiles and birds (Catling, 1992). Options for pest control in such areas are therefore limited and pesticides have to be avoided.

Flooded rice in irrigated and rainfed production systems is vulnerable to a range of aquatic and semi-aquatic organisms as well as terrestrial insects and rodents that damage rice at different growth stages. However, flooding is recognized as important in controlling many weeds (Chauhan et al., 2012). In regions were *Azolla* (*Azolla* spp.: Salviniaceae) can still grow (these are often eliminated by apple snails) rice is further protected because the *Azolla* blocks light from the developing weeds (Biswas et al., 2005). With efforts to increase water productivity, practices such as intermittent flooding and direct seeding that reduce the amount of flooding result in a higher incidence of some weeds (Kürschner, 2010).

In upland and aerobic rice production systems, a range of distinct pests and diseases affect rice production. Insects such as termites (Isoptera), mole crickets (Gryllotalpidae) and beetle larvae (Scarabidae) attack rice roots, and parasitic nematodes are often so problematic that they restrict possibilities for successive rice crops (Litsinger et al., 1987; Prasad, 2011).

# 5 Reducing labour costs

Farmers are increasingly faced with the challenges of high labour costs and labour shortages. In many regions, increasing labour costs have incentivized farmers towards increased mechanization using either farmer-owned machinery or locally rented equipment. Furthermore, an increasing incidence of absentee landowners and migration towards urban centres often leads to consolidation of farms under a single management. Crop establishment and harvesting practices have been most affected by these recent trends (Savary et al., 2005; Timmer, 2005; Takahashi and Otsuka, 2009); but increasing labour costs will also reduce the feasibility of contracting labour for hand-weeding, hand collection of snails or spot pesticide applications.

Traditionally, flooded rice has been transplanted in much of Asia as seedlings of 15–30 days old. This labour-intensive process can effectively reduce damage from apple snails (Yanes-Figueroa et al., 2014) and weeds (Chauhan et al., 2012). A tendency towards labour saving has resulted in 'broadcasting' methods for crop establishment including wet and dry direct seeding and seedling broadcasting, each of which has associated benefits and trade-offs related to pest damage (Chauhan, 2012; Horgan et al., 2014c).

Wet direct seeding at high seeding rates reduces weed competition but is highly vulnerable to snail damage even under shallow flooding (ca. 1 cm) (Horgan et al., 2014c; Horgan 2016). Dry seeding can conserve water use, but is more vulnerable to weeds (Chauhan, 2012); furthermore, both wet and dry direct seeding in many regions are affected by wader (wet seeding) and passerine birds (dry seeding) that directly consume either the broadcast seeds or the developing seedlings (Wilson et al., 1989; Lane et al., 1998).

Seedling broadcasting and machine transplanting have potential to reduce labour costs while at the same time reducing pest problems (Horgan et al., 2014c); however, there is still a need to improve these techniques to enable transplanting of older seedlings, which are more robust and less affected by birds, insects and snails. Under proper seedbed management, some varieties can be transplanted as old as 40 DAS without yield loss (see Yanes Figueroa et al., 2014).

The choice of crop establishment method determines seedling densities and the number of seedlings per hill, both of which can be associated with different crop health syndromes. For example, a higher seedling density can reduce weed problems (Phuong et al., 2005) and has been noted to reduce the incidence of snail damage (Yanes-Figueroa et al., 2014). In a study from Thailand, Ishii-Eiteman and Power (1997) found green leafhoppers to colonize direct seeded fields more quickly than transplanted fields, but were more mobile in the transplanted fields, resulting in less aggregation of eggs and developing nymphs and more opportunities to transmit viruses between rice plants. Clearly, the spatial arrangement of rice seedlings will affect pest behaviours and their consequent damage potentials as well as influencing disease spread; however, few studies offer any empirical evidence to understand these processes in rice systems (see Section 8).

# 6 Increased availability and marketing of agrochemicals

In its August 2013 edition, the journal *Science* coined the term 'pesticide planet' to draw attention to the overwhelming use of pesticides in modern agriculture (Grullón, 2013). Agrochemical industries have thrived in recent years, chemical supply stores are now a common feature of rural communities, and prophylactic-calendar spray programmes are encouraged by agrochemical suppliers at field schools and open days throughout rice-growing regions (Sprangenberg et al., 2015). Pesticide imports in Asia have more than tripled (3.4 ×) since the beginning of the millennium (Fig. 2) (FAO, 2016). However, evidence suggests that pesticides are an increasingly poor solution, if not a major underlying cause, for much of the pest-mediated yield losses in rice.

Figure 2 indicates massive increases in pesticide trade (imports and exports) in the rice producing nations of Asia over the last two decades. Several recent surveys of rice farmers in Asia also indicate a high level of pesticide use and little recourse to alternative pest control methods (Rejesus et al., 2012; Qiao et al., 2012; Abdollahzadeh et al., 2015). Paradoxically, the trend towards an increasing availability of chemical pesticides has not resulted in the expected lower pest and weed incidences, but to area-wide problems of resistant weeds (Powles and Shaner, 2001), sustained outbreaks of planthoppers and leaf folders (Crambidae) (Sprangenberg et al., 2015; Chintalapati et al., 2016), unstable management of snails (Horgan et al., 2014b) and worrisome consequences for environmental and human health (Pingali and Roger, 1995; Qiao et al., 2012), often without clear yield responses to chemical inputs (Pingali and Gerpacio, 1997; Sprangenberg et al., 2015).

## 6.1 Insecticides

Insect herbivores can quickly adapt to commonly used insecticides (Matsumura et al., 2008). In addition, several insecticides are associated with faster growth rates and

**Figure 4** Varying degrees of damage to rice fields by the brown planthopper (*Nilaparvata lugens*) in Central Java (Indonesia) highlight the role of poor crop management in inducing outbreaks. Note that severe 'hopperburn' (plant death caused by planthoppers) is restricted to the central field in this picture from 2015. All fields in this photo were transplanted at the same time either with cv. Ciherang or the closely related cv. IR64, but farmers applied varying amounts and types of insecticide to their plots. Such well-defined spatial patterns to damage are indicative of insecticide-induced resurgence of herbivores.

increased reproduction in pest insects (Fig. 4) (Wang et al., 2005; Azzam et al., 2009; Chintalapati et al., 2016). To further add to the problem, the efficiency of natural enemies such as spiders and predatory wasps is reduced by insecticides either because they are directly killed by the chemicals or because their behaviours are altered (Kenmore et al., 1984; Tanaka et al., 2000; Horgan et al., 2017a). Increased reproduction and reduced mortality leads to exponential population growth rates in high-fecundity species such as planthoppers and leaf folders, a phenomenon known as 'resurgence' (Hardin et al., 1995). This can occur even on resistant rice varieties (Gallagher et al., 1994; Heinrichs, 1994). Over 100 published scientific reports have now described pesticide-induced resurgence in a range of rice pest insects including planthoppers, stem borers and leaf folders, as well as paddy-dwelling species of health concern such as mosquitoes (Culicidae). These studies have focused not only on a range of insecticides, but also on molluscicides, fungicides and herbicides that increase herbivore fitness (e.g. Grigarick et al., 1990; Wu et al., 2001; Wang et al., 2005; Assam et al., 2009; Xin et al., 2012; Horgan et al., 2014b).

In Indonesia, campaigns to reduce pesticide use through IPM and farmer field schools (FFS) and a government ban of several insecticides in 1986 resulted in a noted decline in outbreaks of the brown planthopper during the late 1980s and 1990s (Settle et al., 1996). Reduced pesticide use in several other Asian countries appears to have had similar effects (Pingali and Gerpacio, 1997). However, in recent years as pesticides have become cheaper and more readily available in Asia (Fig. 2), outbreaks of planthoppers have again increased in frequency and severity (Bottrell and Schoenly, 2012).

New insecticide products are made available to rice farmers on a regular basis. Although resistance to insecticides develops quickly, some products have not yet been associated with resurgence. However, insecticide products are rarely removed from markets even when they are known to cause resurgence, and farmers continue to use a variety of older products (Berg and Tam, 2012). The current situation of rapidly increasing insecticide use in Asia can be predicted to cause large-scale instability in insect pest populations and consequent pest outbreaks, and must be recognized as a major threat to rice production in the region.

## 6.2 Herbicides

Herbicide resistance has also been noted as a major constraint to rice farming in many parts of the world: hundreds of weed species are resistant to major classes of herbicide, and many species are resistant to multiple herbicide classes (Heap, 2014). Regions that have practiced direct seeding over several decades have been particularly affected. For example, weeds have become a major cause of yield decline for rice farmers in Sri Lanka who struggle to control grasses and sages using available products (Ranasinghe, 2003).

Management of weedy rice using Clearfield® and other herbicide-tolerant rice varieties has similarly succumbed to herbicide resistance: Clearfield® rice harbours a mutation in the acetohydroxyacid synthase or acetolactate synthase enzyme rendering it tolerant to imidazolinone (IMI) herbicides (Sudianto et al., 2013). Clearfield® rice together with IMI herbicides were expected to reduce weedy rice problems in the Americas, Asia and Europe. Although resistance development was anticipated, the pace of development in tropical regions (particularly in South America) was more rapid than expected. Planting of 2–3 crops a year in warm regions and the absence of die-off of volunteer rice during winter has also facilitated the transfer of the acetohydroxyacid synthase or acetolactate synthase mutagen to weedy rice (Sudianto et al., 2013).

Increasing the use of herbicides due to water and labour-saving practices (hand-weeding had been common in Asian rice paddies), and the need to increase herbicide concentrations or applications to counter weed resistance, leads to higher herbicide concentrations in soils and water. Herbicides can be detrimental to wildlife (Takamura and Yasuno, 1986; Saka, 2010; Liu et al., 2011). Herbicide residues can also kill native vegetation and reduce the yields of successive grain crops, pastures or vegetable crops such that the use of selective herbicides prohibits certain crop rotations (Sudianto et al., 2013). Institutes and extension agencies that promote the adoption of herbicide-reliant rice production systems will need to consider how these systems can be sustained in an environmentally friendly manner as weeds continue to adapt to available chemicals.

## 6.3 Molluscicides

Molluscicide imports into Asia and other rice-growing regions have similarly increased in recent years (Adalla and Magsino, 2006). Snail damage to rice fields following the introduction of apple snails to several rice producing countries has been mainly treated using chemical molluscicides despite effective cultural control methods (Adalla and Magsino, 2006; Yanes-Figueroa et al., 2014). In fields treated with molluscicides in tropical Asia, surviving apple snails can grow rapidly, due to reduced intraspecific competition, representing a more serious threat to developing rice plants than untreated rice plots that have high-density snail populations but smaller individuals (Horgan, 2017). Farmer surveys in Ecuador have indicated that fields treated with molluscicides may also have a greater propensity for insect damage (Horgan et al., 2014b).

# 7 Pest management for rice: scientific approaches

There are three main ingredients for successful IPM, these are:

i Knowledge of the rice production system, its component species (including the rice plant) and the nature of interactions between these species. It is also essential to understand how rice ecosystems, particularly their pest and weed populations, are regulated through negative feedback loops;

ii Tools to ensure that the outcomes of natural regulation favour yield and that ecological pathways (succession) towards damaging crop health syndromes are avoided. Such tools include traditional pest management practices such as flooding fields to prevent weeds, using traps or barriers to exclude rodents or snails, biological control and host plant resistance (HPR). Among these tools, pesticides must be carefully considered, as has been shown, since pesticides are often the underlying cause of pest-mediated damage and threaten environmental and human health;

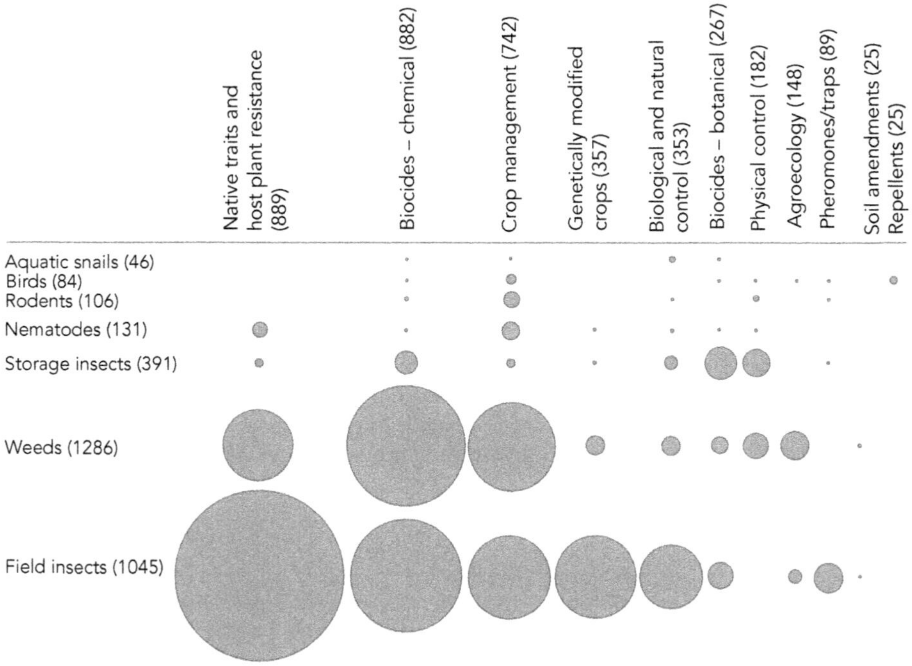


**Figure 5** Focus of peer-reviewed scientific papers concerning rice pests and their management. The area of each circle indicates the number of papers with information relevant to the pest groups indicated in the side column and for broad topics indicated in the top row. The number of relevant records for each topic is indicated in parentheses. Note that a single paper may have information (records) relevant to different topics. Data are based on articles retrieved from Web of Science (1970 to June 2016) using 'rice' together with pest and weed genus names (based on key literature) as search terms.

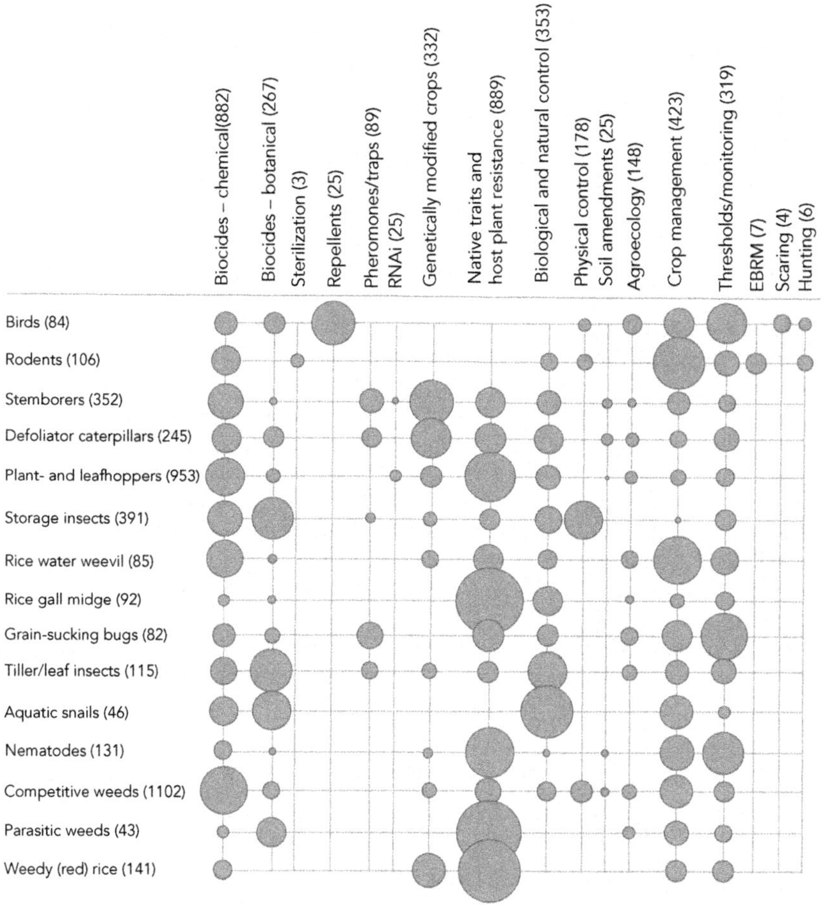


**Figure 6** Focus of peer-reviewed scientific papers concerning rice pests and their management. The area of each circle indicates the proportion of papers focusing on the broad topics indicated in the top row for each pest group (indicated in the first column). The number of relevant records for each topic is indicated in parentheses. Data are from the same source as in Fig. 5. 'EBRM' is ecologically based rodent management.

iii Finally, IPM requires effective communication of knowledge and guidance of farmers towards best management practices. IPM is therefore as much about communication networks, didactic tools, effective extension and agricultural policy as it is about ecology or technology.

Figures 5 and 6 indicate the response by science to pest issues in rice. Peer-reviewed literature rarely addresses traditional (indigenous or farmer-derived) pest management practices, therefore the broad research areas indicated in the figures are not exhaustive of actual farmers' practices; they represent rather the areas of principal scientific interest. It should also be noted that the information in Fig. 5 and 6 represent accumulated scientific literature since 1970 and do not indicate current research emphases. This section focuses on how science has responded to the first two requirements for successful IPM? Issues around education and policy will be addressed in Sections 14 and 15.

# 8 Knowledge of the rice production system

Records of research into crop management and monitoring rank third in rice pest-related scientific literature in Fig. 5 (based on Web of Science retrievals, 1970–2016); however, in the last 20 years, this has represented only about 10% of studies in rice pest management. Early ecological research into rice production systems mainly focused on understanding the autecology of specific rice pests and their effects on rice production. However, despite several studies that have tried to assess damage or develop economic thresholds, there is still little understanding of yield losses due to rice pests: Most estimates present only ranges of damage or yield loss (e.g. 5–60%). However, without knowledge of the frequencies and spatial extent of such losses, such information is of little value. Recent research has attempted to improve the situation by developing technologies to better detect pests including using remote sensing applications (Lamb and Brown, 2001; Yang et al., 2007). It will be necessary to continually update damage thresholds and estimated yield losses as climates and crop management practices change and because different rice varieties often have specific responses to pest attack (see Section 11). It is apparent therefore that, despite calls for changes in crop management, there has been little investment in the science needed to predict responses by pests and diseases. Without such knowledge, IPM will be severely limited.

# 9 Biocides, traps and barriers

## 9.1 Chemical and botanical biocides

According to Fig. 5 and 6, research into biocides has dominated pest management science for rice. Much of this research has included testing and promoting chemical control methods, particularly herbicides. A large body of information has also been generated to explain the mechanisms underlying pest (insect and weed) resistance to chemicals and resurgence in herbivores (see Sections 6.1 and 6.2).

In an attempt to reduce the negative effects of chemical pesticides, several researchers have focused on developing and evaluating botanical pesticides, particularly neem (Lim and Bottrell, 1994). A focus on botanical pesticides is prevalent in research on (i) storage pests, because of the need to avoid chemical contamination of foods (Isman, 2006); (ii) apple snails, because of a need to avoid chemical pesticides in water bodies (Horgan, 2017); and (iii) witchweed (*Striga* spp.: Orobanchaceae) – largely due to the discovery of strigolactones that can stimulate the germination of parasitic weeds and reduce their densities in the seed bank (Yoneyama et al., 2009; Dubey et al., 2010). Botanical biocides are regarded as safer than chemical poisons. However, they can also potentially cause resurgence (although there are still no records of such cases) and they are better applied under the same IPM guidelines as chemical biocides (Lim and Bottrell, 1994).

## 9.2 Pheromones

Research into pheromones and baits was prominent in the 1990s, with research investments declining more recently. Synthetic pheromones are currently available for striped stem borer, *Chilo suppressalis*, ricebug, *Leptocorisa chinensis* Dallas and Indian meal moth, *Plodia interpunctella* (Hubner). These pheromones are rarely employed as sex disruptors but used more commonly in pest monitoring to help predict pest densities or to identify peaks in seasonal occurrence as a guide for management interventions (Kojima et al. 1996; Watanabe et al., 2009; Witzgall et al., 2010; Wilson et al., 2015).

## 9.3 Light traps

Light traps have featured in the management of certain insects, particularly planthoppers: Networks of light traps have operated throughout north-eastern China and Japan to detect waves of migrating planthoppers. During the northern hemisphere spring, brown and white-backed planthoppers migrate several thousand kilometres from South East Asia to the north east of Asia. Light traps have been used to determine the extent, direction and climatic conditions that generate these migrations and to act as an early warning system for rice farmers before planthoppers alight in their fields (Matsumura, 2001; Cheng, 2009). For example, rustic light traps have been operated in Vietnam as part of a successful 'escape strategy' whereby farmers determine rice planting dates based on the occurrence of peak planthopper densities in light traps (planting only after the peak has passed) (Bentley, 2009).

## 9.4 Repellents

Repellents have received considerable research attention as a method to control damage from grain-feeding birds to direct seeded rice, mainly in the United States. Repellents such as methyl anthranilate, pulegone and caffeine; and deterrents including coloured seed coatings have been shown to reduce the consumption of rice seed by birds (i.e. red-winged blackbird, *Agelaius phoeniceus* L: Icteridae) (Avery et al., 1996, 1999, 2005; Pawlina and Proulx, 1999). Simple ideas like seed coatings could help farmers to reap the benefits of birds in consuming dropped grain while at the same time avoiding losses during direct seeding.

## 9.5 Barriers and traps

Innovative physical barriers and traps are a prominent feature of ecologically based rodent and snail management in rice fields, and an essential feature in controlling storage pests including rodents and insects (Fig. 5 and 6). For example, hermetic storage bags reduce damage to stored rice from several storage pests (Guenha et al., 2014). Rustic storage facilities in developing countries are at higher risk from storage pests; here biocontrol agents, pheromone traps, botanical repellents and simple but cheap physical controls can considerably reduce losses from insects (Philips and Throne, 2009). In many cases, traditional exclusion methods including grain store designs and traditional construction methods may be effective against rat damage to stored rice (Sarangi et al., 2009)

## 10 Transgenic and herbicide-resistant rice varieties

Following the development of Bt transgenic crops with resistance against chewing insects, transgenic resistance became the major focus of research into reducing stem borer and leaf folder damage to rice (31% and 26% of all published studies, respectively) (Horgan 2012). Transgenic rice expressing lectins, protease inhibitors and venoms have also been considered to combat rice planthoppers and other rice pests (Murdock and Shade, 2002; Chen et al., 2012). More recently, research into RNAi technology has gained some attention for controlling planthoppers and stem borers (Fig. 6).

Transgenic and gene mutation approaches to developing varieties with novel traits have featured prominently in research into weed management: Glyphosate- and IMI-resistant rice has been developed to prevent damage to rice during herbicide applications and thereby better manage weeds (Sudianto et al., 2013). However, as mentioned above, the sustainability of herbicide-resistant varieties and caution about the transfer of transgenes or mutated genes to weedy rice, especially weedy hybrids has limited the application of these technologies (Chen et al., 2004; Sudianto et al., 2013 and see Section 6.2).

Transgenic rice varieties for pest management have not gained a foothold in rice production systems. This has been largely due to continuing debate over their potential long-term effects and difficulties in managing adaptation by target organisms, particularly in tropical regions where cropping is more intensive (Sudianto et al., 2013; Pu et al., 2014) and in developing nations where farms are small (usually >2 ha) and where area-wide resistance management would be consequently difficult (Teetes, 1994).

## 11 Genetic approaches to resistance and tolerance

Native HPR has dominated research into rice herbivore management for decades. ('Native' refers to resistance derived from rice or its wild relatives.) Furthermore, the application of molecular and biotechnological tools to understanding insect–rice, nematode–rice and witchweed–rice interactions has represented 46% of all studies into rice pest management since 2000. 'Resistance' specifically describes a plant's ability to deter, kill or reduce the fitness of insects or nematodes. However, there is a broad range of possible interactions between pests and their rice host, each with potential benefits or trade-offs for pest management. For example, rice plants can vary in their tolerance to herbivore and parasite attack (Horgan, 2012). Tolerance is the ability of a plant to compensate for damage (Horgan, 2012; Horgan et al., 2016a). Some rice plants may also turn-on ineffective resistance pathways in response to pest attack (Agarrwal et al., 2014). Such responses may be detectable using modern molecular tools, but remain inconsequential in ecological terms.

Because the term HPR has a limited definition, this section focuses on what might be better called research into 'native traits and genes' – these are traits that can be acquired through conventional breeding programmes. Native traits have been the principal focus of research into the management of nematodes (39% of all published nematode-related records), plant- and leafhoppers (42%), witchweed (67%) and rice gall midges (73%) (Fig. 6). Native traits have also featured prominently in research into the management of stem borers (46%) and defoliator caterpillars (41%) as well as rice water weevil (18%). The avoidance of genes involved in grain discolouration has been considered to reduce the impact of grain-sucking bugs (Gravois and Bernhardt, 2000). Furthermore, molecular approaches have been applied to understand shattering and dormancy as traits leading to the invasiveness of weedy rice and to understand gene flow between cultivated and weedy rice (62%). The search for allelopathic rice varieties and molecular approaches to improving rice competition with weeds, although not as prominent as research into other native traits, was topical between 2000 and 2010. More recently, allelopathy as a function of rhizosphere bacteria has gained increasing attention (Lin et al., 2010).

Using native traits as tools for herbivore, nematode, witchweed and disease management has been approached in two principal ways: (i) the screening and elimination of susceptible rice lines during breeding programmes and (ii) the inclusion of major resistance genes into breeding lines and modern varieties.

## 11.1 Screening

Screening involves challenging potential rice breeding lines with important pests or diseases and recording the resulting damage. Screening can be conducted under field conditions, but is more usually conducted in greenhouses or laboratories. Such screening studies are often effective, particularly where they identify varieties that are novel for a target pest population (normally indicated by low damage compared to susceptible rice varieties). A novel variety might include a variety that is phenologically distinct from the main varieties in the region of interest (the target region) and therefore represents a barrier to dispersal and to successful attack by the target pest (Horgan, 2012). Routine screening of this type appears to be successful in reducing damage from rice planthoppers in some parts of Asia (Widawski et al., 1998; Horgan and Crisol, 2013).

Good screening methods are essential to deliver improved varieties to farmers. What is noteworthy here is that excessive yield losses are avoided in the resulting varieties because rice phenotypes with high susceptibility and/or low tolerance are simply excluded from breeding programmes. Unfortunately, routine screening using methods such as standard seedling seedbox tests (SSST) is often unreliable, is difficult to interpret and gives variable results according to conditions in the screening facilities (Fujita et al., 2013). Furthermore, to ensure that screening is useful, test organisms should be derived from recently established herbivore colonies or parasite strains. For example, researchers at the Centro Internacional de Agricultura Tropical (CIAT – Colombia) renew their planthopper, *Sogatodes oryzicola* (Muir), colonies every 2 years when screening for resistance to Rice Hoja Blanca Virus. However, test colonies and strains at many institutes are often deliberately maintained for several years as research 'biotypes' or 'pathotypes' to facilitate laboratory-based breeding programmes (Fujita et al., 2013). Whereas this may hold advantages for gene discovery, it is not recommended for screening studies.

## 11.2 Incorporating major resistance genes in rice

To develop resistant rice varieties, scientists can identify loci containing resistance genes and, using molecular tools such as marker-assisted selection (MAS), ensure that these genes are maintained through the breeding pipeline (Rispail et al., 2010; Chen et al., 2012; Fujita et al.; 2013). The advantage of this approach has been to permit the repeated backcrossing of susceptible HYVs with resistance donors while ensuring that the genes of interest are not eliminated from advanced lines. Increasingly, resistance donors include near-isogenic lines with single or 'pyramided' (including two or more resistance genes in the same line) resistance genes (Fujita et al., 2013).

One major problem associated with this approach to pest management is that resistance (unlike tolerance) exerts selection pressure on the target organisms. Already the development of virulent pest populations has become a serious issue for anti-herbivore resistance: For example, in a recent study, researchers from South and South East Asia evaluated the effectiveness of several donor varieties with known resistance genes against the brown planthopper. Their results indicated that most genes were now ineffective, but that traditional varieties and landraces containing two or more resistance genes remained largely effective (Horgan et al., 2015). This supports the idea that pyramiding resistance genes can improve the durability of anti-herbivore resistance (Fujita et al., 2013). However, ecological theory predicts a limit to pyramiding defence genes without appreciably affecting yield (Brown, 2002). As yet, no published study has tested this hypothesis in the context of resistance against herbivores in rice.

In the face of growing research interest in native traits for pest management, some probing questions need to be asked: For example, with such a large investment in research on insect HPR, why has insecticide use increased so dramatically in Asia in recent years? Such a trend suggests that, if deployed, resistance will simply reduce the profitability of rice farms unless farmers also reduce their pesticide inputs (Pingali and Gerpacio, 1998; Widawsky et al., 1998). Clearly there is a need to better insert HPR and other native traits into IPM strategies; however, there has been a distinct lack of research in this area. In particular, issues around the deployment of resistant varieties need to be more fully examined (Fujita et al., 2013; Horgan et al., 2015). Here, researchers may gain from the rich research into deployment of transgenic crops (Gould, 1998) to better direct deployment of native resistance. However, this will require an improved understanding of rice field ecology as well as understanding the socio-economic realities of rice farmers.

# 12 Enhancing natural regulation

Investments in ecological research for pest management in rice have lagged behind research investments in molecular breeding, biotechnological and biocide approaches (Fig. 5 and 6). Furthermore, among reports from research into the natural pest management of insects indicated in Fig. 5 and 6 are several recent studies on the effects of transgenic crops on specific predatory species from the rice community as well as studies on the mechanisms underlying insecticide-induced resurgence of pest insects (see Section 6.1). Also, much of the research indicated under 'crop management' for weeds in Fig. 5 and 6 has only compared flooding or direct seeding against 'new' herbicide treatments. Studies that clearly focus on ecological approaches to the management of pests and weeds by identifying or testing specific ecologically based interventions are comparatively rare.

## 12.1 Biological control

Biological control agents have been tested for their effects in reducing infestations by herbivorous insects, nematodes and snails in rice. Classical biological control is rarely contemplated even after the introduction of noxious weeds and damaging insects to new regions, although studies do continue to identify suitable biological control agents against introduced aquatic weeds that also occur in rice (Culliney, 2005). Classical biological control works at regional levels and is not specific to rice fields, although rice farmers are identified as beneficiaries of such programmes.

Inundative biological control has been investigated for the control of rice field insects, nematodes and storage pests. These studies have focused on the development of rearing and application methods for *Trichogramma* spp. (Trichogrammatidae) to control stem borers (Ko et al., 2014); the use of *Metarhizium* spp. (Clavicipitaceae) and *Beauveria* spp. (Cordycipitaceae) to control several insect and nematode pests (Loc and Chi, 2005; Meyling and Eilenberg, 2007); and the use of bacteria (i.e. *Pseudomonas fluorescens* [Flügge] Migula and *Bacillus* spp.) to control insects, nematodes and rice diseases (Commare et al., 2002; Padgham and Sikora, 2007). Among the advantages of using fungal and bacterial biocontrol agents are the ease with which they can be reared and the facility with which they can be applied in the field (Meyling and Eilenberg, 2007). However, they do exert selection pressure for pest adaptation if used in excess.

Biocontrol of insects and snails using ducks and fish has gained recent attention as part of integrated farming systems. Ducks in particular have been shown to reduce snail densities. Farmers in many regions where apple snails have been introduced will invite duck herders to direct their ducks through rice fields (Teo, 2001). Successful management of insects and snails by fish has been more difficult to achieve and can have associated problems. For example, the common carp, *Cyprinus carpio* L, clearly reduces snail densities, but carp is an invasive threat in some regions (Ip et al., 2014). In general, the effectiveness of biological control has been difficult to determine. Nevertheless, key areas for research have been largely overlooked, including the discovery and use of entomopathogenic viruses or snail-killing nematodes in rice fields.

## 12.2 Natural pest management

The need to maintain natural enemies in rice fields and allow these to build up populations has been well recognized for decades. A current paradigm for natural enemy dynamics is that generalist predators will consume decomposers at the time of crop establishment and during seedling stages, but later move to consume rice pests as these begin to colonize the young rice plants (Settle et al., 1996). For this reason, it is strongly advised not to spray rice fields with insecticides during the first 40 days after transplanting (Kenmore et al., 1984). The success of this heuristic has been borne out by several studies, and most recently by a convincing study from China (Gurr et al., 2016).

Massive applications of pesticides in recent years (see Section 6) have increased fears that natural enemy numbers have become depleted in Asian rice fields. Therefore, several strategies have been devised to restore natural enemies, increase their efficiency and also improve the resilience of rice ecosystems against perturbations – including resilience against climate change (Gurr, 2009; Gurr et al., 2016; Westphal et al., 2015; Horgan et al., 2016d). For example, in a comprehensive study, Kiritani (1999) indicated that most predators and parasitoids in Japanese rice fields were likely to increase the number of generations they achieved by 2 or 3 under average mean monthly temperature increases of 2°C, whereas most rice pests would only increase their generations by one. This suggests that under increasing temperatures, numerical responses to pest densities by many natural enemies would be enhanced, thereby improving pest regulation. However, spiders are expected to achieve the lowest increases in generations (ca. 0.6 per year) under increasing temperatures (Kiritani 1999), potentially decreasing their effectivity. By creating suitable

**Figure 7** Ecological engineering: Luffa (*Luffa* spp.: Cucurbitaceae) grown on rice bunds is attractive to parasitoids and predatory wasps that feed from its large yellow flowers. Wild birds, including predatory species, use bamboo supports to scan the rice fields for insects and apple snails (see Horgan et al., 2017b). By growing plants of function such as these, farmers can increase the abundance and diversity of the natural enemies of stem borers and leaf folders in their rice fields and, at the same time, increase productivity on their farms.

habitat for predators such as spiders, any potential negative effects of global warming on natural pest regulation could be avoided (Kiritani, 1999).

Strategies that create adequate habitat for natural enemies can be included under the broad title of 'agro-ecology'; however, recently the terms 'ecological intensification' (Gurr et al., 2016) and 'ecological engineering' (Gurr, 2009; Horgan et al., 2016d) have gained prominence. Ecological intensification implies increasing the per unit production of farms through the application of ecological principles while at the same time 'sparing' natural areas for biodiversity conservation. Though the concept is often used in terms of simple intensification practices, ecological intensification is more correctly associated with diversification of crop production in time and space (Bommarco et al., 2013; Gurr et al., 2016). Unfortunately, evidence suggests that intensifying agricultural lands for food and fibre production is unlikely to lead to sparing of natural areas, particularly in developing nations (Phelps et al., 2013). Ecological engineering is a related concept, but stresses the design of habitat based on species interactions and feedback mechanisms that are highly predictable. Ecological engineering goes further by proposing the development of production systems with benefits for both man and nature (Gurr, 2009; Horgan et al., 2016d), in effect 'sharing' spaces with native biodiversity.

Rice systems that diversify habitat by including plants of known function to enhance the services provided by key species in the rice ecosystem (other than direct food production) are examples of ecological engineering. Ecological engineering research in Asia has mainly focused on creating flower strips and vegetable patches as a source of alternative foods and refuge for the natural enemies of planthoppers and stem borers (Fig. 7). Recent reports have indicated that such systems can successfully increase rice yields, reduce pest damage, reduce or eliminate pesticide inputs, increase the biodiversity and function of predators including predatory birds, and increase farm profits and food production by using the space available on rice bunds (levees) or in specially constructed earthen platforms to grow vegetables (Westphal et al., 2015; Gurr et al., 2016; Horgan et al. 2016d, 2017a,b).

Despite such apparent success, there is still a need for further research into ecological engineering and other diversified farming systems. This includes research to avoid problems due to pests such as rats that might benefit from planted bunds or the close cultivation of other crops (Horgan et al., 2016d). Nevertheless, the continual adaptation of such diversified systems by including elements such as fish or ducks, incorporating soil amendments such as biochar or silicon (both of which reduce pest damage [Huang et al., 2015]) and planting high-yielding but pest-resistant rice varieties has enormous potential for achieving both sustainability and food security targets.

## 13 Integrating management practices

The review of pest management research and techniques presented in Sections 7 through 12 indicates a general partitioning of potential management approaches and research efforts. Recent innovations in pest management certainly hold promise in reducing pest damage to rice. However, it is clear that more research attention is required to bring the different 'technologies' together into functioning and sustainable rice production systems. For example, research into ecological engineering and biocontrol has rarely incorporated innovations in HPR into research platforms, and HPR research often ends with the 'resistant variety' or 'resistance locus', without considering deployment and management of pest adaptation, or rice field ecology. Future research might benefit from more successfully integrating such diverse components of pest management.

## 14 Education and extension

A recent study by Parsa et al. (2014) examined reasons for the poor adoption of IPM in different crops through a survey of 163 IPM professionals and practitioners from around the world. The survey identified 51 obstacles to IPM adoption in developing countries. The top 10 reasons for non-adoption included knowledge-related issues (lack of training, little technical support, limited availability of information), structural issues (lack of government policy and support, limited funding), implementation issues (prohibitive costs, limited access to inputs, difficult implementation) and issues related to competing interests (powerful influence of the pesticide industry).

Recent surveys and cognitive studies from Asia suggest that rice farmers are driven to use pesticides, particularly insecticides, based on exaggerated expectations of yield loss (risk aversion) (Palis, 1998; Bandong et al., 2002; Pingali, 2001). A lack of adequate extension support and of basic ecological knowledge means that farmers often initiate pesticide applications when they observe 'insects' or damage, but generally cannot distinguish pests from beneficial organisms and cannot project from their field observations to realistic damage potentials (Bandong et al., 2002). Furthermore, farmers communicate regularly with pesticide vendors that promote prophylactic pesticide applications or calendar sprays without any pest monitoring; herbicides, molluscicides and fungicides are generally applied following calendar applications, an approach that is increasingly common with insecticides (Robinson et al., 2007).

One approach to overcome a lack of farmer knowledge about pests and beneficial organisms has been to directly train farmers through informal courses and campaigns. For example, Indonesia's National Integrated Pest Management Program and the FFS of the Food and Agriculture Organization of the United Nations (FAO) were successful in reducing insecticide use in the 1990s (Matteson, 2000; Thorburn, 2015). The 'three reductions, three gains' (3R3G) campaign promoted through the Ministry of Agriculture and Rural Development in Vietnam after 2000 was similarly successful in reducing pesticide inputs (Huan et al., 2008). The campaign promoted a reduction in seeding rates, fertilizers and pesticides to improve rice profitability (adding that environmental and farmer health would also gain). Campaigns such as these have benefited by exploiting existing extension systems as well as expanding to mass media such as radio and TV (Huan et al., 2008; Bentley, 2009; Westphal et al., 2015). However, education campaigns are expensive and difficult to sustain and evidence suggests that adoption of IPM declines rapidly after education campaigns have ended (Rejesus et al., 2012).

Rice knowledge banks, made available through specialist websites, could sustain the knowledge of farmers and government extension technicians. Knowledge banks are digital information repositories with simplified retrieval interfaces to help users understand rice crop management, pest problems and, occasionally, beneficial organisms. Several national and international institutes maintain such websites. In some cases specialised knowledge 'apps' have been made available to farmers through smartphones (i.e., Rice Knowledge Bank – Bangladesh Rice Research Institute), and several research groups plan to develop interactive 'apps' whereby farmers can receive diagnostic support from remote specialists after answering key questions or uploading images of potential pests encountered in their fields (personal communications with app developers at the National University of Singapore).

As an extension of their knowledge bank, the CABI Plantwise programme has developed a system of plant clinics, including mobile clinics, for rice farmers in Thailand, where farmers receive advice in managing potential pest problems (Leach and Hobbs, 2013). Such systems are designed to support sustainable rice production and promote IPM among rice farmers; however, they are limited by the availability of trained staff and financial support, and if used incorrectly, could encourage unnecessary pesticide applications where information is not followed through with adequate management advice or where knowledge pathways are discontinuous. Government support for such remote extension initiatives will be important to promote their continuity; but it will be essential to ensure healthy partnerships between researchers, funders and infrastructure suppliers during the development and implementation of such schemes if the knowledge they provide is to be used correctly.

# 15 Future trends and conclusion

As discussed throughout this chapter, it is apparent that rice crop management practices are changing in response to key drivers. These changes are often promoted as beneficial to the economic and environmental sustainability of the rice sector. However, although economic advantages can be assessed through traditional cost–benefit analyses using monetary or production indicators, assessing the environmental impact of change has been more difficult. One area of difficulty has been to develop balanced indicators of environmental quality that weigh advantages in one area (e.g. water productivity) against costs in another (e.g. biodiversity loss and water contamination).

Systems approaches can help define and evaluate pest problems associated with different rice production systems and their component practices, as well as determining the best solutions to the problems from among a range of alternatives (Teng and Savary, 1992). As an essential economic activity, it is convenient to examine rice production using value chain analysis (VCA), which attempts to examine in detail the value-added steps in production and marketing, where values are not only monetary but also social (Bolwig et al., 2010). As part of a VCA, farmers' decisions around agronomy and pest management, the costs of management and the potential losses from pests should be addressed. However, only in rare cases can VCA guide researchers towards mediation of environmental impacts (exceptions being where there is a direct cost to production or a yield advantage or penalty due to the environmental consequences of some management intervention).

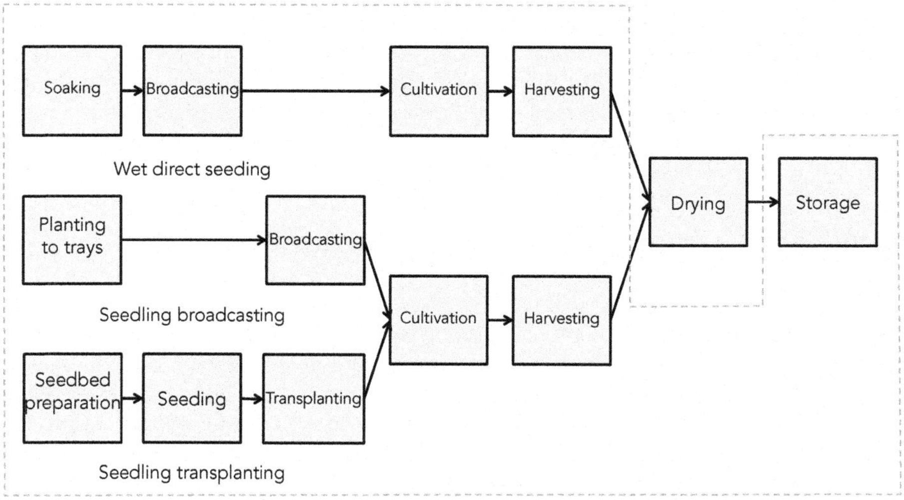


**Figure 8** Life cycle of rice and system boundary (identified inside the dashed box) for topics in this chapter. As an example, the diagram indicates three crop establishment methods for comparison using LCA, highlighting the differences in cultivation and harvesting methods after direct seeding and seedling broadcasting or transplanting. Each stage in the life cycle can be assessed against a series of impact indicators designed to safeguard the environment and human health.

To factor in environmental concerns and manage the environmental impacts of rice production, a range of techniques around life cycle assessments (LCAs) has been developed (i.e. Blengini and Busto, 2009). Although these have initially focused heavily on mitigating greenhouse gas emissions and the energy costs of grain production, there have been recent calls to include assessment of risks to biodiversity and consequently to the ecosystem services provided by animal and plant communities in production landscapes (Teillard et al., 2016).

LCA involves defining the boundaries of the study system, conducting a detailed compilation of all environmental inputs and outputs at each stage of the system and quantifying the relative importance of all impacts using specific indicators (such as water use, eutrophication potential, energy requirements, biodiversity loss and global warming potential) (Fig. 8) (Blengini and Busto, 2009). Whereas VCA focuses mainly on economic and social sustainability, LCA focuses predominantly on issues of environmental sustainability. A natural blending of VCA and LCA might address more fully the essential aspects of sustainable development of the rice sector. However, in general, linking LCA and VCA requires some environmental bookkeeping for which realistic and recognizable cost/value estimates cannot be easily derived.

Without adequate development of LCA techniques that allow a thorough assessment of different rice production systems and practices, researchers might adopt the precautionary principle as they strive for healthy rice ecosystems. Healthy rice ecosystems will produce sufficient quantities of safe food to be economically sustainable, but also function to maintain biodiversity, manage floodwaters, improve rural aesthetics and support cultural identity. The concept recognizes that farmers are themselves affected by environmental degradation and that pesticides are a major health and production challenge (Pingali and Roger, 1995; Horgan et al., 2016d). IPM could gain renewed vigour if inserted into such an integrated framework and assessed using evolving system analytical tools. The major challenge for sustainable futures is to simultaneously respond to a range of global challenges including climate change, food insecurity, increasing poverty and environmental degradation. With such complicated demands on future rice production systems, adequate frameworks based on integrated assessment tools can greatly support policy and direct holistic research approaches.

# 16 Where to look for further information

Global problems require global solutions. Researchers will need to forge links with colleagues from different regions, disciplines and core interests to better grasp the complexity of issues and determine the best solutions to sustainably manage rice systems both as production areas and as a healthy habitat for humans and wildlife. In recent years, a number of researchers have developed agro-ecological approaches to pest management including ecological engineering. In many cases the fruits of their research are only now coming to light. Researchers at AfricaRice and at CIRAD-Agricultural Research for Development have tested plants of function in West Africa and in upland rice areas of Madagascar. Charles Sturt University in Australia (Graham Centre for Agricultural Innovation, School of Agricultural and Wine Sciences) has been a leader in ecological engineering in rice and now has a satellite laboratory in China. Also in Australia, researchers at the Centre for Compassionate Conservation, University of

Technology Sydney, are conducting research in ecological engineering in rice and wildlife-friendly rice farming in Asia and Australia. Researchers at the University of Connecticut (Department of Ecology and Evolutionary Biology) lead efforts in wildlife-friendly rice in the United States.

For information on developing herbivore- and disease-resistant rice using molecular tools, readers can contact researchers at the Faculties of Agriculture of Kyushu University and Saga University in Japan and researchers from the Rice Department of the Government of Thailand. These researchers have greatly contributed to the development of near-isogenic rice lines with several herbivore resistance genes. Researchers at the College of Life Sciences of Wuhan University, China, have also identified and pyramided genes for several resistance traits in modern rice lines and investigated the mechanisms underlying resistance. Researchers at AfricaRice in collaboration with Makerere University (College of Agricultural and Life Sciences) have similarly gained much recent knowledge in control of witchweed using HPR.

Several researchers are developing environmental accounting techniques to improve the sustainability of rice farming. Researchers at the Institute of Physical Geography and Landscape Ecology at Leibniz University in Hannover (Germany) and at the Department of Community Ecology of the Helmholtz-Centre for Environmental Research in Halle (Germany) have developed sustainability indicators for rice ecosystems in Asia. CABI's Plantwise programme has offered a new direction for sustainable pest management and rice health, particularly in Thailand. Together, these researchers continue to work towards the success of IPM and the sustainability of rice production. However, there is considerable scope for expanding research agendas to fill current knowledge gaps and develop workable solutions to emerging rice health syndromes as well as to improve current IPM practices and their levels of adoption.

## 17 References

Abdollahzadeh, G., Sharifzadeh, M. S. and Damalas, C. A. 2015. Perceptions of the beneficial and harmful effects of pesticides among Iranian rice farmers influence the adoption of biological control. *Crop Protection*, 75, 124–31.

Adalla, C. B., Magsino, E. A., Joshi, R. and Sebastian, L. 2006. Understanding the golden apple snail (*Pomacea canaliculata*): biology and early initiatives to control the pest in the Philippines. In Joshi, R. C. and Sebastian, L. S. (eds), *Global Advances in Ecology and Management of Golden Apple Snails*, PhilRice, Nueva Ecija, Philippines, pp. 199–213.

Agarrwal, R., Bentur, J. S. and Nair, S. 2014. Gas chromatography mass spectrometry based metabolic profiling reveals biomarkers involved in rice-gall midge interactions. *Journal of Integrative Plant Biology*, 56, 837–48.

Ampong-Nyarko, K. and Datta, S. K. 1993. Effects of nitrogen application on growth, nitrogen use efficiency and rice-weed interaction. *Weed Research*, 33, 269–76.

Avery, M. L., Decker, D. G., Humphrey, J. S. and Laukert, C. C. 1996. Mint plant derivatives as blackbird feeding deterrents. *Crop Protection*, 15, 461–4.

Avery, M. L., Humphrey, J. S., Decker, D. G. and McGrane, A. P. 1999. Seed color avoidance by captive red-winged blackbirds and boat-tailed grackles. *The Journal of Wildlife Management*, 1003–8.

Avery, M. L., Werner, S. J., Cummings, J. L., Humphrey, J. S., Milleson, M. P., Carlson, J. C., Primus, T. M. and Goodall, M. J. 2005. Caffeine for reducing bird damage to newly seeded rice. *Crop Protection*, 24, 651–7.

Azzam, O. and Chancellor, T. C. B. 2002. The biology, epidemiology, and management of rice tungro disease in Asia. *Plant Disease*, 86, 88–100.

Azzam, S., Wang, F., Wu, J.-C., Shen, J., Wang, L.-P., Yang, G.-Q. and Guo, Y.-R. 2009. Comparisons of stimulatory effects of a series of concentrations of four insecticides on reproduction in the rice brown planthopper *Nilaparvata lugens* Stål (Homoptera: Delphacidae). *International Journal of Pest Management*, 55, 347–58.

Balasubramanian, V., Sie, M., Hijmans, R. J. and Otsuka, K. 2007. Increasing rice production in sub-Saharan Africa: challenges and opportunities. *Advances in Agronomy*, 94, 55–133.

Bandara, R. and Tisdell, C. 2002. Asian elephants as agicultural pests: Economics of control and compensation in Sri Lanka. *Natural Resources Journal*, 42, 491–519.

Bandong, J. P., Canapi, B. L., Dela Cruz, C. G. and Litsinger, J. A. 2002. Insecticide decision protocols: a case study of untrained Filipino rice farmers. *Crop Protection*, 21, 803–16.

Bentley, J. W. 2009. Impact of IPM extension for smallholder farmers in the tropics. In Peshin, R. and Dhawan, A. K. (eds), *Integrated Pest Management: Dissemination And Impact*, Springer Science, pp. 333–46.

Berg, H. and Tam, N. T. 2012. Use of pesticides and attitude to pest management strategies among rice and rice-fish farmers in the Mekong Delta, Vietnam. *International Journal of Pest Management*, 58, 153–64.

Biswas, M., Parveen, S., Shimozawa, H. and Nakagoshi, N. 2005. Effects of *Azolla* species on weed emergence in a rice paddy ecosystem. *Weed Biology and Management*, 5, 176–83.

Blackshaw, R. E., Brandt, R. N., Janzen, H. H., Entz, T., Grant, C. A. and Derksen, D. A. 2003. Differential response of weed species to added nitrogen. *Weed Science*, 51, 532–9.

Blengini, G. A. and Busto, M. 2009. The life cycle of rice: LCA of alternative agri-food chain management systems in Vercelli (Italy). *Journal of Environmental Management*, 90, 1512–22.

Bolwig, S., Ponte, S., Du Toit, A., Riisgaard, L. and Halberg, N. 2010. Integrating poverty and environmental concerns into value-chain analysis: a conceptual framework. *Development Policy Review*, 28, 173–94.

Bommarco, R., Kleijn, D. and Potts, S. G. 2013. Ecological intensification: harnessing ecosystem services for food security. *Trends in Ecology and Evolution*, 28, 230–8.

Bottrell, D. G. and Schoenly, K. G. 2012. Resurrecting the ghost of green revolutions past: The brown planthopper as a recurring threat to high-yielding rice production in tropical Asia. *Journal of Asia-Pacific Entomology*, 15, 122–40.

Brown, J. K. M. 2002. Yield penalties of disease resistance in crops. *Current Opinion in Plant Biology*, 5, 339–44.

Cai, Z., Xing, G., Yan, X., Xu, H., Tsuruta, H., Yagi, K. and Minami, K. 1997. Methane and nitrous oxide emissions from rice paddy fields as affected by nitrogen fertilisers and water management. *Plant and Soil*, 196, 7–14.

Catling, D. 1992. *Rice in Deep Water*, Macmillan Press, London, UK.

Chauhan, B. S. 2012. Weed ecology and weed management strategies for dry-seeded rice in Asia. *Weed Technology*, 26, 1–13.

Chauhan, B. S., Singh, R. G. and Mahajan, G. 2012. Ecology and management of weeds under conservation agriculture: a review. *Crop Protection*, 38, 57–65.

Chen, H., Stout, M. J., Qian, Q. and Chen, F. 2012. Genetic, molecular and genomic basis of rice defense against insects. *Critical Reviews in Plant Sciences*, 31, 74–91.

Chen, L. J., Lee, D. S., Song, Z. P., Suh, H. S. and Lu, B. R. 2004. Gene flow from cultivated rice (*Oryza sativa*) to its weedy and wild relatives. *Annals of Botany*, 93, 67–73.

Cheng, J. 2009. Rice planthopper problems and relevant causes in China. In Heong, K. L. and Hardy, B. (eds), *Planthoppers: New Threats to the Sustainability of Intensive Rice Production Systems in Asia*, International Rice Research Institute, Philippines, pp. 157–78.

Chin, D. V. 2001. Biology and management of barnyardgrass, red sprangletop and weedy rice. *Weed Biology and Management*, 1, 37–41.

Chintalapati, P., Katti, G., Puskur, R. R. and Nagella Venkata, K. 2016. Neonicotinoid-induced resurgence of rice leaffolder, *Cnaphalocrocis medinalis* (Guénee). *Pest Management Science*, 72, 155–61.

Cockfield, G., Mushtaq, S. and White, N. 2012. *Relocation of Intensive Agriculture to Northern Australia: The Case of the Rice Industry*. Technical Report, University of Southern Queensland, Toowoomba, Australia.

Commare, R. R., Nandakumar, R., Kandan, A., Suresh, S., Bharathi, M., Raguchander, T. and Samiyappan, R. 2002. *Pseudomonas fluorescens* based bio-formulation for the management of sheath blight disease and leaffolder insect in rice. *Crop Protection*, 21, 671–7.

Culliney, T. W. 2005. Benefits of classical biological control for managing invasive plants. *Critical Reviews in Plant Sciences*, 24, 131–50.

De Koninck, R., Bernard, S. and Bissonette, J.-F. 2011. *Borneo Transformed: Agricultural Expansion on the Southeast Asian Frontier*, NUS Press, Singapore.

Dela Cruz, C. G. and Litsinger, J. 1988. Insect pests and their natural enemies in ratoon rice. In International Rice Research Institute (ed.), Rice rationing, International Rice Research Institute, Manila, Philippines, pp. 195–208.

Drost, D. C. and Moody, K. 1982. Rat damage in weed control experiments in rainfed transplanted rice. *International Journal of Pest Management*, 28, 295–9.

Dubey, N. K, Shukla, R., Kumar, A., Singh, P. and Prakash, B. 2010. Prospects of botanical pesticides in sustainable agriculture. *Current Science*, 98, 479–80.

Eggum, B. O. and Juliano, B. O. 1973. Nitrogen balance in rats fed rice differing in protein content. *Journal of the Science of Food and Agriculture*, 24, 921–7.

Elphick, C. S. and Oring, L. W. 1998. Winter management of Californian rice fields for waterbirds. *Journal of Applied Ecology*, 35, 95–108.

Food and Agriculture Organization of the United Nations. 2016. *FAOSTAT* [Online]. Available: http://faostat3.fao.org/home/E [Accessed 30 June 2016].

Fujita, D., Kohli, A. and Horgan, F. G. 2013. Rice resistance to planthoppers and leafhoppers. *Critical Reviews in Plant Sciences*, 32, 162–91.

Gallagher, K. D., Kenmore, P. E. and Sogawa, K. 1994. Judicial use of insecticides deter planthopper outbreaks and extend the life of resistant varieties in Southeast Asian rice. In Denno, R. F. and Perfect, T. J. (eds), *Planthoppers: Their ecology and Management*, Springer-Science, pp. 299–614.

Godfray, H. C. J., Beddington, J. R., Crute, I. R., Haddad, L., Lawrence, D., Muir, J. F., Pretty, J., Robinson, S., Thomas, S. M. and Toulmin, C. 2010. Food security: The challenge of feeding 9 billion people. *Science*, 327, 812–18.

Gould, F. 1998. Sustainability of transgenic insecticidal cultivars: integrating pest genetics and ecology. *Annual Review of Entomology*, 43, 701–26.

Gravois, K. A. and Bernhardt, J. L. 2000. Heritability and genotype × environment interactions for discolored rice kernels. *Crop Science*, 40, 314–18.

Grecchi, R. C., Gwyn, Q. H. J., Bénié, G. B., Formaggio, A. R. and Fahl, F. C. 2014. Land use and land cover changes in the Brazilian Cerrado: A multidisciplinary approach to assess the impacts of agricultural expansion. *Applied Geography*, 55, 300–12.

Grigarick, A. A., Webster, R. K, Meyer, R. P, Zalom, F. G. and Smith, K. A. 1990. Effect of pesticide treatments on nontarget organisms in California rice paddies. I. Impact of triphenyltin hydroxide. II. Impact of diflubenzuron and triflumuron. *Hilgardia*, 58(1)

Grullón, G. 2013. Pesticide planet (Infographic). *Science*, 341, 730–1.

Guenha, R., Das Virtudes Salvador, B., Rickman, J., Goulao, L. F., Muocha, I. M. and Carvalho, M. O. 2014. Hermetic storage with plastic sealing to reduce insect infestation and secure paddy seed quality: A powerful strategy for rice farmers in Mozambique. *Journal of Stored Products Research*, 59, 275–81.

Gurr, G. M. 2009. Prospects for ecological engineering for planthoppers and other arthropods in rice In Heong, K. L. and Hardy, B. (eds), *Planthoppers: New Threats To The Sustainability of Intensive Rice Production Systems in Asia*, International Rice Research Institute, Philippines, pp. 371–88.

Gurr, G. M., Lu, Z., Zheng, X., Xu, H., Zhu, P., Chen, G., Yao, X., Cheng, J., Zhu, Z., Catindig, J. L., Villareal, S., Chien, H. V., Cuong, L. Q., Channoo, C., Chengwattana, N., Lan, L. P., Hai, L. H., Chaiwong, J., Nicol, H. I., Perovic, D. J., Wratten, S. D. and Heong, K. L. 2016. Multi-country evidence that crop diversification promotes ecological intensification of agriculture. *Nature Plants*, 2, 16014.

Hafeez, M. M., Bouman, B. A. M., Van de Giesen, N. and Vlek, P. 2007. Scale effects on water use and water productivity in a rice-based irrigation system (UPRIIS) in the Philippines. *Agricultural Water Management*, 92, 81–9.

Hardin, M. R., Benrey, B., Coll, M., Lamp, W. O., Roderick, G. K. and Barbosa, P. 1995. Arthropod pest resurgence: an overview of potential mechanisms. *Crop Protection*, 14, 3–18.

Heap, I. 2014. Global perspective of herbicide-resistant weeds. *Pest Management Science*, 70, 1306–15.

Heinrichs, E. A. 1994. Impact of insecticides on the resistance and resurgence of rice planthoppers. In Denno, R. F. and Perfect, T. J. (eds), *Planthoppers: Their ecology and Management*, Springer-Science, pp. 571–98.

Hinterhuber, A. 2002. Value chain orchestration in action and the case of the global agrochemical industry. *Long Range Planning*, 35, 615–35.

Holt, J., Chancellor, T. C. B., Reynolds, D. R. and Tiongco, E. R. 1996. Risk assessment for rice planthopper and tungro disease outbreaks. *Crop Protection*, 15, 359–68.

Horgan, F. G. 2012. Diversity and defence: Plant–herbivore interactions at multiple scales and trophic levels. In Gurr, G. M., Wratten, S. D., Snyder, W. E. and Read, D. M. Y. (eds), *Biodiversity and Insect Pests: Key Issues for Sustainable Management*, Wiley-Blackwell, Oxford, UK, pp. 241–58.

Horgan, F. G. 2017. Ecology and management of apple snails in rice. In Chauhan, B., Jabran, K. and Gulshan, M. (eds), *Rice Production Worldwide*, Springer, Berlin (in press).

Horgan, F. G. and Crisol, E. 2013. Hybrid rice and insect herbivores in Asia. *Entomologia Experimentalis et Applicata*, 148, 1–19.

Horgan, F. G., Crisol-Martínez, E., Almazan, M. L. P., Romena, A., Ramal, A. F, Ferrater, J. B. and Bernal, C.C. 2016a. Susceptibility and tolerance in hybrid and pure-line rice varieties to herbivore attack: biomass partitioning and resource-based compensation in response to damage. *Annals of Applied Biology*, 169, 200–13.

Horgan, F. G., Felix, M. I., Portalanza, D. E., Sánchez, L., Rios, W. M. M., Farah, S. E., Wither, J. A., Andrade, C. I. and Espin, E. B. 2014b. Responses by farmers to the apple snail invasion of Ecuador's rice fields and attitudes toward predatory snail kites. *Crop Protection*, 62, 135–43.

Horgan, F. G., Naik, B. S., Iswanto, E. H., Almazan, M. L. P., Ramal, A. F. and Bernal, C. C. 2016c. Responses by the brown planthopper, *Nilaparvata lugens*, to conspecific density on resistant and susceptible rice varieties. *Entomologia Experimentalis et Applicata*, 158, 284–94.

Horgan, F. G., Ramal, A. F., Bentur, J. S., Kumar, R., Bhanu, K. V., Sarao, P. S., Iswanto, E. H., Chien, H. V., Phyu, M. H., Bernal, C. C., Almazan M. L. P., Alam, M. Z., Lu, Z. and Huang, S.-H. 2015. Virulence of brown planthopper (*Nilaparvata lugens*) populations from South and South East Asia against resistant rice varieties. *Crop Protection*, 78, 222–31.

Horgan, F. G., Ramal, A. F., Bernal, C. C., Villegas, J. M., Stuart, A. M. and Almazan, M. L. 2016d. Applying ecological engineering for sustainable and resilient rice production systems. *Procedia Food Science*, 6, 7–15.

Horgan, F. G., Ramal, A. F., Villegas, J. M., Almazan, M. L. P., Bernal, C. C., Jamoralin, A., Pasang, J. M., Orboc, G., Agreda, V. and Arroyo, C. 2017b. Ecological engineering with high diversity vegetation patches enhances bird activity and ecosystem services in Philippine rice fields. *Regional Environmental Change*, (in press).

Horgan, F. G., Ramal, A. F., Villegas, J. M., Jamoralin, A., Bernal, C. C., Perez, M. O., Pasang, J. M., Naredo, A. I. and Almazan, M. L. P. 2017a. Effects of bund crops and insecticide treatments on arthropod diversity and herbivore regulation in tropical rice fields. *Journal of Applied Entomology* (in press).

Horgan, F. G., Srinivasan, T. S., Naik, B. S., Ramal, A. F., Bernal, C. C. and Almazan, M. L. P. 2016b. Effects of nitrogen on egg-laying inhibition and ovicidal response in planthopper-resistant rice varieties. *Crop Protection*, 89, 223–30.

Horgan, F. G., Stuart, A. M. and Kudavidanage, E. P. 2014a. Impact of invasive apple snails on the functioning and services of natural and managed wetlands. *Acta Oecologica*, 54, 90–100.
Horgan, F. G., Yanes Figueroa, J. and Almazan, M. L. P. 2014c. Seedling broadcasting as a potential method to reduce apple snail damage to rice. *Crop Protection*, 64, 168–76.
Hu, X.-F., Cheng, C., Luo, F., Chang, Y.-Y., Teng, Q., Men, D.-Y., Liu, L. and Yang, M.-Y. 2016. Effects of different fertilization practices on the incidence of rice pests and diseases: A three-year case study in Shanghai, in subtropical southeastern China. *Field Crops Research*, 196, 33–50.
Huan, N. H, Chien, H. V., Quynh, P. V., Tan, P. S., Du, P. V., Escalada, M. M. and Heong, K. L. 2008. Motivating rice farmers in the Mekong Delta to modify pest management and related practices through mass media. *International Journal of Pest Management*, 54, 339–46.
Huang, W.-K., Ji, H.-L., Gheysen, G., Debode, J. and Kyndt, T. 2015. Biochar-amended potting medium reduces the susceptibility of rice to root-knot nematode infections. *BMC Plant Biology*, 15, 267.
Inthavong, S., Schiller, J. M., Sengsoulivong, V. and Inthapanya, P. 2004. Status of gall midge in Lao PDR. In Bennett, J., Bentur, J. S., Pasalu, I. C. and Krishnaian, K. (eds), *New Approaches To Gall Midge Resistance In Rice*. International Rice Research Institute and Indian Council of Agricultural Research, pp. 77–87.
Ip, K. K. L., Liang, Y., Lin, L., Wu, H., Xue, J. and Qiu, J.-W. 2014. Biological control of invasive apple snails by two species of carp: Effects on non-target species matter. *Biological Control*, 71, 16–22.
Ishii-Eiteman, M. J. and Power, A. G. 1997. Response of green rice leafhoppers to rice-planting practices in northern Thailand. *Ecological Applications*, 7, 194–208.
Isman, M. B. 2006. Botanical insecticides, deterrents, and repellents in modern agriculture and an increasingly regulated World. *Annual Review of Entomology*, 51, 45–66.
Jacob, J., Singleton, G. R., Herawati, N. A. and Brown, P. R. 2010. Ecologically based management of rodents in lowland irrigated rice fields in Indonesia. *Wildlife Research*, 37, 418–27.
Jagadish, S. V. K., Septiningsih, E. M., Kohli, A., Thomson, M. J., Ye, C., Redoña, E., Kumar, A., Gregorio, G. B., Wassmann, R., Ismail, A. M. and Singh, R. K. 2012. Genetic advances in adapting rice to a rapidly changing climate. *Journal of Agronomy and Crop Science*, 198, 360–73.
Jiang, M. X. and Cheng, J. A. 2003. Interactions between the striped stem borer *Chilo suppressalis* (Walk.)(Lep., Pyralidae) larvae and rice plants in response to nitrogen fertilization. *Anzeiger für Schädlingskunde- Journal of Pest Science*, 76, 124–8.
Kenmore, P. E., Carino, F. O., Perez, C. A., Dyck, V. A. and Gutierrez, A. P. 1984. Population regulation of the rice brown planthopper (*Nilaparvata lugens* Stål) within rice fields in the Philippines. *Journal of Plant Protection in the Tropics*, 1, 19–37.
Kiritani, K. 1999. Shift of IPM strategy for rice under global warming in temperate areas. In Zhongshan University (ed.), *Integrated Pest Management In Rice-Based Ecosystem*. Editorial Department, Journal of Zhongshan University, Guangzhou, China, pp. 235–44.
Ko, K., Liu, Y., Hou, M., Babendreier, D., Zhang, F. and Song, K. 2014. Evaluation for potential *Trichogramma* (Hymenoptera: Trichogrammatidae) strains for control of the striped stem borer (Lepidoptera: Crambidae) in the Greater Mekong Subregion. *Journal of Economic Entomology*, 107, 955–63.
Kojima, A., Yamashiro, C. and Arisaka, M. 1996. Regional control threshold of rice stem borer, *Chilo suppressalis* Walker (Lepidoptera: Pyralidae), indicated by adult catch with sex-pheromone trap. *Japanese Journal of Applied Entomology and Zoology*, 40, 279–86.
Kürschner, E., Hensche, C., Hildebrandt, T., Jülich, E., Leineweber, M. and Paul, C. 2010. Water saving in rice production–dissemination, adoption and short term impacts of alternate wetting and drying (AWD) in Bangladesh. *Humboldt-Universitat Zu Berlin SLE Publication Series*, S241.
Lamb, D. W. and Brown, R. B. 2001. PA – precision agriculture: Remote-sensing and mapping of weeds in crops. *Journal of Agricultural Engineering Research*, 78, 117–25.
Lane, S. J., Azuma, A. and Higuchi, H. 1998. Wildfowl damage to agriculture in Japan. *Agriculture, Ecosystems and Environment*, 70, 69–77.

Laurance, W. F., Sayer, J. and Cassman, K. G. 2014. Agricultural expansion and its impacts on tropical nature. *Trends in Ecology and Evolution*, 29, 107–16.

Leach, M. C. and Hobbs, S. L. A. 2013. Plantwise knowledge bank: delivering plant health information to developing country users. *Learned Publishing*, 26, 180–5.

Lim, G. S. and Bottrell, D. G. 1994. *Neem Pesticides in Rice: Potential and Limitations*, International Rice Research Institute, Manila, Philippines.

Lin, W., Fang, C., Chen, T., Lin, R., Xiong, J. and Wang, H. 2010. Rice allelopathy and its properties of molecular ecology. *Frontiers in Biology*, 5, 255–62.

Litsinger, J. A., Barrion, A. T. and Soekarna, D. 1987. Upland rice insect pests: Their ecology, importance and control. IRRI Research Paper Series, 123. International Rice Research Institute, Manila, Philippines.

Liu, W.-Y., Wang, C.-Y., Wang, T.-S., Fellers, G. M., Lai, B.-C. and Kam, Y.-C. 2011. Impacts of the herbicide butachlor on the larvae of a paddy field breeding frog (*Fejervarya limnocharis*) in subtropical Taiwan. *Ecotoxicology*, 20, 377–84.

Loc, N. T. and Chi, V. T. B. 2005. Efficacy of some new isolates of *Metarhizium anisopliae* and *Beauveria bassiana*, against rice earhead bug, *Leptocorisa acuta*. *Omonrice*, 13, 69–75.

Matsumura, M. 2001. The current status of occurrence and forecasting system of rice planthoppers in Japan. *Journal of Asia-Pacific Entomology*, 4, 195–9.

Matsumura, M., Takeuchi, H., Satoh, M., Sanada-Morimura, S., Otuka, A., Watanabe, T. and Van Thanh, D. 2008. Species-specific insecticide resistance to imidacloprid and fipronil in the rice planthoppers *Nilaparvata lugens* and *Sogatella furcifera* in East and South-East Asia. *Pest Management Science*, 64, 1115–21.

Matsuno, Y., Nakamura, K., Masumoto, T., Matsui, H., Kato, T. and Sato, Y. 2006. Propspects for multifunctionality of paddy rice cultivation in Japan and other countries in monsoon Asia. *Paddy Water Environment*, 4, 189–97.

Matteson, P. C. 2000. Insect pest management in tropical Asian irrigated rice. *Annual Review of Entomology*, 45, 549–74.

Mattson, W. J. 1980. Herbivory in relation to plant nitrogen content. *Annual Review of Ecology and Systematics*, 11, 119–61.

Meyling, N. V. and Eilenberg, J. 2007. Ecology of the entomopathogenic fungi *Beauveria bassiana* and *Metarhizium anisopliae* in temperate agroecosystems: potential for conservation biological control. *Biological Control*, 43, 145–55.

Moody, K. and Drost, D. C. 1983. The role of cropping systems on weeds in rice. In International Rice Research Institute (ed.), *Weed Control In Rice: Proceedings of The Conference on Weed Control In Rice, 1983*. International Rice Research Institute, Los Baños, Philippines, pp. 73–86.

Morse, S. and Buhler, W. 1997. IPM in developing countries: the danger of an ideal. *Integrated Pest Management Reviews*, 2, 175–85.

Murdock, L. L. and Shade, R. E. 2002. Lectins and protease inhibitors as plant defenses against insects. *Journal of Agricultural and Food Chemistry*, 50, 6605–11.

Okuma, M., Fukushima, Y. and Tanaka, K. 1994. Feeding habits of apple snail (*Pomacea canaliculata*) to paddy weeds and damage avoidance to rice seedlings. *Weed Research (Tokyo)*, 39, 109–13.

Padgham, J. L. and Sikora, R. A. 2007. Biological control potential and modes of action of *Bacillus megaterium* against *Meloidogyne graminicola* on rice. *Crop Protection*, 26, 971–7.

Palis, F. G. 1998. Changing farmers' perceptions and practices: the case of insect pest control in Central Luzon, Philippines. *Crop Protection*, 17, 599–607.

Parsa, S., Morse, S., Bonifacio, A., Chancellor, T.C.B, Condori, B., Crespo-Pérez, V., Hobbs, S.L.A, Kroschel, J., Ba, M. N., Rebaudo, F., Sherwood, S. G., Vanek, S. J., Faye, E., Herrera, M. A. and Dangles, O. 2014. Obstacles to integrated pest management adoption in developing countries. *Proceedings of the National Academy of Sciences*, 111, 3889–94.

Pathak, M. D. and Khan, Z. R. 1994. *Insect Pests of Rice*. International Rice Research Institute, Manila, Philippines.

Patil, J., Miller, A. J. and Gaur, H. S. 2013. Effect of nitrogen supply form on the invasion of rice roots by the root-knot nematode, *Meloidogyne graminicola. Nematology,* 15, 483–92.
Pawlina, I. M. and Proulx, G. 1996. Study of house sparrow (*Passer domesticus*) feeding preference to natural color and guard coat blue coated seeds. *Crop Protection*, 15, 143–6.
Phelps, J., Carrasco, L. R., Webb, E. L., Koh, L. P. and Pascual, U. 2013. Agricultural intensification escalates future conservation costs. *Proceedings of the National Academy of Sciences*, 110, 7601–6.
Phillips, T. W. and Throne, J. E. 2009. Biorational approaches to managing stored-product insects. *Annual Review of Entomology,* 55, 375.
Phuong, L.T ., Denich, M., Vlek, P.L.G. and Balasubramanian, V. 2005. Suppressing weeds in direct-seeded lowland rice: Effects of methods and rates of seeding. *Journal of Agronomy and Crop Science*, 191, 185–94.
Pingali, P. L. 2001. Environmental consequences of agricultural commercialization in Asia. *Environment and Development Economics*, 6, 483–502.
Pingali, P. L. and Gerpacio, R. V. 1997. Living with reduced insecticide use for tropical rice in Asia. *Food Policy*, 22, 107–18.
Pingali, P. L. and Gerpacio, R. V. 1998. Toward reduced pesticide use for cereal crops in Asia. In Lutz, E., Binswanger, H., Hazell, P. and McCalla, A. (eds), *Agriculture and the Environment: Perspectives on Sustainable Rural Development.* Washington, DC, World Bank, pp. 254–70.
Pingali, P. L. and Roger, P. A. 1995. *Impact of Pesticides on Farmer Health and the Rice Environment.* International Rice Research Institute, Manila, Philippines.
Powles, S. B. and Shaner, D. L. 2001. *Herbicide Resistance and World Grains*, CRC Press.
Prasad, R. 2011. Aerobic rice systems. *Advances in Agronomy,* 111, 207–47.
Price, P. W. 1991. The plant vigor hypothesis and herbivore attack. *Oikos*, 244–51.
Prot, J.-C., Soriano, I. R. S. and Matias, D. M. 1994. Major root-parasitic nematodes associated with irrigated rice in the Philippines. *Fundamental and Applied Nematology,* 17, 75–8.
Pu, D., Shi, M., WuU, Q., Gao, M., Liu, J. F., Rer, S, Yang, F., Tang, P., Ye, G., Shen, Z., He, J., Yang, D., Bu, W., Zhang, C., Song, Q., Xu, D., Strand, M. R. and Chen, X. 2014. Flower-visiting insects and their potential impact on transgene flow in rice. *Journal of Applied Ecology,* 51, 1357–65.
Pucher, J., Steinbronn, S., Mayrhofer, R., Schad, I., El-Matbouli, M. and Focken, U. 2013. Improved sustainable aquaculture systems for small-scale farmers in northern Vietnam. In Fröhlich, H. L., Schreinemachers, P., Stahr, K. and Clemens, G. (eds), *Sustainable Land Use and Rural Development in Southeast Asia: Innovations and Policies for Mountainous Areas*, Springer, pp. 281–317.
Qiao, F., Huang, J., Zhang, L. and Rozelle, S. 2012. Pesticide use and farmers' health in China's rice production. *China Agricultural Economic Review,* 4, 468–84.
Rana, S. and Jain, D. 2013. Impact of modern agricultural practices on population density of Indian peafowl (*Pavo cristatus*) in Haryana, India. *Agricultural Science Digest*, 33, 230–3.
Ranasinghe, L. L. 2003. Evaluation of factors responsible for low efficacy of farmer's weed management practices in rice cultivation. *Annals of Sri Lanka Department of Agriculture*, 5, 199–205.
Reddy, C. S., Laha, G. S., Prasad, M. S., Krishnaveni, D., Castilla, N. P., Nelson, A. and Savary, S. 2011. Characterizing multiple linkages between individual diseases, crop health syndromes, germplasm deployment, and rice production situations in India. *Field Crops Research*, 120, 241–53.
Rejesus, R. M., Mutuc, M. E. M., Yasar, M., Lapitan, A. V., Palis, F. G. and Chi, T. T. N. 2012. Sending Vietnamese rice farmers back to school: Further evidence on the impacts of farmer field schools. *Canadian Journal of Agricultural Economics/Revue Canadienne d'Agroeconomie*, 60, 407–26.
Rispail, N., Dita, M. A., González-Verdejo, C., Pérez-de-Luque, A., Castillejo, M. A., Prats, E., Román, B., Jorrín, J. and Rubiales, D. 2007. Plant resistance to parasitic plants: molecular approaches to an old foe. *New Phytologist,* 173, 703–12.
Robinson, E. J. Z., Das, S. R. and Chancellor, T. B. C. 2007. Motivations behind farmers' pesticide use in Bangladesh rice farming. *Agriculture and Human Values,* 24, 323–32.

Rodenburg, J., Demont, M., Sow, A. and Dieng, I. 2014. Bird, weed and interaction effects on yield of irrigated lowland rice. *Crop Protection*, *66*, 46–52.

Saka, M. 2010. Acute toxicity of rice paddy herbicides simetryn, mefenacet, and thiobencarb to *Silurana tropicalis* tadpoles. *Ecotoxicology and Environmental Safety*, 73, 1165–9.

Salim, M. and Saxena, R. C. 1991. Nutritional stresses and varietal resistance in rice: effects on whitebacked planthopper. *Crop Science*, 31, 797–805.

Santiapillai, C. and Read, B. 2010. Would masking the smell of ripening paddy-fields help mitigate human–elephant conflict in Sri Lanka? *Oryx*, 44, 509–11.

Sarangi, S. K., Singh, R. A. nd Singh, K. A. 2009. Indigenous method of rat proof grain storage by Adi tribes of Arunachal Pradesh. *Indian Journal of Traditional Knowledge*, 8, 230–3.

Savary, S., Castilla, N. P., Elazegui, F. A. and Teng, P. S. 2005. Multiple effects of two drivers of agricultural change, labour shortage and water scarcity, on rice pest profiles in tropical Asia. *Field Crops Research*, 91, 263–71.

Sexsmith, J. J. and Pittman, U. J. 1963. Effect of nitrogen fertilizers on germination and stand of wild oats. *Weeds*, 11, 99–101.

Sogawa, K., Liu, G. J. and Shen, J. H. 2003. A review on the hyper-susceptibility of Chinese hybrid rice to insect pests. *Chinese Journal of Rice Science*, 17, 23–30.

Spangenberg, J. H., Douguet, J.-M., Settele, J. and Heong, K. L. 2015. Escaping the lock-in of continuous insecticide spraying in rice: Developing an integrated ecological and socio-political DPSIR analysis. *Ecological Modelling*, 295, 188–95.

Stafford, J. D., Kaminski, R. M. and Reinecke, K. J. 2010. Avian foods, foraging and habitat conservation in world rice fields. *Waterbirds*, 33, 133–50.

Stewart, W. M., Dibb, D. W., Johnston, A. E. and Smyth, T. J. 2005. The contribution of commercial fertilizer nutrients to food production. *Agronomy Journal*, 97, 1–6.

Stuart, A. M., Palenzuela, A. N., Bernal, C. C., Ramal, A. F. and Horgan, F. G. 2014. Effects of fertiliser applications on survival and recruitment of the apple snail, *Pomacea canaliculata* (Lamarck). *Crop Protection*, 64, 78–87.

Sudianto, E., Beng-Kah, S., Ting-Xiang, N., Saldain, N. E., Scott, R. C. and Burgos, N. R. 2013. Clearfield® rice: Its development, success, and key challenges on a global perspective. *Crop Protection*, 49, 40–51.

Suding, K. N., Lejeune, K. D. and Seastedt, T. R. 2004. Competitive impacts and responses of an invasive weed: dependencies on nitrogen and phosphorus availability. *Oecologia*, 141, 526–35.

Settle, W. H., Ariawan, H., Astuti, E. T., Cahyana, W., Hakim, A. L., Hindayana, D. and Lestari, A. S. 1996. Managing tropical rice pests through conservation of generalist natural enemies and alternative prey. *Ecology*, 77, 1975–88.

Takahashi, K. and Otsuka, K. 2009. The increasing importance of nonfarm income and the changing use of labor and capital in rice farming: the case of Central Luzon, 1979–2003. *Agricultural Economics*, 40, 231–42.

Takamura, K. and Yasuno, M. 1986. Effects of pesticide application on chironomid larvae and ostracods in rice fields. *Applied Entomology and Zoology*, 21, 370–6.

Tanaka, K., Endo, S. and Kazano, H. 2000. Toxicity of insecticides to predators of rice planthoppers: Spiders, the mirid bug and the dryinid wasp. *Applied Entomology and Zoology*, 35, 177–87.

Teetes, G. L. 1994. Adjusting crop management recommendations for insect-resistant crop varieties. *Journal of Agricultural Entomology*, 11, 191–200.

Teillard, F., Maia de Souza, D., Thoma, G., Gerber, P. J. and Finn, J. A. 2016. What does Life-Cycle Assessment of agricultural products need for more meaningful inclusion of biodiversity? *Journal of Applied Ecology*, 53, 1422–9.

Teng, P. S. and Savary, S. 1992. Implementing the systems approach in pest management. *Agricultural Systems*, 40, 237–64.

Teo, S. S. 2001. Evaluation of different duck varieties for the control of the golden apple snail (*Pomacea canaliculata*) in transplanted and direct seeded rice. *Crop Protection*, 20, 599–604.

Thorburn, C. 2015. The rise and demise of integrated pest management in rice in Indonesia. *Insects,* 6, 381–408.

Timmer, C. P. 2005. Food security and economic growth: an Asian perspective. *Asian-Pacific Economic Literature,* 19, 1–17.

Tindall, K. V. and Stout, M. J. 2003. Use of common weeds of rice as hosts for the rice water weevil (Coleoptera: Curculionidae). *Environmental Entomology,* 32, 1227–33.

Van Groenigen, J. W., Burns, E. G., Eadie, J. M., Horwath, W. R. and Van Kessel, C. 2003. Effects of foraging waterfowl in winter flooded rice fields on weed stress and residue decomposition. *Agriculture, Ecosystems and Environment,* 95, 289–96.

Vu, Q., Quintana, R., Fujita, D., Bernal, C. C., Yasui, H., Medina, C. D. and Horgan, F. G. 2014. Responses and adaptation by *Nephotettix virescens* to monogenic and pyramided rice lines with *Grh*-resistance genes. *Entomologia Experimentalis et Applicata,* 150, 179–90.

Wang, A.-H., Wu, J.-C., Yu, Y.-S., Liu, J.-L., Yue, J.-F. and Wang, M.-Y. 2005. Selective insecticide-induced stimulation on fecundity and biochemical changes in *Tryporyza incertulas* (Lepidoptera: Pyralidae). *Journal of Economic Entomology,* 98, 1144–9.

Watanabe, T., Takeuchi, H., Ishizaki, M., Yasuda, T., Tachibana, S.-I., Sasaki, R., Nagano, K., Okutani-Akamatsu, Y. and Matsuki, N. 2009. Seasonal attraction of the rice bug, *Leptocorisa chinensis* Dallas (Heteroptera: Alydidae), to synthetic attractant. *Applied Entomology and Zoology,* 44, 155–64.

Westphal, C., Vidal, S., Horgan, F. G., Gurr, G. M., Escalada, M., Chien, H. V., Tscharntke, T., Heong, K. L. and Settele, J. 2015. Promoting multiple ecosystem services with flower strips and participatory approaches in rice production landscapes. *Basic and Applied Ecology,* 16, 681–9.

Widawsky, D., Rozelle, S., Jin, S. and Huang, J. 1998. Pesticide productivity, host-plant resistance and productivity in China. *Agricultural Economics,* 19, 203–17.

Wilson, B. E., Hardy, T. N., Beuzelin, J. M., Vanweelden, M. T., Reagan, T. E., Miller, R., Meaux, J., Stout, M. J. and Carlton, C. E. 2015. Expansion of the Mexican rice borer (Lepidoptera: Crambidae) into rice and sugarcane in Louisiana. *Environmental Entomology,* 44, 757–66.

Wilson, E. A., Leboeuf, E. A., Weaver, K. M. and Leblanc, D. J. 1989. Delayed seeding for reducing blackbird damage to sprouting rice in southwestern Louisiana. *Wildlife Society Bulletin,* 17, 165–71.

Witzgall, P., Kirsch, P. and Cork, A. 2010. Sex pheromones and their impact on pest management. *Journal of Chemical Ecology,* 36, 80–100.

Wu, J., Xu, J., Yuan, S., Liu, J., Jiang, Y. and Xu, J. 2001. Pesticide-induced susceptibility of rice to brown planthopper *Nilaparvata lugens. Entomologia Experimentalis et Applicata,* 100, 119–26.

Xin, Z., Yu, Z., Erb, M., Turlings, T. C. J., Wang, B., Qi, J., Liu, S. and Lou, Y. 2012. The broad-leaf herbicide 2, 4-dichlorophenoxyacetic acid turns rice into a living trap for a major insect pest and a parasitic wasp. *New Phytologist,* 194, 498–510.

Yanes Figueroa, J., Almazan, M. L. P., Horgan, F. G. 2014. Reducing seed-densities in rice seedbeds improves the cultural control of apple snail damage. *Crop Protection,* 62, 23–31.

Yang, C. -M., Cheng, C. -H. and Chen, R. -K. 2007. Changes in spectral characteristics of rice canopy infested with brown planthopper and leaffolder. *Crop Science,* 47, 329–35.

Yoneyama, K., Xie, X., Yoneyama, K. and Takeuchi, Y. 2009. Strigolactones: structures and biological activities. *Pest Management Science,* 65, 467–70.

Chapter 13

# Control of rodent pests in rice cultivation

*P. R. Brown, CSIRO Agriculture and Food, Australia; B. Douangboupha, National Agricultural and Forestry Research Institute, Lao PDR; N. M. Htwe, International Rice Research Institute, Myanmar; J. Jacob, Julius Kühn Research Institute, Germany; L. Mulungu, Sokoine University of Agriculture, Tanzania; N. T. My Phung, Provincial Department of Agriculture and Rural Development, Vietnam; G. R. Singleton, International Rice Research Institute, The Philippines and University of Greenwich, UK; A. M. Stuart, International Rice Research Institute, The Philippines; and Sudarmaji, Assessment Institute for Agricultural Technology of Yogyakarta, Indonesia*

1 Introduction
2 Rodent impacts on rice
3 Rodent pests of rice and their biology
4 Rodent management methods
5 Case studies of management of rats in rice
6 Other vertebrate pests in rice
7 Future trends and conclusion
8 Where to look for further information
9 References

## 1 Introduction

Rodents remain one of the main nuisances to mankind. For thousands of years they have been causing damage to crops, stored grain and infrastructure, and are reservoirs for devastating human diseases such as plague and typhus. The bones of rats and mice are found along with human bones from the mid-Pleistocene (1–2.5 million years ago). Rats and mice subsequently radiated throughout the world. Rodents continue to cause serious damage to staple food crops such as rice (John 2014), despite advances in methods of control and management techniques.

The order Rodentia is one of the most widely distributed order of animals throughout the world. There are more than 2700 rodent species described, and the Rodentia order accounts for 42% of all mammal species (Aplin et al. 2003; Macdonald 2001). They are primarily seed eaters, but some are insectivorous and some are omnivorous. The success of the pest rodent species can be attributed to their short lifespan, high rates of fecundity (high litter sizes, short gestation period and short duration to sexual maturity), potential to disperse long distances, complex social hierarchy, flexible social systems and their

http://dx.doi.org/10.19103/AS.2016.0003.24

physiology and body structure which allow them to live in a wide range of environments (Prakash 1988; Buckle and Smith 2015). There are five broad suborders of rodents based largely on their jaw structure and gnawing phenology and function: Anomaluromorpha (scaly tailed squirrels, springhares), Castorimorpha (beavers, pocket gophers, kangaroo rats), Hystricomorpha (African mole rats, porcupines, cane rats, chinchilla rats, cavies, guinea pigs, capybara, agoutis, nutria, coypu), Myomorpha (jerboas, muskrats, voles, lemmings, true mice and rats, spiny dormice, mole rats, bamboo rats) and Sciuromorpha (mountain beaver, dormice, squirrels, chipmunks, prairie dogs, marmots).

In some cultures, rodents are held in high regard: In China and Japan, the clever and quick-witted rat is considered a symbol of good luck and wealth. It is the first animal in the Chinese Zodiac. In Vietnam, rodent meat is eaten to bestow good fortune and fertility on the newly-weds. In Myanmar, the rat is one of eight calendar figures – people born on a Thursday carry the rat as their figure and Buddhists visit pagodas to worship their birth figure. In India, there are a few temples where the local rat species is not allowed to be harmed, and people freely feed and provide milk for them. In general, however, rodents are held in poor esteem and considered major pests because of their devastating impacts. However, less than 10% of the species belonging to the Rodentia order are significant agricultural pests; the others generally play an important beneficial role in the ecosystem (Singleton et al. 2007).

Rodents are among the world's most important pests (Prakash 1988; Singleton et al. 1999a; Buckle and Smith 2015), causing significant damage to agricultural production throughout the world (Elias and Fall 1988; Hoque et al. 1988; Lund 1988; Marsh 1988; Fiedler 1994; Singleton 2003; Singleton et al. 2010a). It has been estimated that less than 10% of rodent species are major pest species of agricultural and urban areas (Stenseth et al. 2003; Singleton et al. 2007); however, it is in the poorer developing countries that the impact of rodents is generally greatest. In Indonesia, rodents are considered the most important pre-harvest pest of rice, causing around 15% losses annually (Geddes 1992); and, in Tanzania, rodents cause an estimated 5–15% loss of maize (Leirs 2003). Farmers report yield losses of between 0 and 100%, with occasional outbreaks devastating on smallholder farmers (Singleton et al. 2010b; John 2014). Surveys of rice farmers have revealed that rodents are considered the most important constraint for rice production for 98% of farmers, and have been described as the pest they have least control over (Schiller et al. 1999).

Rodents also cause problems through disease transfer to humans and to their livestock. Rodents are the carriers of human diseases (rodent-borne zoonoses) such as the plague, rat-borne typhus, hantavirus diseases and arenavirus infections (Mills 1999; Meerburg et al. 2009a). Rodents carried parasites that were responsible for the Black Death (*Yersinia pestis*), which killed 20 million people from 1347 to 1350 (between 30 and 50% of Europe's population) (Cantor 2001). Rodents also transmit leptospirosis, a disease with a global distribution that has had major impacts in Indonesia, Thailand, Vietnam, Australia and the Pacific Islands in recent years (e.g. Tangkanakul et al. 2005; Tubiana et al. 2013). In developing countries, mortality rates range from 2 to 10% of those infected, but it is easily treated with antibiotics if diagnosed early. The disease affects many rural communities, especially in developing countries and is often misdiagnosed as malaria or dengue fever (because of the influenza-like symptoms). Most farmers come in contact with the spirochaetes of leptospirosis as they work in their fields. In 2000, there were 14 285 cases of leptospirosis with 362 deaths reported in Thailand (Tangkanakul et al. 2005).

Much is known about the biology, genetics and behaviour of rodents, especially through studies of the laboratory rat (*Rattus norvegicus*) and the laboratory mouse (*Mus domesticus*).

Laboratory mice have advanced our understanding of genetics and reaction systems for mammals and they provide an important model for disease and for drug design and monitoring (Berry and Scriven 2005). Despite this understanding, we are, comparatively, on a fast-learning curve about how wild rodents interact with their environment. More knowledge is required, particularly on rodents that cause damage to rice crops.

Given the important role of rice production in global food security, it is important to manage rodent damage to rice. It is therefore necessary to understand the ecology of the pest species and the nature of the damage they cause, leading to the development of appropriate methods to minimise the damage caused. Rodent management is primarily about controlling damage caused (therefore reducing losses to farmer's rice yield), not killing rodents *per se*.

# 2 Rodent impacts on rice

## 2.1 Pre-harvest losses from rodents

Rodents may damage rice at all stages, from sowing through to ripening and harvest. After sowing, rodents can consume the entire seed or seedling, which leads to the complete removal of a plant. Once tillers have emerged, rodents generally bite off the tillers near the base, but provided there is sufficient plant remaining, it is possible for the plant to compensate for that damage by producing new tillers or regrowing cut tillers. However, following the maximum tillering stage, there will be a yield penalty – the later the damage, the greater the yield penalty (Buckle et al. 1979; Buckle 2015; My Phung et al. 2010). From mid-tillering, rodents cause damage to rice by cutting at an oblique angle near the base of tillers. Rice is most susceptible to rodent damage from the booting (reproductive) stage to harvest.

It is notoriously difficult to assess rodent damage and losses to rice at a crop and landscape scale because damage is typically patchy (e.g. Buckle 2015; Miller et al. 2008). Measurements can be made of the number of rodent-cut tillers on rice plants, but this may not necessarily translate into yield losses. Rodent damage might be easy to see, but vegetative recovery can mask the appearance of damage at harvest (Wood and Singleton 2015). Farmer estimates are subjective and can be wildly inaccurate. Farmers often do not notice damage until it is greater than 5–10%.

The nature of rodent damage to rice can vary, depending on the species of rodent and the type of rice grown. There is no consistent pattern of damage and losses due to rodents in different rice systems in different countries (Table 1). Some species, such as *Bandicota bengalensis*, cut tillers and hoard them in their burrows (see Htwe et al. 2017). Other species feed in the field. Some species eat in small areas and cause localised patches of damage. When abundance is high, some species, in particular *R. argentiventer* and *R. tanezumi*, cause damage towards the centre of fields, away from the field edge. This pattern of damage is known as the 'stadium effect' (Benigno 1980).

There have been a range of methods used to assess rodent damage to rice crops. These include diagonal transects (Rennison 1979), random quadrat sampling (e.g. Salvioni 1991) and parallel transects from the edge of the crop to halfway into a field with sampling points at right angles to set distances from the edge (Aplin et al. 2003; Singleton et al. 2005). The latter method is designed to capture the spatial variation of rodent damage, particularly given there is more damage to rice tillers towards the centre of a field. Miller et al. (2008) were able

**Table 1** The twenty largest rice-producing countries in the world and the corresponding pest species of rodents. Rice statistics from 2015 (IRRI rice statistics: http://ricestat.irri.org:8080/wrs2/entrypoint.htm). There is a lack of information on rodent pests in rice for some countries

| Country | Rice system | Rice area (000 ha) | Rice production (000 tonnes) | Main rodent species | Estimated losses | Nature of losses | Reference |
|---|---|---|---|---|---|---|---|
| China | Irrigated lowland | 30,210 | 208,243 | *Rattus nitidus, Rattus turkestanicus, Bandicota indica, Rattus losea, Rattus rattus* complex, *Microtus fortis, Rattus rattoides, Rattus norvegicus, Mus musculus, Apodemus agrarius* | 5–10% | Severe, varies between years | Deng & Wang (1984), Zhang et al. (1999), Wood & Singleton (2015), Zhao (1996), Singleton (2003) |
| India | Irrigated lowland | 43,000 | 150,015 | *Nesokia indica, Mus booduga, Bandicota bengalensis, Millardia meltada, Bandicota indica, Rattus nitidus, Rattus meltada* | 5–15% | Some outbreaks associated with bamboo masting | Sridhara (1992), Wood & Singleton (2015), Barnett & Prakash (1976), Singh et al. (1994), Parshad (1999) |
| Indonesia | Irrigated lowland | 12,160 | 57,165 | *Rattus argentiventer* | 10–17% in Java | Chronic and occasional outbreaks associated with weather events | Singleton (2003), Singleton et al. (2005), Sudarmaji et al. (2010) |
| Bangladesh | Irrigated and rainfed lowlands | 12,000 | 51,905 | *Rattus rattus* complex, *Bandicota bengalensis, Bandicota savilei, Bandicota indica* | 5–10% (>50% in some districts) | Some outbreaks associated with bamboo masting | Aplin et al. (2003), Islam et al. (1993), Singleton (2003), Belmain et al. (2015) |
| Vietnam | Irrigated lowland | 7,660 | 45,120 | *Rattus argentiventer, Rattus sakaratensis* (formerly considered *R. losea*), *Rattus rattus* complex | 5–30% | 100,000 to 700,000 ha with high annual damage (>10%); outbreaks due to unusual rainfall events | Brown et al. (2005), Singleton (2003), Huan (2010) |
| Thailand | Rainfed and irrigated lowlands | 9,650 | 24,848 | *Rattus argentiventer, Rattus sakaratensis* (formerly considered *R. losea*), *Rattus rattus* complex, *Bandicota indica, Bandicota savilei, Rattus exulans, Mus caroli, Mus cervicolor* | 1–18% | Highest in central region; varies markedly between years | Boonsong et al. (1999), Singleton (2003) |

| | | | | | | | |
|---|---|---|---|---|---|---|---|
| Myanmar | Irrigated and rain-fed lowlands | 6,800 | 19,063 | *Rattus rattus* complex, *Bandicota bengalensis*, *Bandicota savilei* | 5–40% | Some outbreaks associated with bamboo masting or extreme weather events | Aplin et al. (2003), Singleton (2003) Htwe et al. (2013) |
| The Philippines | Irrigated lowland, upland | 4,650 | 18,254 | *Rattus tanezumi*, *Rattus argentiventer*, *Rattus norvegicus*, *Rattus exulans* | District losses 10–20% | Variable, some outbreaks | Fall (1977), Singleton (2003), Singleton et al. (2008); Stuart et al. (2011) |
| Brazil | Irrigated lowland | 2,300 | 11,765 | *Holochilus brasiliensis*, *Sigmodon hispidus* | | Some outbreaks associated with bamboo masting | |
| Japan | Irrigated lowland | 1,610 | 10,852 | Insufficient information | | | No recent published studies |
| Pakistan | Irrigated lowland | 2,850 | 10,351 | *Nesokia indica*, *Bandicota bengalensis*, *Millardia meltada*, *Mus* spp. | 10–25% damage, 2–43% loss (mean 19%) | | Wood & Singleton (2015), Fulk & Akhtar (1981) |
| The United States of America | Irrigated lowland | 1,042 | 8,724 | Insufficient information | | | No recent published studies |
| Cambodia | Rainfed lowland, upland, irrigated lowland | 2,900 | 6,797 | *Rattus argentiventer*, *Bandicota indica*, *Rattus exulans*, *Rattus rattus* complex | Patchy | No national data | Aplin et al. (2003), Singleton (2003) |
| Egypt | Irrigated lowland | 800 | 5,797 | *Arvicanthis niloticus* | | | No recent published studies |
| South Korea | Irrigated lowland | 799 | 5,771 | Insufficient information | | | No recent published studies |
| Sri Lanka | Irrigated lowland | 1,225 | 4,853 | *Bandicota bengalensis*, *Rattus rattus* complex, *Mus cervicolor* | ? | ? | Bambaradeniya et al. (2004) |
| Nepal | Irrigated lowland | 1,560 | 4,655 | Insufficient information | | | No recent published studies |
| Nigeria | Rainfed lowland | 2,400 | 4,300 | *Arvicanthus niloticus*, *Dasymys incomtus*, *Tatera kempii*, *Mastomys natalensis*, *Lemniscomys striatus*, *Uranomys foxi*, *Rattus rattus*, *Mus musculoides* | 4.8–12.6% | | Rabiu & Rose (2004), Funmilayo & Akande (1977), Wood & Singleton (2015), |
| Madagascar | Irrigated lowland, rainfed lowland | 1,450 | 3,700 | *Rattus rattus* | 2.5% yield loss | 86% of farmers report rodents as important pests | Duplantier (1999) |
| Peru | Irrigated lowland | 400 | 3,120 | Insufficient information | | | No recent published studies |

to better capture this spatial variation in rice terraces by estimating areas of low, medium and high rodent damage and then using stratified random assessment of rodent damage.

Rodents are significant pests in all rice cropping systems: irrigated lowland, rainfed lowland, upland rainfed and deep water/tidal wetlands (Singleton 2003; Table 1). In Vietnam, the rodent problem escalated in the 1990s in the Mekong and Red River deltas due to an increase in cropping intensity; two and three rice crops were grown annually where previously only one or two crops were grown (Lan et al. 2003). In Indonesia (Singleton et al. 2003a), Malaysia (Lam 1988), Myanmar (Htwe et al. 2013) and Vietnam (Huan et al. 2010), asynchrony of cropping was reported as a major driver of high losses to rodents. We propose that regardless of the rice agro-ecosystem, there is likely to be a relationship between the level of losses to rodents and both the intensity of cropping and the level of asynchrony of cropping. Intensive lowland irrigated rice is the most intensive cropping system and therefore suffers the highest losses to rodents due to an extended period of increased availability of food that results in rats breeding more often during a year, and for longer when there is asynchronous cropping (Singleton et al. 2010b).

It is important to understand economic thresholds for rodents and yield losses. Much work has been done on insects in rice (see chapter 12 of this volume), but there are only a few examples of this for rodents in rice, and these have been achieved mainly through simulation modelling (Brown et al. 2011; My Phung et al. 2010; Mulungu et al. 2014a).

## 2.2 Post-harvest losses from rodents

Rodent losses to stored rice have been claimed to be higher than rodent pre-harvest losses in South Asia (Parshard 1999). The numbers of studies of rodent post-harvest losses, however, are few. Recent publications have documented losses at 4–14% in Myanmar (Belmain et al. 2015; Htwe et al. 2017), 10% in Laos (Brown et al. 2013) and 3% in Bangladesh (Belmain et al. 2015). The losses by rodents depend on the species composition, storage structures, sanitation around storage areas and the duration of storage. The dominant species in stored grain were those of the *R. rattus* complex in Laos; *R. rattus, R. exulans* and *B. bengalensis* in Myanmar; and *Mus musculus* in Bangladesh. Interestingly, the main pest species to grain stores are often different from those that cause losses in the field with perhaps the exception of Myanmar (Htwe et al. 2017), indicating that often we cannot apply our understanding of the breeding dynamics and spacing behaviour of field rodent species to those that live in and around houses and grain stores.

# 3 Rodent pests of rice and their biology

There are many species of rodents that cause damage to rice. Some have global distributions (e.g. *R. rattus* complex; Aplin et al. 2011), some have wide distributions across much of Southeast Asia (e.g. *R. argentiventer* and *R. sakaratensis* (formerly considered *R. losea*)) and in sub-Saharan Africa, for example *Mastomys natalensis* (Leirs et al. 1994; Mulungu et al. 2011), and still others are found in a relatively more limited range. Some regions have only one or two species, while others can have relatively large rodent communities (e.g. Mekong Delta, Vietnam, with up to ten different species) (Aplin et al. 2003). A summary of the pest rodent species present in the main rice-producing countries is shown in Table 1 and a summary of the nature of the impact is shown in Table 2.

**Table 2** Level of threat posed by selected rodents to rice crops. Key: + = low threat, ++ = moderate threat, +++ = high threat,? = lack of direct evidence. Source: Aplin et al. (2003). There is incomplete knowledge about the rodent species in Brazil, Cambodia, Egypt, Japan, Madagascar, Nepal, Nigeria, Peru, South Korea, Sri Lanka and the United States

| Species | Removal of seeds | Removal of emerging plants | Cutting of tillers | Removal of maturing grain | Remarks |
|---|---|---|---|---|---|
| *Bandicota bengalensis* | ++ | ++ | ++ | ++ | Breeding linked to crop stage, damages all stages, causes most damage near burrows, hoards grain (1.7 kg of rice grain/burrow system) |
| *Bandicota indica* | + | + | ++ | ++ | Breeding not linked to crop stage, only causes damage at moderate to high densities, hoards grain |
| *Bandicota savilei* | ? | ? | ? | ? | Can be locally abundant and prefers lowland rainfed rice with no widespread flooding |
| *Cannomys badius* | ? | ? | ? | ? | Causes some damage to upland rice crops in Laos, tunnels below rice plants and consumes plants from below |
| *Mastomys natalensis* | +++ | +++ | ++ | ++ | Breeding is throughout the year but most female individuals are active during the crop maturity stage, damages all stages (Mulungu at al. 2013, 2015a) |
| *Mus booduga/terricolor* | +? | +? | +? | +? | Can breed all months of the year, can cause damage at any crop stage |
| *Mus caroli/cervicolor* | ? | ? | ? | ? | Common in rice fields, very agile and climb plants to damage developing seeds |
| *Mus cookii* | ? | ? | ? | ? | Often captured in and around rice fields, but little is known of its impact |
| *Mus musculus* | + | + | + | + | Breeding linked to season, damage at all stages, but *Mus* spp. often makes up a small proportion of rodent community |
| *Nesokia indica* | ? | ? | + | +++ | Can cause extensive damage to rice and other crops, in Pakistan it consumes only ripening grain |
| *Rattus argentiventer* | ++ | ++ | +++ | +++ | Key pest in Southeast Asia causing heavy damage at high densities. Breeding linked to crop stage |
| *Rattus exulans* | + | + | ++ | ++ | Usually confined to houses and villages, but can cause some damage to crops |
| *Rattus losea/sakaratensis* | ++ | ++ | ++ | ++ | Breeding linked to crop stage, competes with *R. argentiventer* in most of ranges |
| *Rattus nitidus* | ? | ? | ? | ? | China (Sichuan Province) and Thailand, breeding linked to crop stage |
| *Rattus norvegicus* | + | + | + | + | Mainly pest in urban areas, with scattered populations in Thailand, Vietnam and the Philippines, damage not recorded |
| *Rattus rattus* complex | ++ | ++ | +++ | +++ | Found in village, urban, crop fields and forests, competes with other species, damage only around the edge of fields, breeding linked to crop stage |
| *Rattus tanezumi* | ++ | ++ | +++ | +++ | Found in village, urban, crop fields and forests, damage often concentrated in the centre of rice fields, can breed throughout the year but peaks in breeding linked to crop stage |
| *Rattus turkestanicus* | ? | ? | ? | ? | Breeding linked to maturing rice in Fujian Province and Xiamen Province (China), damage to early ripening rice crops |

The breeding dynamics of some rodent species is linked to the crop stage of rice (Table 2), for example, *R. argentiventer* (My Phung et al. 2011). It has a natural preference for waterlogged areas with grassy habitat, making it ideally suited to rice fields (Aplin et al. 2003). If there is one rice crop per year, *R. argentiventer* will breed once a year. If there are two rice crops per year, they have two breeding seasons. Three breeding seasons are apparent when there are three rice crops per year. *R. argentiventer* commences breeding at maximum tillering stage and continues through to harvest. If the rice crops are not synchronised and the breeding season is extended, then the young from the first litter will be mature enough to breed themselves, thus resulting in an exponential increase in abundance (Leung et al. 1999).

A similar association between rice crop growth and breeding activity is apparent for *R. tanezumi* in the Philippines and *M. natalensis* in Tanzania. However, these species are able to continue breeding at low levels (10–20% adult females breeding) at times when no ripening rice is available (Duque et al. 2008; Htwe et al. 2012; Mulungu et al. 2013; Stuart et al. 2015).

Pest rodent populations have the ability to grow rapidly in response to good environmental conditions or following lethal control because they have high rates of reproduction (many offspring, short gestation period) and/or survival (high growth rates, *r*-selected species; with boom and bust characteristics), compared to species at carrying capacity (slow growth rates, *K*-selected species) (Hein and Jacob 2015). Pest rodents can either be native to a particular area or introduced by humans. Most reach sexual maturity at 2–3 months of age, with the females producing 6–7 young after a gestation period of only 2–3 weeks. Females are also capable of post-partum oestrus, meaning they can become pregnant straight after giving birth to a litter of young, thus allowing another litter to be produced when one litter is weaned.

In intensive lowland-irrigated rice cropping systems, rodents cause chronic losses, although the amplitude will vary from year to year. Probably of greater economic concern for smallholder farmers are the episodic outbreaks that can lead to losses of 50–100% of the rice crop. There have been three types of outbreaks described for rodents in rice cropping systems (see Singleton et al. 2010a for review):

1 Masting events (e.g. bamboo and beech trees) not related to climate or farming systems,
2 Changes in abiotic conditions alone (aseasonal or unusual rainfall event, or major climatic events such as El Niño) that create episodic rodent population outbreaks in response to increases in food availability, and
3 Changes in cropping systems: anthropogenic responses to calamitous events such as typhoons, cyclones and droughts. Anthropogenic responses include delayed or asynchronous planting or increased intensity of cropping.

Rodents build nests in and around rice fields or surrounding vegetation, non-rice habitats or houses/village habitats. In irrigated lowland rice monocultures of the Philippines, female *R. tanezumi* nest along the edges of rice fields and within ripening fields, between the rice tillers (Fall 1977). Whereas, in complex agroecosystems, female *R. tanezumi* predominantly nest in adjacent coconut groves, including within the crowns of coconut trees (Stuart et al. 2012).

A number of studies on the movement and habitat use, relating to changes in the abundance of *R. argentiventer*, have been conducted and show that habitat use of *R. argentiventer* was responsive to rice crop growth stage, rice plant cover (Brown et al. 2001; My Phung et al. 2012) and patches of high-quality habitat (Jacob and Wegner 2005). The movement of *R. argentiventer* is greater in the non-breeding seasons than in breeding seasons, with males travelling further than females in the breeding seasons (Brown et al. 2001). Not enough studies of other species in rice systems have been conducted to draw generalities.

In rice fields, an established population of *R. tanezumi* is restricted to a home range of around 100 m (Alfonso et al. 1985). However, the home range changes at different stages of the rice crop (Singleton et al. unpublished data); just before and after rice harvest, the average home range of *R. tanezumi* individuals was found to be 0.6 ha for males and 0.3 ha for females. During the milky-ripe stage of the rice crop, males had an average home range of 1.8 ha and females had an average home range of 0.8 ha.

Knowledge of foraging behaviour and food preferences is vital to the successful use of rodenticides. Baits placed within rice fields are likely to be more attractive and more readily accepted during the tillering stage of the rice crop because food availability is low and pest rodents move into the rice field from the surrounding habitats during this time (Stuart et al. 2015). When rice plants enter the reproductive stage, the acceptability of rodenticide bait is expected to decline and will be less effective (Buckle and Rowe 1981).

Generally, rodent pest species are considered to be opportunistic omnivores with a dominant component of their diet comprising seed (Bomford 1987), insects (Berry and Bronson 1992; King et al. 1996), crabs, snails (Tann et al. 1991) and macadamia nuts (Horskins et al. 1998). The dietary preference of *R. argentiventer* was mainly rice, followed by green material, insects and mung bean (My Phung et al. 2011).

In a rice field ecosystem of the Philippines, rice has been identified as the main component of the *R. tanezumi* diet during the ripening to harvest stages of the rice crop (Tigner 1972; Htwe et al. 2014). Insects, weeds (e.g. *Digitaria* sp., *Ipomoea aquatica* and *Echinochloa colonum*), snails and crabs have been identified as the dominant food items during the off-season months when rice grains were absent. Other foods included grasses and other vegetative matter (Htwe et al. 2014). *R. tanezumi* also feed on agricultural crops other than rice, such as coconut, maize and banana.

# 4 Rodent management methods

There are a wide range of management options available that can be used to control rodents in rice systems. These are largely designed to kill animals, but there are a few exceptions (e.g. habitat management and biocontrol; detailed below). These methods can be categorised as physical, chemical, biocontrol and others. The ultimate aim of a rodent management programme is to reduce the impact (e.g. reduce rat damage to rice), not kill animals *per se*. The advantages and disadvantages of each management method are presented in Table 3. Farmers often attempt to control rodents when they see damage, which is often too late because rodent population numbers are already high, so a good understanding of the timing of control for the maximum benefit of reducing rice yield loss is necessary (see Section 4.5 for details).

**Table 3** Advantages and disadvantages of different control strategies to manage rodents in rice-based cropping systems

| Method | Advantages | Disadvantages |
|---|---|---|
| Trapping – snap traps | • Easy to set | • Can only kill one animal at a time<br>• Useful in small-scale situations<br>• Not effective when abundance of rodents high<br>• Unwanted non-target effects |
| Trapping – live-traps | • Easy to set | • Small number of animals captured<br>• Need to kill captured animals |
| Barriers – plastic fence | • Protects small areas of crop, for example, rice nurseries | • Takes time and effort to set up<br>• Plastic only lasts 1 or 2 seasons |
| Barriers – trap-barrier system (TBS) | • Can capture a large number of rats<br>• Can protect large areas of rice (10–15 ha) | • Takes time and effort to set up<br>• Needs to be planted 2–3 weeks earlier than surrounding fields<br>• Needs a community-based approach for best results |
| Habitat management | • Increases predation risk | • Needs to be applied at the right time<br>• Takes effort to apply herbicides or slashing<br>• Conflict with conservation |
| Crop synchronisation | • Reduces length of rat breeding season | • Needs coordination among neighbouring farmers (>100 ha) |
| Bund size | • Reduces rodent burrows | • May not be appropriate in all situations |
| Rodenticide – acute | • Easy to apply<br>• See dead rodents<br>• Immediate effect | • Unwanted non-target effects<br>• Often no antidote available<br>• Bait shyness |
| Rodenticide anticoagulant | • Easy to apply<br>• Antidote available<br>• No learnt bait shyness | • Takes longer for effect<br>• Can be expensive<br>• Unwanted non-target effects |
| Fumigation | • Relatively easy to apply | • Need to locate all burrows to be effective<br>• Takes a lot of labour to do properly |
| Repellents | • Limited non-target effect | • No strong evidence for lasting effect |
| Sterility control | • None known | • Bait delivery difficult |
| Reproductive inhibitors | • None known | • Bait delivery difficult |
| Predators | • Relatively cheap | • No definitive proof that predators increase rice yields |
| Parasites and diseases | • Reduce reliance on chemical rodenticide | • Limited availability of commercial products |
| Diversionary feeding | • Cheap | • No evidence it works in rice fields |
| Hunting | • Can be conducted at any time<br>• Large numbers can be killed when hunting as a group<br>• Source habitats can be targeted | • Group hunting takes time and effort to organise |
| Electricity | • No advantages | • Highly dangerous |
| Ultrasound/ electromagnetic devices | • No advantages | • Not effective |
| Ecologically based rodent management (EBRM) | • Positive benefit:cost ratios | • Requires coordination at community scale |

## 4.1 Physical rodent management methods

### 4.1.1 Trapping

*Trapping* is a widespread and common tool for removing rodents. There are many 'village industries' based around the construction and sales of local rat traps. These include wire traps in the uplands of Laos and bamboo traps in Myanmar. Kill traps are generally spring loaded and are designed to rapidly compress the cervical vertebrae for a very quick kill. These traps need to be either attractive to rats via the placement of an attractive bait or are set along runways. The effectiveness of physical trapping depends on the behaviour of the species. For example, the house mice (*M. musculus*) are more likely to be trapped because they are relatively neophilic, whereas some species are notoriously difficult to trap, for example, black rats (*R. rattus*) and Norway rats (*R. norvegicus*) because they are relatively neophobic. Therefore, a pre-baiting strategy may be required to leave the traps unset for 3–5 days before setting, and a good bait is often needed to make traps effective. Other types of traps are live- and multiple-capture cage traps. Resourceful farmers are known to invent an amazing array of trapping devices, such as snaring devices or buckets of water with a slippery surface so the rodents fall into the bucket and ultimately drown.

### 4.1.2 Barriers

*Barriers* can be built around small areas to protect crops. Plastic barriers have been used successfully for nursery rice crops to protect the seedlings from rodent attack before they are planted out in the field (Sudarmaji et al. 2003). Plastic barriers have also been used around rice fields, but they require regular maintenance. Permanent barriers have been built from bricks and concrete, but there are issues about water flow; drains need to be designed that do not provide access points for rodents. Some species of rodents are excellent climbers, so barriers are not always effective for all rodent species.

### 4.1.3 Trap-barrier system

One special type of barrier is the *trap-barrier system* (TBS) or community TBS (CTBS). The earliest application of fences plus traps was in Malaysia, where rodent populations had built up in fallow land adjoining rice crops (Lam et al. 1990). This approach was modified to include a trap crop (also called a lure crop) which is planted about 2–3 weeks earlier than the surrounding fields, and thus is slightly more attractive than the surrounding fields (Singleton et al. 1998). Multiple-capture traps are placed inside the fence, along the sides, to capture rodents trying to access the trap crop. Research has found that trap crops of about 50 x 50 m are best, and that this TBS can protect an area of 10–15 ha of surrounding rice fields (Singleton et al. 2003b). This occurs because rats are attracted to the trap crop from about 200 m away (Brown et al. 2003), so at a village scale, several TBSs can be set up to protect a large area of rice (see Section 5.1 for the case study of CTBS in Vietnam).

Interestingly, the TBS was tested in the uplands of Laos and had limited success (Brown et al. 2007). Instead the farmers were more concerned about rodents invading their grain stores. So they modified the approach and placed the fence plus multiple capture traps (facing outwards) around their grain stores. They were very satisfied with the level of management.

### 4.1.4 Hunting

There are various forms of *hunting* used in different countries. In Laos, rats are hunted using simple bows and arrows or other more sophisticated devices. In Vietnam, farmers work together as a small group, and with the aid of dogs, to locate active rodent burrows, pour water into the burrows and either catch or hit the rats as they emerge from the burrows. In Indonesia, groups of farmers walk in a line through a rice field carrying a string containing tin cans to make noises, and thereby herd rats into a net placed at the other end of the field to catch rats. In Myanmar, rat hunters have made a tool which can produce insect noises to call rats. They kill rats once they come out from their burrows. In the Philippines, rat campaigns may involve a circle of rat hunters that hit rats as they are driven out of a diminishing area of refuge in a fallow field as it is ploughed by a hand tractor.

### 4.1.5 Habitat management

Another type of physical control is through *habitat management*. Effective management can be achieved through treatment of non-crop areas that are sources of habitat for rodents, often undisturbed weedy areas which provide burrows and nesting sites. This can be managed by slashing the weeds to reduce nesting habitat and increase predation risk. At a broad scale, *crop synchrony* is a highly effective form of management. For rodent species that exhibit a peak in breeding activity that coincides with the ripening stage of rice crops, crop synchrony limits the duration of this breeding period by reducing food availability. However, these measures increase uniformity in rice cropping, which may be in conflict with conservation aims to increase buffer zones and diversify crop margins for the benefit of biodiversity.

### 4.1.6 Rice bunds

*Rice bunds* (the small earthen banks surrounding rice field plots) provide an important potential habitat for rodents to construct their burrows. Larger-sized bunds can provide ample space for burrow sites, whereas narrow bunds are not preferred by rodents to build burrows. For *R. argentiventer* and *R. tanezumi*, few rat burrows are found when bunds are less than 30 cm wide and 30 cm high (Brown et al. 2006; Stuart et al. 2015). Thus, minimising the width and depth of bunds to less than 30 x 30 cm is recommended for rodent management.

## 4.2 Chemical rodent management methods

There are a wide range of *rodenticides* available. These can be broadly categorised as acute and anticoagulant. Acute rodenticides are generally for use in field situations, whereas anticoagulants are not normally registered for use in field settings (normally used in warehouses or houses).

### 4.2.1 Acute rodenticides

*Acute rodenticides* cause death from minutes up to 24 hours of ingestion. An acute rodenticide that is widely used by smallholder rice farmers is zinc phosphide ($Zn_3P_2$). It comes as a grey or black powder, and needs to be mixed with a bait substrate. It is usually mixed and coated onto grains, such as broken rice. It has a garlic odour and is toxic to a wide range of rodent pests (Buckle and Eason 2015). It is available as commercial products in many countries in concentrations of 1–5%, although many countries have banned its use. Phosphine gas is given off in the acidic environment of the stomach, and then the gas enters the blood stream, causing heart failure and damage to internal organs (Buckle and Eason 2015). There is no antidote for the poison. In a relative sense, there are few non-target issues with zinc phosphide because phosphine gas rapidly dissipates and most of the toxic product is released; furthermore zinc phosphide does not bioaccumulate, so it poses not only low secondary hazard but also high primary hazard to many vertebrates and invertebrates (Brown et al. 2002). However, inadequate storage can pose a risk to man and domestic animals.

Rats are suspicious of new objects and some species are known to develop an aversion to the acute rodenticides. When rodents consume a sub-lethal dose and get sick, they learn to associate the illness with the rodenticide, so they will not consume any more. These animals are defined as bait shy. Pre-baiting with untreated grain can be used to reduce bait shyness. However, even with pre-baiting, the use of zinc phosphide is unlikely to achieve a high level of success in rats (Buckle 1999).

## 4.2.2 Anticoagulant rodenticides

*Anticoagulant rodenticides* were developed to overcome bait shyness, and they work by blocking the recycling of the active form of vitamin K that is essential for blood clotting. The animal dies of internal bleeding (haemorrhage) several days after ingesting the rodenticide. Thus, it does not associate any sickness with the bait, which allows sufficient feeding to occur. For humans, an advantage of anticoagulant rodenticides is that vitamin K can be administered as an antidote. Warfarin was originally developed for the treatment of human thrombosis in the 1930s, but was found to be an effective rodenticide in the 1940s and became commercially available from 1950s.

The difference between first- and second-generation anticoagulants is that the first-generation anticoagulants require multiple feeds over many days until a sufficient quantity of poison has been ingested. First-generation anticoagulants include pindone, diphacinone, warfarin and coumatetralyl. Rodents generally die within 10 days of ingestion of the bait. Second-generation anticoagulants require only a single feed, but death occurs between 3 and 7 days. Second-generation anticoagulant rodenticides include bromadiolone, brodifacoum and difenacoum. Different baiting strategies are therefore required for the different types of anticoagulants. If using a first-generation anticoagulant, the bait should be available continuously, whereas, if using a second-generation anticoagulant, a pulse baiting strategy should be used. There are significant secondary poisoning hazards for all types of anticoagulants because of the relatively long persistence in tissues of rodents and the way in which anticoagulants can bioaccumulate through the food chain (Eason et al. 2002).

Pulse baiting involves leaving bait in stations for at least three nights and assessing the amount of bait taken. The bait stations should then be removed. After a week, the bait stations should be replenished and left for another three nights, replenishing each night if required. When there is little bait taken, baiting should cease, because most rodents should have been killed. This method reduces the chance of rodents eating excess amounts of bait once they have received a lethal dose, therefore reducing the risk of non-target poisoning. However, this technique requires more effort in checking bait stations; this may incur a greater labour cost, but is much cheaper in terms of the quantity of rodenticide used.

Pre-baiting may be required to obtain effective control of rats because they generally avoid new objects and foodstuffs (neophobia). Pre-baiting allows rats and mice to get used to feeding at a known site and on a particular bait. This ensures that a lethal dose of poison is consumed before illness develops and feeding stops. A sub-lethal dose can lead to 'poison or bait shyness'. The use of the second-generation anticoagulants reduces the need for pre-feeding.

Three types of resistance have been described – operational, toxicological and genetical. Furthermore, the behaviour of rodents, such as neophobia (fear of new objects), and conditioned or unconditioned aversion to the bait base or rodenticide, may facilitate avoidance of ingestion of a fatal dose of a rodenticide. This may explain why application of rodenticides may fail that cannot be accounted for by physiological resistance. Avoidance behaviour, which could be heritable, can and does reduce the efficacy of rodenticides and may also enhance the effects of physiological resistance. There are few documented cases of resistance to anticoagulants in rice-based systems.

Nearly all rodenticides are now out of patent, and there is little work in developing new types of rodenticides (although, see Section 8 on future research). Most research effort is on developing unique attractive bait bases or combining known compounds in bait. Commercial formulations include wax blocks, extruded pellets or bait coated on attractive grain or seed.

### 4.2.3 Fumigation

*Fumigation* is used in some countries to gas burrows. In Indonesia, for example, granules of sulphur are mixed with rice straw, placed into a hand-operated fumigation device, set alight, and then air is pumped through the chamber by turning a hand pump and the smoke gets pumped into the rodent burrows (Singleton et al. 1998). Where smoke comes from holes, they are filled with wet mud. The pressure of the pumping pushes the sulphur gas deep into the burrow complex to kill rodents in the burrow. Other types of fumigants include pellets or granules containing aluminium or magnesium phosphide, which are inserted into rodent burrows and sealed with mud.

### 4.2.4 Repellents

Many repellent compounds have been used for studying the interaction of herbivores and plants using small rodents as model species. Only some studies were aimed at identifying suitable repellents in an applied context. Often repellent compounds work in controlled environments such as laboratories and enclosures but their effects in the field are negligible. As a result only few rodent repellents are registered for controlling rodent damage.

## 4.3 Biocontrol of rodents

Methods for the biocontrol or rodents include sterility control, reproductive inhibitors, predators and parasites, and diseases.

### 4.3.1 Sterility control

Currently, there are no effective *sterility control* or *reproductive inhibitors* available for use in field settings. Curcumol and triptolide are registered sterilants for rodent management in China (Huang 2014). It is not clear if they are effective in reducing damage to rice in field situations. A potential chemical sterilant is 4-vinylcyclohexene diepoxide (VCD), which rapidly depletes the follicle population in the ovaries in rats and mice (Jacob et al. 2008). VCD is being considered for registration in the United States.

'BIORAT' is marketed in Vietnam and the Philippines. It is a combination of warfarin sodium 0.02% + *Salmonella enteritidis* var. Danysz Lysine (Painter et al. 2004). There are no published data in mainstream literature to support or refute the use of BIORAT. There are, however, two publications that caution the use of BIORAT because of human health concerns (Friedman et al. 1996; Painter et al. 2004).

### 4.3.2 Predators

*Predators* are frequently suggested to reduce the impact of rodents, yet there are very few studies actually demonstrating the impact of increasing predation and increasing rice yields or the yield of other crops. Nest boxes and perches have been suggested as devices for encouraging owls and predatory birds, but no well-designed replicated study demonstrates their effectiveness. The risk of predation can be enhanced through habitat management, which has been demonstrated through giving up densities. Therefore, habitat management is more likely to be effective through increasing predation risk than through increasing perching poles or nest boxes for owls in rice systems.

### 4.3.3 Parasites and diseases

A number of *parasites and diseases* have been trialled against rodents in rice systems. In general, few have demonstrated increases in rice yields and few are commercially available. A rodent-specific parasite *Sarcocystis singaporensis* has been trialled effectively in rice fields in Thailand and resulted in reduced yield loss and with positive benefit–cost ratios, similar to that obtained with conventional control techniques (Jäkel et al. 2006). Positive results were also obtained when this form of biocontrol was applied as part of an integrated ecologically based rodent management (EBRM) approach in the northern uplands of Laos (Jäkel et al. 2016). The parasitic protozoan needs two hosts to maintain its lifecycle: a snake (*Python reticulatus*) and rodents of the genera *Rattus* or *Bandicota*. Sporocysts of *S. singaporensis* need to be isolated and then mass-produced and inoculated into baits (see Jäkel et al. 2006 for details). The parasite does not infect humans. Currently, this protozoan bait is commercially available only in a few countries.

## 4.4 Other management strategies

There are some other management strategies that are not strictly physical or chemical, so are grouped here as 'others'.

### 4.4.1 Diversionary feeding

*Diversionary feeding* could be used in a situation where high-value produce could be protected by providing an alternative feed source. There are, however, no effective studies demonstrating the benefits of such an approach in rice-based systems.

**Table 4** The principles of ecologically based rodent management (EBRM) and attributes of successful EBRM for management of rodent pests in rice crops and how they compare in different country contexts

| Principles of EBRM | Southeast Asia (rats in rice fields of Vietnam and Indonesia) | Laos (upland rice systems) | Tanzania |
|---|---|---|---|
| (1) The management actions are environmentally sound | Yes – combination of community trap-barrier system, field sanitation prior to maximum tillering, reduce bund size, synchronise planting and harvesting, destroy rat burrows, conduct community rat campaigns at key times successful for management of rats in rice fields in Vietnam and Indonesia | Partially – combination of targeted trapping in key habitats, targeted community rat campaigns, crop synchrony in lowland pockets of the upland ecosystem, but no widespread effective management strategy has been designed and tested. Rodenticides increasingly available and used | Yes – combination of physical trapping (use of locally made traps), buckets of water buried to the soil surface, field sanitation prior to transplanting, reduce bund size, synchronise planting and harvesting, destroy rat burrows in rice fields |
| (2) They are cost effective | Yes – beneficial benefit:cost ratios determined | Yes – beneficial benefit:cost ratios determined | No – not sufficiently evaluated |
| (3) The actions are sustainable | Yes – integrated into government policy | Yes – there is some integration into government policy and at village level | Yes – integrated into village cropping calendar |
| (4) They are applied at a large scale | Yes – applied at the community scale; management units >100 ha | No – most activities are conducted individually | No – applied at the individual scale; management units <2.5 ha |
| (5) They are politically advantageous | Yes – integrated into government policy | Partially – some issues highlighted in government policy, but not all | No – integrated into only village cropping calendar |
| (6) They are socially acceptable | Yes – farmers see the benefit of working together to reduce rat problems | Partially – many farmers are still working individually | No – farmers are still working individually, there is a need for more synthetisation on a change in knowledge, attitudes and practices to farmers for working together to reduce rat problems |

| Attributes of successful EBRM | Southeast Asia (rats in rice fields of Vietnam and Indonesia) | Laos (upland rice systems) | Tanzania |
|---|---|---|---|
| Understanding of the taxonomy, biology and ecology (behaviour and life history characteristics) of the pest species within the agricultural system | Good understanding of *R. argentiventer* and other species in Vietnam and Indonesia through long-term studies | Reasonable understanding of *R. rattus* complex, but more breeding data and studies on movements and the like are required. More basic information about biology and ecology of rodents involved in bamboo masting outbreaks (*nuu khii*) | Good understanding of *M. natalensis* in eastern Tanzania through long-term studies |
| Monitoring systems to determine thresholds for management (linked to socio-economic thresholds) | Rodent populations and damage monitored routinely | Monitoring systems not established across the country. Hampered by poor communication and remoteness | Rodent populations and damage monitored routinely by farmers and scientists |
| Practicing rodent biologists in the region? | Rodent ecologists employed by research and extension agencies | Rodent ecologists employed, but few in number, and few resources | Rodent ecologists and extensionists employed by university and Ministry of Agriculture and food security at rodent control centre agencies, respectively |
| Experimental field studies used to evaluate management strategies and test hypotheses about rodent population dynamics | Replicated manipulative studies conducted | Observational studies have been conducted, and some treatments have been implemented, but need widespread replicated manipulative studies | Replicated manipulative studies conducted |
| Consider a range of management strategies – do not rely on rodenticides alone | A range of management strategies are recommended and supported by government initiatives | A range of management strategies are recommended and supported by government initiatives, but more emphasis is needed during widespread *nuu khii* outbreaks | A range of management strategies are recommended by scientists and supported by village local government initiatives |

Based on Brown & Khamphoukeo (2010), Brown et al. (2006), Brown et al. (2007), Jacob et al. (2010), Mulungu et al. (2013 & 2015b), Singleton & Brown (1999), Singleton (1997), and Singleton et al. (1999a).

### 4.4.2 Electricity

*Electricity* has been used in some countries, but it is a highly dangerous technique for humans. Farmers have been known to tap into the main electricity supply poles through their village and set up a wire 2–3 cm off the ground so that rats are electrocuted. Significant human problems occur, for example, when children play in the vicinity or when adults walk across fields at night. This practice can also lead to livestock deaths and power outages. It is a highly dangerous practice and is not recommended. There are no safe commercially available electrical devices for killing rats in field situations.

### 4.4.3 Ultrasound and electromagnetism

Some commercial products that emit *ultrasound and/or electromagnetic* pulses are available, but they are designed for use in commercial, warehouse or domestic situations, not for use in large open spaces, such as rice fields. There is little evidence these devices are actually effective.

## 4.5 Ecologically based rodent management

EBRM is essentially a combination of the control methods mentioned above, but conducted in a manner most amenable to the ecology and biology of the particular rodent pest species and the particular rice agro-ecosystem. EBRM has evolved from an integrated pest management paradigm, but there is more emphasis on specific design of strategies. The initial impetus for EBRM came from Singleton (1997), and has since been further developed. There are six key principles for EBRM, but some additional attributes should also be considered (outlined in Table 4). To develop a successful rodent management strategy, it is important to identify the rodent species of concern and understand its ecology in the specific ecosystem. A few key issues are the timing of control (to take advantage of stage of breeding and population abundance) and undertaking management over sufficiently large areas to reduce the chance of reinvasion after treatment. Case study examples for three countries (Vietnam, Laos and Tanzania) are provided in Section 5.

# 5 Case studies of management of rats in rice

## 5.1 EBRM in Vietnam

Rodents are one of the top three rice pests in Vietnam. Farmers describe rodents as the pest they have least control over. Rodents affect households that are dependent on rice production for their livelihoods and impact on poor farming communities which have few resources. Traditionally, farmers have relied heavily on the use of rodenticides, electrocution and spreading sump oil mixed with insecticides onto flooded rice fields to manage the rodent problem, but these can be expensive, are often applied individually by farmers in an uncoordinated manner after significant damage has already occurred, and have negative environmental consequences.

A series of research projects to learn more about the rodent problems were conducted in lowland irrigated rice systems of the Red River Delta and the Mekong River Delta of

**Table 5** Ecologically based rodent management strategies that have proven successful to reduce the damage that rats cause to rice crops in Vietnam (based on Huan et al. 2010)

| Control action | Reason for action | Timing for management |
|---|---|---|
| Community actions | Get the farming community to work together over large areas to identify rodent burrows and destroy them systematically | At land preparation stage up until tillering stage |
| Synchrony of cropping | Synchronises planting of crops (within 2 weeks) and limits the length of the breeding season of the rats | Needs to be set up prior to land preparation |
| Rat campaign | Concentrate on source habitats | Before planting |
| Small bund size in fields | Rats do not dig burrows in bunds smaller than 30 cm wide | Needs to be established before cropping is done |
| Field sanitation (field hygiene) | Clearing long grass and weeds around irrigation canal banks and other non-crop areas to reduce nesting habitat as well as refuge habitat for rodents when no crop is present | From land preparation to harvest |
| Community trap-barrier system (CTBS) | To reduce damage to crops through capture of rats | Needs to be established 3 weeks prior to transplanting of surrounding rice fields for best results |
| Linear trap-barrier system (LTBS) | Plastic fence set up with multiple capture traps to intercept rats moving between source and sink habitats | Can be set up at any time, depending on where rats are residing and where damage is occurring |

Vietnam. They were designed to understand the rodent problems in lowland rice cropping systems, the species involved, the nature of damage and the testing of management techniques at a village scale. These projects lead to the development of recommendations for rodent management (see Brown et al. 2006 and Huan et al. 2010 for details). A number of studies were conducted to understand the biology and ecology of the rodent pest species, and it was found that the breeding and behaviour of the main rodent pest species (particularly *R. argentiventer* and *R. sakaratensis*) were linked to the development of the rice crops. Rodent problems increased as the number of rice crops per year increased (up to three rice crops per year), and the level of damage increased through successive rice crops (Brown and My Phung 2011).

Farming communities were trained and supported in implementing EBRM through 'training of trainers' of local extension staff which was built up and expanded over the course of the project. Modules were integrated into national training programmes. Training and supporting activities expanded from core sites each year to neighbouring villages and districts over each subsequent year.

A range of community-based rodent control options were trialled and were found to be relatively inexpensive to implement and resulted in reductions in yield losses. These include community actions such as synchronised cropping, rat control campaigns at key times, field hygiene and the use of CTBS when damage is expected to be high (Table 5).

A coordinated approach was required across communities and villages, and research and extension agencies at the provincial and national levels to implement effective rodent management.

The majority of farmers adopted community actions as a successful rodent control strategy (Brown et al. 2006). The adoption of CTBS occurred only on sites where government subsidies were available to farmers. After implementing EBRM, rodent damage was reduced by 33–50% (reduced by up to 88% in Ha Nam Province), rice yields were increased by 2–5%, rodenticide use was reduced by 62–90% and the use of electrocution was reduced by 95%. There was a strong shift away from individual actions to group or community actions. Key findings to ensure sustainable EBRM include the need to have good coordination between civic and government agencies to enable farmer participation, to have strong, effective leadership of farmer groups and for management to be conducted early in the growth of the rice crop before rodent populations commence breeding.

An additional participatory study was set up with farmers that included the use of a simulation model of rodent abundance and impact on rice yields (My Phung et al. 2013). The model adequately predicted farmer's rice yields and was an integral component of the learning cycle of farmers and extension workers in achieving a clear understanding of the rationale for changing their traditional rodent control strategies. Farmers subsequently timed their rodent control operations earlier than they had practiced before the experiment (targeting the tillering rice crop stage rather than later when rat damage had already occurred) (My Phung et al. 2013). Farmers liked the idea of community cooperation, but it was more difficult to set up and coordinate than undertaking control themselves. They wanted to be able to show they had captured many rats, but they did not want to be embarrassed about returning from rat hunts with very few captured rats.

## 5.2 EBRM in Laos

In the upland environment, rodents are considered one of the most important pests of upland rice, maize, Job's tear (sorghum) and other crops with mean yield losses estimated at 20% (Douangboupha et al. 2010). Upland rice farmers generally rate them as being second only to weeds as the overall most important constraint to upland rice cultivation (Schiller et al. 1999).

The frequency and duration of rodent outbreaks vary markedly from one province to another. Bamboo masting and rodent (*nuu khii*) outbreaks are episodic, but such population outbreaks occur in many parts of Laos. These are sometimes responsible for extreme crop losses (50–100% losses), occasionally leading to localised or widespread famine. In 2008, severe food shortages due to *nuu khii* outbreaks were reported in seven upland provinces. The main causes of these outbreaks appear to be bamboo masting events and changes in cropping patterns. A number of rodent species are involved in these outbreaks.

The main rodent pest in the upland farming system of Laos is the black rat, *R. rattus* complex. It is found throughout the upland farming system, inhabiting upland crops, lowland systems (in the small pockets of lowland rice along river valleys), and in and around villages. The rodents follow the available food sources between these habitats through the different seasons. It was therefore important to develop rodent management strategies that targeted the rodent populations in the different habitats at key times to reduce damage in the field and damage to produce after harvest.

Farmers in upland environments control rats through traditional methods that include using snap traps, hunting with dogs and sticks, shooting with catapults or arrows, and

**Table 6** List of final recommended rodent management strategies for upland rice farming systems in Laos (based on Brown et al. 2007)

| Upland | Lowland/garden | Village |
|---|---|---|
| • Trap continuously<br>• Use pitfall traps (1 m deep, 0.5 m wide at opening)<br>• Set up bait trap-barrier system (TBS)<br>• Work together to hunt rats in field stores<br>• Work together to hunt rats after harvest | • Trap continuously<br>• Work together to hunt rats after harvest<br>• Set up pitfall traps<br>• Dig burrows and hunt with dogs | • Establish rules/ regulations for sustainable management of rodents (stop use of rodenticide, stop eating predator of rat, promote village campaign and work together)<br>• Raise cats and dogs<br>• Conduct sanitation throughout village<br>• Set up grain store TBS<br>• Get school children working together to trap rats<br>• Conduct village campaign at key times |

guarding the field at night with a small fire (Brown and Khamphoukeo 2007). Poisons such as zinc phosphide were also used. Farmers could purchase rodenticides from local retailers, but the product did not have instructions in local Lao language, and the active ingredient was unknown. Rodenticides were applied in the field only when rodent numbers were high, and heavy crop damage had already been observed.

A range of rodent management strategies were designed with farmers, which were then trialled by farmers in 'Treatment' sites through training to implement the recommended activities. Five 'Treatment' and five 'Reference' sites (without management intervention) were established in Luang Prabang and Luang Namtha provinces. Farmers in Treatment sites in Luang Prabang spent significantly more time trapping rats, but overall spent less money applying rodent management practices compared to those in Reference sites. In contrast, farmers in Treatment sites in Luang Namtha spent less time controlling rats but spent more money (one site applied rodenticides that were expensive) compared to those in Reference sites.

Treatment farmers conducted rice store TBS, pitfall traps, sanitation, bait TBS and TBS in the field, and there was an 80% reduction in use of rodenticides in Treatment sites (some villages banned the use of rodenticides) when compared to Reference sites (Table 6).

Rodent damage to upland and lowland rice was relatively low throughout the project, while damage increased from booting to flowering to harvesting stages. There was no significant difference between Treatment and Reference sites, with crop damage for corn 7% (range 0–26%), for lowland rainfed rice 3% (0–16%), for lowland dry season rice 1% (1–2%) and for upland rainfed rice 1% (0–7%).

Moderate levels of breeding in the main pest species (*R. rattus* complex) were evident throughout the year (15–25% adult females pregnant each month). Therefore, management of rodents in villages and neighbouring areas needs to be continuous.

Changes in knowledge, attitudes and practices (Post-KAP) were evaluated at the end of the study (Brown and Khamphoukeo 2010).

- Rodents were a significant pest on farms (88%).
- Effective control methods were trapping (50%), rodenticides (21%), cats (8%) and digging burrows (8%).
- Farmers would continue to use trapping (56%), rodenticides (15%), cats (11%) and digging burrows (7%).
- 78% of farmers wanted to spend less time and money controlling rats.
- 77% of farmers thought that if they used the recommended rodent management strategies they would spend less time and money controlling rats.
- 71% of farmers said they would work together in the future.

## 5.3 EBRM in Tanzania

In Tanzania, rice production increased dramatically from 985 000 t in 2002 to 2 248 000 t in 2011, owing to expansion of farming areas from 566 000 ha in 2002 to 1 119 000 ha in 2011 (FAOSTAT 2013). Yields, however, have not increased correspondingly (Ching'ang'a 1985) due to several complex factors; one of the most important is food loss due to crop pests, including rodents (Mulungu et al. 2015a). Farmers reported that rodent pests caused 20–60% crop losses in both fields and stores each year (Mulungu et al. 2015b). They often damage crops throughout the growing season, from germination to harvest, causing an estimated 5–11% loss at pre-harvest stages during the wet and dry seasons, respectively (Sixbert 2014). The level of damage, however, was not uniform throughout the growth stages of rice. At planting, for example, rodents dug up and ate the planted rice seeds in nurseries or in fields that were directly planted, and consequently necessitated repeated late replanting and ultimately resulted in lower yields. The rice crop was able to partially compensate for damage, therefore making it difficult to characterise when the damage occurred and its severity. Significant compensation was noted at the transplanting (14 days after sowing (DAS)) and vegetative (45 DAS) stages (Mulungu et al. 2014a).

In irrigated systems in Tanzania, farmers produce rice crops twice per year, one during the rainy season and another during the dry season exclusively under irrigation, and the fields were arranged in a rice–fallow matrix landscape. The recruitment of new rodents into the population was higher in the rice fields compared with fallow land. This could imply that rodents from surrounding areas were attracted to the rice field. Such dispersal into rice fields is most apparent at transplanting and harvesting time where recruitment is highest. Interestingly, recruitment was observed to be low at the booting stage, which is different from what is reported in Southeast Asia. High-quality food resources in rice fields could increase the survival chances of newly born and older rodents during the population increase phase. The multi-mammate rat, *M. natalensis,* had higher survival rates in rice fields than in fallow land, arguably because of enhanced food quality (Leirs et al. 1994; Mulungu et al. 2014b).

*M. natalensis* is the most abundant rodent pest species in rice crop fields in Tanzania. This genus has been recorded at high densities in disturbed landscape in agricultural fields throughout sub-Saharan Africa (Leirs 1995; Leirs et al. 1996). In irrigated rice fields in Tanzania, its population density fluctuates markedly between months with the highest population peak reported during the dry season (Mulungu et al. 2013). The pest is sexually active throughout the year, although it reaches the highest level when the rice crop is at the maturity stage. Together, this suggests that breeding is highly influenced by the rice production systems, which is different from maize-dominated mosaic habitats. In a maize-dominated habitat, the occurrence of rodent outbreaks is reportedly influenced by rainfall pattern (Linn 1991; Leirs 1995). In irrigated rice agroecosystems, water and food are not limiting factors. Breeding occurs over a longer period and population size is generally higher, with a much weaker link with rainfall (although breeding is still most prominent in the rainy seasons). More juvenile rats are recorded in August and September, indicating that the main breeding season is during the rainfall season (Mulungu et al. 2013). Therefore, the rodents are not influenced by rainfall *per se*, but rather by the quantity and quality of their food, which is dependent on the phenology of the rice crops and surrounding vegetation.

In irrigated rice fields, vegetative plant materials (leaves, stems and seeds) are the most abundant components of the diet of *M. natalensis*, while other food types (invertebrates, fruits) are consumed only in low quantities (Mulungu et al. 2014b). Agricultural cropping patterns in Tanzania typically consist of a relatively small-scale matrix of agricultural fields and fallow land (Odhiambo et al. 2005). Habitat quality for rodent pest species will likely vary according to such changes in land use, and it is expected that the population dynamics of resident animals will exhibit important spatio-temporal differences that can potentially affect crop damage patterns and severity. An understanding of the dispersion patterns of a pest is an important pre-requisite for developing an effective management programme for the pest in question. *M. natalensis* in irrigated rice fields generally exhibited an aggregated spatio-temporal distribution (Mulungu 2015a). Heat maps of trapping grids visually confirmed this dispersal pattern, indicating the clumped nature of captured rodents in irrigated rice crop fields at different crop growth stages (Mulungu et al. 2015a).

Movement of *M. natalensis* in rice fields thus seems to be driven by food availability and flooding status, which can be attributed to land use practices. Adult *M. natalensis* have smaller home ranges than subadults in rice fields, indicating that rice fields are suitable for breeding (Mulungu et al. 2015c). However, travel distances are larger in rice fields, especially at the transplanting stage, during which rice fields are flooded and provide less food. Rats move into neighbouring fallow fields leading to temporary high densities in fallow land. A decrease in travel distance was observed in rice fields when the rice crop is ripening, which can be explained by higher food availability and a more suitable, non-flooded situation.

A majority of farmers (80–90%) cultivate paddy in small fields ranging from 0.5 to 1.5 acres. Therefore, with these small fields, rodent damage to crop is high for small-scale farmers. Current management in rice fields in Tanzania consists mainly of the use of acute rodenticides (53%), trapping and physical measures such as putting polythene around nurseries as rodent barriers (47%) (Mulungu et al. 2015b). Farmers conduct rodent control as individuals. Control actions are reactive, with very little planning and poor safety precautions for the use of rodenticides. The rodenticides in the irrigated rice systems are distributed along the banks of the rice in indiscriminate amounts and with questionable results.

## 5.4 Principles of EBRM

The six principles of EBRM are compared across the three case studies (Table 4). In order to have effective sustainable rodent management, all six principles of EBRM need to be addressed. EBRM seems to be far more effective in Vietnam than in Laos or Tanzania, mainly related to the longer length of time that research has been conducted in Vietnam. An additional five attributes of successful EBRM are described, and again we see that the attributes are applied better in Vietnam than in Laos and Tanzania. This comparison allows an examination of where the relative strengths and weaknesses exist, and perhaps what could be done to address them. This approach could be applied for any rodent pest in any country.

# 6 Other vertebrate pests in rice

The other vertebrate pests of rice, such as birds, are not well documented. In Myanmar, munias (*Lonchura* spp.; known locally as rice birds) are considered the main rice field pest according to their foraging habit in rice fields. Other bird pests include sparrows (*Passer* spp.) and parrots (species unknown), especially where crops are ripening earlier or later than surrounding fields. Elephants (*Elephas maximus*) can sometimes become pests in some parts of the Ayeyarwady region and Rakhine State, Myanmar. In Sri Lanka, reported vertebrate pests of rice include elephants, peacocks (*Pavo cristatus*), monkeys (various species) and wild pigs (*Sus scrofa*). In Thailand, munias and weavers (*Ploceus* spp.) are considered to be the main bird pests of rice, with occasional reports of wild pigs and monkeys damaging upland rice (Boonpramuk, U. pers. comm.). In the Philippines, four species of birds are considered pests of rice, including the Eurasian tree sparrow (*Passer montanus*), and three species of munia. The pest status of these birds is not well documented as the only reliable method available to quantify bird damage in rice is the use of bird exclusion nets (Rodenburg et al. 2014). When farmers in Aurora province, northern Luzon, the Philippines, were asked to rank the top three rice pests, birds were ranked as the second most important pest during the previous dry season crop (Stuart et al. 2011).

Bird pests of rice are perhaps best documented in Africa, where the red-billed quelea (*Quelea quelea* L. subsp. *quelea*) can cause severe losses to rice crops over a relatively short period of time due to their gregarious and migratory behaviour (de Mey and Demont, 2013). Based on farmer surveys across 20 countries in Africa, birds are considered to be the second most important pest of rice after weeds (IRRI 2010). Bird damage is most prominent during the dry season, when there are less alternative food sources available, such as wild grass seeds (Ruelle and Bruggers 1982). Preventative measures used to manage bird pests in rice include synchronous cropping; weed control (as weeds attract birds during the early grain filling stage) and nest destruction using avicides, explosives or flamethrowers (de Mey and Demont 2013). Protective measures include the use of repellent substances; protecting fields or nurseries with nets or wires; and bird scaring using humans, noise-making devices, flags or scarecrows. However, aside from weed management that has been shown to reduce bird visitation rates to rice fields (Rodenburg et al. 2014), evidence that clearly demonstrates the effectiveness of these methods is limited.

# 7 Future trends and conclusion

For several years, no new rodenticidal compounds have been registered for use in crop protection or in the biocidal sector, and this may be the case for years to come.

This is partially due to the availability of effective compounds (anticoagulants) for the management of commensal rodents, which internationally represents a larger market than crop protection. There is some work to identify new compounds such as cellulose-based bait or sodium selenite that seem unpalatable and ineffective (Jokić et al. 2014; Schmolz 2010), methaemoglobin (Rennison et al. 2013) and plant toxins (Pauling et al. 2009; Yuan et al. 2014). Another strategy to improve the efficacy of products is to use a combination bait that includes two anticoagulants, or an anticoagulant and cholecalciferol (Endepols et al. 2016; Baldwin et al. 2016).

Studies on implementing EBRM for rodents in rice cropping systems have provided favourable findings. The studies demonstrate favourable economic returns for the farmers (e.g. Singleton et al. 2005; Brown et al. 2006; Jacob et al. 2010); yet, farmers often do not always continue to practice EBRM. Therefore, more research is required to examine why farmers do not implement such efficient methods, although they know they are effective.

There are opportunities to rigorously explore the synergistic effects of management strategies that have multiple benefits for rice production, for example concurrent weed and rodent management. Entomologists recommend growing flowering plants along the edges of rice fields to promote beneficial predatory arthropods and parasitoids (see chapter 14 in this volume), but this unfortunately provides high-quality food and shelter for rats (Horgan et al. 2016). Bringing together expertise from different fields could be used to explore management practices that have multiple beneficial effects.

Improved monitoring systems and forecasting systems are urgently needed. Most monitoring is conducted too late after damage has already occurred and there are very few predictive models available. Routine monitoring is conducted in some countries (e.g. in Vietnam, Brown and My Phung 2011), but most estimates of rodent impact and damage collected from the field by extension staff are not collected rigorously. There is strong interest in developing automated remote sensing equipment for rodents (in all crops, not just in rice). The challenge is making them reliable and cheap, and not too data intensive. This is a rapidly advancing field. There are also mobile phone applications to record farmer observations, and these will be valuable for region-wide monitoring (e.g. the *MouseAlert* system for mice in Australian cropping systems: www.mousealert.org.au).

Rodents are clearly a significant problem in rice agro-ecosystems in many parts of the globe. The data on post-harvest losses are limited but the published information on pre-harvest losses emphasise the importance of rodent impacts on regional food security (John 2014). One analysis calculated that what rodents eat and spoil globally could feed 280 million people in developing countries for a year (Meerburg et al. 2009b). If rodent damage could be reduced from 10 to 5%, then more rice would be available for human consumption.

It is possible to manage rodent pests and to reduce the damage caused to growing rice crops and therefore increase yields (see case study examples). This outcome occurs primarily through a thorough understanding of the pest species and their breeding dynamics and behaviour, and then a well-supported management strategy that does not rely on a single control technique. It is possible to manage rodents without relying entirely on rodenticides. That is why a good understanding of the rodent ecology and breeding dynamics is required. A range of management methods include crop synchrony, habitat management, management of bund sizes, community campaigns at key times, field sanitation and various trapping methods. Management needs to be linked to a monitoring system and control should be applied early, before significant damage occurs. Finally, management needs to be conducted over a large area. In developing countries that are dominated by smallholder farmers (holdings less than 2 ha), community action is essential for effective management of rodents. We emphasise that food security is a strong driver for EBRM at a village level and also at the national level (see also John 2014).

There are now many examples of well-implemented rodent management strategies, particularly in Southeast Asia. It is possible to learn from these and adapt the processes to other situations. However, more could be done: the rodent problem will not go away; constant effort is required to keep on top of the problem and to continually improve all aspects, including management methods, institutional support and capacity building of extension staff, researchers and farmers.

# 8 Where to look for further information

Some key resources (web sites, book chapters, etc.) are provided below.

- Singleton (2003) *Impacts of Rodents on Rice Production in Asia* (Published by IRRI): Reviewed pre-harvest rodent impacts for 11 Asian countries.
    - Available at: https://books.google.com.au/books?id=c_nlNaRP2XMC&pg=PA1&lpg=PA1&dq=Impacts+of+Rodents+on+Rice+Production+in+Asia&source=bl&ots=vAH5rCNr8a&sig=YJbXHYMimWmtkSgVU31WmrkHgZk&hl=en&sa=X&ved=0ahUKEwjjjM2fq8_MAhVKv5QKHedGAq8Q6AEIPjAF#v=onepage&q=Impacts%20of%20Rodents%20on%20Rice%20Production%20in%20Asia&f=false.
- Aplin et al. (2003) *Field methods for rodent studies in Asia and the Indo-Pacific* (Published by ACIAR): Described many key methods and approaches needed to undertake studies on rodents. It also contains a description of key rodent pests for the Asia/Pacific region.
    - Available at: http://aciar.gov.au/publication/mn100.
- Singleton et al. (2010a) *Rodent Outbreaks: Ecology and Impacts* (Published by IRRI): Presents chapters from a small conference looking at rodent outbreaks.
    - Available at: http://irri.org/resources/publications/books/rodent-outbreaks-ecology-and-impacts
- *International Conference on Rodent Biology and Management* (ICRBM). There have been five ICRBMs since 1998.
    - All conferences and links to papers are available at http://www.icrbm.org/
- EU-funded projects on rodent management in Africa (StopRats, EcoRat, RatZooMan, StapleRat).
    - More information and links available at http://projects.nri.org/stoprats/background
- *IRRI Knowledge Bank* contains information, fact sheets and links about pre-harvest and post-harvest management of rodents.
    - Pre-harvest available at: http://www.knowledgebank.irri.org/training/fact-sheets/pest-management/rats/rodent-control-non-chemical-in-lowland-irrigated-rice;
    - Post-harvest available at: http://www.knowledgebank.irri.org/step-by-step-production/postharvest/storage/storage-pests/rodents-as-storage-pest.
- Buckle and Smith (2015): *Rodent Pests and Their Control* (2nd Edition) (Published by CAB International): This is the 2nd edition of the classic book. It has been updated and republished.
    - Available at: http://www.cabi.org/bookshop/book/9781845938178.

# 9 References

Alfonso, P. J., Fieldler, L. A., and Sumangil, J. P., (1985). Rodent ecology, population dynamics and behaviour. In Sanchez, F. F. and Benigno, E. A. (Eds), *Rodent Biology and Control (with special reference to the Philippines)*, pp. 25–47. The National Crop Protection Center: Los Baños, Philippines.

Aplin, K. P., Brown, P. R., Jacob, J., Krebs, C. J., and Singleton, G. R. (2003). *Field Methods for Rodent Studies in Asia and the Indo-Pacific*. Australian Centre for International Agricultural Research: Canberra, p. 223.

Aplin, K. P., Suzuki, H., Chinen, A. A., Chesser, R. T., ten Have, J., Donnellan, S. C., Austin, J., Frost, A., Gonzalez, J. P., Herbreteau, V., Catzeflis, F., Soubrier, J., Fang, Y.-P., Robins, J., Matisoo-Smith, E., Bastos, A. D. S., Maryanto, I., Sinaga, M. H., Denys, C., Van Den Bussche, R. A., Conroy, C., Rowe, K., and Cooper, A. (2011). Multiple geographic origins of commensalism and complex dispersal history of black rats. *PLoS ONE* 6, e26357.

Baldwin, R. A., Meinerz, R., and Witmer, G. W. (2016). Cholecalciferol plus diphacinone baits for vole control: a novel approach to a historic problem. *Journal of Pest Science* 89, 129–35.

Bambaradeniya, C. N. B., Edirisinghe, J. P., Silva, D. N., Gunatilleke, C. V. S., Ranawana, K. B., and Wijekoon, S. (2004). Biodiversity associated with an irrigated rice agro-ecosystem in Sri Lanka. *Biodiversity & Conservation* 13, 1715–53.

Barnett, S. A., and Prakash, I. (1976). *Rodents of Economic Importance*. Heinemann: London)

Belmain, S. R., Htwe, N. M., Kamal, N. Q., and Singleton, G. R. (2015). Estimating rodent losses to stored rice as a means to assess efficacy of rodent management. *Wildlife Research* 42, 132–42.

Benigno, E. A. (1980). Rat damage distribution in rice and corn fields. *Philippine Agriculturist* 63, 239–50.

Berry, R. J. and Bronson, F. H. (1992). Life history and bioeconomy of the house mouse. *Biological Review* 67, 519–50.

Berry, R. J., and Scriven, P. N. (2005). The house mouse: a model and motor for evolutionary understanding. *Biological Journal of the Linnaen Society* 84, 335–47.

Boonsong, P., Hongnark, S., Suasa-ard, K., Khoprasert, Y., Promkerd, P., Hamarit, G., Nookarn, P., and Jäkel, T. (1999). Rodent management in Thailand. In Singleton, G. R., Hinds, L. A., Leirs, H., and Zhang, Z. (Eds), *Ecologically-based Management of Rodent Pests*. Vol. ACIAR Monograph No. 59, pp. 338–58. Australian Centre for International Agricultural Research: Canberra.

Bomford, M. (1987). Food and reproduction of wild house mice. II. A field experiment to examine the effect of food availability and food quality on breeding in spring. *Wildlife Research* 14, 197–206.

Brown, P. R., and Khamphoukeo, K. (2007). Farmers' knowledge, attitudes, and practices with respect to rodent management in the upland and lowland farming systems of the Lao People's Democratic Republic. *Integrative Zoology* 2, 165–73.

Brown, P. R., and Khamphoukeo, K. (2010). Changes in farmers' knowledge, attitudes and practices after implementation of ecologically-based rodent management in the uplands of Lao PDR. *Crop Protection* 29, 577–82.

Brown, P. R., and My Phung, N. T. (2011). Pattern and dynamics of rodent damage to lowland irrigated rice crops in An Giang, Vietnam. *International Journal of Pest Management* 57, 67–76.

Brown, P. R., and Tuan, N. P. (2005). Compensation of rodent pests after removal: control of two rat species in an irrigated farming system in the Red River Delta, Vietnam. *Acta Oecologica* **28**, 267–79.

Brown, P. R., Chambers, L. K., and Singleton, G. R. (2002). Pre-sowing control of house mice (*Mus domesticus*) using zinc phosphide: efficacy and potential non-target effects. *Wildlife Research* 29, 27–37.

Brown, P. R., Khamphoukeo, K., Sonekham, Invanh, Simone, Vison, Chanthi, Khamphamy, Lanfeuang, McWilliam, A., and Kingkeo, S. (2007). Development of rodent management strategies for the upland farming system in Lao PDR. *Lao Journal of Agriculture and Forestry* 15, 124–45.

Brown, P. R., Leung, L. K.-P., Sudarmaji, and Singleton, G. R. (2003). Movements of the ricefield rat, *Rattus argentiventer*, near a trap-barrier system in rice crops in West Java, Indonesia. *International Journal of Pest Management* 49, 123–9.

Brown, P. R., McWilliam, A., and Khamphoukeo, K. (2013). Post-harvest damage to stored grain by rodents in village environments in Laos. *International Biodeterioration & Biodegredation* 82, 104–9.

Brown, P. R., My Phung, N. T., and Gaydon, D. S. (2011). Rats in rice: linking crop and pest models to explore management strategies. *Wildlife Research* 38, 560–7.

Brown, P. R., Singleton, G. R., and Sudarmaji (2001). Habitat use and movements of the rice-field rat, *Rattus argentiventer*, in West Java, Indonesia. *Mammalia* 65, 151–66.

Brown, P. R., Tuan, N. P., Singleton, G. R., Ha, P. T. T., Hoa, P. T., Hue, D. T., Tan, T. Q., Tuat, N. V., Jacob, J., and Muller, W. J. (2006). Ecologically-based management of rodents in the real world: application to a mixed agro-ecosystem in Vietnam. *Ecological Applications* 16, 2000–10.

Buckle, A. P. (1999). Rodenticides – their role in rodent pest management in tropical agriculture. In Eds Singleton, G. R., Hinds, L. A., Leirs, H., and Zhang, Z. (Eds), *Ecologically-based Management of Rodent Pests*. Vol. ACIAR Monograph No. 59, pp. 163–77. Australian Centre for International Agricultural Research: Canberra.

Buckle, A. P. (2015). Damage assessment and damage surveys. In Eds Buckle, A. P., and Smith, R. H. (Eds), *Rodent Pests and Their Control*, 2nd Edition, pp. 209–30. CAB International: Oxford, UK.

Buckle, A. P., and Eason, C. T. (2015). Control methods: chemical. In Buckle, A. P., and Smith, R. H. (Eds), *Rodent Pests and Their Control*, 2nd Edition, pp. 123–54. CAB International: Oxford, UK.

Buckle, A. P., and Rowe, F. P., (1981). Overseas Development Administration/Department of Agriculture, Malaysia. Rice field rat project, Malaysia. Technical Report. 1977–1980. Centre for Overseas Pest Research, London, UK, p. 99.

Buckle, A. P., and Smith, R. H. (2015). *Rodent Pests and their Control*, 2nd Edition. CAB International: Oxford, UK.

Buckle, A. P., Yong, Y. C., and Rowe, F. P. (1979). Yield response of the rice variety Improved Mashuri to simulated rat damage. *Malaysian Agriculture Journal* 52, 135–44.

Cantor, N. F. (Ed.) (2001). *In the Wake of the Plague: The Black Death and the World it Made*. Simon & Schuster UK Ltd: London.

Ching'ang'a, H. M. (1985). Rice in Tanzania. Rice Improvement in Eastern, Central and Southern Africa, Proceedings of the International Rice Workshop, 9–19 April 1984, Lusaka, Zambia, International Rice Research Institute (Philippines), pp. 97–106.

de Mey, Y., and Demont, M. (2013). Evidence and control. In Wopereis, M. C., Johnson, D. E., Ahmadi, N., Tollens, E., and Jalloh, A. (Eds), *Realizing Africa's Rice Promise*, pp. 241–9. CAB International: Oxfordshire, UK.

Deng, Z., and Wang, C.-X. (1984). Rodent control in China. In Clark, D. O., Marsh, R. E., and Beadle, D. E. (Eds), *Proceedings of the Eleventh Vertebrate Pest Conference*, pp. 47–53. University of California, Davis, California: Red Lion Motor Inn, Sacramento, California.

Douangboupha, B., Singleton, G. R., Brown, P. R., and Khamphoukeo, K. (2010). Rodent outbreaks in the uplands of the Lao PDR. In Singleton, G. R., Belmain, S. R., Brown, P. R., and Hardy, B. (Eds), *Rodent Outbreaks: Ecology and Impacts*, pp. 99–111. International Rice Research Institute: Los Baños, Philippines.

Duplantier, J., and Rakotondravony, D. (1999). The rodent problem in Madagascar: agricultural pest and threat to human health. In Singleton, G. R., Hinds, L. A., Leirs, H., and Zhang, Z. (Eds), *Ecologically-based Management of Rodent Pests*. Vol. ACIAR Monograph No. 59, pp. 441–59. Australian Centre for International Agricultural Research: Canberra.

Duque, U. G., Joshi, R. C., Singleton, G. R., Marquez, L. V., Florague, M. A., and Sebastian, L. S. (2008). Biology and management of rodent communities in intensive irrigated rice cropping systems in Luzon Island. In Joshi, R. C., Singleton, G. R., Sebastian, L. S. (Eds), *Philippine Rats: Ecology and Management*, p. 5. Philippine Rice Research Institute, Science City of Muñoz, Nueva Ecija.

Eason, C. T., Murphy, E. C., Wright, G. R. G., and Spurr, E. B. (2002). Assessment of risks of brodifacoum to non-target birds and mammals in New Zealand. *Ecotoxicology* 11, 35–48.

Elias, D. J., and Fall, M. W. (1988). The rodent problem in Latin America. In Prakash, I. (Ed.), *Rodent Pest Management*, pp. 13–28. CRC Press: Boca Raton, Florida.
Endepols, S., Klemann, N., Richter, D., and Matuschka, F. R. (2016). The potential of coumatetralyl enhanced by cholecalciferol in the control of anticoagulant resistant Norway rats (*Rattus norvegicus*). *Pest Management Science* (in press).
Fall, M. W. (1977). *Rodents in Tropical Rice*. Technical Bulletin No. 36, Rodent Research Center. Rodent Research Centre, University of Philippines: Los Baños, Philippines.
FAOSTAT (2013) Country Profile, United Republic of Tanzania. Available at http://www.fao.org/countryprofiles/index/en/?iso3=TZA. Accessed 13 May 2013.
Fiedler, L. A. (1994). Rodent pest management in eastern Africa. FAO, Plant Production and Protection Paper No. 123, Rome.
Friedman, C. R., Malcolm, G., Rigau-Perez, J. G., Arambulo, P., and Tauxe, R. V. (1996). Public health risk from Salmonella-based rodenticides. *Lancet* 347, 1705–6.
Fulk, G. W., and Akhtar, M. T. (1981). An investigation of rodent damage and yield reduction in rice. *Tropical Pest Management* 27, 116–20.
Funmilayo, O., and Akande, M. (1977). Vertebrate pests of rice in southwestern Nigeria. *Pest Articles and News Summaries (PANS)* 23, 38–48.
Geddes, A. W. M. (1992). *The relative importance of pre-harvest crop pests in Indonesia*. Natural Resource Institute Bulletin 47. Natural Resources Institute: Chatham, UK.
Hein, S., and Jacob, J. (2015). Recovery of small rodent populations after population collapse. *Wildlife Research* 42, 108–18.
Hoque, M. M., Sanchez, F. F., and Benigno, E. A. (1988). Rodent problems in selected countries in Southeast Asia and islands in the Pacific. In Prakash, I. (Ed.), *Rodent Pest Management*, pp. 85–99. CRC Press: Boca Raton, Florida.
Horgan, F. G., Ramal, A. F., Bernal, C. C., Villegas, J. M., Stuart, A. M., and Almazan, M. L. P. (2016). Applying ecological engineering for sustainable and resilient rice production systems. *Procedia Food Science* 6, 7–15.
Horskins, K., White, J., and Wilson, J. (1998). Habitat usage of *Rattus rattus* in Australian macadamia orchard systems: implications for management. *Crop Protection*, 174, 359–64.
Htwe, N. M., and Singleton, G. R., (2014). Is quantity or quality of food influencing the reproduction of rice-field rats in the Philippines? *Wildlife Research* 41, 56–63.
Htwe, N. M., Singleton, G. R., and Maw, P. P. (2017). Post-harvest impacts of rodents in Myanmar; how much rice do they eat and damage? *Pest Management Science* 73, 318–24.
Htwe, N. M., Singleton, G. R., and Nelson, A. D. (2013). Can rodent outbreaks be driven by major climatic events? Evidence from cyclone Nargis in the Ayeyawady Delta, Myanmar. *Pest Management Science* 69, 378–85.
Htwe, N. M., Singleton, G. R., Hinds, L. A., Propper, C. R., Sluydts, V. (2012). Breeding ecology of rice field rats, *Rattus argentiventer* and *R. tanezumi* in lowland irrigated rice systems in the Philippines. *Agriculture, Ecosystems & Environment* 161, 39–45.
Huan, N. H., Nga, V. T. Q., Brown, P. R., My Phung, N. T., and Singleton, G. R. (2010). Rodent impacts in lowland irrigated intensive rice systems in Vietnam. In Singleton, G. R., Belmain, S. R., Brown, P. R., and Hardy, B. (Eds), *Rodent Outbreaks: Ecology and Impacts*, pp. 139–52. International Rice Research Institute: Los Baños, Philippines.
Huang, E. (2014). Approved and banned rodenticides in China. Chemlinked web site Available from https://agrochemical.chemlinked.com/approved-and-banned-rodenticides-china
IRRI, AfricaRice and CIAT (2010). Global Rice Science Partnership (GRiSP). International Rice Research Institute, Manila, Philippines; Africa Rice Center, Cotonou, Benin; and International Center for Tropical Agriculture, Cali, Colombia
Islam, Z., Morton, R. G., and Jupp, B. P. (1993). The effects of rat damage on deepwater rice yields in Bangladesh. *International Journal of Pest Management* 39, 250–4.
Jacob, J., and Wegner, R. E. (2005). Does continuous removal of individuals separate high- and low-quality ricefield rats? *Journal of Wildlife Management* 69, 821–6.

Jacob, J., Singleton, G. R., and Hinds, L. A. (2008). Fertility control of rodent pests. *Wildlife Research* 35, 487–93.

Jacob, J., Sudarmaji, R., S. G., Rahmini, Herawati, N. A., and Brown, P. R. (2010). Ecologically based management of rodents in lowland irrigated rice fields in Indonesia. *Wildlife Research* 37, 418–27.

Jäkel, T., Khoprasert, Y., Promkerd, P., and Hingnark, S. (2006). An experimental field study to assess the effectiveness of bait containing the parasitic protozoan *Sarcocystis singaporensis* for protecting rice crops against rodent damage. *Crop Protection* 25, 773–80.

Jäkel, T., Mouaxengcha, K., Nuber, U., and Douangboupha, B. (2016). Integrated rodent management in outbreak-prone upland rice growing areas of Northern Laos. *Crop Protection* 79, 34–42.

John, A. (2014). Rodent outbreaks and rice pre-harvest losses in Southeast Asia. *Food Security* 6, 249–60.

Jokic, G., Vuksa, M., Dedovic, S., and Kljajic, P. (2014). Laboratory testing of wood mouse and common vole sensitivity to bromadiolone, sodium selenite, and cellulose. *Journal of Pest Science* 87, 309–14.

King, C. M., Innes, J. G., Flux, M., and Kimberly, M. O. (1996). Population biology of small mammals in Pureora Forest Park: 2. The feral house mouse (*Mus musculus*). *New Zealand Journal of Ecology* 20, 253–69.

Lam, Y. M. (1988). Rice as a trap crop for the rice field rat in Malaysia. In A. C. Crabb and R. E. Marsh (Eds), *Proceedings of the 13th Vertebrate Pest Conference*, pp. 123–8.

Lam, Y. M., Supaad, M. A., Chang, P. M., Mohamed, M. S., Goh, C. S., and Radzi, H. (1990). An innovative approach for protecting rice against severe rat depredation. *Proceedings of the 3rd International Conference on Plant Protection in the Tropics*, pp. 1–23.

Lan, L. P., Aplin, K. P., Hung, N. M., Quoc, N. V., Chien, H. V., Sang, N. D., and Singleton, G. R. (2003). Rodent communities and historical trends in rodent damage in the Mekong Delta of Vietnam: establishing an ecological basis for effective pest management. In Singleton, G. R., Hinds, L. A., Krebs, C. J., and Spratt, D. M. (Eds), *Rats, Mice and People: Rodent Biology and Management*. Vol. ACIAR Monograph 96, pp. 290–6. ACIAR: Canberra.

Leirs, H. (2003). Management of rodents in crops: the Pied Piper and his orchestra. In Singleton, G. R., Hinds, L. A., Krebs, C. J., and Spratt, D. M. (Eds), *Rats, Mice and People: Rodent Biology and Management*. Vol. ACIAR Monograph 96, pp. 183–90. ACIAR: Canberra.

Leirs, H. (1995). Population ecology of *Mastomys natalensis* (Smith 1834): implication for rodent control in East Africa. Agricultural Edition No. 35, Belgian Administration for Development Cooperation, Brussels, Belgium, pp. 1–268

Leirs, H., Verhagen, R., and Verheyen, W. (1994). The basis of reproductive seasonality in *Mastomys* rats (Rodentia, Muridae) in Tanzania. *Journal of Tropical Ecology* 10, 55–66.

Leirs, H., Verhagen, R., Verheyen, W., Mwanjabe, P., and Mbise, T. (1996). Forecasting rodent outbreaks in Africa: an ecological basis for Mastomys control in Tanzania. *Journal of Applied Ecology*, 33, 937–43.

Leung, L. K.-P., Singleton, G. R., Sudarmaji, and Rahmini (1999). Ecologically-based population management of the rice-field rat in Indonesia. In Singleton, G. R., Hinds, L. A., Leirs, H., and Zhang, Z. (Eds), *Ecologically-based Management of Rodent Pests*. Vol. ACIAR Monograph No. 59, pp. 305–18. Australian Centre for International Agricultural Research: Canberra.

Linn, I. J. (1991). Influence of 6-methoxybenzoxazolinone and green vegetation on reproduction of Multimmamate rat (*Mastomys coucha*). *South African Journal of Wildlife Research* 21, 33–7.

Lund, M. (1988). Rodent problems in Europe. In Prakash, I. (Ed.), *Rodent Pest Management*, pp. 29–33. CRC Press: Boca Raton, Florida.

Macdonald, D. W. (Ed.) (2001). *The New Encyclopaedia of Mammals*. Andromeda: Oxford, UK.

Marsh, R. E. (1988). Rodent problems on the North American continent. In Prakash, I. (Ed.), *Rodent Pest Management*, pp. 1–11. CRC Press: Boca Raton, Florida.

Meerburg, B. G., Singleton, G. R., and Kijlstra, A. (2009a). Rodent-borne diseases and their risks for public health. *Critical Reviews in Microbiology* 35, 221–70.

Meerburg, B. G., Singleton, G. R., and Leirs, H. (2009b). The year of the rat ends - time to fight hunger! *Pest Management Science* 65, 351–2.

Miller, R. W., Stuart, A. M., Joshi, R. C., Banks, P. B., and Singleton, G. R. (2008). Biology and management of rodent communities in complex agroecosystems – rice terraces. In Singleton, G. R., Joshi, R. C. and Sebastian, L. S. (Eds), *Philippine Rats: Ecology and Management*, pp. 25–36. Science City of Muñoz, Neuva Ecija: Philippine Rice Research Institute.

Mills, J. N. (1999). The role of rodents in emerging human disease: examples from the hantaviruses and arenaviruses. In Singleton, G. R., Hinds, L. A., Leirs, H., and Zhang, Z. (Eds), *Ecologically-based Management of Rodent Pests*. Vol. ACIAR Monograph No. 59, pp. 134–60. Australian Centre for International Agricultural Research: Canberra.

Mulungu, L. S., Sixbert, V., Ngowo, V., Mdangi, M., Katakweba, A. S., Tesha, P., Mrosso, F. P., Mchomvu, M., Kilonzo, B. S., and Belmain, S. R. (2015a). Spatio-temporal patterns in the distribution of the multimammate mouse, *Mastomys natalensis*, in rice crop and fallow land habitats in Tanzania. *Mammalia* 79, 177–84.

Mulungu, L. S., Borremans, B., Ngowo, V., Mdangi, M., Katakweba, A. S., Tesha, P., Mrosso, F. P., Mchomvu, M., and Kilonzo, B. S. (2015c). Comparative study of the movements of *Mastomys natalensis* in irrigated rice and fallow fields in eastern Tanzania. *African Journal of Ecology* 53, 473–9.

Mulungu, L. S., Lagwen, P. P., Mdangi, M. E., Kilonzo, B. S., Belmain, S. R. (2014a). Impact of spatio-temporal simulations of rat damage on yield of rice (*Oryza sativa* L.) and implications for rodent pest management. *International Journal of Pest Management*, 60, 269–74.

Mulungu, L. S., Mahlaba, T. A., Massawe, A. W., Kennis, J., Crauwels, D., Eiseb, S., Monadjem, A., Makundi, R. H., Katakweba, A. A. S., Leirs, H., and Belmain, S. R. (2011). Dietary differences of the multimammate mouse, *Mastomys natalensis* (Smith 1834), across different habitats and seasons in Tanzania and Swaziland. *Wildlife Research* 38, 640–6.

Mulungu, L. S., Mlyashimbi, E. C. M, Ngowo, V., Mdangi, M., Katakweba, A. S., Tesha, P., Mrosso, F. P., Mchomvu, M., Kilonzo, B. S., and Belmain, S. R. (2014b). Food preferences of the Multi-mammate mouse, *Mastomys natalensis* (Smith 1834) in irrigated rice field in eastern Tanzania. *International Journal of Pest Management*, 60, 1–8.

Mulungu, L. S., Mrosso, F. P., Katakweba, A. A. S., Mdangi, M. E., Tesha, P. P. H., Ngowo, V., Mchomvu, M., and Kilonzo, B. S. (2015b). Farmer's knowledge, attitude and practice on rodent management in lowland irrigated rice in Central-eastern, Tanzania. *International Research Journal of Plant Science*, 6, 7–14.

Mulungu, L. S., Ngowo, V., Mdangi, M., Katakweba, A. S., Tesha, P., Mrosso, F. P., McHomvu, M., Sheyo, P. M., and Kilonzo, B. S., (2013). Population dynamics and breeding patterns of multimammate mouse, *Mastomys natalensis* (Smith 1834), in irrigated rice fields in eastern Tanzania. *Pest management Science* 69, 371–7.

My Phung, N. T., Brown, P. R., and Leung, L. K.-P. (2011). The diet of the female ricefield rat, *Rattus argentiventer*, influences their breeding performance in a mixed rice cropping ecosystem in An Giang province, the Mekong Delta, Vietnam. *Wildlife Research* 38, 610–6.

My Phung, N. T., Brown, P. R., and Leung, L. K.-P. (2013). Use of computer simulation models to encourage farmers to adopt best rodent management practice in lowland irrigated rice systems in An Giang, the Mekong Delta, Vietnam. *Agricultural Systems* 116, 69–76.

My Phung, N. T., Brown, P. R., Leung, L. K.-P. and Tuan, L. M. (2010). The effect of simulated rat damage on irrigated rice yield and compensation. *Crop Protection* 29, 1466–71.

My Phung, N. T., Brown, P. R., Leung, L. K.-P. and Tuan, L. M. (2012). Changes in population abundance, reproduction and habitat use of the rice-field rat, *Rattus argentiventer*, in relation to rice-crop growth stage in a lowland rice agroecosystem in Vietnam. *Wildlife Research* 39, 250–7.

Odhiambo, R. O., Makundi, R. H., Leirs, H., and Verhagen, R. (2005). Community structure and seasonal abundance of rodents of maize farms in southwestern Tanzania. *Belgian Journal of Zoology* 135: 113–18.

Painter, J. A., Molbak, K., Sonne-Hansen, J., Barrett, T, Wells, J. G., and Tauxe, R. V. (2004). Salmonella-based rodenticides and public health. *Emerging Infectious Diseases* 10, 985–7.

Parshad, V. R. (1999). Rodent control in India. *Integrated Pest Management Reviews* 4, 97–126.

Pauling, C., Ogilvie, S. C., Miller, A., Ataria, J. M., Waiwai, J., Doherty, J., and Eason, C. T. (2009). Matauranga rakau paitini – naturally occurring toxins in New Zealand plants with potential for vertebrate pest control. Report prepared for Nga Pae o te Maramatanga, Lincoln University Wildlife Management Report No. 50, pp 1–30

Prakash, I. (1988). *Rodent Pest Management*. CRC Press: Boca Raton, Florida.

Rabiu, S., and Rose, R. K. (2004). Crop damage and yield loss caused by two species of rodents in irrigated fields in northern Nigeria. *International Journal of Pest Management* 50, 323–6.

Rennison, B. D. (1979). Monitoring rat damage in West Malaysia: a consultant's report and recommendations for the Malaysia/UK Rice Rat Control Project. Rodent Pests Department, Ministry of Agriculture, Fisheries and Food, UK.

Rennison, D., Conole, D., Tingle, M. D., Yang, J., Eason, C. T., and Brimble, M. A. (2013). Synthesis and methemoglobinemia-inducing properties of analogues of para-aminopropiophenone designed as humane rodenticides. *Bioorganic & Medicinal Chemistry Letters* 23, 6629–35.

Rodenburg, J., Demont, M., Sow, A., and Dieng, I. (2014). Bird, weed and interaction effects on yield of irrigated lowland rice. *Crop Protection* 66, 46–52.

Ruelle, P., and Bruggers, R. L. (1982). Traditional approaches for protecting cereal crops from birds in Africa. In *10th Vertebrate Pest Conference*, 23–25 February 1982. Monterey: California

Salvioni, M. (1991). Evaluation of rat (*Rattus-rattus*) damage in rice fields in Madagascar. *Tropical Pest Management* 37, 175–8.

Schiller, J. M., Boupha, B. D., and Bounnaphol, O. (1999). Rodents in agriculture in the Lao PDR - a problem with an unknown future. In Singleton, G. R., Hinds, L. A., Leirs, H., and Zhang, Z. (Eds), *Ecologically-based Management of Rodent Pests*. Vol. ACIAR Monograph No. 59, pp. 372–87. Australian Centre for International Agricultural Research: Canberra.

Schmolz, E. (2010). Efficacy of anticoagulant-free alternative bait products against house mice (*Mus musculus*) and brown rats (*Rattus norvegicus*). *Integrative Zoology* 5, 44–52.

Singleton, G. R. (1997). Integrated management of rodents: A Southeast Asian and Australian perspective. *Belgian Journal of Zoology* 127, 157–69.

Singleton, G. R. (2003). Impacts of Rodents on Rice Production in Asia. International Rice Research Institute, IRRI Discussion Paper Series No. 43. Los Baños: Philippines.

Singleton, G. R., and Brown, P. R. (1999). Management of mouse plagues in Australia: integration of population ecology, bio-control and best farm practice. In Cowan, D. P., and Feare, C. J. (Eds), *Advances in Vertebrate Pest Management*, pp. 189–203. Filander-Verlag: Berlin.

Singleton, G. R., Belmain, S. R., Brown, P. R., and Hardy, B. (Eds) (2010a). *Rodent Outbreaks: Ecology and Impacts*. International Rice Research Institute: Los Baños, Philippines.

Singleton, G. R., Belmain, S. R., Brown, P. R., Aplin, K. P., and Htwe, N. M. (2010b). Impacts of rodent outbreaks on food security in Asia. *Wildlife Research* 37, 355–9.

Singleton, G. R., Brown, P. R., Jacob, J., Aplin, K. P., and Sudarmaji (2007). Unwanted and unintended effects of culling: A case for ecologically-based rodent management. *Integrative zoology* 2, 247–59.

Singleton, G. R., Hinds, L. A., Leirs, H., and Zhang, Z. (Eds) (1999a). *Ecologically-Based Management of Rodent Pests*. (Australian Centre for International Agricultural Research: Canberra)

Singleton, G. R., Joshi, R. C., and Sebastian, L. S. (Eds) (2008). *Philippine Rats: Ecology and Management*. Science City of Muñoz. Neuva Ecija: Philippine Rice Research Institute, p. 215.

Singleton, G. R., Kenney, A. J., Tann, C. R., Sudarmaji, and Hung, N. Q. (2003a). Myth, dogma and rodent management: good stories ruined by data? In Singleton, G. R., Hinds, L. A., Krebs, C. J., and Spratt, D. M. (Eds), *Rats, Mice and People: Rodent Biology and Management*. Vol. ACIAR Monograph 96, pp. 554–60. ACIAR: Canberra.

Singleton, G. R., Leirs, H., Hinds, L. A., and Zhang, Z. (1999b). Ecologically-based management of rodent pests - re-evaluating our approach to an old problem. In Singleton, G. R., Hinds, L. A., Leirs, H., and Zhang, Z. (Eds), *Ecologically-based Management of Rodent Pests*. Vol. ACIAR Monograph No. 59, pp. 17–29. Australian Centre for International Agricultural Research: Canberra.

Singleton, G. R., Sudarmaji and Brown, P. R. (2003b). Comparison of different sizes of physical barriers for controlling the impact of the rice field rat, *Rattus argentiventer*, in rice crops in Indonesia. *Crop Protection* 22, 7–13.

Singleton, G. R., Sudarmaji, and Suriapermana, S. (1998). An experimental field study to evaluate a trap-barrier system and fumigation for controlling the rice field rat, *Rattus argentiventer*, in rice crops in West Java. *Crop Protection* 17, 55–64.

Singleton, G. R., Sudarmaji, Jacob, J., and Krebs, C. J. (2005). Integrated management to reduce rodent damage to lowland rice crops in Indonesia. *Agriculture, Ecosystems and Environment* 107, 75–82.

Singh, Y. P., Kumar, D., and Gangwar, S. K. (1994). Status paper on rodents in North-Eastern hill region and their management. *Rodent Newsletter* 18, 3–18.

Sixbert, V. (2014). Demographic changes in rodent population and damage estimate in irrigated rice in Mvomero District, Morogoro, MSc. Dissertation, Sokoine University of Agriculture, Morogoro: Tanzania, p. 72.

Sridhara, S. (1992). Rice. In Prakash, I., and Ghosh, P. K. (Eds), *Rodents in Indian Agriculture, Vol. 1*, pp. 211–30. Scientific Publishers: Jodhpur, India.

Stenseth, N. C., Leirs, H., Skonhoft, A., Davis, S. A., Pech, R. P., Andreassen, H. P., R., S. G., Lima, M., Machangu, R. M., Makundi, R. H., Zhang, Z., Brown, P. R., Shi, D., and Wan, X. (2003). Mice, rats and people: the bio-economics of agricultural rodent pests. *Frontiers in Ecology and the Environment* 1, 367–75.

Stuart, A. M., Prescott, C. V., and Singleton, G. R., (2012). Natal nest locations of the Asian house rat (*Rattus tanezumi*) in lowland rice-coconut cropping systems: a coconut penthouse or rice bunds with water frontage? *Wildlife Research* 39, 496–502.

Stuart, A. M., Prescott, C. V., Singleton, G. R., and Joshi, R. C., (2011). Knowledge, attitudes and practices of farmers on rodent pests and their management in the lowlands of the Sierra Madre Biodiversity Corridor, Philippines. *Crop Protection* 30, 147–54.

Stuart, A. M., Singleton, G. R., and Prescott, C. V., (2015). Population ecology of the Asian house rat (*Rattus tanezumi*) in complex lowland agroecosystems in the Philippines. *Wildlife Research* 42, 165.

Sudarmaji, Rochman, Singleton, G. R., Jacob, J., and Rahmini (2003). The efficacy of a trap-barrier system for protecting rice nurseries from rats in West Java, Indonesia. In G. R. Singleton, L. A. Hinds, C. J. Krebs and D. M. Spratt (Eds), *Rats, Mice and People: Rodent Biology and Management*, pp. 306–8. ACIAR: Canberra.

Sudarmaji, Singleton, G. R., Brown, P. R., Jacob, J., and Herawati, N. A. (2010). Rodent impacts in lowland irrigated intensive rice systems in West Java, Indonesia. In Singleton, G. R., Belmain, S. R., Brown, P. R., and Hardy, B. (Eds), *Rodent Outbreaks: Ecology and Impacts*, pp. 115–27. International Rice Research Institute: Los Baños, Philippines.

Tangkanakul, W., Smits, H. L., Jatanasen, S., and Ashford, D. A. (2005). Leptospirosis: an emerging health problem in Thailand. *Southeast Asian Journal of Tropical Medicine and Public Health* 36, 204–8.

Tann, C. R., Singleton, G. R., and Coman, B. C. (1991). Diet of the house mouse (*Mus domesticus*) in the mallee wheatlands of north-western Victoria. *Wildlife Research* 18, 1–12.

Tigner, J. R. (1972). Seasonal food habits of *Rattus rattus mindanensis* (The Philippine Ricefield Rat). University of Colorado: Colorado, p. 66.

Tubiana, S., Mikulski, M., Becam, J., Lacassin, F., Lefèvre, P., Gourinat, A.-C., Goarant, C., and D'Ortenzio, E. (2013). Risk factors and predictors of severe Leptospirosis in New Caledonia. *PLOS Neglected Tropical Diseases* 7, e1991.

Wood, B. J., and Singleton, G. R. (2015). Rodents in Agriculture and Forestry. In Buckle, A. P., and Smith, R. H. (Eds), *Rodent Pests and their Control*, 2nd Edition, pp. 33–80. CAB International: Oxford, UK.

Yuan, J. B., Huang, Q., Ren, G., Shi, M., Chen, L., Yang, W. L., Chen, H. F., Yang, M., Yang, B., Yang, G. H., and Luo, X. Q., (2014). Acute and subacute toxicity of the extract of *Aristolochiae fructus* and honey-fried *Aristolochiae fructus* in rodents. *Biological & Pharmaceutical Bulletin* 37, 387–93.

Zhang, Z., Chen, A., Ning, Z., and Huang, X. (1999). Rodent pest management in agricultural systems in China. In Singleton, G. R., Hinds, L. A., Leirs, H., and Zhang, Z. (Eds), *Ecologically-based Management of Rodent Pests*. Vol. ACIAR Monograph No. 59, pp. 261–84. Australian Centre for International Agricultural Research: Canberra.

Zhao, G. (1996). Rodent pest control and plant protection. In Wang, Z. W., and Zhang, Z. B. (Eds), *Theory and Practice of Rodent Pest Management*, pp. 19–37. Science Press: Beijing.

Chapter 14

# Integrated weed management techniques in rice

*Simerjeet Kaur and Gulshan Mahajan, Punjab Agricultural University, India; and Bhagirath S. Chauhan, The University of Queensland, Australia*

1 Introduction
2 Weeds in rice crops: the problem
3 The need for IWM in rice production
4 IWM: an overview
5 IWM techniques: preparation and control of the growing environment
6 IWM techniques: establishing the rice crop
7 IWM techniques: managing the rice crop
8 The integration of weed management techniques
9 Summary
10 Where to look for further information
11 References

## 1 Introduction

The world's population currently stands at more than 7 billion, and it is expected to reach 8.5 billion by the year 2025, causing a rise in global demand for food. It is estimated that to meet this demand, global food production needs to be increased by over 40% by 2030 and 70% by 2050 (FAO 2009). This poses a great challenge for agricultural scientists, because limited options exist for horizontal expansion. Vertical expansion is therefore the only choice for sustaining the needs of the growing population.

Rice (*Oryza sativa* L.) is a principal source of food for more than half of the world's population. It is particularly important in South and South-East Asia, and more than 90% of rice is grown and consumed in Asia. During the era of the green revolution in Asia, semi-dwarf, photoperiod-insensitive, fertilizer-responsive, non-lodging varieties of rice and improved production and protection techniques helped to enhance rice production and productivity. Further enhancement of rice productivity depends on proper recognition of biological constraints such as weeds, and the alleviation of these problems through evolving and implementing appropriate management strategies.

http://dx.doi.org/10.19103/AS.2016.0003.25

In this chapter, we review the problem of weeds in rice crops and the need for integrated weed management (IWM) to address this issue. After a brief overview of the IWM approach and its advantages, three sections offer detailed discussion of the wide variety of techniques which may be incorporated into an IWM strategy, including preventive methods, tillage, crop rotation and choice of cultivar, water management, use of fertilizers and herbicides, and mechanical and biological control of weeds. We then review examples of how these techniques can be combined and integrated, and offer recommendations for future IWM research.

# 2 Weeds in rice crops: the problem

Weeds are one of the major biological constraints, causing an annual loss of about 45% of the annual losses in agricultural produce (Rao 2000). Despite continuous research in weed science and extension efforts to apply new knowledge, weeds continue to cause considerable losses to farm productivity. Weeds are unwanted, undesirable plants, which interfere with a crop's utilization of land and water resources by competing for nutrients, water and sunlight. They therefore adversely affect paddy yield and efficiency of harvesting.

Weeds also raise the cost of production and intensify pest and disease problems by serving as alternate hosts. High weed density provides a habitat for the growth of various pest organisms (e.g. insects, nematodes and disease pathogens), which adversely affect rice production. For instance, *Echinochloa* spp. is an alternate host for fungal pathogens that affect rice. Therefore, effective management of the weed not only controls weeds, but also helps in managing the sheath blight disease of rice (Kaur and Singh 2015a). Living or decaying weeds can also secrete toxic root exudates or leaf leachates that depress the normal growth of rice plants (Xuan et al. 2005).

The competition between weeds and rice depends on a number of factors, including the relative growth stages of the rice and weeds, the nature of stand establishment (transplanting vs. direct seeding), plant density (both plant population and geometry), the rice cultivar being grown, and moisture and nutrient availability. Weed composition in rice also varies with the rice ecology. Rice environments can be divided on the basis of the hydrology, into irrigated lowland rice, rain-fed lowland rice, flood-prone rice and upland rice. Irrigated lowland rice is grown in bunded fields for one or more crops per year, and involves assured irrigation, which maintains 5–10 cm of standing water in the field. Rain-fed lowland rice is grown in bunded fields that are flooded with rainwater for at least part of the cropping season, to water depths that exceed 100 cm for no more than ten days. In both irrigated and rain-fed lowlands, fields are predominantly puddled.

Transplanting rice seedlings either manually or mechanically is the conventional method of crop establishment; weed seeds then germinate after rice transplantation and weeds have to compete with the well-established rice seedlings. Hence, the crop–weed competition is low. The extent of yield reduction in direct-seeded rice due to weeds is high: 51–74% in rain-fed lowland rice, compared to only 15–20% in irrigated lowland rice. The critical period of crop–weed competition extends up to 60 days after sowing (DAS).

In flood-prone environments, fields suffer from excess water and uncontrolled deep flooding of more than 50–400 cm water. The deep-water rice and floating rice found in these environments provide poor yields of about 1.0 to 1.5 t $ha^{-1}$ (DRD 2014). In such environments, rice plants face severe weed pressure in the early stages of their development, and perennial and floating type rice weeds cause high yield reductions at later stages. Unchecked weeds cause 18–59% loss in grain yield in this type of rice crop (Sharma 1997).

The upland rice ecosystem consists of rain-fed fields with no surface water accumulation. The fields are prepared and seeded (broadcast, drill or dibble) when dry. Upland rice competes poorly with weeds and uncontrolled weed growth often results in negligible or zero yield. There is larger variability in weed species composition in upland rice than in other production systems. This depends upon ecology, cropping system and management practice.

To obtain a weed-free spectrum and ensure maximum crop productivity and profitability in rice, it is important to manage weeds effectively and economically. A weed-free period for the first 25–45 DAS is required to avoid any loss in yield in dry direct-seeded rice (Chauhan and Johnson 2011) and if weeding is delayed beyond 20 days after rice emergence, irreparable damage will be done by weeds. During early establishment, weeds complete 20–30% of their growth while the crop accomplishes only 2–3% of its growth. Reductions in yield due to weeds are 10–15%, 35–45% and 20–25% in transplanted rice, upland direct-seeded rice and wet-seeded rice, respectively. Significant amounts of nutrients (24.7 kg $ha^{-1}$ nitrogen, 5.8 kg $ha^{-1}$ phosphorus and 63.4 kg $ha^{-1}$ potassium) are taken from the soil by weeds in one season (Sharma 2007). Apart from loss of macronutrients, micronutrients such as iron (2.5 kg $ha^{-1}$), zinc (61 g $ha^{-1}$), copper (15 g $ha^{-1}$) and manganese (168 g $ha^{-1}$) are also lost due to weed infestation (Pandey and Thakur 1988).

Azmi and Baki (1995) estimated that the rice grain yield losses caused by grasses [mainly *E. crus-galli* (L.) P. Beauv.], broadleaved weeds and sedges were 41, 28 and 10%, respectively. Weedy rice/red rice is the weed with the highest impact for lowland rice in Brazil, with losses estimated at around US$ 300 $ha^{-1}$ (Burgos et al. 2008). Rice grain production in India suffers a yearly loss of 15 million tonnes due to weed competition (Chatterjee and Maity 1981) and the trend is similar in other Asian countries.

Weed–crop competition increases as growth progresses, particularly in direct-seeded rice, since weeds germinate at the same time and compete for light and nutrients with the rice seedlings. There is little problem with weed control in manual and mechanically transplanted rice. However, weeds (especially red/weedy rice) are a very significant problem for dry direct-seeded rice, and, to a slightly lesser extent, for wet direct-seeded rice. The conversion from transplanted to direct-seeded rice results in more competitive weed flora and, therefore, more comprehensive weed management approaches are required for effective weed control (Singh et al. 2009). In transplanted rice conditions (paddies), water-loving weeds like *Echinochloa crus-galli, Echinochloa colona* (L.) Link, *Leptochloa chinensis* (L.) Nees., *Cyperus iria* L., *Cyperus difformis* L., *Eclipta prostrata* (L.) L. and *Sphenoclea zeylanica* Gaertn. infest the crop. With the shift in crop establishment method from transplanting to direct seeding, various other weeds are also found to infest rice fields, including aerobic grasses such as *Dactyloctenium aegyptium* (L.) Willd., *Digitaria sanguinalis* (L.) Scop., *Eragrostis diarrhena* (Schult. & Schult. f.) Steud., *Eleusine indica* (L.) Gaertn., *Acrachne racemosa* (Roxb.) Lindl. Ex Chiov., and *Commelina benghalensis* L.; sedges such as *Cyperus compressus* L. and *Cyperus rotundus* L.; and broadleaved weeds such as *Digera arvensis* Forsk., *Phyllanthus niruri* L., *Amaranthus viridis* L. and *Trianthema portulacastrum* L. (Malik et al. 2014).

# 3 The need for IWM in rice production

Due to the rising cost of labour, labour shortages and the drudgery of manual weed control in rice, herbicides have assumed great significance as a weed control technique, and are viewed as the main solution for the weed problem in rice production. Chemical weed control tactics play a significant role in modern weed management, and they contribute to maintaining the productivity of diverse cropping systems, reducing yield losses and facilitating conservation tillage systems. In India, for example, herbicide use is estimated to be 0.6 kg $ha^{-1}$, in comparison to the global average of 3 kg $ha^{-1}$. The 25% of the total pesticide market in India is directed towards protecting rice, wheat and soybean crops. In 2007, relative rates of herbicide use were estimated at 3571, 4122, 12159, 17164 and 200488 metric tons of active ingredients in Bangladesh, India, the United Kingdom, Germany and the United States, respectively (FAO 2015). The herbicide revolution in these countries allowed good weed control and caused significant improvement in crop productivity.

However, with the intensive use of herbicides, multiple problems such as weed shifts, environmental pollution and herbicide resistance are appearing, and these problems have global impact (Heap 2015). Continuous applications of a particular cropping practice, such as the use of a particular herbicide, can inadvertently contribute to a shift in the dominance and distribution of weeds. Controlling weeds with only one or two techniques gives some weeds a chance to adapt to those practices. For example, the use of herbicides with the same mechanism of action, year after year, has resulted in herbicide-resistant weed chemotypes. Repeated herbicide use may result in weed flora shifts towards more persistent perennials, as well as a build-up of herbicide residues in the soil and in consumable products. Moreover, weeds which share a similar genetic heritage with rice crops (e.g. weedy rice/red rice) are not successfully controlled with the selective herbicides currently in use.

Herbicides will continue to play a significant role as an important tool for weed control in rice. However, weed science must also develop and incorporate low-input/low-cost/non-monetary practices to create IWM systems that reduce selection pressure and environmental degradation. IWM, besides controlling existing weed problems, puts greater emphasis on preventing weed reproduction and establishment, and on minimizing weed competition with the crop. It emphasizes integrating techniques and knowledge in a way which combats the causes of weed infestation, rather than just responding to problems after they occur. The goal of IWM is to optimize crop production through the intensive use of preventive methods, scientific knowledge, management skills, monitoring procedures, and the efficient use of diverse control practices.

At present, hand weeding still occupies more than 40% of small farmers' time in the least developed countries, and this practice does not completely prevent crop losses caused by weeds. On larger farms, the use of herbicides is more common, but numerous studies show the ill effects of the extensive (indiscriminate) use of herbicides on human health (devastating, life-long diseases and deformities in children and unborn foetuses) and the environment (groundwater pollution due to runoff and deep percolation) (Antle and Pingali 1994; Kolpin et al. 1998). IWM has the potential to improve outcomes on farms both large and small, in developing countries and in the developed world.

# 4 IWM: an overview

The FAO defines integrated pest management as a method whereby all economically, ecologically and toxicologically justifiable methods are employed to keep harmful organism(s) below the threshold level of economic damage, keeping in the foreground the continuous employment of natural limiting factors. IWM practices ensure that weeds are in a constantly changing system that uses many different control practices, and thus limits their ability to adapt.

Adoption of IWM is essential for economical and effective management of weeds, minimizing the size of the weed seed bank over time, and management of herbicide resistance. IWM integrates different agronomic, ecological, cultural and chemical practices to manage weeds, so that reliance on any one weed control technique is reduced. Effective and sustainable weed management involves the combined use of preventive, cultural, mechanical, chemical and biological weed control techniques in an effective and economical way. No tool can offer complete control, but the integration of these tools in suitable combinations works efficiently. Reducing the reliance on one or two specific weed control techniques or tools extends the life of these methods for future use. The objective of IWM is to maintain weed densities at a manageable level, preventing shifts to weeds which will be more difficult to control and managing weed biodiversity.

IWM still allows some use of herbicides, but it combines this with various non-monetary, low-input, low-cost resource conservation techniques such as zero tillage, establishment methods, planting patterns, good water and fertilizer management, stale seedbed, crop residue incorporation, mulching, green manuring and timeliness of operations. These combinations of techniques are being used to reduce the herbicide load in ecosystems. It is important to consider current, location-specific farming activities when making recommendations for production practices for weed management. A better understanding of the mechanisms of inter- and intra-specific interaction can lead to more effective weed management practices. IWM involves the concept of multiple tactics of weed management to maintain weed population below a certain level while conserving environment quality.

Monitoring and regulating shifts in weed flora and the appearance of resistant weeds and preventing introduction of new weeds are major components of IWM, as well as an emphasis on low-cost agronomic/ecological, biological and biotechnological methods for environmental safety. Site-specific IWM strategies have to be evaluated considering the available resources in a situation and the hydrology/environment in which rice is being grown. A successful IWM strategy enhances farmer's profitability, ensures environmental protection and responds to consumer preference. The integration of different methods of weed management can increase the efficiency of applied herbicides. Crop competitiveness against weeds can also be improved by integration of improved production and protection practices, their timeliness, optimum fertilization and water management and incorporation of crop residues in the soil (Chauhan et al. 2012b).

The weed scenario present within crops is the end consequence of different agronomic and ecological weed management practices. Good weed control practices should increase weed species diversity, while maintaining their biomass below a critical threshold level for desired crop growth and yield (Miyazawa et al. 2004). Rice is a very important crop, especially in Asia, and sustainable weed control is possible only when rice production makes use of effective weed management practices.

In the next three sections, we review the large number of IWM practices that may be useful to rice growers, along with their potential advantages and disadvantages.

# 5 IWM techniques: preparation and control of the growing environment

## 5.1 Preventive methods

'One year's seeding makes seven years' weeding' is a saying used to highlight the danger of allowing weeds to grow and seed themselves (Cousens and Mortimer 1995). Weed prevention includes all measures which check the entry and establishment of weeds in an area. It also includes farm hygiene to prevent the yearly production of seeds, tubers and rhizomes by the weed species already present on the farm. It deals with all those practices that help to prevent weeds from becoming a problem over time.

Physical, mechanical or chemical methods which prevent weeds from producing viable seeds are considered a part of weed prevention. Quarantine laws, which aim to prevent weeds from infesting a new habitat, are also included. Non-cropped areas like irrigation channels, drainage channels, ditches, fence lines, farm boundaries, bunds and other similar non-cropped areas are often neglected by farmers, but should be considered as a part of weed prevention strategy. These places act as a source of weed infestation to the cropped areas. Such areas can be kept free from weeds with the use of contact herbicides or physical methods (Rao 2000).

Weeds spread to new areas through contamination; crop seeds contaminated with weed seeds can be a huge problem. Rice crop seed that is to be used for next year's sowing must be harvested at full maturity and stored safely. The contamination of crop seed can be prevented in two ways: by producing a weed-free crop, or by removing the seeds of weeds before storage or at sowing time. Seed cleaning devices such as sieves, magnetic separators, velvet rollers, dented plates and/or salt solution separation can be used to clean crop seed. The rice nursery should be free from weeds that mimic rice plants. Seeds of the Clearfield rice variety/hybrid (non-transgenic, modified with traditional plant breeding methods) can be used in combination with broad-spectrum weed control, especially of weedy rice with imazethapyr, quinclorac and imazamox herbicides. If rice seed needs to be brought in from an outside source, then the source should be reliable, and the seeds should be certified as being for the desired cultivar.

Weedy rice has spread in several regions through the use of contaminated rice seeds and contaminated machinery. Therefore, the use of clean rice seeds and machines can greatly reduce the spread of weedy rice to new areas. There is a dire need to increase sanitary awareness among growers to avoid weedy rice infestations.

## 5.2 Field levelling and tillage

Precision land levelling or using a laser leveller to create regular, sloping fields enables appropriate water management, and this has a profound effect on weed growth and the bioefficacy of herbicides. A well-levelled field is essential for good weed control: fields with low and high areas contain varying water depths, which are too deep for the rice in the low spots and too shallow for weed control in the high spots. If the field is not levelled, the rice seedlings cannot establish in the low spots, resulting in stunted plants with low tiller production, reducing crop competitiveness.

The stale seedbed practice has been found very effective in reducing infestation of weedy/red rice, volunteer rice plants from the previous season, and *Cyperus rotundus* in the crop (Mahajan et al. 2013). In this technique, one or two flushes of weeds can be destroyed before planting the crop. Weeds are allowed to germinate after a light irrigation or rainfall and the emerging seedlings are killed using either a non-selective herbicide or tillage operations. Using a herbicide is to be preferred, since it avoids bringing buried seeds back to the soil surface, and the crop can then be planted without any further tillage. Although the stale seedbed practice helps in reducing seed banks of weedy rice and other weeds, the extent of efficacy depends on the degree of dormancy. The stale seedbed technique is time consuming and the delay to sowing the rice crop translates into poor yield. It exhausts only 1–2 cm of the shallow soil layer of non-dormant weed seeds, which can be replenished with tillage. Climate change is a potential threat in areas where weeds are controlled with the stale seedbed technique, because an increase in the $CO_2$ level in the atmosphere may increase the tolerance of weeds for glyphosate, reducing glyphosate efficiency in the future. The stale seedbed technique is a powerful tool for weed management in organic agriculture systems.

Among the several objectives of tillage, weed control is one of the most important. Jethro Tull (1674–1741), the father of tillage, developed various implements for weed control. However, tillage serves only as a temporary means of weed control because the soil contains many dormant weed seeds. Moreover, intensive cultivation could potentially result in increased surface soil degradation and erosion losses. Ploughing may bury weed seeds at a depth that prevents germination, but may also expose others from deeper soil layers to the surface where conditions are more suitable for germination.

Tillage affects weeds through cutting, burial, uprooting and removal of weeds; moreover, it changes the soil microclimate and as a consequence alters weed germination, emergence and establishment. Under zero-till conditions, 60–90% of the weed seeds were located in the top 5 cm of the soil, depending upon the soil type. This differential vertical distribution of seeds in the soil affects weed population composition as different soil depths differ in the availability of moisture/air, temperature, light, microfauna and predator activity (Chauhan et al. 2006a,b, 2007, 2012a). In conservative tillage systems, weed seeds present in the top soil zone are forced to germinate and can be killed with any contact herbicide, and crops can be sown without any tillage operations.

Thorough land preparation can significantly decrease the incidence of weeds in rice by killing weeds, weed seeds and vegetative propagules and provide weed-free conditions at the time of planting; a conducive environment for the rapid growth of rice seedlings. Weed management is significantly influenced by any variation in the intensity or occurrence of tillage, as tillage affects weed growth by changing the soil conditions needed for seed germination through uprooting/dissecting/burying them deep enough to prevent emergence and shifting seeds both vertically and horizontally in the soil. With reduced tillage, the density of some annual and perennial weeds can increase.

Weed species that require light for germination grow more successfully in zero-till systems, as in these systems most weed seeds remain on or close to the soil surface (Chauhan and Johnson 2009a), whereas the majority of weed seeds are buried 10–15 cm below the surface in tilled fields. Germination of *E. colona* increased in the presence of light and emergence was greater from shallow soil layers, which suggests that this weed can be a dominant weed in conservation agriculture systems (Chauhan and Johnson 2009b).

The tillage system used affects perennial weeds more than annual weeds; conservation agriculture may promote the germination and emergence of newly shed weed seeds that stay on or near the soil surface. Perennial weeds are very difficult to manage under zero tillage as these produce vegetative reproductive parts like rhizomes, stolons and tubers. When there are good reasons to employ low or zero-tillage systems, other effective weed control practices should therefore be employed to manage weeds successfully.

Flooding and exquisite puddling is an excellent weed management tool in lowland rice. Summer ploughing and cultivation of crops during post-rainy periods is helpful in reducing weed infestation. Lowering bulk density to 1.8 g $cc^{-1}$ through soil compaction is a very effective tool for minimizing weed pressure in direct-seeded lowland rice. Good field drainage and good water management are prerequisite for wet-seeded rice establishment and for reduction of herbicide phytotoxicity on non-target flora.

The changes in tillage practices for rice cultivation in the Indian Punjab have resulted in major shifts in weed flora in the summer season, from the traditional summer season weeds to water-loving weeds like *Echinochloa crus-galli*, *Paspalum distichum* and *Cyperus iria*. New weeds like *Ischaemum rugosum*, *Cyperus difformis* and *Fimbristylis tenera* have also invaded the rice fields. Further, continuous cultivation of rice under flooded conditions changed the soil environment in favour of semi-aquatic and few aquatic weeds such as *Leptochloa chinensis*, *Ammannia baccifera*, *Caesulia axillaris*, *Sphenoclea zeylanica*, *Sagittaria guyanensis*, *Ipomoea aquatica* and *Oxalis* species. Shifts in weed flora occur due to changes in rice ecology from paddies to aerobic conditions (Rao et al. 2007; Mahajan et al. 2009). The above literature suggests that research is needed to evaluate the effects of both tillage system and residue management on weed dynamics.

## 5.3 Cover crops

Cover crops and mulches can reduce weed problems by preventing weed seed germination or by suppressing the growth of emerging seedlings. Cover crops play an important role in suppressing perennial weeds if tillage and smother crop approaches are followed. An effective smother crop leaves the ground weed-free at its harvest. Mulches can be organic or inorganic in nature, consisting of living plant ground cover, loose particles spread over soil or sheets of artificial/natural materials laid on the surface. Selection of a cover crop depends upon growth duration, rooting behaviour and nitrogen-fixing ability. Choices include *Azolla*, a green manure crop like *Sesbania*, sannhemp and others. Another important consideration in selecting a cover crop is that it should not compete with the main crop for common resources.

# 6 IWM techniques: establishing the rice crop

## 6.1 Establishment methods

Rice establishment methods affect the level of weed infestation and the species composition in the crop. The transplanting method is primarily adopted for good weed control, because puddling creates anaerobic conditions. However, there has been a shift from transplanting to direct seeding in several countries of South-East Asia in recent years. Due to receding water levels, the rising cost of labour and deteriorating soil structure due to puddling, direct-seeded rice is gaining popularity.

In developing countries, direct seeding has been the principal method of rice establishment since the 1950s. Farmers with different resource levels, land development, infrastructure and weather conditions follow different direct seeding practices in the Indo-Gangetic Plains (Pandey and Velasco 2005). Direct seeding has many benefits, including reduced labour requirements, less drudgery, earlier crop maturity, more efficient water use, higher tolerance of water deficit and fewer methane emissions. However, weeds are the number one problem in direct-seeded rice. IWM techniques are more relevant and important in direct-seeded rice systems than in transplanted rice systems, because the more diverse weed flora means that a single control method cannot provide good weed control.

The weedy rice problem in direct-seeded rice systems can be greatly reduced by either shifting to the manual transplanting of seedlings or by broadcasting seedlings (aka parachute planting) in flooded fields. The drudgery of manual transplanting can be avoided through the use of mechanical rice transplanters, which now allow transplanting in both puddled and zero-tilled fields (Kamboj et al. 2013).

## 6.2 Sowing methods

The random transplanting method is the most common method followed in Asian countries. However, line sowing is very useful in fields that are experiencing severe problems with weedy rice. It is difficult to distinguish weedy rice plants from cultivated rice plants at an early stage if the rice is sown using the broadcasting method. When rice is sown in lines or rows, it is easier to identify and remove weedy rice seedlings between the rows. Moreover, both manual and mechanical weeding are much easier to perform in row-seeded rice than in broadcast rice (Chauhan, 2013).

A dry-seeded rice crop can be sown in rows using a seed drill fitted to a tractor, while wet-seeded rice can be sown in lines using a drum seeder. There is a need to increase awareness among growers regarding the importance of line sowing, and to increase adoption of seeders among farmers. There is also a need to develop seeders fitted to tractors, which can easily sow rice seeds in puddled conditions.

## 6.3 Crop rotation and rice cultivars

A key component of weed management in traditional farming was the rotation of different crops with different life cycles. Crop rotation is defined as a system of growing different kinds of crops in succession on the same land. With each crop, certain typical weeds appear, and these weeds increase rapidly if a favourable crop is grown continuously. With time, these weed species become hard to control (Chauhan et al. 2012b).

Including crop rotation in conservation agriculture is a successful approach to reducing weed pressure. Crop rotations are effective in controlling crop-associated weeds. Crop rotation with crops requiring different management practices is known to disrupt the growth cycle of weeds, including weedy rice (Chauhan and Johnson 2010b; Chauhan 2013). The long-term weed population dynamics will be affected by the choice and sequencing of crops. Growing a non-rice crop allows the use of different cultural practices and alternative, effective herbicides. In rice monoculture systems (e.g. in eastern and south India), one rice crop (preferably the dry season crop) can be rotated with an upland crop. Mixed cropping of rice–*Azolla* and rice–green manure can help in reducing the weed infestation. Inclusion of a short-duration crop (e.g. green gram/black gram) between two rice crops can also help in decreasing the population of weedy rice. Summer cowpea for fodder or *Sesbania* for green manure in between the rice–wheat rotation after the wheat crop can provide fair weed control, thus minimizing the use of herbicides.

In a study in Italy, the rotation of soya bean for a year led to greater than 90% reduction in the weedy rice seed bank. Some crops such as sorghum, pearl millet and maize may drastically suppress the weed population and reduce their biomass through allelopathic interferences. Pearl millet may exhibit residual weed suppression in the following crop, so it should be included as a fodder crop in rotation. Adoption of rotation crops, however, will depend on their market prices and yield stability.

Rice plants release toxic allelochemicals (either as root exudates or sourced from decaying plant materials) that can suppress and kill weeds. The potential of some allelopathic rice cultivars to inhibit weed growth is up to 40%. An IWM programme in rice should attempt to effectively exploit the competitive ability of crops in suppressing weed growth. Different crops and cultivars can reduce weed biomass from 4 to 83% during a full season of competition (Minotti and Sweet 1981). The competitive ability of the cultivar depends upon both weed tolerance and weed-suppressive abilities. Weed tolerance is the crop's ability to give a high yield despite weed competition, while weed-suppressive ability is the ability of the crop to restrain weed growth through competition. The differential morphological character of cultivars plays a vital role in crop–weed competition. Mahajan et al. (2015) reported that fast early growth may be the fundamental attribute of a weed-suppressive genotype but a strong weed-suppressive ability is not always linked to high-yielding ability. The competitive ability of rice cultivars is associated with light interception-related traits (Chauhan 2012a,b). Therefore, the use of competitive cultivars is an important IWM strategy (Mahajan et al. 2013).

There has been less emphasis on weed-competitive traits in transplanted rice due to the reduced weed problem in puddled conditions (Mahajan and Chauhan 2013); however, weed competitive cultivars can be exploited in direct-seeded rice systems (Chauhan 2012a; Mahajan and Chauhan 2011). Generally, tall cultivars compete more vigorously with weeds but they tend to lodge and have lower yield potential than dwarf-statured cultivars. Intermediate height may be a more appropriate trait for direct seeding in suppressing weeds, along with early vigour, more tillers, inclined leaves, more biomass accumulation at the early crop stage, and high leaf area index and specific leaf area during vegetative growth for rapid ground cover by the canopy. Early-maturing rice cultivars and rice hybrids also have a smothering effect on weeds due to their improved vigour and early canopy cover (Chauhan and Johnson 2010a).

The effectiveness of competitive cultivars in weed suppression can be increased with agronomic manipulations. Changing the plant spacing pattern and time of sowing might be helpful in providing additional weed control when herbicide doses are decreased (Mahajan and Chauhan 2011). In sub-Saharan Africa, cultivars of rice (cv. Nipponbare) that exhibit strong post-attachment resistance to the root hemi-parasitic weed *Striga hermonthica* have been identified and these may be used in breeding programmes to enhance host resistance (Gurney et al. 2006).

Rice cultivars having early vigour are always more competitive against weeds, including weedy rice. However, weed competitive variety of rice can only suppress weeds to a limited extent if used in isolation. Therefore, there is a need to integrate other weed management strategies when evaluating the competitive advantage of rice cultivars. There is a need to identify suitable cultivars for dry-seeded rice with desirable traits like good mechanical strength of seedlings to facilitate good crop establishment, early seedling vigour, an efficient root system and yield stability. Rice cultivars with coloured (e.g. purple) leaves and stems may help in reducing weedy rice. Weedy rice emerging in fields planted with such rice cultivars can be easily distinguished and pulled out. A purple-leaf rice cultivar may have low yield potential and lower market values, but this may be better than abandoning the field due to a severe weedy rice problem. The utility of this approach should be explored under different weedy rice scenarios.

## 6.4 Plant population

Good quality seeds/seedlings and plant protection measures should be used for maintaining optimum population of the crop. In an agricultural ecosystem, by maintaining narrow row spacing and high seeding rates, a crop provides a more smothering effect on weeds as less space is available for weeds to flourish. With the help of a high crop density per unit area, weeds can be suppressed by maintaining the dominant position of the crop plants over weeds through a modification in the canopy structure. Plant population and row arrangement also affect weed growth by cutting light availability at the ground level. Higher plant population can put pressure on the availability of space for weed growth and make the crop more competitive against weeds (Mahajan et al. 2006; Phuong et al. 2005).

Narrower row spacing improves the competitiveness of crops with weeds by developing faster canopy cover. The rice crop's ability to compete against weeds for light can therefore be increased with cultural management practices such as reduced crop row spacing (Chauhan and Johnson 2010a). High seeding rates could partly control weeds, depending upon the biology of weeds and rice cultivars present in the field. In a transplanted crop, dry matter accumulation by weeds can be reduced by increasing plant density from 33 to 44 plants $m^{-2}$ (hills). Weed growth is higher in a thin-populated as compared to a thick-populated crop. In direct-seeded rice, high seeding rates are generally used to help control weeds by smothering, as well as to compensate for damage by rats and birds or for poor crop establishment. Higher plant population is of importance, especially in low-input and organic production systems or where herbicide resistance has developed in weeds.

In weedy rice-infested areas (e.g. in Malaysia), the use of high seeding rates has helped to reduce the problem of weedy rice (Chauhan 2013). In a study on direct-seeded rice, it was observed that *E. colona* density was not influenced by seed rate; however, sedges density was influenced by seed rate. Sedge density decreased by 35% when the seed rate was increased from 15 to 240 kg $ha^{-1}$ (Mahajan et al. 2010). Although increasing the seed rate beyond the optimum level had no influence on weed density, weed biomass decreased linearly with increasing seed rates. The increase in seeding rate from 15 to 125 kg $ha^{-1}$ led to rapid canopy closure, thus reducing weed competition and significantly decreasing weed biomass (Chauhan et al. 2011). Narrower row spacing also resulted in early crop canopy cover, allowing less light penetration through the rice crop leaves, and thus smothering weeds and preventing them from competing effectively with the crop (Chauhan and Johnson 2011b). As a result, it gave rise to higher productive tillers and grain yield (Mahajan and Chauhan 2011).

In transplanted rice, seedlings are traditionally transplanted in a random manner by migrant/farm labourers. However, a change in plant geometry (e.g. transplanting in a triangular/square manner or in paired rows) may help cultivars to suppress weeds more effectively. Favourable conditions for crop growth are achieved with a higher plant density and narrow row spacing, enabling the rice crop to compete more effectively with weeds and so produce a higher yield (Phuong et al. 2005).

The twin planting pattern reduced weed biomass in dry-seeded rice and resulted in an increase in yield (Mahajan and Chauhan 2011). More weed biomass was produced using a row spacing of 30 cm, compared with that produced using 15-cm spacing or 10-20-10-cm paired rows (Chauhan and Johnson 2011). Weed biomass was at its highest level at a low seed rate of 20 kg $ha^{-1}$ and decreased significantly with each successive increase in seed rate from 20 kg to 50 kg $ha^{-1}$, while weed biomass increased significantly with each increase in row spacing from 15 to 30 cm (Kaur and Singh 2014). In another study, it was observed that grain yield remained similar at different row spacings (20 cm or 30 cm) in weed-free environments (Chauhan 2012a,b). However, in weedy or partially weedy conditions, narrow row spacing provided a significantly higher grain yield than did wider row spacing. In crops planted in narrow rows, the critical period for weed–crop competition was usually less than that for crops planted in wider rows; rice crops sown with wider spacing were vulnerable to weed competition for a longer period, and this may provide an explanation for the greater weed biomass (Chauhan and Johnson 2010a, 2011b).

A higher than an optimum plant population may have benefits in weed control, but it is often accompanied by reduced tillering, increased proportion of ineffective tillers, more lodging problems, increased chance of rat damage, insect attack and disease infection, reduced panicle size, more spikelet sterility and increased nitrogen demand. Therefore, it is necessary to evaluate the effect of high seeding rates on weed management and rice productivity in cases where herbicide use is limited, for example in organic production systems or where herbicides are less effective.

# 7 IWM techniques: managing the rice crop

## 7.1 Water management

Weeds reduce water-use efficiency in rice by 12–46% (Mahajan et al. 2015). Although water productivity in the form of evapotranspiration is low, a huge amount of water is utilized for land preparation/paddies in puddled, transplanted rice systems. Many weeds cannot germinate or grow in flooded soils, making flooding an effective tool for controlling those weeds. In addition to forming a hard pan and preparing slurry into which seedlings or pre-germinated seeds can be sown, puddling buries weeds in the lower layers of mud where they decompose by an anaerobic action to form ammonium compounds, which can then be used directly by the crop. However, deep ponding of water in early growth stages should be avoided, because developing rice seedlings can be killed or their growth can be retarded due to lack of oxygen and accumulation of toxic concentrations of applied herbicides. When the transplanted seedlings have established themselves (approximately one week after transplanting), the field should be completely flooded to a depth of 7.5–10 cm to inhibit weed growth. As the rice grows, the depth of the water can be gradually increased to 15 cm.

The efficacy of rice herbicides increases significantly with an increase in the ponding period from three to nine days after transplanting. The problem of red/weedy rice can be reduced by water seeding and use of good and clean crop seeds. Flooding is commonly used as the primary cultural weed control method to suppress weeds in puddled, transplanted rice; however, the effect of flooding on weeds is species specific (Chauhan and Johnson 2010b). It can be effectively used to suppress the emergence and growth of weedy rice, but some weed species, namely *Leptochloa chinensis* and *Eleusine indica*, may be encouraged by alternate wetting and drying in puddled as well as dry-seeded rice. The soil moisture content in the upper 0–15 cm soil layer affects the emergence patterns of weeds.

Differential competition is also observed with water stress: grass weeds provide substantial competition to rice crops under stressed conditions (Mahajan et al. 2015). Due to canal irrigation systems, weeds such as *Ageratum conyzoides* L., usually typical of hilly areas, have invaded the rice fields, watercourses and bunds of the fields in certain regions. Development of semi-aquatic conditions through the continuous cultivation of rice under flooded conditions has encouraged the weed species of marshy areas like *Scirpus roylei* (Nees) Parker, *Eleocharis acicularis* (L.) R. & S., *Najas arguta* Kunth. and *Hydrilla verticillata* (L. f.) Royle.

In direct-seeded rice systems, there is a need to introduce rice cultivars capable of emerging under anaerobic or flooded conditions to avoid weedy rice/red rice infestation. The use of such rice cultivars with appropriate flooding time and depth can greatly help to reduce weedy rice, especially in regions where water is plentiful. This method will help greatly in reducing chemical load.

## 7.2 Fertilizer use

Fertilizer use has significant effects on weed emergence, growth, dormancy, hardiness and persistence, and on weed dynamics. Soil fertility and use of fertilizers both affect weed diversity and crop–weed competition in rice crops; some weeds favour low-fertility soils, whereas others predominate in well-fed soils. Manipulation of crop fertilization, especially nitrogen, is a promising agronomic practice in reducing weed interference in crops (Cathcart et al. 2004). However, under high weed pressure, crop fertilization with various nutrients helps weeds more than the crop because nutrient absorption is faster and higher in weeds than in crop plants.

Increasing N application rate up to 150 kg $ha^{-1}$ caused significant improvement in grain yield of direct-seeded rice when weeds were well controlled; however, under poor weed control conditions, it resulted in a drastic reduction in yield (Mahajan and Timsina 2011). Thus, without other weed management techniques, crop fertilization alone is not helpful in achieving higher net returns. There should be more emphasis on the application of fertilizers using band placement methods along crop rows, at the most effective times and in optimum quantities. To date, limited research has been done in this direction, particularly for the specific case of rice crops.

## 7.3 Mechanical control of weeds

Mechanical control consists of methods that kill or suppress weeds through physical interruption. Such methods include pulling, digging, discing, ploughing and mowing. Different hand tools have been used to kill and destroy weeds. In the eighteenth century, Jethro Tull invented the first animal-drawn weeding implement, the horse hoe. The rice rotary weeder and cono weeder can be employed for weed control in rice. *Halod* and *bushening* is employed in some parts of India to control weeds (Angiras and Rana 1998).

In transplanted rice, a rotary weeder can be used four times starting from 15 days after transplanting at a ten-day interval to manage weeds. It saves labour by reducing hand weeding, aerates the soil and root zone, and enhances the root activity, which ultimately translates into higher yields. In upland dry-seeded rice, repeated inter-cultivation by passing a hoe or small blade harrow 2–3 times between 20 and 45 DAS can effectively control weeds. In rain-fed rice, two weedings are carried out. The first weeding can be done between 15 and 21 days after germination, and the second weeding may be done 30–45 days after first weeding.

## 7.4 Hand weeding/pulling

In the past, hand weeding was synonymous with weed management due to abundant labour availability, the cheaper cost of labour and the nature of agriculture as the major occupation. Hand weeding is time consuming and tedious. When weeds are large enough to be gripped, they are pulled out of the soil and discarded. Smaller weeds can be hand-pulled. Early hand weeding is better since any delay will enable the weeds to absorb nutrients. Hand hoeing is used as a very handy method of weed control, particularly where line planting is practised. Hand hoeing is faster than hand weeding and even creeping perennials can be controlled.

Integration of weed control methods is very important for solving long-term weed problems and so hand weeding and hoeing have a place even in some modern agricultural systems. The escaped weed plants after the use of herbicides must be uprooted before seed setting. These escaped plants may be of a resistant population or may be due to lack of proper spraying technology. Therefore, to reduce the weed seed bank, weak/left over plants should not be allowed to form seeds. Germinated weeds should be pulled twice or thrice in order to eliminate their competition.

## 7.5 Biological control

Biological control of weeds is broadly defined as 'the use of an agent, a complex of agents, or biological processes to bring about weed suppression'. All forms of microbial organisms can be considered as biological control agents. Biological control makes use of living organisms including insects, herbivorous fish, animals, disease-causing organisms and competitive plants to limit weed infestation. This method cannot eradicate weeds, but weed population can be reduced to a significant degree.

Examples in the case of rice include *Hirschmanniella spinicaudata*, a rice root nematode that controls most upland rice weeds; *Bactra verutana*, a moth that destroys *Cyperus rotundus*; *Altica cyanea* (the steel blue beetle) that completely destroys *Ludwigia parviflora* and the rust fungus *Puccinia canaliculata*, which controls *Cyperus esculentus* L.

Integrated rice–duck farming (Aigamo Method), a 500-year-old tradition from Japan, makes use of the mutually beneficial relation between ducks and rice crops to increase rice productivity, reduce pests and weeds in rice fields and provide nutrients to rice plants (Li et al. 2012).

Seed predation by granivore fauna, such as ants and other insects, could be used as an important tool in weed management systems where newly produced weed seeds remain on the soil surface, mostly under zero-till systems (Chauhan et al. 2010). By integrating seed predation with other weed control measures, herbicide use and the risks associated with it can be reduced, as can the need for costly manual labour. The size of the weed seed bank can also be reduced significantly by seed predation. Weed seed predators can be encouraged by retaining crop residues in the field, since these serve as forage for the predators. Since no additional cost is required to apply such approaches, these environment-friendly, safe and economical approaches for weed control can be combined with existing practices as a component of an IWM package.

The potential of biocontrol agents should be harnessed. However, it must be recognized that there may be a very low abundance of these natural enemies to weeds at the particular times required to control the weeds in a specific agricultural situation.

## 7.6 Chemical approaches

The importance of herbicide use is closely related to the cost and availability of labour. Herbicides are one of the first labour-saving technologies to be adopted and have proved to offer superior weed control. They are more energy- and labour-efficient than manual or mechanical methods of weed management. Weeds which mimic crops and those with asexual propagation methods can be more effectively controlled with chemical measures than with mechanical methods (Chauhan 2012a,b). The success of direct seeding of rice is dependent on the use of herbicides, because without their use weeds grow extremely rapidly in the early stages before the fields can be flooded, so that even manual means of control are often not feasible. Herbicides are also used in transplanted rice systems. The costs involved with herbicide use are likely to remain a major constraint to their widespread adoption. However, increased labour wages have led to adoption of chemical weed control alone or as a component of IWM by many farmers in Asia during recent times. As a consequence, the use of herbicides varies considerably between countries.

Herbicides may be classified as non-selective or selective, and pre-, early post- or post-emergence. Most herbicides used in rice production are selective, having narrow or broad spectrum, while having a limited effect on the crop. Selectivity is dependent upon the compounds included in a herbicide, its dose, and the time and method of application. It is therefore important to follow the recommendations for selective and effective weed control. Non-selective herbicides such as glyphosate are sometimes used before establishing rice on weed infestations such as wild/weedy rice, which are difficult to control with selective herbicides.

Pre-emergence herbicides are applied to the soil and control weeds before they emerge, while post-emergence are applied to weeds after they emerge. Among the amide group are the herbicides butachlor, pretilachlor and propanil. Butachlor can be applied either as a pre-emergence or early post-emergence measure to give control over a wide range of annual grasses and some broadleaf weeds. Several pre-emergence herbicides including butachlor, thiobencarb, pendimethalin, oxadiargyl, pretilachlor, anilofos, pyrazosulfuron ethyl, oxadiazon, oxyfluorfen and nitrofen provide a fair degree of weed control when applied alone or supplemented with hand weeding (Chauhan 2012a; Chauhan and Opeña 2013; Mahajan and Chauhan 2011; Kaur and Singh 2015b). However, pre-emergence herbicides have to be applied within a short timeframe in adequate soil moisture, preferably moist/standing water, to be effective.

Selective herbicides are not available to manage weedy rice in conventional rice cultivars. However, some studies have reported that pre-plant herbicides (e.g. oxadiazon and metolachlor) may provide effective control of weedy rice. However, these herbicides should be applied before rice planting to avoid damage to cultivated rice. In direct-seeded rice systems, especially in dry direct-seeded systems where a broad range of weeds is present, it is necessary to use mixtures of different compatible herbicides (Chauhan 2012b) or sequential sprays of pre-emergence herbicides followed by post-emergence herbicides, in order to ensure effective weed management (Mahajan et al. 2009).

However, some weed species are not controlled effectively even after using herbicide mixtures. Moreover, some weed species keep emerging throughout the crop season because of their high degree of seed dormancy. Injudicious and continuous use of a single herbicide over a long period of time may result in the development of resistant biotypes, shifts in weed flora and negative effects on the succeeding crop; soil/aerial environments including soil microflora and fauna; and residues in crop produce (Chauhan et al. 2012a). In Indian Punjab, continuous use of butachlor for years has resulted in a shift in weed population from *Echinochloa* spp. to *Ischaemum rugosum*, *Cyperus iria* and *Caesulia axillaris;* and continuous use of pretilachlor has caused a weed population shift from *Echinochloa* spp., *Cyperus iria* and *Caesulia axillaris* to *Ischaemum rugosum* (Walia 2010). Such situations demand IWM strategies. Further, to improve the efficacy of applied herbicides and to reduce the cost of weed management, low-cost and highly efficient herbicide application technologies including spray equipment and nozzles, herbicide carriers and adjuvants should be adopted along with ensuring the timeliness of applications of herbicide. Moreover, further research is needed on herbicide mixtures to delay resistance development, to reduce the cost of weed management and to provide more options in the weed control spectrum.

Herbicide-tolerant rice is an effective option in the IWM programme and also promotes adoption of resource conservation technologies. Imidazolinone-tolerant rice has been developed through chemically induced seed mutagenesis and conventional breeding, while glufosinate- and glyphosate-tolerant rice has been created through transgenic technologies to contribute to solving the problem of managing weeds, especially weedy/red rice, perennial and parasitic weeds (Gealy et al. 2003). The development of a non-transgenic herbicide-tolerant cultivar (e.g. imidazolinone-tolerant) can serve as an effective means of weed management through the application of new-generation herbicides that are highly effective, non-toxic and rapidly biodegradable (Chauhan et al. 2012b; Mahajan and Chauhan 2013). Commercial mixtures of both imazethapyr + imazapic and imazapyr + imazapic used in Clearfield® rice technology have a residual activity in soil with stronger effects on succeeding ryegrass and the next-season non-tolerant rice (Santos et al. 2014; Marchesan et al. 2010). These technologies can be adopted by large-scale farmers; however, the costs and investment risks of high-quality certified seeds associated with unpredictable environmental variations pose a real barrier for small and marginal farmers (Delmer, 2005). In addition, although herbicide-tolerant rice currently allows selective chemical control, outcrossing (0.003–0.008%) between Clearfield™ rice and red rice (Shivrain et al. 2007), between Mediterranean transgenic rice to conventional rice (0.036%) and red rice (0.086%) (Messeguer et al. 2004) and between glufosinate-tolerant rice and red rice (0.3–1%) (Zhang et al. 2003) is a major concern.

# 8 The integration of weed management techniques

## 8.1 Examples of technique integration

Hand weeding can be integrated with herbicide use to minimize the effects of the burgeoning cost of labour and its non-availability (Singh et al. 2001; Rao and Nagamani 2007; Raoet al. 2007), and to achieve optimum rice productivity and profitability. In dry-seeded rice, control techniques should be mixed so that reliance on herbicides could be reduced by following IWM approaches. Crop–weed competition is greatly affected by the cultivar being grown and the planting pattern. The rice cultivar and planting pattern can interact to influence weed biomass due to differential competitive ability of cultivars under different planting patterns. Rice cultivars respond differently to different planting patterns. Paired row planting of the PR-115 cultivar, for example, resulted in less weed biomass than when sown in uniform rows (Mahajan and Chauhan 2011). By exploiting the weed competitiveness of cultivars (IET-21214) in a paired-row pattern, the intensive use of herbicides (i.e. sequential sprays of pre- and post-emergence herbicides in dry-seeded rice) could be reduced to a single application of either one of the herbicides (Mahajan et al. 2014). Increasing N fertilizer to 150 kg $ha^{-1}$ resulted in significant increase in grain yield when the weeds were controlled by pre- and post-emergence herbicides. However, N addition resulted in a drastic reduction in yield under poor weed control conditions (Mahajan and Timsina 2011). The authors of this study suggested that greater understanding is required on the ecology of weeds in response to varying levels of N and weeds. Seed rate is inversely correlated to weed interference; higher seed rates caused significant reductions in weed biomass, whereas higher than optimum seed rate (15–30 kg $ha^{-1}$) caused reduction in yield (Mahajan et al. 2010). Therefore, optimum seed rate in direct-seeded rice may minimize the effect of weed competition on rice growth and grain yield and thus can reduce herbicide use, thereby promoting IWM.

## 8.2 Developing new approaches in IWM

Research in IWM must cover various factors in crop production in a systematic manner. To find the best IWM approach for a crop, it is necessary to look into the role of zero/conservation tillage, weed biology, autoecology and syncology; the critical period of crop–weed competition; methods of weed control; weed-suppressive and weed-tolerant cultivars; modelling of crop–weed interference; effect of crop rotation and weed seed bank dynamics; and 'lab-to-land' services. The complexity involved in addressing these issues requires a multi-disciplinary approach (Swanton and Weise 1991).

Below, we suggest for critical areas for future IWM research:

1. As weeds are dynamic, continuous monitoring and refinement of management strategies are essential for alleviating the adverse effects of weeds on agricultural productivity and environmental health. Weed surveillance in a crop gives an insight to the extent of the weed problem and knowledge of the associated weed composition, which then helps in devising weed management practices before the weed problem reaches alarming proportions. Remote sensing can be used as a tool for agricultural weed detection, and can potentially provide a solution to the current logjam in precision weed management (Ramanjit-Kaur et al. 2011; Lamb and Brown 2001; Thorp and Tian 2004).
2. Understanding weed biology and ecology is a prerequisite for developing strategies for the cultural management of any weed. With adequate knowledge of the germination requirements of weed seeds, weeds can be controlled by stimulating germination at a time when their seedlings can be easily killed or by providing an environment that induces very low germination (Chauhan and Johnson, 2010b). Information such as that 92% of *E. crus-galli* seeds germinated on the soil surface; that a burial depth of only 0.4 cm reduced seedling emergence by 50%; that emergence and plant biomass were reduced by the addition of 6 ton $ha^{-1}$ of rice residue to the soil surface, and that deep flooding in early growth stages significantly suppressed

growth of *E. crus-galli* could contribute to improved weed control approaches. For example, these pieces of information might inform a decision to implement soil inversion by tillage to bury weed seeds to deeper layers from where their emergence is negligible; and/or addition of crop residue on the soil surface as mulch and early flooding of the crop for managing *E. crus-galli* and other weeds with similar ecological requirements (Chauhan and Johnson 2011a).

3. A better understanding of weed seed bank dynamics is critical for the development of more efficient weed management strategies and to regulate communities of weed species. The long-term goal of IWM strategies is to reduce the size of the weed seed bank in different cropping systems. Weeds should be uprooted before they produce seeds or even flower to avoid further addition to the weed seed bank.
4. Weed species distribution and prevalence will be affected by changes in atmospheric $CO_2$, rainfall and temperature under the impact of climate change. Fourteen of the world's worst weeds are $C_4$ plants, while 76% of the harvested crop area is planted with $C_3$ crops. *E. colona*, *E. crus-galli* and *C. iria* have the $C_4$ photosynthetic pathway and as such have greater potential growth rates and light use efficiencies than plants with the $C_3$ pathway, such as rice. The effect of climate change is species specific: studies on the effect of $CO_2$ enrichment on weed species revealed that *Echinochloa colona* responded to elevated $CO_2$, but *Cyperus rotundus* L. and *Eleusine indica* did not; weedy rice responds more strongly to increasing levels of $CO_2$ than does cultivated rice. Given climate change, $C_3$ crops would benefit more from elevated $CO_2$ than $C_4$ weeds, and so losses due to $C_4$ weeds might decrease. However, it is not clear how the effects of an increase in $CO_2$, increasing temperature and more frequent drought conditions will interact with each other to affect overall plant growth/yield. Very little effort has so far been made to study the impact of climate change on weeds, weed ecology and their response to weed management practices including herbicides. Future research efforts are required to find solutions to these new challenges.

# 9 Summary

The formulation of packages of IWM strategies for different rice ecosystems requires adequate knowledge of the agronomic, soil, genetic, biological, ecological, physiological, biochemical, environmental, toxicological and economic aspects of plant growth, behaviour and production. Comprehensive integration of multiple techniques and modified tools is necessary to ensure system productivity and sustainability, long-term weed management, food security and environmental safety. It is therefore essential to adopt best management practices, which include applying multiple herbicides with different modes of action, rotating crops, adopting the best cultural weed management practices, use of weed-free seeds, scouting fields routinely, cleaning equipment, non-cropped areas and water channels to reduce the transmission of weeds to other fields. An approach that integrates different strategies allows us to manage weeds effectively and may help delay the evolution of herbicide resistance. It is necessary to develop and popularize IWM techniques, which take into account the ecological, geographic, climatic and agronomic aspects of cropping systems for different agro-ecological regions of India. To achieve effective, long-term and sustainable weed control, IWM strategies that target the prevention of weed invasion, recruitment and reproduction need to be developed.

The IWM package should be effective enough for long-term maintenance of natural resources and agricultural productivity with minimal adverse environmental impact and adequate economic returns to farmers.

# 10 Where to look for further information

Readers may look the following books for further information related to integrated weed management in direct-seeded rice:

Bay-Petersen, J. (1994). Integrated management of paddy and aquatic weeds in Asia. FFTC book series no. 45. *Proceedings of the International Seminar 'Biological Control and Integrated Management of Paddy and Aquatic Weeds in Asia'* held in Tsukuba, Japan, 19–25 October 1992. Food and Fertilizer Technology Center.

Chauhan, B. S. and Mahajan, G. (eds) (2014). *Recent Advances in Weed Management*, Vol. 1. Springer-Verlag, New York, 411p.

Singh, Y., Singh, V. P., Chauhan, B., Orr, A., Mortimer, A. M., Johnson, D. E. and Hardy, B. (eds) (2008). *Direct Seeding of Rice and Weed Management in the Irrigated Rice-Wheat Copping System of the Indo-Gangetic Plains*. International Rice Research Institute, Los Baños (Philippines) and Directorate of Experiment Station, G. B. Pant University of Agriculture and Technology Pantnagar (India), 272p.

# 11 References

Angiras, N. N. and Rana, S. S. (1998). Integrated weed management in direct seeded puddled-sprouted rice (*Oryza sativa* L.). *Indian J. Agron.* 43(4): 644–9.

Antle, J. M. and Pingali, P. L. (1994). Pesticides, productivity, and farmer health: A Philippine case study. *Am. J. Agr. Econ.* 76(3): 418–30.

Azmi, M. and Baki, B. B. (1995). The succession of noxious weeds in tropical Asian rice fields with emphasis on Malaysian rice ecosystem. 15th Asian Pacific Weed Science Society Conference, Tsukuba, Japan, pp. 51–67.

Bay-Petersen, J. (1994). Integrated management of paddy and aquatic weeds in Asia. FFTC book series no. 45. *Proceedings of the International Seminar 'Biological Control and Integrated Management of Paddy and Aquatic Weeds in Asia'* held in Tsukuba, Japan, 19–25 October 1992. Food and Fertilizer Technology Center for the Asian and Pacific Region Taipei, Republic of China on Taiwan, 230p.

Burgos, N. R., Norsworthy, K. J., Scott, R. C. and Smith, K. L. (2008). Red rice (*Oryza sativa*) status after 5 years ofimidazolinone-resistant rice technology in Arkansas. *Weed Technol.* 22: 200–8. http://dx.doi.org/10.1614/WT-07-075.1.

Cathcart, R. J., Chandler, K. and Swanton, C. J. (2004). Fertilizer N rate and the response of weeds toherbicide. *Weed Sci.* 52: 291–6.

Chatterjee, B. N. and Maity, S. K. (1981). *Principles and Practices of Rice Growing*, 2nd edn. Oxford and IBH Publications, New Delhi, p. 293.

Chauhan, B. S. (2013). Strategies to manage weedy rice in Asia. *Crop Prot.* 48: 51–6.

Chauhan, B. S., and Johnson, D. E. (2009a). Seed germination ecology of junglerice (*Echinochloa colona*): A major weed of rice. *Weed Sci.* 57: 235–40.

Chauhan, B. S. and Johnson, D. E. (2010a). Implications of narrow crop row spacing and delayed *Echinochloa colona* and *Echinochloa crus-galli* emergence for weed growth and crop yield loss in aerobic rice. *Field Crops Res.* 117: 177–82.
Chauhan, B. S. and Johnson, D. E. (2010b). The role of seed ecology in improving weed management strategies in the tropics. *Adv. Agron.* 105: 221–62.
Chauhan, B. S. and Johnson, D. E. (2011a). Ecological studies on *Echinochloa crus-galli* and the implications for weed management in direct-seeded rice. *Crop Prot.* 30: 1385–91.
Chauhan, B. S. and Johnson, D. E. (2011b). Row spacing and weed control timing affect yield of aerobic rice. *Field Crops Res.* 121: 226–31.
Chauhan, B. S. and Johnson, D. E. (2009b). Influence of tillage systems on weed seedling emergence pattern in rainfed rice. *Soil Tillage Res.* 106(1): 15–21.
Chauhan, B. S. and Mahajan, G. (eds) (2014). *Recent Advances in Weed Management*, Vol. 1. Springer-Verlag, New York, 411p.
Chauhan, B. S. and Opeña, J. (2013). Weed management and grain yield of rice sown at low seeding rates in mechanized dry-seeded systems. *Field Crops Res.* 141: 9–15.
Chauhan, B. S., Gill, G. and Preston, C. (2006a). Influence of tillage systems on vertical distribution, seedling recruitment and persistence of rigid ryegrass (*Lolium rigidum*) seed bank. *Weed Sci.* 54: 669–76.
Chauhan, B. S., Gill, G. and Preston, C. (2007). Effect of seeding systems and dinitroaniline herbicides on emergence and control of rigid ryegrass (*Lolium rigidum*) in wheat. *Weed Technol.* 21: 53–8.
Chauhan, B. S., Gill, G. and Preston, C. (2006b). Tillage system effects on weed ecology, herbicide activity and persistence: A review. *Aust. J. Exp. Agric.* 46: 1557–70.
Chauhan, B. S., Mahajan, G., Sardana, V., Timsina, J. and Jat, M. L. (2012b). Productivity and sustainability of the rice-wheat cropping system in the Indo-Gangetic Plains of the Indian subcontinent: Problems, opportunities, and strategies. *Adv. Agron.* 117: 315–69.
Chauhan, B. S., Migo, T., Westerman, P. R. and Johnson, D. E. (2010). Post-dispersal predation of weed seeds in rice fields. *Weed Res.* 50: 553–60.
Chauhan, B. S., Kaur, P., Mahajan, G., Randhawa, R. K., Singh, H. and Kang, M. S. (2014). Global warming and its possible impacton Indian agriculture. *Adv. Agron.* 123: 65–112. doi:10.1016/B978-0-12-420225-2.00002-9.
Chauhan, B. S., Singh, R. G. and Mahajan, G. (2012a). Ecology and management of weeds under conservation agriculture: A review. *Crop Prot.* 38: 57–65.
Chauhan, B. S. (2012a). Weed ecology and weed management strategies for dry-seeded rice in Asia. *Weed Technol.* 26: 1–13.
Chauhan, B. S. (2012b). *Weed Management in Direct-seeded Rice Systems.* International Rice Research Institute, Los Baños (Philippines), 20p.
Chauhan, B. S., Singh, V. P., Kumar, A. and Johnson, D. E. (2011). Relations of rice seeding rates to crop and weed growth in aerobic rice. *Field Crops Res.* 21: 105–15.
Cousens, R. and Mortimer, M. (1995). *Dynamics of Weed Populations.* Cambridge University Press, Cambridge, UK, 332p.
Delmer, D. P. (2005). Agriculture in the developing world: Connecting innovations in plant research to downstream applications. *Proc. Nat. Acad. Sci. U.S.A.* 102(44): 15739–46.
DRD (2014). Rice- A status paper. Govt. of India, Directorate of Rice Development, 250A Patliputra Colony, Patna, Bihar, 144p.
FAO (2009). www.fao.org/fileadmin/templates/wsfs/docs/expert_paper/How_to_feed_the_world_in_2050.pdf (accessed 28 December 2015).
FAOSTAT (2015). http://faostat3.fao.org/home/inputs.htmlDOWNLOAD (accessed 28 December 2015).
Gealy, D. R., Mitten, D. H. and Rutger, J. N. (2003). Gene flow between red rice (*Oryzasativa*) and herbicide-resistant rice (*O. sativa*): Implications for weed management. *Weed Technol.* 17: 627–45.

Gurney, A. L., Slate, J., Press, M. C. and Scholes, J. D. (2006). A novel form of resistance in rice to the angiosperm parasite *Striga hermonthica*. *New Phytologist*. 169(1): 199–208.

Heap, I. (2015). The International survey of herbicide resistant weeds. Online as cited on December 29. Available www.weedscience.org.

Kamboj, B. R., Yadav, D. B., Yadav, A., Goel, N. K., Gill, G., Malik, R. K. and Chauhan, B. S. (2013). Mechanized transplanting of rice (*Oryza sativa* L.) in nonpuddled and no-till conditions in the rice-wheat cropping system in Haryana, India. *Am. J. Plant. Sci.* 4: 2409–13.

Kaur, S. and Singh, S. (2014). Influence of crop density on weeds, growth and yield of direct-seeded rice. *Indian J. Weed Sci.* 46(4): 318–21.

Kaur, S. and Singh, S. (2015a). Impact of crop establishment methods and weed control on weeds, insect-pest and disease infestation in rice in North-Western Indo-Gangetic Plains. *Inter. J. Agri. Sci.* 7(4): 487–91.

Kaur, S. and Singh, S. (2015b). Bioefficacy of different herbicides for control of weeds in direct seeded rice.*Indian J. Weed Sci.* 47(2): 1–4.

Kaur, R., Mahey, R. K., Mukherjee, J. and Mahajan, G. (2011). Weed management in rice-wheat cropping system through remote sensing-a review. *J. Agric. Sci. and Technol.* 5(3): 366–84.

Kolpin, D. W., Thurman, E. M., and Linhart, S. M. (1998). The environmental occurrence of herbicides: The importance of degradates in ground water. *Arch. Environ. Contam. Toxicol.* 35: 385–90.

Lamb, D. W. and Brown, R. B. (2001). PA – precision agriculture: Remote-sensing and mapping of weeds in crops. *J. Agric. Eng. Res.* 78(2): 117–25.

Li, S. S., Wei, S. H., Zuo, R. L., Wei, J. G. and Qiang, S. (2012). Changes in the weed seed bank over 9 consecutive years of rice-duck farming. *Crop Prot.* 37: 42–50.

Mahajan, G. and Chauhan, B. S. (2011). Effects of planting pattern and cultivar on weed and crop growth in aerobic rice system. *Weed Technol.* 25: 521–5.

Mahajan, G. and Chauhan, B. S. (2013). The role of cultivars in managing weeds in dry-seeded rice production systems. *Crop Prot.* 49: 52–7.

Mahajan, G. and Timsina, J. (2011). Effect of nitrogen rates and weed control methods on weeds abundanceand yield of direct-seeded rice. *Arch. Agron. Soil Sci.* 57(3): 239–50.

Mahajan, G., Chauhan, B. S. and Gill, M. S. (2013). Dry-seeded rice culture in Punjab state of India: Lessons learned from farmers. *Field Crops Res.* 144: 89–99.

Mahajan, G., Chauhan, B. S. and Johnson, D. E. (2009). Weed management in aerobic rice in Northwestern Indo-Gangetic Plains. *J. Crop Improv.* 23: 366–82.

Mahajan, G., Gill, M. S. and Singh, K. (2010). Optimizing seed rate to suppress weeds and to increase yield in aerobic direct-seeded rice in Northwestern Indo-Gangetic Plains. *J. New Seeds* 11(3): 225–38.

Mahajan, G., Poonia, V. and Chauhan, B. S. (2014). Integrated weed management using planting pattern, cultivar, and herbicide in dry-seeded rice (*Oryza sativa* L.) in northwest India. *Weed Sci.* 62: 350–9. doi:10.1614/WS-D-13-00115.1.

Mahajan, G., Ramesha, M. S. and Chauhan, B. S. (2014). Response of rice genotypes to weed competition in dry direct-seeded rice in India. *Sci. World J.* http://dx.doi.org/10.1155/2014/641589.

Mahajan, G., Sardana, V., Brar, A. S. and Gill, M. S. (2006). Effect of seed rates, irrigation intervals and weed pressure on productivity of direct-seeded rice (*Oryza sativa*). *Indian J. Agric. Sci.* 76 (12): 756–9.

Mahajan, G., Ramesha, M. S. and Chauhan, B. S. (2015). Genotypic differences for water-use efficiency and weed competitiveness in dry direct-seeded rice. *Agron. J.* 107(4): 1573–83.

Malik, R. K., Kumar, V., Yadav, A. and McDonald, A. (2014). Conservation agriculture and weed management in South Asia: Perspective and development. *Indian J. Weed Sci.* 46(1): 31–5.

Marchesan, E., Santos, F. M., Grohs, M., Avila, L. A., Machado, S. L. O., Senseman, S. A., Massoni, P. F. S. and Sartori, G. M. S. (2010). Carryover of imazethapyr and imazapic to nontolerant rice. *Weed Technol.* 24: 6–10.

Messeguer, J., Marfa, V., Catala, M. M., Guiderdoni, E. and Mele, E. (2004). Afield study of pollen-mediated gene flow from Mediterranean GM riceto conventional rice and the red rice weed. *Mol. Breed* 13: 103–12.

Minotti, P. L. and Sweet, R. D. (1981). Role of crop competition in limiting losses from weeds. In D. Pimentel (ed.), *CRC Handbook of Pest Management in Agriculture*, Vol. II. CRC Press, Boca Raton, FL, pp. 351–67.

Miyazawa, K., Tsuji, H. and Yamagata, M. (2004). Response of weed flora to combinations of reduced tillage, biocide application and fertilization practices in a 3-year crop rotation. *Weed Biol. Manage* 4: 24–34.

Pandey, J. and Thakur, K. N. (1988). Effect of herbicides and nitrogen on nutrient removal by weeds in transplanted rice. *Indian J. Agron.* 33: 259–363.

Pandey, S. and Velasco, L. (2005). Trends in crop establishment methods in Asia and research issues. In *Rice is Life: Scientific Perspectives for the 21st Century*. Proceedings of the World Rice Research Conference, 4–7 November 2004, Tsukuba, Japan, pp. 178–81.

Phuong, L. T., Denich, M., Vlek, P. L. G. and Balasubramanian, V. (2005). Suppressing weeds in direct-seeded lowland rice: Effects of methods and rates of seeding. *J. Agron. Crop Sci.* 191: 185–94.

Rao, A. N. and Nagamani, A. (2007). Available technologies and future research challenges formanaging weeds in dry-seeded rice in India. In *Proceedings of the 21stAsian Pacific Weed Science Society Conference* from 2 to 6 October, held at Colombo, Sri Lanka.

Rao, A. N., Mortimer, A. M., Johnson, D. E., Sivaprasad, B. and Ladha, J. K. (2007). Weed management in direct-seeded rice. *Adv. Agron.* 93: 155–257.

Rao, V. S. (2000). *Principles of Weed Science*, 2nd edn. Science Publishers Inc, New Hampshire, USA, 557p.

Santos, L. O., Pinto, J. J. O., Piveta, L. B., Noldin, J. A., Galon, L. and Concenço, G. (2014). Carryover effect of imidazolinone herbicides for crops following rice. *Am. J. Plant Sci.* 5: 1049–58.

Sharma, A. R. (1997). Effect of integrated weed management and nitrogen fertilization on the performance of rice under flood-prone lowland conditions. *J. Agric. Sci.* 129(4): 409–18.

Sharma, R. (2007). Integrated weed management in wheat and rice crop. *Ind. Farming* 11: 29–34.

Shivrain, V. K., Burgos, N. R., Anders, M. M., Rajguru, S. N., Moore, J. and Sales, M. A. (2007). Gene flow between Clearfield™ rice and red rice. *Crop Prot.* 26: 349–56.

Singh, S., Chhokar, R. S., Gopal, R., Ladha, J. K., Gupta, R. K., Kumar, V. and Singh, M. (2009). Integrated weed management: A key to success for direct-seeded rice in the Indo-Gangetic Plains. In J. K. Ladha, Yadwinder-Singh, O. Erenstein and B. Hardy (eds), *Integrated Crop and Resource Management in the Rice-wheat System of South Asia*. International Rice Research Institute, Los Baño: Philippines, pp. 261–78.

Singh, V. P., Singh, G. and Singh, R. K. (2001). Integrated weed management in direct seededspring sown rice under rainfed low valley situation of Uttaranchal. *Indian J. Weed Sci.* 33: 63–6.

Singh, Y., Singh, V. P., Chauhan, B., Orr, A., Mortimer, A. M., Johnson, D. E. and Hardy, B. (eds) (2008). *Direct Seeding of Rice and Weed Management in the Irrigated Rice-wheat Cropping System of the Indo-Gangetic Plains.* International Rice Research Institute, Los Baños (Philippines), and Directorate of Experiment Station, G. B. Pant University of Agriculture and Technology Pantnagar (India). 272p.

Swanton, C. J. and Weise, S. F. (1991). Integrated weed management: The rationale and approach. *Weed Technol.* 5(3): 657–63.

Thorp, K. R. and Tian, L. F. (2004). A review on remote sensing of weeds in agriculture. *Precis. Agric.* 5: 477–508.

Walia, U. S. (2010). *Weed Management,* 3rd edn. Kalyani Publishers, Ludhiana, 373p.

Xuan, T. D., Tawata, S., Khanh, T. D. and Chung, I. M. (2005). Biological control of weeds and plant pathogens in paddy rice by exploiting plant allelopathy: An overview. *Crop Prot.* 24: 197–206.

Zhang, N. Y., Linscombe, S., Oard, J. and Zhang, N. Y. (2003). Out-crossing frequency and genetic analysis of hybrids between transgenic glufosinate herbicide-resistant rice and the weed, red rice. *Euphytica* 130: 35–45.

# Index


Acute rodenticides 354–355
Aerobic rice cultivation 113–114
good agricultural practices (GAP) in
fertilizer management 116
irrigation 116–118
land preparation 115
pest control (nematodes) 115–116
seeds and seed treatment 115
variety selection 114
weed management 115
as water-saving measures 7–10
benefits 8–9
description 7–8
risks 9–10
Agrochemicals
and IPM
herbicides 319
insecticides 317–319
molluscicides 319
and resource-use efficiency 218
Algal weeds 94–95
Alternate wetting and drying (AWD) 10–13
benefits 11–13
description 10
risks 13
Analysis of variance (ANOVA) 49
ANOVA. *see* Analysis of variance (ANOVA)
Anticoagulant rodenticides 355–356
Augmentative biological control 279–280

Barriers
and rodents 353
and traps 323
BCR. *see* Benefit–cost ratio (BCR)
Benefit–cost ratio (BCR) 90
Best farmers' yield
definition of 158
quantification of yield gaps 165–169
'Best management practices' (BMP) 202–204
Biochar production and utilization 140–141
Biocontrol, of rodent pests
parasites and diseases 357
predators 357
sterility control 356–357
Biological control
augmentative 279–280
of IPM 276–277, 326–327
of IWM 390–391
Biology, of rodents 348–351
BIORAT 357
BMP. *see* 'Best management practices' (BMP)
Botanical biocides 322
BPH. *see* Brown planthopper (BPH)
Brown planthopper (BPH) 250

CA. *see* Conservation agriculture (CA)
CABI Plantwise programme 330
Central Rice Research Institute (CRRI) 79
Central Soil Salinity Research Institute (CSSRI) 79
Chemical approaches, of IWM 391–392
Chemical biocides 322
Chemical rodent management
acute rodenticides 354–355
anticoagulant rodenticides 355–356
fumigation 356
repellents 356
*Chilo suppressalis* (Walker) 242–245
Chlorophyll meters
dynamic SPAD meter threshold value approach 29
fixed SPAD meter threshold value approach 28–29
Climate change
and SRI 194–196
and sustainability 218–220
Climate/weather, in Sundarbans region 74–75
Coastal acid saline soils 77–78
Coastal saline soils, rice cultivation in
alternative livelihood options 99
and dry season rice 86–90
income from rainfed lowlands 96
integrated plant nutrient supply (IPNS) 91–95
algal weeds in rainfed lowland rice fi elds 94–95
and micronutrients 93–94
nitrogen management in rice fields 91–93
P and K fertilizers 93
*Kharif* rice 98
management of 77–78
managing toxic gases 90–91
mulching/cover crops 95
overview 69–72
*Rabi*/summer/*boro* rice 98–99
removing excess soluble salts 91
salt- and flood-tolerant rice varieties 78–82
Sundarbans region 72–77
climate and weather 74–75
contingency plan for cyclone affected area of 96–98
crops and cropping pattern 72–74
rice field management 85–86
soil types 75–77
and wet season rice 82–85
Combustion
for paddy drying 136–138
power plant 138
Complex farming systems 192–193

Conservation agriculture (CA) 204–205
Controlled-release N fertilizers 36–39
Cover crops 384
and mulching 95
'Crop health syndromes' 310–312
Cropping intensity 314–315
Crop rotation 385–387
Crops/cropping pattern, in Sundarbans region 72–74
CRRI. *see* Central Rice Research Institute (CRRI)
CSSRI. *see* Central Soil Salinity Research Institute (CSSRI)
Cultivation season 108
Cultural practices, of IPM
fertilizer management 275–276
host plant resistance 274–275
overview 273–274

DAS. *see* Days after sowing (DAS)
Days after sowing (DAS) 378
Deep placement, of N fertilizers 42–44
Direct-seeding, for crop establishment 198–199
Diseases, of rodent pests 357
Dissemination
for IPM 283–285
of SRI 187–188
Diversionary feeding 357–359
Dry direct seeded rice cultivation 105–107
good agricultural practices (GAP) in
cultivars 108
cultivation season 108
fertilizer use 111
harvesting and threshing 112
irrigation 110–111
overview 107–108
roguing 112
seed rate and sowing depth 109
seed treatment 108–109
soil and field preparation 108
sowing 109–110
thinning 111
weed management 111–112
Dry season rice, and coastal saline soils 86–90
Dynamic SPAD meter threshold value approach 29

Earth obseRvation Model based ricE information Service (ERMES) 297
Ecologically based rodent management (EBRM) 360
in Laos 362–364
principles of 365–366
in Tanzania 364–365
in Vietnam 360–362
Economic action thresholds 281
Economic yield 158
Education and extension, of IPM 329–330
Electricity, and rodents 360
Electromagnetism, and rodents 360
Energy production
energy balance analysis for anaerobic digestion 138–139
overview 133–136
ERMES. *see* Earth obseRvation Model based ricE information Service (ERMES)
ESP. *see* Exchangeable sodium percentage (ESP)
Ethical, social and behavioural issues, in PPPs
labour force demographics 299
overview 298–299
risk perception and knowledge of proper use procedures 299–300
European Space Agency Sentinel-1A 297
European sustainable use of pesticides (SUD) 300–302
Evolution, of SRI
fundamentals 186
practices 185
principles 185–186
and soil biology 186
EXAMS. *see* Exposure Analysis Modeling System (EXAMS)
Exchangeable sodium percentage (ESP) 70
Experimental yield
definition of 158
quantification of yield gaps 164–165
Exposure Analysis Modeling System (EXAMS) 297

Farmer Field Schools (FFS) 283, 302, 318
Farmer involvement, in IPM 302–303
Farmers' fertilizer practice (FFP) 31
Farmyard manure (FYM) 47, 83
Fertilizer management 275–276, 313
aerobic rice cultivation 116
in dry direct seeding 111
and IWM 389
N fertilizers
controlled-release and slow-release 36–39
deep placement of 42–44
FFP. *see* Farmers' fertilizer practice (FFP)
FFS. *see* Farmer field schools (FFS)
Field levelling and tillage 382–384
Fixed SPAD meter threshold value approach 28–29
Fixed-time variable rate dose approach 31–33
Foliage feeders 255–260
Fumigation 356
FYM. *see* Farmyard manure (FYM)

Geographical information systems (GIS) 297
GHG. *see* Greenhouse gases (GHG)
GIS. *see* Geographical information systems (GIS)
Global warming potential (GWP) 196

Global Yield Gap and Water Productivity Atlas (GYGA) 163–164
Good agricultural practices (GAP)
in aerobic rice cultivation
fertilizer management 116
irrigation 116–118
land preparation 115
pest control (nematodes) 115–116
seeds and seed treatment 115
variety selection 114
weed management 115
in dry direct seeding
cultivars 108
cultivation season 108
fertilizer use 111
harvesting and threshing 112
irrigation 110–111
overview 107–108
roguing 112
seed rate and sowing depth 109
seed treatment 108–109
soil and field preparation 108
sowing 109–110
thinning 111
weed management 111–112
Greenhouse gases (GHG) 122, 196–197
Green Revolution 187, 293
GWP. *see* Global warming potential (GWP)
GYGA. *see* Global Yield Gap and Water Productivity Atlas (GYGA)

Habitat management
and IPM 278–279
of rodents 354
Hand weeding/pulling 390
Harvesting, and threshing 112
Health and Environmental Risks: Organisation, Integration and Cross-fertilisation of Scientific Knowledge (HEROIC) 298
Herbicides 319
HEROIC. *see* Health and Environmental Risks: Organisation, Integration and Cross-fertilisation of Scientific Knowledge (HEROIC)
Higher-tier risk assessments 297
High-yielding rice varieties (HYVs) 78, 313–314
Host plant resistance 274–275
HTC. *see* Hydrothermal carbonization (HTC)
Hunting, of rodents 354
Hydrothermal carbonization (HTC) 141
HYVs. *see* High-yielding rice varieties (HYVs)

In-field rice residue processing
incorporation 128–129
open-field burning 126–128
INM. *see* Integrated nutrient management (INM)
Insecticides 317–319
and IPM
economic action thresholds 281
resistance 282–283
selectivity against natural enemies 281–282
Integrated nutrient management (INM). *see* Integrated plant nutrient supply (IPNS)
Integrated pest management (IPM)
and agrochemicals
herbicides 319
insecticides 317–319
molluscicides 319
and augmentative biological control 279–280
barriers and traps 323
chemical and botanical biocides 322
concepts and options for 272
'crop health syndromes' 310–312
cultural practices
fertilizer management 275–276
host plant resistance 274–275
overview 273–274
dissemination mechanisms for 283–285
effects of insecticide application 277–278
and habitat management 278–279
increasing rice production
areas of 312–313
cropping intensity 314–315
fertilizer use 313
high-yielding rice varieties (HYVs) 313–314
and insecticides
economic action thresholds 281
resistance 282–283
selectivity against natural enemies 281–282
knowledge of rice production system 322
and labour costs 316–317
and light traps 323
natural regulation
biological control 276–277, 326–327
education and extension 329–330
natural pest management 327–329
overview 271–272, 309–310
and pheromones 323
and PPPs
and European SUD 300–302
farmer involvement in 302–303
and repellents 323
resistance and tolerance
incorporating genes 325–326
overview 324–325
screening 325
scientific approaches 320–322
transgenic and herbicide-resistant rice varieties 324
and water productivity 315–316
Integrated plant nutrient supply (IPNS) 77, 91–95

algal weeds in rainfed lowland rice fields 94–95
and micronutrients 93–94
nitrogen management in rice fields 91–93
organic resources and mineral fertilizers 47–50
P and K fertilizers 93
Integrated rice – duck farming (Aigamo method) 390
Integrated weed management (IWM)
biological control 390–391
chemical approaches 391–392
cover crops 384
crop rotation and rice cultivars 385–387
developing new approaches in 393–394
establishment methods 384–385
examples of technique integration 392–393
fertilizer use 389
field levelling and tillage 382–384
hand weeding/pulling 390
mechanical control 389–390
overview 377–378, 381–382
plant population 387–388
preventive methods 382
purpose of 380
sowing methods 385
water management 388–389
weeds in rice crops 378–379
International Organisation for Biological and Integrated Control of Noxious Animals and Plants (IOBC-OILB) 301
International Panel on Climate Change (IPPC) 218
International Rice Research Institute (IRRI) 79, 283, 303
International Water Management Institute (IWMI) 191
IOBC-OILB. *see* International Organisation for Biological and Integrated Control of Noxious Animals and Plants (IOBC-OILB)
IPM. *see* Integrated pest management (IPM)
IPNS. *see* Integrated plant nutrient supply (IPNS)
IPPC. *see* International Panel on Climate Change (IPPC)
IRRI. *see* International Rice Research Institute (IRRI)
Irrigation techniques
in aerobic rice cultivation 116–118
in dry direct seeding 110–111
overview 3–4
water-saving measures
aerobic rice as 7–10
alternate wetting and drying (AWD) 10–13
saturated soil culture (SSC) 13–15
scale-dependency of 5–7
in southeast Australia 15–17
water productivity 4–5
IWM. *see* Integrated weed management (IWM)
IWMI. *see* International Water Management Institute (IWMI)

*Kharif* rice 98

Labour costs, and IPM 316–317
Labour force demographics 299
Labour saving, and SRI 197–198
Land preparation. *see* Soil/field/land preparation
Laos, EBRM in 362–364
LCAs. *see* Life cycle assessments (LCAs)
LCC. *see* Leaf colour charts (LCCs)
Leaf colour charts (LCCs) 29–33, 116
fixed-time variable rate dose approach 31–33
real-time threshold values 29–31
Leafhoppers 249–255
*Leptocorisa acuta* (Hemiptera: Alydidae) 260–264
Life cycle assessments (LCAs) 332
*Lissorhoptrus oryzophilus* (Kuschel) 237–239
Livestock feed, rice straw for 143
Long-Term Continuous Cropping Experiment (LTCCE) 303
LTCCE. *see* Long-Term Continuous Cropping Experiment (LTCCE)

Marker-assisted selection (MAS) 325
MAS. *see* Marker-assisted selection (MAS)
Mass Media Campaign (MMC) 283
MD. *see* Mekong Delta (MD)
Mechanical control, of IWM 389–390
'Mediterranean-Rice' (MED Rice) 296
Mekong Delta (MD) 130
Micronutrients 46–47
and IPNS 93–94
Mineral fertilizers 47–50
MMC. *see* Mass Media Campaign (MMC)
Molluscicides 319
Mulching/cover crops 95
Multiple simulation models 296
Mushroom production, rice straw for 141–143

Natural pest management 327–329
Near-infrared (NIR) spectral response 33
N fertilizers
controlled-release and slow-release 36–39
deep placement of 42–44
*Nilaparvata lugens* (Stål) 250–255
NIR. *see* Near-infrared (NIR) spectral response
Nitrification and urease inhibitors 39–42
Nitrogen management, in rice fields 91–93
Nutrient management
chlorophyll meters
dynamic SPAD meter threshold value approach 29

fixed SPAD meter threshold value approach 28–29
leaf colour charts (LCCs) 29–33
fixed-time variable rate dose approach 31–33
real-time threshold values 29–31
micronutrients 46–47
N fertilizers
controlled-release and slow-release 36–39
deep placement of 42–44
optical sensors 33
organic resources and mineral fertilizers 47–50
overview 25–27
phosphorus and potassium 44–46
real-time site-specific 27–33
site-specific 33–36
urease and nitrification inhibitors 39–42

Off-field rice residue processing
collection 131–133
overview 129–130
OPE. *see* Open Pan Evaporimeter (OPE)
Open-field burning 126–128
Open Pan Evaporimeter (OPE) 117
Operation mechanization 199–200
Optical sensors 33
ORC. *see* Organic Rankin Cycle (ORC)
Organic Rankin Cycle (ORC) 126, 136
Organic resources, and mineral fertilizers 47–50
*Orseolia oryzae* (Wood-Mason) 245–249
Overview 294–296

PADDY. *see* Pesticide paddy field model (PADDY)
P and K fertilizers 93
Panicle feeders 260–264
Parasites, of rodent pests 357
PCR. *see* Polymerase chain reaction (PCR)
PECs. *see* Predicted environment concentrations (PECs)
Performance indicators, for SRP 223–225
Personal protective equipment (PPE) 299
Pest control (nematodes) 115–116
Pesticide paddy field model (PADDY) 296
Pheromones, and IPM 323
Phosphorus 44–46
Physical rodent management
barriers 353
habitat management 354
hunting 354
rice bunds 354
trap-barrier system (TBS) 353
trapping 353
Planthoppers 249–255
Plant population, and IWM 387–388
Plant protection products (PPPs)
ethical, social and behavioural issues in
labour force demographics 299
overview 298–299
risk perception and knowledge of proper use procedures 299–300
IPM
and European SUD 300–302
farmer involvement in 302–303
overview 293–294
and risk assessment
higher-tier risk assessments 297
limitations of 298
multiple simulation models 296
overview 294–296
tier-1 assessments 296–297
Polymerase chain reaction (PCR) 249
Post-harvest losses, and rodents 348
Potassium 44–46
Potential yield
definition of 158
quantification of yield gaps 161–162
PPE. *see* Personal protective equipment (PPE)
Predators, of rodent pests 357
Predicted environment concentrations (PECs) 295
Pre-harvest losses, and rodents 345–348
Production and consumption, of rice
projections to 2050 215–217
strategic importance 214–215

Quantification, of yield gaps
Global Yield Gap and Water Productivity Atlas (GYGA) 163–164
overview 160–161
rice modelling 162–163
using best farmers' yield 165–169
using experimental yield 164–165
using potential yield 161–162
Quantitative Evaluation of the Fertility of Tropical Soils (QUEFTS) model 34
QUEFTS. *see* Quantitative Evaluation of the Fertility of Tropical Soils (QUEFTS) model

*Rabi*/summer/*boro* rice 98–99
Rainfed lowland rice fields
algal weeds 94–95
and rice cultivation 96
rice production 191–192
Real-time site-specific nutrient management 27–33
chlorophyll meters
dynamic SPAD meter threshold value approach 29
fixed SPAD meter threshold value approach 28–29
leaf colour charts (LCCs) 29–33

fixed-time variable rate dose approach 31–33
real-time threshold values 29–31
optical sensors 33
overview 25–27
Real-time threshold values 29–31
Repellents
and IPM 323
and rodents 356
Resistance and tolerance, of IPM
incorporating genes 325–326
overview 324–325
screening 325
Resource-use efficiency
agrochemicals 218
climate change impacts 218–220
water 217–218
RGSV. see Rice grassy stunt (RGSV)
Rice bunds, of rodents 354
Rice by-products/rice residues
biochar production and utilization 140–141
for energy production
energy balance analysis for anaerobic digestion 138–139
overview 133–136
in-field processing
incorporation 128–129
open-field burning 126–128
off-field processing
collection 131–133
overview 129–130
overview 121–123
rice husks 123–126
combustion for paddy drying 136–138
rice straw 123–126
combustion power plant 138
for livestock feed 143
for mushroom production 141–143
Rice cultivars 385–387
in dry direct seeding 108
Rice cultivation. *see also* Good agricultural practices (GAP)
aerobic 113–114
in coastal saline soils
alternative livelihood options 99
contingency plan for cyclone affected area 96–98
and dry season rice 86–90
income from rainfed lowlands 96
integrated plant nutrient supply (IPNS) 91–95
*Kharif* rice 98
management of 77–78
managing toxic gases 90–91
mulching/cover crops 95
overview 69–72
*Rabi*/summer/ *boro* rice 98–99
removing excess soluble salts 91
rice field management 85–86
salt- and flood-tolerant rice varieties 78–82
Sundarbans region 72–77
and wet season rice 82–85
dry direct seeded 105–107
overview 105
Rice-feeding insects
foliage feeders 255–260
leafhoppers and planthoppers 249–255
overview 235–236
panicle feeders 260–264
rice gall midges 245–249
root and stem feeders 237–239
stem borers 240–245
Rice fields
management 85–86
nitrogen management in 91–93
Rice gall midges 245–249
Rice grassy stunt (RGSV) 253
Rice husks
availability of 123–126
combustion for paddy drying 136–138
Rice modelling 162–163
Rice production
and IPM
areas of 312–313
cropping intensity 314–315
fertilizer use 313
high-yielding rice varieties (HYVs) 313–314
knowledge of 322
Rice ragged stunt (RRSV) 253
Rice straw
availability of 123–126
combustion power plant 138
for livestock feed 143
for mushroom production 141–143
RICEWQ model 297
Risk assessment, and PPPs
higher-tier risk assessments 297
limitations of 298
multiple simulation models 296
overview 294–296
tier-1 assessments 296–297
Risk perception 299–300
Rodents
biocontrol of
parasites and diseases 357
predators 357
sterility control 356–357
biology of 348–351
chemical rodent management
acute rodenticides 354–355
anticoagulant rodenticides 355–356
fumigation 356
repellents 356
diversionary feeding 357–359

ecologically based rodent management (EBRM) 360
in Laos 362–364
principles of 365–366
in Tanzania 364–365
in Vietnam 360–362
and electricity 360
impacts on rice
post-harvest losses 348
pre-harvest losses 345–348
overview 343–345
physical rodent management
barriers 353
habitat management 354
hunting 354
rice bunds 354
trap-barrier system (TBS) 353
trapping 353
ultrasound and electromagnetism 360
vertebrate pests in rice 366
Root and stem feeders 237–239
Roguing, in dry direct seeding 112
RRSV. *see* Rice ragged stunt (RRSV)

Saline Alkali Tolerant Varietal (SAVT) 79
Salt- and flood-tolerant rice varieties 78–82
Saturated soil culture (SSC) 13–15
benefits 14
description 13
risks 14–15
SAVT. *see* Saline Alkali Tolerant Varietal (SAVT)
Scale-dependency, of water-saving measures 5–7
SCI. *see* System of Crop Intensification (SCI)
Scoring system, for SRP 227
Screening, and IPM 325
Seed rate, and sowing depth 109
Seeds/seed treatment
in aerobic rice cultivation 115
in dry direct seeding 108–109
Site-specific nutrient management (SSNM) 33–36, 128
Slow-release N fertilizers 36–39
Soil biology, of SRI 186
Soil/field/land preparation
in aerobic rice cultivation 115
in dry direct seeding 108
Soil organic matter (SOM) 127
Soil types, in Sundarbans region 75–77
SOM. *see* Soil organic matter (SOM)
Sowing
depth, and seed rate 109
in dry direct seeding 109–110
methods 385
SRI. *see* System of rice intensification (SRI)
SRP. *see* Standard recommended practices (SRP); Sustainable Rice Platform (SRP)
SSC. *see* Saturated soil culture (SSC)
SSNM. *see* Site-specific nutrient management (SSNM)
Standard recommended practices (SRP) 193
Stem borers 240–245
Sterility control, of rodents 356–357
SUD. *see* Sustainable use of pesticides (SUD)
Sundarbans region, and rice cultivation 72–77
climate and weather 74–75
contingency plan for cyclone affected area of 96–98
crops and cropping pattern 72–74
rice field management 85–86
soil types 75–77
Sustainability
overview 213–214
production and consumption
projections to 2050 215–217
strategic importance 214–215
resource-use efficiency
agrochemicals 218
climate change impacts 218–220
water 217–218
Sustainable Rice Platform (SRP)
for assessment 225–227
description 222
objectives 227–228
performance indicators 223–225
scoring system 227
target groups and associated benefits 228
technologies 222–223
voluntary sustainability standards (VSS) 220–221
Sustainable Rice Platform (SRP)
for assessment 225–227
description 222
objectives 227–228
performance indicators 223–225
scoring system 227
target groups and associated benefits 228
technologies 222–223
voluntary sustainability standards (VSS) 220–221
Sustainable use of pesticides (SUD) 300–302
SWI. *see* System of Wheat Intensification (SWI)
System of Crop Intensification (SCI) 193
System of rice intensification (SRI)
and 'best management practices' (BMP) 202–204
and climate change 194–196
and complex farming systems 192–193
conservation agriculture (CA) 204–205
direct-seeding for crop establishment 198–199
dissemination of 187–188
evolution of
fundamentals 186
practices 185

principles 185–186
and soil biology 186
and greenhouse gases (GHG) 196–197
and labour saving 197–198
mechanization of operations 199–200
protean nature of 183–184
and rain-fed rice production 191–192
value-chain development 200
yields
assessing 200–202
and income 188–191
System of Wheat Intensification (SWI) 193

Tanzania, EBRM in 364–365
TBS. *see* Trap-barrier system (TBS)
Thinning, in dry direct seeding 111
Threshing, and harvesting 112
Tier-1 assessments 296–297
Toxic gases, and rice cultivation 90–91
TOXic substances in Surface WAters (TOXSWA) 297
TOXSWA. *see* TOXic substances in Surface WAters (TOXSWA)
Transgenic and herbicide-resistant rice varieties 324
Trap-barrier system (TBS) 353
Trapping, of rodents 353
*Trichispa sericea* (Guerin-Meneville) 257–260

Ultrasound, and rodents 360
UNEP. *see* United Nations Environment Programme (UNEP)
United Nations Environment Programme (UNEP) 221
Urease and nitrification inhibitors 39–42

Value chain analysis (VCA) 331
Value-chain development 200
VCA. *see* Value chain analysis (VCA)
Vertebrate pests, in rice 366
Vietnam, EBRM in 360–362
Voluntary sustainability standards (VSS) 220–221

Water management, and IWM 388–389
Water productivity 4–5
and IPM 315–316
scale-dependency of 5–7
Water resource-use efficiency 217–218
water-saving measures
aerobic rice as 7–10
benefits 8–9
description 7–8
risks 9–10
alternate wetting and drying (AWD) 10–13
benefits 11–13
description 10
risks 13
saturated soil culture (SSC) 13–15
benefits 14
description 13
risks 14–15
scale-dependency of 5–7
in southeast Australia 15–17
water productivity 4–5
Water use efficiency (WUE) 197
WBPH. *see* Whitebacked planthopper (WBPH)
Weeds
algal 94–95
management
in aerobic rice cultivation 115
in dry direct seeding 111–112
in rice crops 378–379
Wet season rice, rice cultivation in 82–85
Whitebacked planthopper (WBPH) 250
WUE. *see* Water use efficiency (WUE)

Yield gaps
causes of
crop simulation models 169–171
expert/household/on-farm field survey 171–174
on-farm experiments 174
overview 157–159
quantification of
Global Yield Gap and Water Productivity Atlas (GYGA) 163–164
overview 160–161
rice modelling 162–163
using best farmers' yield 165–169
using experimental yield 164–165
using potential yield 161–162
Yields, SRI
assessing 200–202
and income 188–191

CPSIA information can be obtained
at www.ICGtesting.com
Printed in the USA
LVOW02*2320040617
536940LV00010B/442/P

9 781786 760289